The Qualitative Theory of Optimal Processes

CONTROL AND SYSTEMS THEORY

A Series of Monographs and Textbooks

Editor

JERRY M. MENDEL

University of Southern California
Los Angeles, California

Associate Editors

Karl J. Åström
Lund Institute of Technology
Lund, Sweden

Michael Athans
Massachusetts Institute of Technology
Cambridge, Massachusetts

David G. Luenberger
Stanford University
Stanford, California

Volume 1: Discrete Techniques of Parameter Estimation: The Equation Error Formulation, JERRY M. MENDEL

Volume 2: Mathematical Description of Linear Systems, WILSON J. RUGH

Volume 3: The Qualitative Theory of Optimal Processes, R. GABASOV *and* F. KIRILLOVA

OTHER VOLUMES IN PREPARATION

208. Sarachik, P. E., and Kreindler, E. "Controllabilitz "Controllability and Observability of Linear Discrete-Time Systems," Intern. J. Control, Vol. 1, No. 5, 1965.

209. Scorza - Dragoni, G. Un Teorema Sulle Funzioni Continue Rispetto ad Una e Misurabili Rispetto as Un'altra Variable. Rend Sem. Mat. Univ. Padova, 17, 1948.

210. Shemer, J. E., and Gupta, S. C. "Applications of Butkovskii's Form of Discrete Maximum Principle," ISA Trans., Vol. 5, No. 4, 1966.

211. Silva, L. M. "Predictor Control Optimizes Control System Performance," Trans. ASME, Vol. 77, No. 8, 1955.

212. Silverman, L. M., and Meadows, H. E. "Controllability and Time-Variable Unilateral Networks," IEEE Trans. on Circuit Theory, Vol. CT-12, No. 3, 1965.

213. Sivan, R. On the Controllability of the Heat Equation. Third Congress IFAC, Abstracts, London, 1966.

214. Snow, D. R. "Singular Optimal Controls for a Calss of Minimum Effort Problems," J. Soc. Industr. and Appl. Math., A2, No. 2, 1964.

215. Tarnove, I. "A Controllability Problem for Non-Linear Systems," Math. Theory Control, New York - London, Acad. Press, 1967.

216. Tou, J. T.

a) "Optimum Control of Discrete Systems Subject to Saturation," Proc. IEEE. Vol. 52, No. 1, 1964.

b) "Die Zeitoptimale Reglung Diskontinuierlicher Systeme mit Begrenzung des Regelsignals." Regelungstechnik, Bd. 12, No. 11, 1964.

217. Tung, F., and Striebel, C. T. "A Stochastic Optimal Control Problem and its Applications," J. Math. Anal. and Appl., Vol. 12, No. 2, 1965.

218. Vogt, W. G., and Cullen C. G. "The Minimum Number of Inputs Required for the Complete Controll-ability of a Linear Stationary Dynamical System," IEEE Trans. Automat. Control. Vol. AC-12, No. 3, 1967.

219. Walt, F. M. "An Engineering Approach. Hierarchical Optimization Criteria," IEEE Trans. Automat. Control, Vol. 12, No. 2, 1967.

220. Wang, P. K. C. "Invariance, Uncontrollability, and Unobservability in Dynamical Systems," IEEE Trans. Automat. Control, Vol. 10, No. 3, 1965.

221. Warga, J.

a) "Relaxed Variational Problems," J. Math. Anal. and Appl. Vol. 4, No. 1, 1962.

b) "Necessary Conditions for Minimum in Relaxed Variational Problems," J. Math. Anal. and Appl. Vol. 4 No. 1, 1962.

222. Wazewski, T. "Systemes de Commande et equations au Contingent," Bull. Acad. polon. sci., Ser. math. astr., phys., Vol. 9, No. 3, 1961.

223. Wells, C. H. "Minimum Norm Control of Discrete Systems," IEEE Internat. Convent. Res., Vol. 15, No. 3. 1967.

224. Witsenhausen, H. S. "On the Sensitivity of Optimal Control Systems," IEEE Trans. Automat. Control, Vol. 10, No. 4, 1965.

225. Wonham. W. M. "Stochastic Problems in Optimal Control," IEEE Internat. Convent. Rec. Vo. 11, No. 2, 1963.

INDEX

The Qualitative Theory of Optimal Processes

R. Gabasov F. Kirillova

Mathematics Institute of BSSR
Minsk, USSR

Translated by John L. Casti
IIASA, Laxenburg, Austria

MARCEL DEKKER, INC. New York and Basel

 (The original version, Качественная теория оптимальных процессов, was first published by Nauka, Moscow, in 1971.)

MARCEL DEKKER, INC.
270 Madison Avenue, New York, New York 10016

LIBRARY OF CONGRESS CATALOG CARD NUMBER: 76-41471

ISBN: 0-8247-6545-1

Current printing (last digit):
10 9 8 7 6 5 4 3 2 1

PRINTED IN THE UNITED STATES OF AMERICA

To the Pleasant Memory of

Evgenia Alexseevicha Barbashina

PREFACE

The theory of optimal processes has been the subject of fundamental monographs by leading specialists. In this work, we have tried not to repeat methods and results which are already well known. The basic material of the book is founded upon the authors' investigations and the work of their colleagues, which is reflected in both the choice of material and in the character of the bibliography.

R. Gabasov
F. Kirillova

CONTENTS

PART II

THE MAXIMUM PRINCIPLE IN THE THEORY OF OPTIMAL PROCESSES

Chapter VIII. Computational Problems of Optimal Control

Chapter IX. Toward a Theory of Optimal Processes for Discrete Systems

INTRODUCTION

§1. Basic Problems in the Theory of Optimal Processes

The theory of optimal processes has arisen from problems of modern science and technology. Control problems have played a decisive role in originating a new branch of the calculus of variations; a principal source of optimal control problems is space navigation. Currently, it is difficult to find a domain of human activity in which, in one way or another, we are not faced with scientific questions based upon concepts of control.

To illustrate the basic problems of the theory of optimal processes, we consider the problem of minimal time. We are given an object which changes its state under the action of deterministic forces. Based upon the available information about the current state of the object, it is required to find inputs such that the object is transferred from the initial state to a terminal state in minimal time. It is clear that before beginning a theoretical investigation of this problem, it is necessary to establish its mathematical model. Questions connected with the mathematical description of the object form the problem of identification of the object.

We postulate that the behavior of the object under investigation may be described by the equation

$$\frac{dx}{dt} = f(x,u,t) \quad , \qquad x(t_o) = x_o \quad . \tag{1}$$

Here $x = \{x_1,\ldots,x_n\}$ is a vector characterizing the state of the object, t is the time, $u = \{u_1,\ldots,u_r\}$ is a vector denoting those parameters of the object which may be at our disposal. To each initial state x_0 and choice of function u(t), t $\geq$ 0, called the control, there corresponds a unique solution (trajectory, motion) x(t), $t \geq t_0$, of Eq. (1). Usually, the function $x(t) = \{x_1(t),\ldots,x_n(t)\}$ is not available for direct measurement. Information about the current state of the object is contained in the function $z(t) = \{z_1(t),\ldots,z_m(t)\}$, $m \leq n$, connected with x(t) by the relation

$$z(t) = h(x(t),t)$$

(in control theory, the functions $u_1(t),\ldots,u_r(t)$ are also called the input perturbations (inputs), while the functions $z_1(t),\ldots,z_m(t)$ are the output functions (outputs)). The problem of observability consists in the establishment of the current (or initial) state of the object by known functions of the output. If these two problems are solvable for the object under investigation, then the above minimal time problem may be given in the following form: two vectors x_0,x_1 are given; by a control u(t), it is required to transfer the trajectory x(t) of the system (1) from x_0 to x_1 in minimal time. At once there arises the question: does there exist even one control transferring the trajectory x(t) from the initial state x_0 to the terminal state x_1? This basic question is the problem of controllability.

Thus, the minimal time problem has been reduced to the determination of the best control among those which generate trajectories of the system (1) passing between x_0 and x_1. For the solution of this problem (and partially for the problem of controllability) it is clear that two conditions need be stipulated: 1) the class of admissible control functions; 2) the criterion function of the process (in the minimal time problem, this is time). The problem of existence of optimal controls may be formulated in the following way: among the admissible controls, does there exist an optimal control which minimizes the chosen criterion function?

The next step in the solution of the minimal time problem is to narrow the set of admissible controls by isolating a subset containing the optimal control (or controls). Such a subset is isolated by imposing new conditions characterizing optimal controls on the admissible controls. The discovery of conditions which optimal controls must satisfy leads into the circle of questions surrounding problems of necessary conditions for optimality. If, in the minimal time problem, there exists an optimal control and the necessary conditions for optimality single out a unique admissible control, then clearly the latter is the (unique) optimal control. However, if the necessary conditions single out several controls, then there arises the question of their optimality, the problem of sufficient conditions for optimality.

For the final solution of the minimal time problem, it remains to realize methods for isolating optimal controls by computational means. Each realization assumes a definite algorithm. The solution of the problem of computational algorithms for optimization problems determines the successful solution of the preceding problems.

Thus, in general there are four basic problems arising in the optimization of controlled objects. Each of the enumerated problems contains, in itself, many other problems about which we do not speak. Some of these are the problems of optimizing stochastic systems, optimal adaptive systems, problems of statistical prediction, problems of invariance, autonomy, the theory of chattering regimes, the problem of synthesizing optimal controls (optimal feedback control), questions of uniqueness of optimal controls, the well-posed statement of optimization problems, and the theory of optimization for discrete systems. The problems stated above acquire many specific features in optimizing under situations of conflict type, when it is necessary to take into account the noncooperative interests of several sides. In recent years, this circle of questions has been formalized in the theory of differential games.

§2. A Brief Survey of the Current State of the Theory of Optimal Processes

The theory of identification is a science as ancient as mathematics. In modern times, aspects of the identification problem have been studied in [150b, 153, 166, 207]. A more restricted problem is the parametric identification of objects when the equations of the object are considered to be known to within parameters, which are sought such that the motion generated by the mathematical model (1), (2) be in some sense close to the motion given by the output of the object. Both deterministic and stochastic formulations of this problem are possible. In general form, these problems may be considered as problems of optimization with respect to parameters [18b]. Different methods have been successfully applied for the solution of similar questions [18d,184].

The theories of controllability and observability have been formalized as independent divisions of the theory of optimal processes following the work of R. Kalman [58a]. Different problems of these theories had been solved earlier [38b].

According to Kalman, a dynamical system

$$\frac{dx}{dt} = Ax + Bu \quad , \qquad x(t_o) = x_o \tag{3}$$

(A is an nxn matrix, B an nxr matrix) is called completely controllable if for each state x_o there is a number t_1 and a piecewise-continuous control $u(t)$, $t_o \le t \le t_1$, such that the trajectory $x(t)$ of the system (3) satisfies the condition $x(t_1) = 0$. In order that the system (3) be completely controllable, it is necessary and sufficient that the rank $[B, AB, \ldots, A^{n-1}B] = n$. This result for $r = 1$ was obtained in [58a] and in the general case in [185]. Nonstationary linear systems

$$\frac{dx}{dt} = A(t)x + B(t)u \quad , \qquad x(t_o) = x_o \quad , \tag{4}$$

from the point of view of controllability, have been considered in [74j]. It is known [74j,n] that the system (4) is completely controllable if for some $t \ge t_o$

$$\left.\begin{aligned} \text{rank} \quad \{Q_o(\bar{t}), Q_1(\bar{t}), \ldots, Q_{n-1}(\bar{t})\} = n, \\ Q_o(t) = B(t), \; Q_k(t) = A(t)Q_{k-1}(t) - \dot{Q}_{k-1}(t), \\ k = 1, \ldots, n-1. \end{aligned}\right\} \tag{5}$$

Necessary and sufficient conditions for the controllability of the system (4) are expressed through the fundamental matrix introduced in [74n]. Controllability in the small (for the initial state x_o in a sufficiently small neighborhood of the origin) for the nonlinear system

$$\frac{dx}{dt} = f(x,u,t) \quad , \quad x(t_o) = x_o \quad , \quad f(0,0,t) = 0 \quad , \tag{6}$$

is proven in [64d, 74c, 1] under the assumption that the linear model (4), with $A(t) = \frac{\partial f(0,0,t)}{\partial x}$ $B(t) = \frac{f(0,0,t)}{\partial u}$, satisfies the condition (5). Cases of (6) when the linear approximation (3) or (4) is not controllable are called <u>critical</u>. Critical cases of controllability for the second order system

$$\frac{dx}{dt} = f(x) + bu \quad ,$$

are studied in [59]. Questions of controllability of ordinary dynamical systems (6) are also considered in [23, 61, 100, 137b, 146, 157, 170, 193, 200a, 208, 215]. A close relative of the system (6) is the equation with a delayed argument

$$\frac{dx(t)}{dt} = f(x(t), x(t - h), u(t), t) \quad , \tag{7}$$

where h is the delay. The initial state $x_o(\cdot)$ of this system (the minimal information uniquely determining the future motion of the system) is characterized by the vector function $x_o(\cdot) = \{\Phi(\tau),\ t_o - h \leq \tau \leq t_o\}$. In this connection, in addition to the concept of controllability introduced above (which is called relative controllability below), of basic interest for (7) is the property of complete controllability, for which the trajectory x(t) not only reaches the origin but also remains there: $x(t) \equiv 0,\ t_1 \leq t \leq t_1 + h$. Implicit conditions for controllability of the system

$$\frac{dx(t)}{dt} = A(t)x(t) + A_1(t)x(t - h) + B(t)u(t) \tag{8}$$

have the same form as for the ordinary systems (3),(4). Explicit (expressed through A, A_1, B) conditions for relative controllability are introduced in [66a,b]. The problem of complete controllability is considered in [74n]. In the general case, conditions for the complete controllability of the system (8) are still unknown. Questions of controllability for systems other than (6) have been little developed. Only scattered results are known. Partial differential equations are considered in [213,220], while Eq. (3) in a Banach space is considered in [154a,b].

The basic result of the theory of observability for linear systems was obtained in [58a]. According to

Kalman [58a], the dynamical system

$$\dot{x} = Dx \quad , \quad x(t_o) = x_o \quad , \tag{9}$$

is called completely observable with respect to the output

$$z(t) = c'x(t) \quad , \quad t_o \le t \le t_1 \quad , \tag{10}$$

if for any vector p, the value of $p'x_o$ may be reconstructed for each point x_o by the measurements $z(t)$, $t_o \le t \le t_1$. A necessary and sufficient condition for the complete observability of the system (9), relative to the output (10), is the requirement

$$\text{rank} \quad \{c, D'c, \ldots, [D']^{n-1}c\} = n \quad .$$

Generalization of this result to the case of several outputs and to nonstationary systems follows the same scheme as in the analagous problems of controllability. This is explained by the duality principle [58a] between controllability and observability: the system (3) is completely controllable if and only if the system (9),(10) with $D = -A'$, $c = b$ is completely observable. Questions of the observability of nonlinear systems have been almost totally neglected in the literature. Only the result of [3] is known about observability in the linearized approximation (see also [196]). Effective criteria for the observability of systems differing from the usual dynamical system are, at present, unknown.

The first results concerning existence of optimal controls for the minimal time problem (3) appeared in [38a]. In this work, the set of measurable functions with values in the cube

$$|u_i| \leq 1 \quad , \qquad i = 1,\ldots,r \tag{11}$$

is chosen to be the set of admissible controls. In [108a] the cube (11) is replaced by a convex polytope. Generalizations of the results of [38a] to the nonstationary system (4) and to a set of admissible controls representable by a sphere in a standard normed space are given in [74a,b].

Questions of the existence of optimal controls in the minimal time problem for the nonlinear system (1) were first studied in the work [125]. Instead of (11), let there be given a compact set U in r-dimensional space consisting of admissible controls u(t), $t_o \leq t \leq t_1$, transferring the point x_o of the system (1) to the origin in a time no greater than $t_1 - t_o$. If the following conditions are satisfied, then there exists an optimal minimal time control: 1) the trajectory x(t) of the system (1) is uniformly bounded for all admissible controls, $t_o \leq t \leq t_1$; 2) the set $f(x,U,t) = \{f(x,u,t)\colon u \varepsilon U\}$ is compact. The method of proof for this result [221a] is not difficult to extend to optimization problems in which the criterion has, for example, the form

$$J(u) = \int_{t_o}^{t_1} f_o(x,u,t)dt \quad . \tag{12}$$

Other systems for obtaining existence theorems are presented in [205]. Analagous results for the particular case $f(x,u,t) = g(t,x) + h(t,x)u$, where $u \varepsilon U$, U a compact, convex set, appear in [187].

At the foundation of the proofs, we find Fillipov's lemma on implicit functions. This lemma has been successively generalized in succeeding investigations [148, 173, 209]. It is not difficult to prove that, in general, each of the basic functions involved in the formulation of Fillipov's theorem are essential. However, for particular cases of system (1), special boundary conditions, and special criteria, they may turn out to be unnecessary. In fact, in [194b] it is shown that the theorem on existence of optimal minimal time controls for the system

$$\frac{dx}{dt} = A(t)x + b(u,t) \quad , \qquad u \varepsilon U \quad ,$$

with continuous $A(t)$, $b(u,t)$ and compact set U can be proven without the additional assumption of convexity. Weakened conditions for Fillipov's theorem also appear in the works [145,159]. In general, the problem of minimizing a functional along trajectories of a dynamical system has no solution even when the set of admissible controls is extended to the class of measurable functions. For validity of an existence theorem for optimal controls, a more radical extension of the class of admissible controls is required. This problem was solved by R.V. Gamkrelidze in the theory of optimal chattering regimes.

It is known that it is always possible to construct a sequence of controls along which the sequences of values of the functional converge to a lower bound. If the sequence of trajectories uniformly converges to some curve satisfying the boundary conditions, then the limiting curve is called an optimal chattering regime [26,38e]. To study optimal chattering regimes, R.V. Gamkrelidze used the following construction.

Consider the problem of minimizing the functional (12) along the trajectory $x(t)$, $t_o \leq t \leq t_1$, of the system (1) using measurable functions $u(t)$ with values in a given compact set U. Instead of this problem, we look at the problem of minimization of the functional

$$J(u,\alpha) = \int_{t_o}^{t} \sum_{i=1}^{n+2} \alpha_i(t) f_o(x,u_i,t)\,dt \qquad (13)$$

along trajectories of the system

$$\frac{dx}{dt} = \sum_{i=1}^{n+2} \alpha_1(t) f(x, u_i, t) \tag{14}$$

with controls $\alpha_i(t)$, $u_i(t)$, $i = 1,2,\ldots,n+2$, satisfying the conditions

$$\alpha_i(t) \geq 0 \quad , \qquad \sum_{i=1}^{n+2} \alpha_i(t) = 1 \quad , \qquad u_i(t) \varepsilon U \quad .$$

If the trajectories of the system (14) are uniformly bounded, then the last problem satisfies the conditions of the existence theorem for optimal controls since the set

$$\left\{ \begin{matrix} \sum \alpha_i(t) f_o(x, u_i, t) \\ \sum \alpha_i(t) f(x, u_i, t) \end{matrix} \;\middle|\; \begin{matrix} \alpha_i(t) \geq 0 \;, \; \sum \alpha_i(t) = 1 \\ u_i(t) \varepsilon U \end{matrix} \right.$$

is convex in (n+1)-dimensional space. By means of the optimal controls $\alpha_i^o(t)$, $u_i^o(t)$, a sequence of controls realizing an optimal chattering regime for (1) is constructed [38e, 190b, 221b]. The question of existence of optimal controls and chattering regimes has also been studied in [45b, 74e,f, 78, 222].

Necessary conditions for optimality are currently the most well-developed area in the theory of optimal processes. The first results for linear equations were obtained in [86, 124a, 142, 169, 211]. A fundamental event in the theory of optimal processes, signifying the opening of a new branch of the calculus of variations, was the announcement in 1956 of the maximum principle [19] by L.S. Pontryagin, V.G. Boltyanskii, and R.V. Gamkrelidze.

According to the maximum principle, an optimal control $u^o(t)$ (in the two-point minimal time problem for (1)) possesses the property that there exists a nontrivial solution $\psi(t)$ of the system

$$\frac{d\psi}{dt} = -A'\psi \quad , \qquad A = \frac{\partial f(x^o(t),u^o(t),t)}{\partial x} \quad ,$$

along which the function

$$H(x^o(t),\psi(t),u,t) = \psi'(t)f(x^o(t),u,t)$$

assumes its maximum over all elements $u \in U$:

$$H(x^o(t),\psi(t),u^o(t),t) = \max_{u \in U} H(x^o(t),\psi(t),u,t) \quad . \tag{15}$$

This result, proved initially for time-optimal linear systems, was obtained for the general case of the functional (12) by V.G. Boltyanskii in [18a]. The basic construction, using needle-variations and the theorem on separation of convex sets, allows us to overcome difficulties with the methods of the calculus of variations connected with the closed set U. An analagous construction was developed by R.V. Gamkrelidze for problems with state variable constraints [38c,d]. Later, A.Ya. Dubovitskii, A.A. Miliotin, R.V. Gamkrelidze, L. Neustadt, and others presented [38f, 50b, 77a,b, 156, 194c] new schemes of proof for the maximum principle, which gave rise to an essential broadening of the circle of problems solvable by this method.

After the appearance of the maximum principle, attempts were made to generalize the classical calculus of variations to optimization and control problems. It turned out that the prerequisites for completion of this goal already existed in old works on the calculus of variations. The modern works of R. Kalman, A.I. Lur'e, V.A. Troitskii and others [87, 92, 94, 121, 130a, 165, 176, 191] have shown the possibility of reducing a large number of problems to problems of the classical calculus of variations.

Use of dynamic programming in the theory of necessary conditions for optimality has been impeded by the requirement of smoothness of the Bellman function. A fundamental contribution here was made by V.G. Boltyanskii [18c]. Effective application of the dynamic programming method was obtained in the linear case (3) for the problem of minimizing the quadratic functional

$$J(u) = \int_0^{t_1} (x'Mx + u'Ru)dt , \quad t \leq \infty , \quad M > 0 ,$$

which was considered in detail by A.M. Letov [87a]. In the theory of necessary conditions for optimality, the method of increments, developed in the work of L.I. Rozonoer [114b] (see also [29]), has turned out to be fruitful.

The set of methods based on theorems of functional analysis which are applicable for optimization of linear control systems, have been given the name "the functional-analytic approach". A pioneering work in this direction is the investigation of R. Bellman and his co-workers [138], where the theorem about the separation of convex sets is explicitely used. The problem of moments (in early works it is called the L-problem of moments) was introduced into the arsenal of tools for the theory of optimal processes by N.N. Krasovskii [74a,b].

The explosion of literature in the theory of necessary conditions for optimality for ordinary dynamical systems (1) (the basic mass of literature on optimal processes currently contains several thousand or more articles related to the theory of necessary conditions for optimality) has now reached the state where it is possible to write a maximum principle (or analagous condition) for any optimization problem.

It was noted above that use of constructive work [38e] allows us to be relieved of responsibility with respect to the existence of optimal controls in the problem (1),(12). Instead, one must consider the problem (13),(14). However, there is one difficulty on this path: we are totally deprived of information about the form of the optimal control which is usually supplied by the maximum principle. Actually, the maximum principle for (13), (14) leads to the condition

$$H(x^o(t),\ \psi(t),\ u_1^o(t),\dots,u_{n+1}^o(t),\ \alpha_1^o(t),\dots,\alpha_{n+1}^o(t),t)$$

$$= \max_{\substack{u_i \varepsilon U \\ \alpha_i \geq 0,\ \sum \alpha_i = 1}} H(x^o(t),\ \psi(t),\ u_1,\dots,u_{n+1},\alpha_1,\dots,\alpha_{n+1},t)$$

$$= \max_{\substack{\alpha_i \geq 0 \\ \sum \alpha_i = 1}} \sum \alpha_i \max_{u_i \varepsilon U} \psi'(t)f(x^o(t),u_i,t) = \max_{u \varepsilon U} \psi'(t)f(x^o(t),u,t)\ ,$$

i.e. the function (15) does not depend upon the controls α_i, $i = 1,\dots,n+1$. This is a particular case of the situation where, along some control $u^*(t)$, the function (15) does not depend upon the control parameters. Such a control $u^*(t)$ is called singular in [114b]. Many problems of space navigation [89] lead to singular controls. All of this serves as motivation for a thorough investigation of singular controls. In the simplest cases, singular controls are easily found from their definition [114n]. For complex cases, a deeper investigation is required [17, 24, 45a, 50c, 54, 174, 178a, 214].

The first results for the necessary conditions of optimality were based upon the controls obtained by H. Kelly [63a]. This success was, to a great degree, determined by use of a new type of control variation differing from the traditional needle-shaped variations.

Further generalizations, based on new control variations, were made by R. Kopp, H. Moyer, and H. Robbins [71,113]. The case of many singular controls appeared in the investigations of I.B. Vapnyarskii [26].

A series of control system optimization problems is connected with the best choice of parameters. The problem of optimal control with parameters, considered by V.G. Boltyanskii [18b] (see also [18d, 164, 182, 184, 224]), was formulated in the following form. Among q-dimensional vectors w and r-dimensional measurable functions u(t) with values in the set U, choose $w^o, u^o(t)$ for which

$$J(u^o, w^o) \leq J(u,w) = \int_0^{t_1} f_o(x,u,w)\,dt \quad ,$$

when the system dynamics have the form

$$\frac{dx}{dt} = f(x,u,w) \quad , \qquad x(0) = x_o \quad .$$

Necessary conditions for optimality of parameters and controls are obtained for this problem in [18d]. Progressively, results obtained in the theory of necessary conditions for optimality for the ordinary dynamical system (1), have been transferred to objects described by more complex equations. Systems with delays are investigated in the series of works [72, 74m, 104a, 127, 179, 195, 203]. With complexity of the system, the description of the maximum principle appreciably loses its original compactness. Necessary conditions for optimality in distributed parameter systems have been studied by A.G. Butkovskii, Yu.V. Egorov, K.A. Lur'e, T.K. Sirazetdinov, and others [21a,d, 22, 47, 52a, 53a, 93, 118]. Here, except for integral equations in [21d,30], different partial differential equations [52,53a] are considered. There have also appeared works transferring the maximum principle to equations in Banach spaces [53b, 137a, 151, 158, 160c, 162]. In the latter case there arise difficulties and the maximum principle in general form has not been proven.

The theory of sufficient conditions for optimality developed up to now, has been based either on dynamic programming or its extensions (cf. the works of V.G. Boltyanskii, N.N. Krasovskii, V.F. Krotov [18c, 74g,h, 77c]). A typical result has the following form. If a smooth function $S(x,t)$ is given, satisfying the equation

$$\frac{\partial S(x,t)}{\partial t} = \max_{u \in U} \left[- \frac{\partial S'(x,t)}{\partial x} f(x,u,t) - f_o(x,u,t) \right] \tag{16}$$

and the condition $S(x,t_1) = \psi(x)$, then the control $u(x,t)$, chosen according to the rule

$$- \frac{\partial S'(x,t)}{\partial x} f(x,u(x,t), t) - f_o(x,u(x,t), t)$$

$$= \max_{u \in U} \left[- \frac{\partial S'(x,t)}{\partial x} f(x,u,t) - f_o(x,u,t) \right] , \tag{17}$$

provides the minimum value $S(x_o,t_o)$ of the functional

$$J(u) = \psi(x(t_1)) + \int_{t_o}^{t_1} f_o(x,u,t)dt$$

along trajectories of the system (1), with measurable controls $u(t)$ having values in some set U. In this approach to optimal control problems, the basic difficulty is the solution of Eq. (16), but the result (17), expressing the optimal control as a function of the state x and time t, in many cases turns out to be preferable to the result from the maximum principle where the optimal control is obtained in the form of a function of time.

The requirement of smoothness of the Bellman function may be relaxed [18c]. Using dynamic programming,

V.G. Boltyanskii [18d] proved sufficiency of the maximum principle for regulator synthesis.

Another effective method for obtaining sufficient conditions of optimality for nonlinear systems is the method of increments [114a,b]. In the work of L.I. Rozonoer [114a], local optimality of controls satisfying the maximum principle is proved. The method of increments allows us to directly use peculiarities of the problem under consideration (linearity, convexity, etc.). Sufficiency of the maximum principle is proved by functional analytic methods for a series of linear optimization problems in [50a,74d].

The important algorithms for computation of optimal controls differ as to the basic idea used and also the choice of space upon which the iteration is constructed. Gradient methods are based upon necessary conditions for optimality. Among algorithms using iteration in control space, the algorithms presented in [10, 20, 48a, 49, 63b, 132, 133, 141, 152] are typical. In [7, 31, 84, 197] relations using Lagrange multipliers (or penalty functions) are superimposed upon the control system. In [133] a procedure is described for successively decreasing the residuals in either the side relations or the value of the functional. Algorithms of this type lead, generally speaking, to a local minimum.

Algorithms based on iteration in the control system state space proceed on the basis of dynamic programming. Such schemes are very convenient for problems where the basic constraints are on the states of the system. Among algorithms of this sort are discrete dynamic programming schemes [13], the method of successive analysis of variants [95], the method of telescoping, and the method of local variations [6, 98b, 130]. It is noted in [13] that algorithms based on dynamic programming [9, 11b, 14, 37, 98a] lead to optimal controls but require a large high-speed computational memory for accurate solutions. A series of algorithms are known using iteration in the space of dual variables [5, 43, 57, 67, 79, 101, 150a, 181, 199]. Basically, these algorithms apply to optimization problems which are linear in the state variables. These algorithms lead to the determination of optimal controls.

Under the term "discrete" control system is meant a system whose dynamics are described by a recurrence relation (or difference equation)

$$x(k + 1) = f(x(k), u(k), k) \quad . \tag{18}$$

Dynamic programming is a universal method for the investigation of discrete systems. However, its application to concrete problems [122] may turn out to be less effective than utilization of the maximum principle. Thus, the theory of optimal processes for discrete systems is a real problem. A series of problems in optimal processes for continuous systems have a trivial solution for the discrete analogue. However, problems connected with explicit calculation are complicated and their solution does not have such an attractive form as for the continuous case.

Basic works on the theory of optimal processes in discrete systems, with rare exceptions ([58a]) for controllability, [21b,d] for sufficient conditions of optimality), are devoted to the problem of obtaining necessary conditions for optimality. The final goal of these works is to prove a discrete analogue of the maximum principle. N.N. Krasovskii [74b] has proved one variant of the maximum principle for minimal time problems for the system

$$x(k + 1) = Ax(k) + bu(k) \ , \quad |u| \leq 1 \ , \quad x(0) = x_o \ ,$$

in which minimum norm controls are sought. We formulate this result.

If the control $u^o(k)$, $|u^o| \leq \mu$, $0 \leq k \leq N - 1$, transfers the point x_o to the origin in a minimal time N and there does not exist a control $u^*(k)$, solving this problem with $|u^*(k)| < \mu$, then the optimal control necessarily satisfies the condition

$$\psi'(k)bu^o(k) = \max_{|u| \leq \mu} \psi'(k)bu ,$$

where $\psi(k)$ is a nontrivial sequence of n-vectors satisfying the adjoint relation

$$\psi(k - 1) = A'\psi(k) . \tag{19}$$

L.I. Rozonoer [114b] proved the maximum principle for the problem of minimizing the functional

$$J(u) = c'x(N) \tag{20}$$

along trajectories of the system

$$x(k + 1) = A(k)x(k) + b(u,k) , \quad x(0) = x_o ,$$

with a free right endpoint and a control sequence $\{u(k)\}$, each element of which takes values from a set U. This result takes the following form: along the optimal control $u^o(k)$, the maximum condition

$$\psi'(k)b(u^o(k),k) = \max_{u \in U} \psi'(k)b(u,k) ,$$

is satisfied, where $\psi(k)$ is the solution of Eq. (19) with initial condition $\psi(N-1) = c$.

A.G. Butkovskii [21b] constructed an example of a discrete system whose optimal control does not satisfy a maximum condition. This proved that the many attempts to transfer the maximum principle to general discrete systems were doomed to failure. The original problem now assumed a restricted form: to find classes of discrete systems whose optimal control may be found with the help of a maximum principle. In the works [110, 160b, 161], different methods of proof that the maximum principle

$$\psi'(k)f(x^o(k),u^o(k),k) = \max_{u\varepsilon U} \psi'(k)f(x^o(k),u,k) \quad , \tag{21}$$

$$\psi(k-1) = \frac{\partial f'(x^o(k),u^o(k),k)}{\partial x}\psi(k) \quad , \tag{22}$$

is satisfied for minimal time problems (and in problems of minimization of the function (20)) are given for the system (18) if the set

$$\{f(x,u,k): \quad u\varepsilon U\} \quad , \tag{23}$$

is convex.

The convexity requirement (23) may be replaced [110, 167, 168] by a one-sided convexity condition (convexity in a direction [167b]) if a functional of the form

$$J(u) = \sum_{k=o}^{N-1} f_o(x(k),u(k),k)$$

is considered. In this case, the set

$$\left\{\begin{matrix} f(x,u,k) \\ f_o(x,u,k) \end{matrix} \middle| \, u\varepsilon U \right\}$$

is considered in n+1 dimensional space.

By virtue of the example from [21b], many investigations have directed their efforts to a localization of the maximum principle in discrete systems. Different assumptions, replacing the global maximum condition of (21) by a local maximum condition and conditions of stationarity are formulated in [21b,d, 110, 144, 171, 172, 175, 177, 201, 202, 204, 210, 216, 223]. Another approach to necessary conditions of optimality, which in some sense preserves the global character of these conditions, is given in [34j].

Completing this brief survey of basic results in the theory of optimal processes, we note a series of very interesting results related to other parts of this theory. These are the works in stochastic control problems, optimal adaptive processes and, finally, differential games. For these questions, we suggest consulting the publications [1, 11c, 41, 42, 44, 51, 55, 62, 68, 69, 70, 73, 74i, 76, 81b, 97a,b, 99, 108b,d, 120, 124b,c, 136, 155, 183, 188, 192, 206, 217, 225]. We find still more material in the surveys [88a, 117, 139b].

§3. Contents of the Monograph

The basic problem which motivated the authors to write this book is the demonstration of two solution methods for the main problems in the theory of optimal processes. The first method, which is systematically used, is called the method of increments and is a development for a wide circle of problems of the analogous method in the work [114b]. These problems arise as particular cases of the theory of necessary conditions of optimality. As a foundation of the second method of investigation of optimal processes, we use several theorems of functional analysis. Besides the examples used in [21d, 74n, 80], here we use a new development. Each of the methods used has a characteristic region of effective application. Roughly speaking, the first method is used for the study of nonlinear problems, the second for the more restricted class of problems connected with linear systems. In order to eliminate possible questions we immediately note that, although the second method has a more restricted region of application, several results obtained by the methods of functional analysis in this domain cannot be obtained by the more general first method. Thus, these methods complement each other.

The basic objects of study in this book are the ordinary dynamical systems (1) and systems with a delay, along with their linear versions. It seems to us that these two classes of systems sufficiently reflect the essence of many systems and are simple enough to allow us to carry out investigations with a minimal amount of computation. In order not to fill the text with tedious computations, the authors have decided on the basic idea of illustration by simple cases. The reasoning behind this is that generalization of the proved results does not entail any great difficulties. Often the general case follows without proof.

Each of the methods used may be successfully applied to the study of optimization problems for systems differing from (1). As an exception, in Chapter VI results on the maximum principle are deduced for broader classes of systems. This is done only to illustrate the effectiveness of a new form of the maximum principle.

It is known that a bibliography on the theory of optimal processes must number several thousand entries. A proper survey of this mass of work requires a large special effort of its own. In this monograph, the bibliography contains only those works which are cited in the text.

We pass to a brief survey of the contents of the book.

The monograph consists of two parts. In the first part we study questions which are not directly connected with problems of optimization of control systems but which sometimes play a decisive role for the investigation of related problems.

The first two chapters are devoted to the problem of controllability. In the first chapter, the concept of controllability according to Kalman is introduced, while in the second investigations are begun with a new concept of controllability which is simpler than the first. The known results from Chapter I are deduced along the way for completeness of the study.

In Chapter III the observability of a dynamical system is defined as the concept strictly dual to the controllability concept introduced in Chapter II. In this connection, we have replaced Kalman's terminology "costate" by the more transparent (to our way of thinking) word "direction", having in mind the known duality between a "point" and a "direction". The duality principle formulated in this chapter differs from the analogous result of Kalman. In retrospect, it is clear that the circle of questions considered in Chapters II-IV are one and the same. To a great degree, this is explained by the method of investigation used in these chapters. New formulations of the problems from Chapters II-IV allow us to directly and effectively apply the method of increments for the solution of questions which, up until now, had not been considered anywhere. In each of these chapters, methods of functional analysis are cited and schemes for their application are given.

Chapter IV is closely connected with the preceding ones both as to the problem statements and the solution methods. Out of respect for tradition, we call the basic problem of this chapter the problem of identification but, of course, it is really a problem of observability but now not relative to the state of the object, but to its parameters. A duality principle is formulated in this chapter showing the relationship between the questions considered here and the material of Chapter I. Problems from Chapters II-IV admit a statistical formulation (sometimes more natural than that actually given here), but these questions for a series of reasons are not considered in this monograph.

The connecting element between the first and second parts of the book is Chapter V. Specialists in applied problems characteristically are not concerned with general questions of the existence of solutions, but similar problems in the theory of optimal processes play an important and basic role . In this chapter, which is mainly based on the ideas of A.F. Fillipov and R.V. Gamkrelidze, a series of sufficient conditions for the existence of optimal controls are presented. Finally, consideration of specific problems allows us to find both new conditions and new methods of proof for existence theorems. Here are given hypotheses about connections between questions of the existence of optimal controls in continuous sytems and the possibility of extending the maximum principle to discrete systems.

The second part of the monograph is devoted to the maximum principle in its different aspects. This part (Chapters VI-IX) exceeds the first in volume of material and contains the traditional questions in the theory of optimal processes.

Chapter VI covers a wide circle of questions connected with necessary conditions for optimality and, in the first place, with a statement of the Pontryagin Maximum Principle. First of all, a simple proof of this result is given with the help of the method of increments. Next it is transferred to systems with a delay. Analysis of the method of proof reveals that the maximum principle may be given new form for a large class of systems and is invariant within this class.

In Chapter VI a great deal of attention is given to singular controls, differing from others in that along them the maximum principle is no longer effective. The method of studying singular controls is based upon a subsequent development of the method of increments [114b]. Also in this chapter are introduced necessary conditions defined on controls satisfying the Pontryagin maximum principle that, in many cases, allow us to restrict the set of controls which are candidates for optimality. As a second method for reducing the number of pretenders to the optimal control, we present the reduction of the original problem to a problem of optimization with parameters. The latter problems are of independent interest and, therefore, in Chapter VI we present the basic necessary conditions of optimality for these problems. The remaining part of the chapter is devoted to an application of functional analysis for the investigation of global properties of optimal systems. This question was already considered in the monographs [21d, 74n, 80]. However, our path differs in an essential way from that taken in these books. A first difference is that in the earlier monographs information necessary for the proofs of the facts from functional analysis which are used does not follow from elementary analysis and linear algebra. Another difference is that here we use a systematic application of a series of theorems of functional analysis (among them the L-problem of moments, which in [21d, 74n, 80] is placed at the basis of the investigation). At the basis of our investigations are the minimax theorem and the theorem about the separation of convex sets. In a series of cases, this allows us to by-pass difficulties connected with application of the L-problem.

In Chapter VII results are given about the question of sufficient conditions for optimality, on the uniqueness of optimal controls, and on the well-posed statement of several problems in the theory of optimal processes. Here also are considered optimization problems with a hierarchical criterion function.

Chapter VIII has as its only goal the illustration of the principal possibilities for application of the methods of this monograph to computational problems of optimal control. A detailed description of algorithms, their experimental testing, and other questions connected with this line of investigation will be considered elsewhere.

The material of Chapters VIII and Chapter IX on discrete systems concludes the monograph. The theory of optimization of discrete systems is important in control problems and is connected with continuous systems. In Chapter IX results of the authors are collected concerning the basic necessary conditions for optimality.

These remarks give an initial impression about the contents of the book.

§4. Basic Notation

The majority of operations in this book are carried out in finite-dimensional spaces. Infinite-dimensional spaces are conveniently introduced with the help of several symbols. No special knowledge in functional analysis or operator theory is required of the reader. Several delicate questions of function theory are explained with elementary tools, making no pretense to rigor. Conditions on the type of smoothness of the functions which is necessary to make the operations valid, are not always stated. Such an approach saves redundant and completely obvious conditions which, if stated for every concrete case, would only expand the text.

The finite-dimensional (n-dimensional) vector space [46,112] is denoted by the symbol E_n. The dual space to E_n is denoted by E_n'. Thus, the elements of E_n are column vectors, while the elements of E_n' are row vectors. Row vectors are denoted with a prime. For example x, $x \varepsilon E_n$, is an n-column vector, while x' is an n-row vector. The symbol $g'x$ denotes the scalar product of the vectors g and x: $g'x = \sum_{i=1}^{n} g_i x_i$. For denoting the lengths (norms) of vectors x in E_n, we use the symbol $||x||$. The actual meaning of this symbol may be different in different cases. Typical (standard) forms of the norm are:

$$||x|| = \left(\sum_{i=1}^{n} x_i^2 \right)^{1/2} = \text{euclidean norm}$$

$$||x|| = \left(\sum_{i=1}^{n} |x_i|^p \right)^{1/p} , \quad p \geq 1 ,$$

$$||x|| = \max_{1 \leq i \leq n} |x_i| ,$$

(the last form is obtained from the preceding for $p \to \infty$). The norm of an element $g \varepsilon E_n'$ is obtained from the norm in E_n according to the rule

$$||g|| = \max_{||x|| \leq 1} g'x .$$

The norms $||g||$ corresponding to the above norms $||x||$ have the form

$$||g|| = \left(\sum_{i=1}^{n} g_i^2 \right)^{1/2} ,$$

$$||g|| \quad \left(\sum_{i=1}^{n} |g_i|^q \right)^{1/q} , \quad \frac{1}{q} + \frac{1}{p} = 1 ,$$

$$||g|| = \sum_{i=1}^{n} |g_i| .$$

If $||g||$ is a norm associated with $||x||$, then it is not difficult to show that

$$||x|| = \max_{||g|| \leq 1} g'x \quad .$$

In many cases it is sufficient to restrict attention to the euclidean norm. Then the associated norm will also be euclidean. To denote a linear system of n-vector functions x(t) defined on a set T, we use the symbol $E_n(T)$. Sometimes, the simpler symbol $E_n(\cdot)$ is also used; elements of $E_n(\cdot)$ are denoted by $x(\cdot)$. The family of continuous functions having n-continuous derivatives is denoted by the symbol C^n; C is the family of continuous functions.

Different classes of admissible controls are used in this book. The class of piecewise-continuous, r-vector functions with values in a bounded subset $U \subset E_r$ is denoted by D; D_1 is the class of piecewise-continuous and piecewise-smooth functions with values from U; F is the class of measurable, r-dimensional vector functions with values in U.

We define piecewise-continuous functions to be those functions defined on some set, having a finite number of points of discontinuity of the first kind on each bounded subset. The meaning of the term "measurable function" is the following: u(t) is measurable on a set $T = [t_o, t_1]$ if and only if for any $\varepsilon > 0$ it is possible to find a continuous function v(t) and a set σ of intervals from T whose total measure is less than ε, such that $u(t) \equiv v(t)$ for all $t \in T - \sigma$. Hence, it is clear that continuous functions are measurable, as are piecewise-continuous functions and functions having a finite or countable number of discontinuities on T.

For some classes of admissible controls, we use the special notations

$$U_\infty^L(\cdot) = U_\infty^L(T) = \{u(t): \quad u(t)\varepsilon F \quad , \quad \operatorname*{vrai\,max}_{t\varepsilon T} ||u|| \le L\} \quad ,$$

$$U_p^L(\cdot) = U_p^L(T) = \left\{u(t): \quad u(t)\varepsilon F \quad , \int_{t_0}^{t_1} ||u(t)||^p \, dt \le L\right\} \quad ,$$

$U(\cdot) = U(T)$ is the given family of controls. Here $\operatorname*{vrai\,max}_{t\varepsilon T} ||u(t)||$ is calculated in the following way: we select a set w of zero measure in T and find $\sup_{t\varepsilon T-w} ||u(t)||$. Then $\operatorname*{vrai\,max}_{t\varepsilon T} ||u(t)||$ is taken to be $\inf_{w} \sup_{t\varepsilon T-w} ||u(t)||$, where w is taken over all possible null sets in T. Clearly, continuous and piecewise continuous functions satisfy

$$\operatorname*{vrai\,max}_{t\varepsilon T} ||u(t)|| = \max_{\varepsilon} ||u(t)|| \quad .$$

For given functions $u(t)$, $t\varepsilon T$, $x(t)\varepsilon E_n(T)$, the expression

$$f(x(\cdot)) = \int_{t_0}^{t_1} u'(t)x(t) \, dt \quad ,$$

forming a number from these functions is called a linear functional. It is not difficult to see that

$$\left.\begin{aligned} \sup_{u(\cdot)\varepsilon U_\infty(\cdot)} f(x(\cdot)) &= \int_{t_0}^{t_1} ||x(t)|| \, dt \quad , \\ \sup_{u(\cdot)\varepsilon U_\infty(\cdot)} f(x(\cdot)) &= \left(\int_{t_0}^{t_1} ||x(t)||^q \, dt\right)^{1/q} \end{aligned}\right\}, \tag{24}$$

where $||x||$ corresponds to the meaning of $||u||$. In general, any operation of composition of the given function which yields a number is called a functional (nonlinear). When the function is a vector, then we speak of an operator. Let $u(t)$, $t\varepsilon T$, be a function, x a vector. If there is given a rule by which to each $u(\cdot)$ it is possible to find an x, then we write $x = S(u(\cdot))$. We will have occasion to deal with operators of the form

$$x \equiv Su(\cdot) = \int_{t_0}^{t_1} S(t_1,t)u(t)\,dt \quad , \tag{25}$$

$$x \equiv S(u(\cdot)) = \int_{t_0}^{t_1} S(t_1,t,u(t))\,dt \quad ,$$

where $S(t_1,t)$ is an nxr matrix function of two arguments, $S(t_1,t,u)$ is an n-vector function of three arguments. For the linear operator (25), we will often use the symbols $g'S$, $||g'S||$:

$$||g'S|| = \left(\int_{t_0}^{t_1} ||g'S(t_1,t)||^q dt\right)^{1/q} \quad ,$$

$$\text{if } U(\cdot) \equiv U_p(\cdot) \quad .$$

Here the norm in the integrand is consistent with $||u||$.

We note now some traditional symbols: $o(\varepsilon)$: $\frac{o(\varepsilon)}{\varepsilon} \to 0$ as $\varepsilon \to 0$; $O(\varepsilon)$: $O(\varepsilon) \leq k$; int X is the union of all interior points of the set X (points lying in X together with some neighborhood); $\overline{X}$ is the closure of the set X (the union of X and all of its limit points); conv X is the convex hull of the set X (the smallest convex set containing X); det A is the determinant of the matrix A; Sp A is the spur (trace) of the matrix A (the sum of the elements lying on the main diagonal); E is the unit matrix; δ_{ij} is the Kronecker symbol ($\delta_{ij} = 0$, $i \neq j$, $\delta_{ii} = 1$);

$$\text{sign } \alpha = \begin{cases} 1 \ , & \alpha > 0 \ , \\ -1 \ , & \alpha < 0 \ , \end{cases}$$

The sign 0 is not defined (if it is not specifically mentioned); [a] is the integer part of the number a.

Throughout the book capital letters, as a rule, denote matrices, lower case letters are vectors. In those cases where coordinate form is used, the summation convention over repeated indices is used.

In the general case, the ordinary dynamical system is given by the equation

$$\frac{dy}{dt} = g(y,\xi,\eta,t) \quad , \qquad y(t_o) = y_o = y_o\ (\zeta) \quad . \tag{26}$$

Here $y = \{y_1,\ldots,y_n\}$ is the state vector of the system, $\xi = \{\xi_1,\ldots,\xi_r\}$ is the vector of external perturbations, $\eta = \{\eta_1,\ldots,\eta_q\}$ is the parameter vector of the system, and $\zeta = \{\zeta_1,\ldots,\zeta_p\}$ is the parameter vector of the initial state. To each choice of the function $\xi(t)$ and the vectors η,ζ, there corresponds a unique solution $y(t)$, $t \geq t_o$, of Eq. (26). We will call this motion unperturbed (programmed) motion, while generation of the function $\xi(t)$ and the vectors η,ζ is the control program. As usual, the programmed (unperturbed) characteristics of the system will be compared with perturbations which we denote by $\tilde{y}(t)$, $\tilde{\xi}(t)$, $\tilde{\eta}$, $\tilde{\zeta}$. We introduce quantitites representing the deviation of the perturbed from the unperturbed systems:

$$\begin{aligned} x(t) &= \tilde{y}(t) - y(t) \quad , & u(t) &= \tilde{\xi}(t) - \xi(t) \quad , \\ w &= \tilde{\eta} - \eta \quad , & v &= \tilde{\zeta} - \zeta \quad . \end{aligned} \tag{27}$$

The function $x(t)$, $t \geq t_o$ we call the controlled motion, the function $u(t)$, $t \geq t_o$ is the control, while w and v retain their earlier names. The differential equation of controlled (perturbed) motion has the form

$$\frac{dx}{dt} = f(x,u,w,t) \quad , \qquad x(t_o) = x_o(v) \quad , \tag{28}$$

where $f(x,u,w,t) = g(y(t) + x, \xi(t) + u, \eta + w,t) - g(y(t), \xi(t),\eta,t)$, $f(0,0,0,t) \equiv 0$, $x_o(v) = y_o(\zeta + v) - y_o(\zeta)$, $x_o(0) = 0$. Only the equation (28) for perturbed motion is studied in this monograph, since passage to the unperturbed case is clear by (27).

The following form of Eq. (28) is often used in the sequel ($w = 0$):

$$\frac{dx}{dt} = A(t)x + B(t)u + C(t)ux + \frac{1}{2}D(t)xx + \frac{1}{2}G(t)uu + o \quad .$$

Here we use the notations

$$[A(t)]_{ij} = \frac{\partial f_i(0,0,t)}{\partial x_j} \quad , \quad [B(t)]_{iv} = \frac{\partial f_i(0,0,t)}{\partial u_v} \quad ,$$

$$[C(t)]_{ijv} = \frac{\partial f_i(0,0,t)}{\partial x_j \partial u_v} \quad , \quad [D(t)]_{ijk} = \frac{\partial f_i(0,0,t)}{\partial x_j \partial x_k}$$

$$[G(t)]_{iv\mu} = \frac{\partial f_i(0,0,t)}{\partial u_v \partial u_\mu} \quad , \quad i,j,k = 1,\ldots,n \quad ; \quad v,u = 1,\ldots,r \quad ,$$

$$[C(t)ux]_i = \sum_{j=1}^{n} \sum_{v=1}^{r} [C(t)]_{ijv} u_v x_j \quad ,$$

$$[D(t)xx]_i = \sum_{k=1}^{n} \sum_{j=1}^{n} [D(t)]_{ijk} x_k x_j \quad ,$$

$$[G(t)uu]_i = \sum_{v=1}^{r} \sum_{\mu=1}^{r} [G(t)]_{iv\mu} u_\mu u_v \quad ,$$

$$o = o(||x||^2 + ||x|| \; ||u|| + ||u||^2) \quad .$$

In cases when $w \neq 0$, but $u = 0$, the above notation is retained substituting w for u.

The system (28) is called stationary when the right side does not depend on the time t; otherwise we speak about nonstationary systems. By a linear system, we understand

$$\frac{dx}{dt} = A(t)x + B(t)u \quad .$$

Letting the r-vector u denote the input (r-dimensional), we speak about single-input ($r = 1$) systems, about systems with several inputs ($r > 1$), and about linear systems with nonlinear inputs if

$$\frac{dx}{dt} = A(t)x + b(u,t) \quad .$$

The quantity

$$z = h(x,t) \quad , \quad h(0,t) = 0 \quad ,$$

connected with the state x of the system (28), is called the output of the system, where we speak about m outputs if the dimension of the vector z equals m.

The following system of references is used in this book. Formulas, theorems, lemmas, and examples in each chapter have an autonomous (single) number. For reference to a result of the given chapter, a single number is used. If a result relates to a different chapter, then two numbers are used, where the first number refers to the chapter.

PART I: GENERAL QUESTIONS IN THE THEORY OF THE CONTROL OF DYNAMICAL SYSTEMS

CHAPTER I

Controllability of Dynamical Systems

§1. Statement of the Problem. Definitions. The Idea of the Method of Increments in the Theory of Controllability.

We consider the dynamical system

$$\frac{dx}{dt} = f(x,u,t), \quad x(t_0) = x_0, \quad f(0,0,t) = 0 \ . \qquad (1)$$

whose motion is defined on the interval $T = [t_0, t_1]$, $t_1 - t_0 > 0$. Here $x = \{x_1,\ldots,x_n\}$ is the system state vector, $u = \{u_1,\ldots,u_r\}$ is the perturbation vector, and t is the time.

Let some piecewise continuous perturbation $u(t)$, $t \varepsilon T$, give rise to the motion $x(t)$, $t \varepsilon T$, $x(t_0) = x_0$, of the dynamical system (1). We call the perturbation $u(t)$, $t \varepsilon T$, the control, while the motion $x(t)$ is called the controlled motion.

Problem. To find a control under which the controlled motion passes between the points x_0 and $x_1 = 0$: $x(t_0) = x_0$, $x(t_1) = x_1$.

Definitions.

1. The state x_0 of the dynamical system (1) is called TL-controllable is there exists a piecewise continuous function $u(t)$, $||u(t)|| \leq L$, $t \varepsilon T$, such that the corresponding motion $x(t)$ satisfies the condition $x(t_1) = 0$.

2. A dynamical system is called TL-controllable if there exists a number $\alpha = \alpha(T,L)$ such that all states

of the set $\{x = ||x|| \leq \alpha\}$ are TL-controllable.

3. If the state x_o (of a dynamical system) is TL-controllable for some T and L, then it is called controllable.

4. A dynamical system is called completely controllable if it is TL-controllable for any T $(t_1 < t_o)$ and $L > 0$.

5. A dynamical system is called (completely) controllable in the large if it is controllable (completely controllable) and $\sup_{T,L} \alpha(T,L) = \infty$.

The first approach to the study of controllability in this book is based upon the investigation of auxiliary functions. The idea of using auxiliary functions for studying controllability is the following: on the TL-controllable initial states x_o, the system (1) defines a vector function $\phi(x_o)$. With the help of increments considered along trajectories of the system (1), the vector $\phi(x_o)$ is expressed by $u(t)$, $t \in T$, and the parameters of the system (1). Conditions are found under which the set $\{\phi(x_o)\}$ is open. If the auxiliary functions $\phi(x)$ are given by a nonsingular transformation $x \to \phi(x)$, then the conditions obtained are conditions for controllability.

§2. The Formula for Increment of a Vector Function Along Trajectories of a Dynamical System.

We introduce the vector function $\phi(x) = \{\phi_1(x), \ldots, \phi_\ell(x)\}$, $\phi(0) = 0$, admitting the expansion

$$\phi(x) = Vx + o(x) \quad . \tag{2}$$

For a differentiable nxn matrix function $\Psi(t)$ we clearly have the identity

$$\frac{d}{dt}\Psi(t)x(t) = \frac{d\Psi(t)}{dt}x(t) + \Psi(t)\frac{dx(t)}{dt} \quad .$$

Thus,

$$\Psi(t)x(t)\Big|_{t_0}^{t_1} = \int_{t_0}^{t_1} \dot{\Psi}(t)x(t)dt + \int_{t_0}^{t_1} \Psi(t)\dot{x}(t)dt \ . \qquad (3)$$

Let x_0 be a TL-controllable state, i.e. $x(t_1) = 0$. Then from (2) and (3) we have

$$\phi(x_0) = \int_{t_0}^{t_1} \Psi(t)\dot{x}(t)dt + \int_{t_0}^{t_1} \dot{\Psi}(t)x(t)dt + 0(x_0) \ .$$

Conversely, if this equality is satisfied then $\Psi(t_1)x(t_1) = 0$. Instead of $\dot{x}(t)$ we substitute its expression from (1), while for the function $\phi(t)$ we define the equation on T

$$\dot{\Psi}(t) = -\Psi(t)A(t) \quad , \qquad \Psi(t_0) = -V \ .$$

As a result, we obtain the formula

$$\phi(x_0) = \int_{t_0}^{t_1} \Psi(t)B(t)u(t)dt + \eta \quad , \qquad (4)$$

where

$$\eta = \eta_1 + \eta_2 + \eta_3 \quad , \qquad \eta_1 = \int_{t_0}^{t_1} \Psi(t)C(t)u(t)x(t)dt \quad ,$$

$$\eta_2 = \int_{t_0}^{t_1} \Psi(t)o(||x||^2 + ||x||\,||u|| + ||u||^2)dt \quad ,$$

$$\eta_3 = 0(x_0) \ .$$

§3. The Controllability of Stationary Linear Systems

1. Single-Input Systems. We consider a control system for which the equation of controlled motion has

the form

$$\frac{dx}{dt} = Ax + bu \quad , \tag{5}$$

where A is a constant nxn matrix, b is a constant n-vector, and u is a scalar. If in (4) the function $\phi(x)$ is chosen to be the identity transformation $\phi(x) = x$, then we obtain the following representation for the controllable states:

$$x_o = \int_{t_o}^{t_1} \Psi(t)bu(t)dt \tag{6}$$

Here the function $\Psi(t)$, $t\varepsilon T$, satisfies the equation

$$\dot{\Psi} = -\Psi A \quad , \qquad \Psi(t_o) = -E \quad .$$

We integrate the right side of (6) by parts:

$$x_o = -b \int_{t_o}^{t_1} u(t)dt - \int_{t_o}^{t_1} \Psi(t)Ab \int_{t}^{t_1} u(\tau)d\tau dt \quad .$$

We apply this operation to the second term of the last expression, then to the last term of the new expression, and so forth. As a result, after n steps we obtain

$$x_o = \sum_{\beta=0}^{n-1} (-1)^{\beta+1}A^{\beta}b \int_{t_o}^{t_1} \frac{(t - t_o)^{\beta}}{\beta!} u(t)dt$$

$$+ (-1)^n \int_{t_o}^{t_1} \Psi(t)A^n b \int_{t}^{t_1} \frac{(\tau - t)^{n-1}}{(n - 1)!} u(\tau)d\tau dt \, . \tag{7}$$

We show that the vector function $p(t) = \Psi(t)A^n b$ admits the expansion

$$p(t) = \sum_{\beta=0}^{n-1} \lambda_{\beta}(t)A^{\beta}b \tag{8}$$

Actually, since by virture of the Cayley-Hamiltion Theorem [39] each square matrix A satisfies its own characteristic equation

$$\lambda^n + \alpha_1 \eta^{n-1} + \cdots + \alpha_{n-1}\lambda + \alpha_n = 0 \quad .$$

Then we have the equation

$$A^n + \alpha_1 A^{n-1} + \cdots + \alpha_{n-1}A + \alpha_n E = 0 \quad . \tag{9}$$

Multiplying both sides of this equation on the left by $\Psi(t)$, on the right by $A^n b$, and noting that $p^{(k)}(t) = (-1)^k \Psi(t) A^{n+k} b$, we obtain the differential equation for $p(t)$:

$$p^{(n)} - \alpha_1 p^{(n-1)} + \cdots + (-1)^{n-1}\alpha_{n-1}p + (-1)^n \alpha_n p = 0 \quad . \tag{10}$$

We assume that (8) is false, i.e. there exists a time $t = \bar{t}$ and a vector d such that

$$d'p(t) \neq 0, \; d'A^\beta b = 0, \; \beta = 0,1,\ldots,n-1 \quad . \tag{11}$$

By virtue of (10) the function $\nu(t) = d'p(t)$ satisfies the equation

$$\nu^{(n)} - \alpha_1 \nu^{(n-1)} + \cdots + (-1)^{n-1}\alpha_{n-1}\nu + (-1)^n \alpha_n \nu = 0 \tag{12}$$

with the initial conditions

$$\nu(t_o) = -d'A^n b, \; \dot{\nu}(t_o) = d'A^{n+1}b, \ldots, \nu^{(n-1)}(t_o) = (-1)^n d'A^{2n-1}b.$$

But $A^n b = \sum_{\beta=0}^{n-1} \delta_\beta A^\beta b$ and thus from (11) it follows that

$$\nu(t_o) = 0,\ \nu(t_o) = 0, \ldots, \nu^{(n-1)}(t_o) = 0 \quad . \tag{13}$$

The homogeneous equation (12), with the conditions (13), has the unique solution $\nu(t) \equiv 0$, which contradicts assumption (11): $\nu(t) \not\equiv 0$. The asserted expansion (8) is proved.

We substitute (8) into (7):

$$x_o = \sum_{\beta=0}^{n-1} A^{\beta} b \int_{t_o}^{t_1} \left[(-1)^{\beta+1} \frac{(t - t_o)^{\beta}}{\beta!} + (-1)^n \int_{t_o}^{t} \lambda_{\beta}(\tau) \frac{(t - \tau)^{n-1}}{(n - 1)!} d\tau \right] u(y)dt \ . \tag{14}$$

Let μ_{β}, $\beta = 0,\ldots,n-1$, be arbitrary numbers. We show that for any $\tau = t_1 - t_o > 0$, there exists a piecewise continuous function satisfying the equation

$$\mu_{\beta} = \int_{t_o}^{t_1} \left[(-1)^{\beta+1} \frac{(t - t_o)^{\beta}}{\beta!} + (-1)^n \int_{t_o}^{t} \lambda_{\beta}(\tau) \frac{(t - \tau)^{n-1}}{(n - 1)!} d\tau \right] \cdot u(t)dt \quad . \tag{15}$$

For this it suffices to prove (cf. §6.11) that the functions

$$\ell_{\beta}(t) = (-1)^{\beta+1} \frac{(t - t_o)^{\beta}}{\beta!} + (-1)^n \int_{t_o}^{t} \lambda_{\beta}(\tau) \frac{(t - \tau)^{n-1}}{(n - 1)!} d\tau \ ,$$

$$\beta = 0,\ldots,n-1; \quad t_o \leq t \leq t_1 \ ,$$

are linearly independent. We assume the contrary: there exist numbers g_{β}, $\beta = 0,\ldots,n-1$, $||g|| = 1$, such that $\Delta(t) = \sum_{\beta=0}^{n-1} g_{\beta} \ell_{\beta}(t) \equiv 0$. Then $\Delta^{(\beta)}(t_o + 0) = 0$, $\beta = 0,\ldots,n-1$. However, direct calculation gives $\Delta^{(\beta)}(t_o + 0) = g_{\beta}(-1)^{\beta+1}$, which contradicts the assumption of the linear dependence of the functions $\ell_\beta(t)$, $t \in T$, $\beta = 0,\dots,n-1$.

Thus, for any μ_β the problem (15) has a solution relative to $u(t)$, $t \in T$.

This result, in conjunction with (14), allows us to formulate the following proposition.

Theorem 1. The set of controllable states of the system (5), and only those, have the form

$$x_0 = \sum_{\beta=0}^{n-1} \mu_\beta A^\beta b \quad , \qquad -\infty < \mu_\beta < \infty \quad .$$

Corollary. For controllability (complete controllability) in the large of the stationary linear system (5), it is necessary and sufficient that

$$\text{rank } [b, Ab, \dots, A^{n-1}b] = n \quad .$$

2. Systems with Several Inputs. A stationary linear system with r inputs has the form

$$\frac{dx}{dt} = Ax + Bu \quad , \tag{16}$$

where A,B are constant nxn, nxr matrices, and $u = \{u_1, \dots, u_r\}$ is a vector. We denote the columns of the matrix B by b^ν, $\nu = 1,\dots,r$. Then the system (16) may be written in the form

$$\frac{dx}{dt} = Ax + \sum_{\nu=1}^{r} b^\nu u_\nu \quad .$$

By virtue of the superposition principle (this may be verified by a direct calculation), we obtain the following result for the system (16).

Theorem 2. The set of controllable states for the system (16), and only those, have the form

$$x_o = \sum_{\nu=1}^{r} \sum_{\beta=0}^{n-1} \mu_\beta^\nu A^\beta b^\nu \quad , \qquad -\infty < \mu_\beta^\nu < \infty$$

<u>Corollary.</u> <u>The stationary linear system (16) is controllable (completely controllable) in the large if, and only if,</u>

$$\text{rank } [B, AB, \ldots, A^{n-1}B] = n \quad .$$

<u>Example 1.</u> The 4<u>th</u> order control system with 4 inputs

$$\left.\begin{aligned}
\frac{dx_1}{dt} &= dx_1 + x_2 + u_1 + 2u_2 + 5u_4 \quad , \\
\frac{dx_2}{dt} &= 4x_1 - x_2 + u_2 + 4u_4 \quad , \\
\frac{dx_3}{dt} &= 6x_1 - x_2 + 2x_3 + x_4 + u_3 + 3u_4 \quad , \\
\frac{dx_4}{dt} &= -14x_1 - 5x_2 - x_3 \quad ,
\end{aligned}\right\} \qquad (17)$$

is completely controllable since in this case

$$A = \begin{pmatrix} 3 & 1 & 0 & 0 \\ -4 & -1 & 0 & 0 \\ 6 & -1 & 2 & 1 \\ -14 & -5 & -1 & 0 \end{pmatrix} , \qquad B = \begin{pmatrix} 1 & 2 & 0 & 5 \\ 0 & 1 & 0 & 4 \\ 0 & 0 & 1 & 3 \\ 0 & 0 & 0 & 0 \end{pmatrix}$$

and the vectors $b^1 = \{1,0,0,0\}$, $b^2 = \{2,1,0,0\}$, $b^3 = \{0,0,1,0\}$, $Ab^3 = \{0,0,2,-1\}$ are linearly independent.

3. The Minimal Number of Inputs of a Controllable System

Let it be possible to choose the nxr matrix B in the control system

$$\frac{dx}{dt} = Ax + Bu \quad , \tag{18}$$

where A is a fixed nxn matrix. The number r characterizes the number of inputs to the system. For synthesis of a control system, the following question may arise: what is the minimal number r of inputs for which the system is controllable for a fixed matrix A? Generally speaking, the number r may assume any value from 1 to n.

Example 2. The control system described by the nth order equation

$$x^{(n)} + a_n x^{(n-1)} + \cdots + a_1 x = bu$$

$$(x, a_i, b, u = \text{scalar})$$

is always controllable with a single input. Actually, in this example after introducing the variables $x_1 = x$, $x_2 = \dot{x}, \ldots, x_n = x^{(n-1)}$, the system assumes the form

$$\dot{x}_1 = x_2 \quad ,$$

$$\cdots\cdots\cdot$$

$$\dot{x}_{n-1} = x_n \quad ,$$

$$\dot{x}_n = -a_1 x_1 - \cdots - a_n x_n + bu \quad ,$$

$$A = \begin{pmatrix} 0 & 1 & 0 & \cdots & \cdots & 0 \\ 0 & 0 & 1 & 0 & \cdots & 0 \\ \vdots & & & & & \\ 0 & \cdots & \cdots & \cdots & 0 & 1 \\ -a_1 & -a_2 & \cdots & \cdots & -a_{n-1} & -a_n \end{pmatrix}, \qquad b = \begin{pmatrix} 0 \\ \vdots \\ 0 \\ b \end{pmatrix}$$

Clearly, the vectors

$$b = \begin{pmatrix} 0 \\ \vdots \\ 0 \\ b \end{pmatrix}, \qquad Ab = \begin{pmatrix} 0 \\ \vdots \\ 0 \\ b \\ -a_n b \end{pmatrix}, \ldots, \quad A^{n-1}b = \begin{pmatrix} b \\ -a_n b \\ \vdots \\ \vdots \end{pmatrix}$$

are linearly independent for any a_i and $b \neq 0$.

<u>Example 3</u>. In (18), if the matrix A has the form

$$\begin{pmatrix} a_1 & 0 & \cdots & \cdots & \cdots & 0 \\ 0 & a_2 & 0 & \cdots & \cdots & 0 \\ \cdots & \cdots & \cdots & \cdots & \cdots & \cdots \\ 0 & \cdots & \cdots & \cdots & a_{n-1} & 0 \\ 0 & \cdots & \cdots & \cdots & 0 & a_n \end{pmatrix},$$

then it is clear that a necessary and sufficient condition for controllability of the system (18) is that the matrix B have dimension nxn and be nonsingular.

Combining these two examples, it is possible to construct a control system with any number of inputs.

We now consider the general case of the problem stated above. For this we recall some facts from linear algebra [39]. For an arbitrary $x \varepsilon E_n$, the series

$$x,\ A_x,\ A_x^2, \ldots,$$

contains only a finite number p of linearly independent vectors where

$$A^p x = -\nu_1 A^{p-1} x - \cdots - \nu_p x \quad .$$

The polynomial

$$\lambda^p + \nu_1 \lambda^{p-1} + \cdots + \nu_p$$

is called *the minimal polynomial of the vector x.* The least common multiple of the minimal polynomial of a basis of the space E_n is called *the minimal polynomial of the space E_n.* We denote it by $\psi_1(\lambda)$:

$$\psi_1(\lambda) = \lambda^{p_1} + \nu_1^1 \lambda^{p_1 - 1} + \cdots + \nu_{p_1-1}^1 \lambda + \nu_{p_1}^1 \ . \tag{19}$$

If $p_1 < n$, then there exists a minimal polynomial

$$\psi_2(\lambda) = \lambda^{p_2} + \nu_1^2 \lambda^{p_2 - 1} + \cdots + \nu_{p_2-1}^2 \lambda + \nu_{p_2}^2$$

of the space E_n mod I_1, where I_1 is a subspace with basis

$$b^1, Ab^1, \ldots, A^{p_1-1} b^1 \tag{20}$$

and b^1 is an element of E_n, the minimal polynomial of which coincides with (19). There always exists a vector b^2, the minimal polynomial of which coincides with $\psi_2(\lambda)$. Thus, the vectors

$$b^2, Ab^2, \ldots, A^{p_2-1} b^2 \tag{21}$$

are linearly independent. But, since $\psi_2(\lambda)$ is the minimal polynomial for b^2 mod I_1, then $p_1 + p_2$ of the vectors from (20), (21) are linearly independent. If $p_1 + p_2 < n$, then there exists a minimal polynomial

$$\psi_3(\lambda) = \lambda^{p_3} + \nu_1^3\lambda^{p_3-1} + \cdots + \nu_{p_3-1}^3\lambda + \nu_{p_3}^3$$

of the space E_n mod $(I_1 + I_2)$, where I_2 is a subspace with basis (21). By virtue of the finiteness of n, this process stops at some step r. As a result, we obtain the sequence

$$b^1, Ab^1, \ldots, A^{p_1-1}b^1,\ b^2, Ab^2, \ldots, A^{p_2-1}b^2, \ldots$$

$$\ldots, b^r, Ab^r, \ldots, A^{p_r-1}b^r \tag{22}$$

of linearly independent vectors, where the number r is constructed minimally. Thus, for obtaining (22) it is sufficient to construct polynomials $\psi_\nu(\lambda)$ and vectors b^ν for which these polynomials are minimal.

In linear algebra it is proved that the polynomials $\psi_1(\lambda), \ldots, \psi_r(\lambda)$ are uniquely determined: they coincide with the <u>nontrivial</u> (differing from unity) <u>invariant factors</u> of the matrix A. Thus, we have proved the following theorem.

<u>Theorem 3.</u> <u>The minimal number of inputs for controllability of the system (18) equals the number of nontrivial invariant factors of the matrix A.</u>

The invariant factors $i_1(\lambda), \ldots, i_n(\lambda)$ of a matrix are determined by the formulas

$$i_1(\lambda) = \frac{D_n(\lambda)}{D_{n-1}(\lambda)}, \qquad i_2\ (\lambda) = \frac{D_{n-1}(\lambda)}{D_{n-2}(\lambda)},$$

$$\cdots\cdots\cdots\cdots\cdots\cdots\cdots\cdots\cdots\cdots$$

$$i_n(\lambda) = \frac{D_1(\lambda)}{D_0(\lambda)}$$

$$(D_0(\lambda) \equiv 1),$$

where $D_j(\lambda)$ is the greatest common divisor of all j<u>th</u> order minors of the characteristic matrix $\lambda E-A$ (in $D_j(\lambda)$ the highest coefficient is always taken to be one). For practical calculations of the factors $i_j(\lambda)$, several procedures are known. We describe one of them.

There always exist square polynomial matrices $P(\lambda)$, $Q(\lambda)$ with constant, nonzero determinants such that the matrix

$$P(\lambda)[\ E - A]\ Q(\lambda)$$

has the structure

$$\left\{\begin{matrix} i_{n'}(\lambda) & 0 & 0 & \cdots & \cdots & 0 \\ 0 & i_{n'-1}(\lambda) & 0 & \cdots & \cdots & 0 \\ \cdots & \cdots & \cdots & \cdots & \cdots & \cdots \\ 0 & 0 & \cdots & i_1(\lambda) & 0 & \cdots 0 \\ \vdots & \vdots & \vdots & \vdots & \vdots & \vdots \\ 0 & 0 & \cdots & \cdots & \cdots & 0 \end{matrix}\right\}, \quad (23)$$

where n' is the rank of the matrix $\lambda E-A$. The matrix $P(\lambda)$ is the product of the matrices of left elementary operations required to reduce the matrix $\lambda E-A$ to the form (23). <u>Left elementary operations</u>: 1) multiplication of each row by a number $c \neq 0$; 2) addition of the <u>ith</u> row to the <u>jth</u> row after multiplication by an arbitrary polynomial $b(\lambda)$; 3) interchange of any two rows. Analagous operations with the columns are called <u>right elementary operations.</u> The matrix $Q(\lambda)$ is the product of the matrices of right elementary operations needed to reduce $\lambda E-A$ to the form (23).

The rule for determination of the invariant factors of the matrix A follows from all that has been said above. We form the matrix $\lambda E-A$ and then successively apply left and right elementary operations. These operations may always be applied so that from $\lambda E-A$ we are led to a matrix of the form (23).

<u>Example 4.</u> We find the invariant factors of the matrix A of the system (17). We have

$$\lambda E - A = \begin{pmatrix} \lambda - 3 & -1 & 0 & 0 \\ 4 & \lambda + 1 & 0 & 0 \\ -6 & -1 & \lambda - 2 & -1 \\ 14 & 5 & 1 & \lambda \end{pmatrix} \longrightarrow$$

λ

$$\longrightarrow \begin{pmatrix} \lambda - 3 & -1 & 0 & 0 \\ 4 & \lambda + 1 & 0 & 0 \\ 6 & -1 & \lambda - 2 & -1 \\ 14 - 6 & 5 - \lambda & \lambda^2 - 2\lambda + 1 & 0 \end{pmatrix} \longrightarrow$$

λ − 2

−1

−6

$$\longrightarrow \begin{pmatrix} \lambda - 3 & -1 & 0 & 0 \\ 4 & \lambda + 1 & 0 & 0 \\ 0 & 0 & 0 & -1 \\ 14 - 6 & 5 - \lambda & \lambda^2 - 2\lambda + 1 & 0 \end{pmatrix} \longrightarrow$$

(row operations: $\lambda+1$, $5-\lambda$; column operation: $\lambda-3$)

$$\longrightarrow \begin{pmatrix} 0 & -1 & 0 & 0 \\ \lambda^2 - 2\lambda + 1 & 0 & 0 & 0 \\ 0 & 0 & 0 & -1 \\ -\lambda^2 2\lambda - 1 & 0 & \lambda^2 - 2\lambda + 1 & 0 \end{pmatrix} \longrightarrow$$

(row operation: -1)

$$\longrightarrow \begin{pmatrix} 0 & 1 & 0 & 0 \\ \lambda^2 - 2\lambda + 1 & 0 & 0 & 0 \\ 0 & 0 & 0 & 1 \\ 0 & 0 & \lambda^2 - 2\lambda + 1 & 0 \end{pmatrix}.$$

After interchanging rows and columns, we obtain

$$\lambda E - A \longrightarrow \begin{pmatrix} 1 & 0 & 0 & 0 \\ 0 & 1 & 0 & 0 \\ 0 & 0 & \lambda^2 - 2\lambda + 1 & 0 \\ 0 & 0 & 0 & \lambda^2 - 2\lambda + 1 \end{pmatrix}.$$

The invariant factors of the matrix A are $i_1 = \lambda^2 - 2\lambda + 1$, $i_2 = \lambda^2 - 2\lambda + 1$, $i_3 = 1$, $i_4 = 1$ of which only the first two $i_1(\lambda)$, $i_2(\lambda)$, are nontrivial. Conclusion: in the system (17), the minimal number of inputs necessary for its controllability is two.

From knowledge of the polynomials $\psi_\nu(\lambda)$ it is possible to construct vectors b^ν for which these polynomials are minimal. It suffices to demonstrate this procedure for $\psi_1(\lambda)$. We factor $\psi_1(\lambda)$ into irreducible factors over the real number field

$$\psi_1(\lambda) = [\phi_1(\lambda)]^{c_1}[\phi_2(\lambda)]^{c_2} \cdots [\phi_s(\lambda)]^{c_s} ,$$

where $\phi_1(\lambda),\ldots,\phi_s(\lambda)$ are distinct irreducible polynomials whose highest coefficient is unity. Let E^1 be the union of all vectors x satisfying the equation $\phi_1(A)x = 0$. Analagously, we define $E^2,\ldots,E^s$. In the space E^1 we consider the basis $e_1,\ldots,e_{k_1}$ and we denote those elements whose minimal polynomial is $\phi_1(\lambda)$ by e^1. Analagously, we find the elements e^2 in E^2, and so on. As a result, we obtain vectors $e^1,e^2,\ldots,e^s$. The vector $b^1 = e^1 + e^2 + \cdots + e^s$ has $\psi_1(\lambda)$ as its minimal polynomial.

We illustrate these operations in Example 1. The minimal polynomials $\psi_1(\lambda) \equiv \psi_2(\lambda)$ for this example have factors $\psi_1(\lambda) \equiv \psi_2(\lambda) = (\lambda-1)^2$. Since $\psi_1(A) = 0$, the equation $\psi_1(A)x = 0$ is satisfied for all vectors from E_4. We compute the minimal polynomials for elements of the standard basis $e_1 = (1,0,0,0)$, $e_2 = (0,1,0,0)$, $e_3 = (0,0,1,0)$, $e_4 = (0,0,0,1)$. The minimal polynomial of the element e_1 is $\psi_1(\lambda)$. Thus we obtain $b^1 = e_1$. The minimal polynomial of the space E_4 mod I_1 is $\psi_2(\lambda) = (\lambda-1)^2$. Here I_1 is the subspace with basis e_1, Ae_1. All vectors of the subspace E_4 mod I_1 satisfy the equation $\psi_2(Ax) = 0$. The vectors $e_3 = (0,0,1,0)$, $e_4 = (0,0,0,1)$ are a basis for E_4 mod I_1. The minimal polynomial of the vectors e_3, e_4 is $\psi_2(\lambda)$. Thus we may set $b^2 = e_3$ (or $b^2 = e_4$).

4. Control Systems with Constantly Acting Disturbance

Let the control system have the form

$$\frac{dx}{dt} = Ax + Bu + \gamma(t) \quad , \tag{24}$$

where $\gamma(t)$ is a vector function characterizing the constantly acting disturbances. From the formulas (4) for $\phi(x) = x$, we have

$$x_o = \int_{t_o}^{t_1} \Psi(t)Bu(t)\ dt + \int_{t_o}^{t_1} \Psi(t)\gamma(t)\ dt \quad .$$

Therefore, by virtue of the results following from the transformation (6), we obtain

Theorem 4. The set of controllable states of the system (24), and only those, are expressed by the formula

$$x_o = \sum_{\nu=1}^{r} \sum_{\beta=0}^{n=1} \mu_\beta^\nu A^\beta b^\nu + \int_{t_o}^{t_1} \Psi(t)\gamma(t)\ dt \ , \quad -\infty < \mu_\beta^\nu < \infty \ .$$

Corollary. For controllability (complete controllability) in the large of the system (24), it is necessary and sufficient that

$$\text{rank}\ [B, AB, \ldots, A^{n-1}B] = n \quad .$$

5. A Control Problem from the Theory of Pursuit/Evasion

Let the system (18) describe the motion of the pursuing object, while the equation of the evading object has the form

$$\frac{dx^1}{dt} = Ax^1 + Dv \quad ,$$

where x^1 is an n-vector of states, v is a q-vector of controls, and A,D are constant nxn, nxq, matrices. We say that the problem of pursuit/evasion has an admissible solution if for any vectors x_o, x_o^1 and piecewise continuous function v(t), $t\varepsilon T$, it is possible to find a piecewise continuous control u(t), $t\varepsilon T$, such that $x(t_1) = x^1(t_1)$. Here $x(t_o) = x_o$, $x^1(t_o) = x_o^1$, x(t), $t\varepsilon T$, is the trajectory of the system (18) under the control u(t), and $x^1(t)$ is the trajectory of the evading system under the function v(t).

From Theorem 2 we have the assertion

Theorem 5. The pursuit/evasion problem has an admissible solution if and only if the space spanned by the vectors

$$d^1,\ldots,d^q,Ad^1,\ldots,Ad^q,\ldots,A^{n-1}d^1,\ldots,A^{n-1}d^q \quad ,$$

is a subspace of the space spanned by the vectors

$$b^1,\ldots,b^r,b^1,\ldots,Ab^r,\ldots,A^{n-1}b^1,\ldots,A^{n-1}b^r \quad .$$

§4. Positive, Relative, and Conditional Controllability of Linear Stationary Systems

1. Positive Controllability

Definition 6. The dynamical system

$$\frac{dx}{dt} = Ax + bu \tag{25}$$

is called positively controllable in the large if for any initial state x_o, there is a moment $t = \bar{t}$ and a piecewise continuous, non-negative control u(t), $u(t) \geq 0$, $t\varepsilon[t_o,\bar{t}]$, such taht the trajectory x(t) of the dynamical system transfers the initial state x_o $(x(t_o) = x_o)$ to the origin at time $\bar{t}$ ($x(\bar{t}) = 0$) under the control $u(t)$.

The other definitions of controllability introduced in §1 may be analagously reformulated in terms of non-negative controls. It is clear that if a dynamical system with r controls is controllable, then the same system is positively controllable with 2r controls. Investigation into positive controllability may be carried out by the scheme of §3.

<u>Theorem 6.</u> <u>If the system (25) is positively controllable then the controllable states x_o have the form</u>

$$x_o = \sum_{\nu=1}^{r} \sum_{\beta=0}^{\infty} \mu_\beta^\nu A^\beta b^\nu \quad , \quad (-1)^{\beta+1} \mu_\beta^\nu > 0 \quad ,$$

$$\nu = 1,\ldots,r; \quad \beta = 0,1,\ldots,$$

We introduce the definition

<u>Definition 7.</u> The sequence of vectors

$$a_1,\ldots,a_N \tag{26}$$

forms a <u>positive (non-negative) basis</u> if we may represent any n-vector x in the form

$$x = \sum_{j=1}^{N} \mu_j a_j \quad , \quad u_j > 0 \quad , \quad (\mu_j \geq 0) \quad . \tag{27}$$

By virtue of this definition and Theorem 6, for positively controllable systems the sequence $(-1)^{\beta+1} A^\beta b^\nu$, $\nu = 1,\ldots,r$, $\beta = 0,1,\ldots,$ forms a positive basis. It is not difficult to show that if the sequence (26) forms a non-negative basis, then it also forms a positive basis. Actually, if for some x we find among the μ_j corresponding to it, a $\mu_{j_1} = 0$, then we decompose the vector

$$-a_{j_1} = \sum_{j=1}^{N} \mu_j^1 a_j$$

in the basis (26) and add the identity $a_{j_1} + \sum_{j=1}^{N} \mu_j^1 a_j = 0$ to the right side of the expansion (27). Clearly, in the new expansion the coefficient of a_{j_1} will be positive.

For practical verification that a given sequence forms a non-negative basis, the following lemmas may prove useful:

Lemma 1. In order that the sequence (26) form a non-negative basis, it is necessary and sufficient that each element of the sequence

$$c_1, \ldots, c_{N_1} \quad , \tag{28}$$

forming a non-negative basis, be expressible in the form

$$c_k = \sum_{j=1}^{N} \mu_j^k a_j \quad , \quad \mu_j^k \geq 0 \quad . \tag{29}$$

Lemma 2. In order that the sequence (26) form a non-negative basis in E_n, it is necessary and sufficient that

1) the vectors $a_1, \ldots, a_{N-1}$ form a basis in E_n,

2) $$a_N = - \sum_{i=1}^{N-1} \alpha_i a_i \quad , \quad \alpha_i > 0 \quad . \tag{30}$$

Proof of Lemmas 1 and 2. Necessity of the condition of Lemma 1 follows from Definition 7. Let (29) be satisfied with (28) a non-negative basis. We show that (26) is a non-negative basis. Actually, for each x we have

$$x = \sum_{j=1}^{N_1} \mu_j^o c_j \quad , \quad \mu_j^o \geq 0 \quad .$$

Substituting (29) here, we obtain an expansion of x by elements (26) with non-negative coefficients. Lemma 1 is proved.

Necessity of the conditions of Lemma 2. The first condition is obvious. For proof of the second condition, we decompose the element a_N by (26):

$$a_N = \sum_{i=1}^{N} \beta_i a_i \quad , \qquad \beta_i \geq 0 \quad .$$

Hence,

$$a_N = - \frac{1}{1 + \beta_N} \sum_{i=1}^{N-1} \beta_i a_i \quad .$$

Passage from $\beta_i \geq 0$ to $\alpha_i > 0$ is as above.

Sufficiency. In the expansion

$$x = \sum_{i=1}^{N-1} \mu_i a_i \quad , \tag{31}$$

let the numbers μ_{i_k}, $k = 1,\ldots,m$ be negative. We find $\mu_{i_k} a_{i_k}$ from (30):

$$\mu_{i_k} a_{i_k} = - \frac{\mu_{i_k}}{\alpha_{i_k}} [a_N + \alpha_1 a_1 + \cdots + \alpha_{i_k} - 1] a_{i_k}$$

We substitute this into (31). In the new expansion of the vectors $a_1,\ldots,a_N$ there will clearly be no negative coefficients. Lemma 2 is proved.

It is clear that each positively controllable system is controllable. However, the converse is false.

Example 5. The system

$$\dot{x} = u \quad , \quad (x \text{ a scalar})$$

is controllable but it is not positively controllable. States x_o which may be transferred to the origin by non-negative controls lie on the negative real axis.

Example 6. The controllable system

$$\ddot{x} = u \quad ,$$

also is not positively controllable. The initial states x_o which may be transferred to the origin by non-negative controls are given by the expressions

$$x_o = -\mu_o b + \mu_1 Ab \quad , \quad \mu_o > 0 \quad , \quad \mu_1 > 0 \quad ,$$

where $b = \begin{pmatrix} 0 \\ 1 \end{pmatrix}$, $Ab = \begin{pmatrix} 1 \\ 0 \end{pmatrix}$. Thus, only interior points of the fourth quadrant are positively controllable.

Both controllability and positive controllability of a dynamical system are based on properties of the sequence

$$B, AB, \ldots, A^{n-1}B, \ldots . \tag{32}$$

For investigation of the controllability of a system it suffices to check no more than the first n terms of the sequence (the remainder are linearly dependent on the first n), but for studying positive controllability it is necessary to check no less than the first n+1 terms and it may turn out to be necessary to check an arbitrarily large number of terms.

Example 7. In the positively controllable system

$$\dot{x}_1 = x_1 \cos \theta + x_2 \sin \theta \quad ,$$

$$\dot{x}_2 = -x_1 \sin \theta + x_2 \cos \theta + u \quad ,$$

$$0 < 0 \neq \pi k \quad , \qquad k = 1,2,\ldots,$$

the number of terms of the sequence (32) forming a positive basis approaches ∞ as $\theta \to 0$. Necessary and sufficient conditions for positive controllability from the viewpoint of functional analysis are studied in §9. Here we only note that the conditions of Theorem 6 are not sufficient.

Example 8. Let $\dot{x} = x + u$. This equation is not positively controllable but the conditions of Theorem 6 are satisfied: $-1, +1, -1, \ldots,$ is a positive basis.

2. Relative Controllability

Definition 8. The dynamical system (25) is called *controllable relative to the subspace K* $[Kx = 0]$ (*relatively controllable*) if for each state x_o there exists a number $\bar{t} < \infty$, and a piecewise continuous control $u(t)$, $\bar{t}_o \leq t \leq \bar{t}$, such that $Kx(t) = 0$.

Substituting $\phi(x) = KF(\bar{t})x$ into (4), in complete analogy with Theorem 2 we obtain a representation for the relatively controllable states

$$KF(\bar{t})x_o = \sum_{\nu=1}^{r} \sum_{\beta=0}^{n-1} \mu_\beta^\nu KF(\bar{t})A^\beta b^\nu \quad .$$

Theorem 7. The dynamical system (25) is relatively controllable in the large if and only if

$$\operatorname{rank}\{KB, KAB, \ldots, KA^{n-1}B\} = \operatorname{rank}\{K\} \quad . \tag{33}$$

Remark. Instead of "relative controllability", it is possible to write the words "relative complete controllability", "relative TL-controllability", and so on, the sense of which is completely clear.

Condition (33) for the case of an (n-1)-dimensional subspace K: $k'x = 0$ assumes an especially simple form: not all of the numbers $k'A^{\beta}b^{\nu}$, $\beta = 0,1,\dots,n-1$, $\nu = 1,2,\dots,r$ are equal to zero.

3. Conditional Controllability

Definition 9. The dynamical system (25) is called controllable in the subspace M (conditionally controllable) if each initial state from the subspace M ($xo = my$, $y \varepsilon E_n$) is controllable.

The criterion for conditional controllability of the considered systems follows from the representation of controllable states contained in Theorem 2.

Theorem 8. The dynamical system (25) is conditionally controllable if and only if

$$\text{rank}\ \{M,B,AB,\dots,A^{n-1}B\} = \text{rank}\ \{B,AB,\dots,A^{n-1}B\} \quad .$$

Example 9. In the system (17), let the initial state lie in the plane $x_3 = x_4 = 0$. It is required to determine whether this system is conditionally controllable separately in the first control.

In this case, the subspace M is given by the expression $x = My$, $y \varepsilon E_4$, where

$$M = \begin{pmatrix} 1 & 0 & 0 & 0 \\ 0 & 1 & 0 & 0 \\ 0 & 0 & 0 & 0 \\ 0 & 0 & 0 & 0 \end{pmatrix} .$$

For the vector $b' = (1,0,0,0)$ we have $Ab' = (3,-4,6,-14)$. The conditions of Theorem 8 are satisfied meaning that each state from M may be transferred to the origin by the control u_1. However, all states from the plane $x_1 = x_2 = 0$ may not be transferred to the origin by this control. In this subspace the system is controllable by the third control.

§5. Controllability of Nonstationary Linear Systems

Let there be given the differential equation

$$\frac{dx}{dt} = A(t)x + B(t)u \tag{34}$$

with sufficiently smooth matrix functions $A(t)$, $B(t)$.

Substituting $\phi(x) = x$ from (4), we obtain

$$x_o = \int_{t_o}^{t_1} \Psi(t)B(t)u(t)\, dt \quad , \tag{35}$$

where $\Psi(t)$ satisfies the equation

$$\dot{\Psi} = -\Psi A(t) \quad , \qquad \Psi(t_o) = -E \quad .$$

Integrating the right side of (35) by parts in analogy with (7), we obtain

$$x_o = \sum_{\beta=0}^{n-1} (-1)^{\beta+1} Q_\beta(t_o) \int_{t_o}^{t_1} \frac{(t - t_o)^\beta}{\beta!} u(t)\, dt$$

$$+ (-1)^n \int_{t_o}^{t_1} \Psi(t)Q_n(t) \int_{t}^{t_1} \frac{(\tau - t)^{n-1}}{(n-1)!} u(\tau)\, d\tau dt \quad . \tag{36}$$

Here

$$Q_o(t) = B(t), \; Q_{\beta+1}(t) = A(t)Q_\beta(t) - \dot{Q}_\beta(t) \quad , \tag{37}$$
$$\beta = 0,\ldots,n-1 \quad .$$

The following assertion arises from the representation (36):

Theorem 9. Let $A(t) \in C^{n-2}$, $B(t) \in C^{n-1}$. Then the nonstationary dynamical system (34) is completely controllable if

$$\text{rank}\ \{Q_1(t_o),\dots,Q_{n-1}(t_o)\} = n \quad . \tag{38}$$

The results of §§3,4 on sufficient conditions for controllability may be transferred to nonstationary systems. Clearly, this is explained by the property that when condition (38) is satisfied all controllable states of the system (34) may be written as

$$x_o = \sum_{\nu=1}^{r} \sum_{\beta=0}^{n-1} \mu_\beta^\nu q_\beta^\nu(t_o) \quad ,$$

by virtue of (36). Here $q_\beta^\nu(t)$ is the νth column of the matrix $Q_\beta(t)$.

§6. Conditions for Controllability of Linear Dynamical Systems with a Nonlinear Input

In this paragraph we extend the class of admissible controls to the family of measurable functions.

1. Stationary Systems

Let the control system be described by the equation

$$\frac{dx}{dt} = Ax + b(u) \quad , \tag{39}$$

where A is a constant nxn matrix, and b(u) is an n-vector function, continuous with respect to the r-vector u. Investigation into the controllability of the system (39) is based on the following fact. The set of values assumed by the trajectory x(t) at any time $\bar{t} \geq t_o$, under all possible measurable functions u(t), $t_o \leq t \leq \bar{t}$ with values in a compact set U, is compact and coincides with the set of values assumed by the trajectory y(t), $t_o \leq t \leq \bar{t}$, at time $\bar{t}$, where y(t) satisfies

$$\frac{dy}{dt} = Ay + q(t) \quad .$$

Here q(t) represents all possible measurable n-vector functions with values from the convex hull $\Omega = \text{conv } b(U)$ of the set b(U). The proof of this result is given in [194b] and is based upon a theorem of A.A. Lyapunov [160a] on the region of values of vector-valued functions.

We introduce the number ρ_L:

$$\rho_L = \max [\rho: \{x: ||x|| \leq \rho\} \cap \text{conv } \{b(u): ||u|| \leq L\} \subset \text{conv } \{b(u): ||u|| \leq L\}] \quad ,$$

which is equal to the radius of the largest sphere centered at the origin whose intersection with the convex set $\{b(u): ||u|| < L\}$ is contained in the convex hull of this set. Let $b^1, \ldots, b^r$ be a basis for the minimal subspace containing

$$\text{conv } \{b(u): ||u|| \leq L\} \quad . \tag{40}$$

We let B denote the matrix $[b^1, \ldots, b^r]$.

Theorem 10. *If the following conditions are satisfied:*

1) *for some L, the origin lies in the core of the set (40);*

2) *rank $[B, AB, \ldots, A^{n-1}B] = n$,*

then for any T, $t_1 < \infty$, the dynamical system (39) is TL-controllable.

For proof of this theorem, it suffices to show that the controllable states of the system

$$\frac{dx}{dt} = Ax + Bu$$

for sufficiently small controls $||u|| \leq \alpha$ are, by virtue of the above fact and condition 1) of the theorem, controllable for the system (39). Conditions 1),2) of the theorem insure the TL-controllability of any state from some neighborhood of the origin for any T (Theorem 1). These remarks are sufficient for the following converse of Theorem 10.

Theorem 11. If the system (30) is controllable, then condition 2) of Theorem 10 is satisfied.

It is not difficult to prove the following assertion.

Theorem 12. If, in addition to the conditions of Theorem 10, the matrix A is asympototically stable ($\rho_L > 0$ for $L > 0$, $\sup_L \rho_L = \infty$), then the system (39) is controllable (completely controllable) in the large.

Remark. The theorem may be strengthened. The system (39) is TL-controllable for any $t_1 > 0$ if there can be found non-zero vectors $c^1, \ldots, c^m$ from (40) such that 1) $\gamma_1 c^1 + \cdots + \gamma_m c^m = 0$, $\gamma_i > 0$, 2) rank $[C, AC, \ldots, A^{n-1}C]$ $= n$. Here $C = [c^1, \ldots, c^m]$ is an nxm matrix.

2. Use of the Conditions of Positive Controllability of Linear Systems

By definition, the convex hull of the set

$$\{b(u): u \in U\}$$

has the following representation

$$\sum_{i=1}^{n+1} \alpha_i b(u_i), \alpha_i \geq 0, \sum_{i=1}^{n+1} \alpha_i = 1, u_i \in U .$$

Thus, investigation of the controllability of a system with controls having values in a given set U is equivalent to studying the controllability of the class of systems

$$\frac{dx}{dt} = Ax + \sum_{i=1}^{n+1} c^i v_i ,$$

where $c^i(t) = b(u_i(t))$, $u_i(t)$ are measurable functions with values in U, and $v_i = v_i(t)$ are measurable controls satisfying the conditions $v_i(t) \geq 0$, $\sum_{i=1}^{n+1} v_i(t) = 1$. We express one of the v_i (for definiteness v_{n+1}) by means of the others. Then $v_{n+1} = 1 - \sum_{i=1}^{n} v_i$ and the original system assumes the form

$$\frac{dx}{dt} = Ax + \sum_{i=1}^{n} d^i(t)v_i + c^{n+1}(t) , \qquad (41)$$

where $d^i(t) = c^i(t) - c^{n+1}(t)$ and the controls v_i, $i = 1,\ldots,n$, are subject only to the conditions $0 \leq v_i \leq 1$. Choosing different constant vectors u_i from U, we obtain constant vectors d^i, c^{n+1}. Applying now the conditions for positive controllability to the system (41), we obtain sufficient conditions for the controllability of the system (39).

Clearly, the second approach to studying controllability of the system (39) is less general. Its value is that conditions for the controllability of (39) with a given set U are reduced to checking controllability conditions for the simple system (41) with simple bounds on the controls.

3. Nonstationary Systems.

Dynamical systems of the form

$$\frac{dx}{dt} = A(t)x + b(u,t) \quad , \tag{42}$$

where A(t), b(u,t) are sufficiently smooth functions of t, are studied from the point of view of controllability in complete analogy with the preceding case. Basic facts about the equivalence of the system (42) to the system

$$\frac{dy}{dt} = A(t)y + q(t) \quad , \quad q(t) \in \operatorname{conv}\{b(u,t) : u \in U\}$$

occur here. Clearly, different results will be obtained conditioned upon the dependence of the functions A(t), b(u,t) on t. However, similar questions have already been discussed in the preceding paragraph so in this case the necessary calculations are elementary and are omitted.

Of great interest for applications are controllability criteria not using the derivatives of A(t), b(u,t). Regrettably, there are almost no results in the literature along these lines.

§7. A Theorem on the Controllability of Dynamical Systems in the Linear Approximation

We linearize the differential equation of controlled motion

$$\frac{dx}{dt} = f(x,u,t) \quad , \quad f(0,0,t) = 0 \quad , \tag{43}$$

with respect to x along $x \equiv 0$:

$$\frac{dx}{dt} = A(t)x + b(u,t) + h(x,u,t) \quad .$$

Here

$$A(t) = \frac{\partial f(0,0,t)}{\partial x} \quad , \quad b(u,t) = f(0,u,t) \quad ,$$

$$h(x,u,t) = \left[\frac{\partial f(0,u,t)}{\partial x} - \frac{\partial f(0,0,t)}{\partial x}\right] x + o(||x||)a(u,t) \quad .$$

Let $b^1(t),\ldots,b^r(t)$ be a basis for the minimal subspace containing the set

$$\Omega(t) = \text{conv}\,\{b(u,t) : ||u|| \leq L\} \quad .$$

Theorem 13. Assume the following conditions are satisfied:

1) $A(t) \varepsilon C^{n-2}$, $b^\nu(t) \subset C^{n-1}$, $\nu - 1,\ldots,r$; $r \geq t_o$;

2) the functions $\frac{\partial f(0,u,t)}{\partial x}$, $\frac{\partial^2 f(0,u,t)}{\partial u^2}$ are continuous in u;

3) $0 \,\varepsilon\, \text{int}\, \Omega(t)$, $t \geq t_o$, $L > 0$;

4) for some $\bar{t}$, $\bar{t} \geq t_o$,

$$\text{rank}\,\{Q_o(\bar{t}),\ldots,G_{n-1}(\bar{t})\} = n \quad ,$$

where $Q_0(t) = \{b^1(t),\ldots,b^r(t)\}$, $Q_{i+1}(t) = A(t)Q_i(t) - \dot{Q}_i(t)$, $i = 0,\ldots,n-2$.

Then the dynamical system (43) is completely controllable.

Proof. By virtue of (4), the controllable states of the system (43) may be written as

$$x_0 = \int_{t_0}^{t_1} \Psi(t)b(u,t)\ dt + \int_{t_0}^{t_1} \Psi(t)h(x,u,t)\ dt \quad .$$

We consider controls satisfying the conditions $||u|| \leq L$. For L sufficiently small, the trajectories x(t) corresponding to such controls satisfy the inequality

$$||x(t)|| \leq kL \quad ,$$

where k is independent of L. Therefore by definition the function h satisfies

$$||h(x,u,t)|| \leq m(L)L \quad ,$$

where $m(L) \to 0$ as $L \to 0$. On the other hand, from conditions 3),4) of the theorem, it follows that the set

$$\left\{ \int_{t_0}^{t_1} \Psi(t)b(u,t)dt : \ ||u|| \leq L \right\}$$

contains a sphere $||x|| \leq \rho L$, where ρ is independent of L. These facts, together with the scheme presented in the monograph [18d, p. 96], suffice to prove the theorem.

Remark. Condition 3) of the theorem may be weakened just as was done in the remark to Theorem 12.

§8. Methods of Functional Analysis in the Theory of Controllability

1. Application of the Minmax Theorem

We consider the dynamical system whose time variation is given by

$$x(t) = s(t,x_o) + \int_{t_o}^{t} S(t,\tau,u(\tau))\, d\tau \quad , \quad t \in T = [t_o,t_1] \quad .$$

Here $S(t,\tau,u)$, $s(t,x)$ are continuous matrix functions. We will assume that the control functions $u(t)$, $t\varepsilon T$, are measurable and assume values from a compact set $U(L)$, depending on a parameter $L \geq 0$. First of all, we find conditions under which, among the admissible controls, there exists a control under which for any initial state $x(t_o) = x_o$, the trajectory of the system (44) satisfies the condition $x(t_1) = 0$. In other words, we will study controllability properties in the sense of the definitions of §1 but with additional conditions on the function $u(t)$.

For the n-vector $x(t_1)$, the distance of the point $x(t_1)$ from the origin may be written (in an appropriate norm) as

$$||x(t_1)|| = \max_{||g|| \leq 1} g'x(t_1) \tag{45}$$

The set R of vectors $x(t_1)$, corresponding to all possible admissible controls $u(t) \varepsilon U(L)$, $t\varepsilon T$, is called the domain of attainability of the system (44). Under the given assumptions, the set R is convex and closed. By virtue of (45), the minimal distance δ from the origin to R equals

$$\delta(L) = \min_{x\varepsilon R} \max_{||g|| \leq 1} g'x$$

Using the min max theorem (§6.9), we interchange the operations min and max:

$$\delta(L) = \max_{||g|| \leq 1} \min_{x \varepsilon R} g'x \quad . \tag{46}$$

Points x of the set R, corresponding to the control u(t), $t \varepsilon T$, have the form

$$x = x(t_1) = \int_{t_0}^{t_1} S(t_1, \tau, u(\tau))\, d\tau + s(t_1, x_0) \quad .$$

Thus, (46) is equivalent to the expression

$$\delta(L) = \max_{||g|| \leq 1} \left[g's(t_1, x_0) + \int_{t_0}^{t_1} \min_{u(\tau) \varepsilon U(L)} g'S(t_1, \tau, u(\tau))\, d\tau \right] .$$

We introduce the notation

$$\mu(L, g, t, \tau) = \min_{u \varepsilon U(L)} g'S(t, \tau, u) \quad .$$

Then

$$\delta(L) = \max_{||g|| \leq 1} \left[g's(t_1, x_0 + \int_{t_0}^{t_1} \mu(L, g, t_1, \tau)\, d\tau \right] \quad .$$

Thus, in order that the state x_0 be controllable, it is necessary and sufficient that for some T, L, $t_1 - t_0 > 0$, $L > 0$, the following condition is satisfied:

$$\max_{||g|| \leq 1} \left[g's(t_1, x_0) + \int_{t_0}^{t_1} \mu(L, g, t_1, \tau)\, d\tau \right] = 0 \quad . \tag{47}$$

In the particular case $S(t, \tau, u) = S(t, \tau)u$, $U = \{u: |u_\nu| \leq L\}$, condition (47) assumes the form

$$\max_{||g|| \leq 1} \left[g's(t_1, x_0) - L \int_{t_0}^{t_1} \sum_{\nu=1}^{r} |[g'S(t_1, \tau)]_\nu|\, d\tau \right] = 0 \quad . \tag{48}$$

Let the transformation $x \to s(t_1,x)$ be nonsingular. Then in order that condition (48) be satisfied, it is necessary and sufficient that

$$[g'S(t_1,\tau)]_\nu \not\equiv 0 \quad , \quad t\varepsilon T \quad , \quad \text{for some } \nu \qquad (49)$$

for all g, $||g|| = 1$. Indeed, let (48) be satisfied but for some g_1 assume we have $[g_1'S(t_1,\tau)]_\nu \equiv 0$ for $t\varepsilon T$, $\nu = 1,\ldots,r$. By virtue of the nonsingularity of the transformation $s(t_1,x)$, we find an x_o such that $g_1's(t_1,x_o) > 0$, i.e. the left side of (48) is positive. This contradiction proves the necessity of condition (49).

For proof of sufficiency, we set

$$\mu = \min_{||g||=1} \int_{t_o}^{t_1} \sum_{\nu=1}^{r} |[g'S(t_1,\tau)]_\nu| d\tau \quad .$$

From (49) it follows that $\mu > 0$. Let

$$\lambda = \max_{||g||=1} g's(t_1,x_o) \quad .$$

Then for $L > \lambda/\mu$, the left side of (48) becomes negative but, since it is continuous in L and is positive for $L = 0$, there is an $\bar{L}$, $0 \leq \bar{L} \leq \lambda/\mu$, such that (48) is satisfied.

Remark. If the transformation $s(t,x_o)$ is singular or if the considered points x_o do not all belong to E_n, then from the preceding discussion it follows that condition (49) is not necessary for the controllability of (44).

2. The Problem of Controllability and the Theorem on Separation of Convex Sets

The set $R_L = \{x: \; x = x(t_1)\}$, where $x(t_1)$ is the position of the state $x(t)$ of the system (44) at time $t = t_1$, is convex and closed if the controls $u(t)$ $t \varepsilon T$, are all possible measurable functions with values in $U(L)$. For some L,T let the system (44) be uncontrollable. Then the origin and the set R_L have intersection and it is possible to strictly separate them by some hyperplane (cf. §6.10). Now it is easy to prove that condition (48) is sufficient for the controllability of the system (44).

In fact, condition (48) means that

$$g's(t_1,x_o) + \int_{t_o}^{t_1} \mu(L,g,t_1,\tau)\, d\tau \leq 0 \quad , \tag{50}$$

for all g, $||g|| = 1$. Let condition (50) be satisfied but the system (44) not be completely controllable. Then the set R_L does not contain the origin and there exists a vector $\bar{g}$, $||\bar{g}|| = 1$, such that

$$g's(t_1,x_o) + \int_{t_o}^{t_1} g'S(t_1,\tau,u(\tau))\, d\tau > 0 \quad .$$

This inequality must be satisfied for all admissible controls. Thus,

$$\bar{g}'s(t_1,x_o) + \int_{t_o}^{t_1} \min_{u \varepsilon U(L)} \bar{g}'S(t_1,\tau,u(\tau))\, d\tau$$

$$= \bar{g}'s(t_1,x_o) + \int_{t_o}^{t_1} \mu(L,\bar{g},t_1,\tau)\, d\tau > 0 \quad ,$$

which contradicts (50). Necessity of condition (50) is obvious: from $x(t_1) = 0$ in (44), it follows that

$$-g's(t_1,x_o) = \int_{t_o}^{t_1} g'S(t_1,\tau,u(\tau))\, d\tau \geq \int_{t_o}^{t_1} \mu(L,g,t_1,\tau)\, d\tau \quad ,$$

for any g, $||g|| = 1$.

3. The Problem of Moments in the Theory of Controllability

The condition $x(t_1)$ is, by virtue of (44), equivalent to the system of equalities

$$-[s(t_1,x_o)]_i = \int_{t_o}^{t_1} [S(t_1,\tau,u(\tau))]_i d\tau \ , \quad i = 1,\ldots,n \ , \tag{51}$$

where the symbol $[\ ]_i$ denotes the ith coordinate of the vector [ ]. If the system (44) is linear in u, i.e.

$$S(t,\tau,u) = S(t,\tau)u \ ,$$

then (51) may be treated as a problem of moments. For simplicity of formulation, we confine our attention to one-dimensional controls. The numbers $[s(t_1,x_o)]_i$ are "moments" of the function $u(t)$, $t\varepsilon T$, by the system of functions $[S(t_1,\tau)]_i$. A necessary and sufficient condition for solvability of the problem of moments is (cf. §6.11) the linear independence of the functions $[S(t_1,\tau)]_i$, $i = 1,\ldots,n$, i.e.

$$g'S(t_1,\tau) \not\equiv 0 \ , \quad t\varepsilon T \ ,$$

for any g, $||g|| = 1$, which coincides with (49).

§9. Positive Controllability of Dynamical Systems

We use the general scheme presented in §8 to study the positive controllability of the system

$$\frac{dx}{dt} = Ax + Bu \ .$$

1. Single-Input Systems

We begin with the simplest case of a single input $(r=1)$:

$$\frac{dx}{dt} = Ax + bu \quad , \quad x(t_0) = x_0 \quad . \tag{52}$$

The solutions of this linear inhomogeneous equation are expressed by means of the fundamental matrix of solutions $F(t)$ of the corresponding homogeneous equation in the form

$$x(t) = F(t)F^{-1}(t_0)x_0 + \int_{t_0}^{t_1} F(t)F^{-1}(\tau)bu(\tau)\, d\tau \quad . \tag{53}$$

Here the matrix $F(t)$ satisfies the equation

$$\frac{dF(t)}{dt} = AF(t) \quad , \quad F(t_0) = E \quad .$$

The formula (53) may be verified by substituting (53) into (52) or proved by the scheme of §2 ($\phi(x) = x$, $\psi = F^{-1}$). Setting $s(t_1,x_0) = F(t_1)F^{-1}(t_0)x_0$, $S(t_1,\tau,u) = F(t_1)F^{-1}(\tau)bu$, we are led to the system (44). We take $U(L)$ to be the set $U(L) = \{u\colon 0 \le u \le L\}$. Under these conditions we have

$$\mu(L,g,t_1,\tau) = \min_{u \varepsilon U(L)} g'S(t_1,\tau,u) = L[g'F(t_1)F^{-1}(\tau)b]_- \quad ,$$

where the symbol $[a(t)]_-$ denotes the negative part of the function $a(t)$:

$$[a(t)]_- = \begin{cases} a(t) \quad , & \text{if } a(t) < 0 \quad , \\ 0 \quad , & \text{if } a(t) \ge 0 \quad . \end{cases}$$

Repeating the arguments used for proof of the controllability condition (49), it is not difficult to see that a necessary and sufficient condition for the positive controllability of the system (52) is the requirement

$$[g'F(t_1)F^{-1}(\tau)b]_- \not\equiv 0 \quad , \quad t\varepsilon T \quad , \tag{54}$$

for all g, $||g||=1$. This condition is trivially not satisfied if $t_1 - t_0$ is sufficiently small. We transform (53) to a simpler form. Differentiating the identity $F(\tau)F^{-1}(\tau) = E$, we obtain

$$\frac{dF^{-1}}{d\tau} = -F^{-1}A,\ldots,\frac{d^nF^{-1}}{d\tau^n} = (-1)^nF^{-1}A^n \tag{55}$$

under the following conditions at t_0:

$$F^{-1}(t_0) = E,\ldots,\frac{d^nF^{-1}(t_0)}{d\tau^n} = (-1)^nA^n \quad . \tag{56}$$

Since each square matrix A satisfies its own characteristic equation

$$A^n + \gamma_1A^{n-1} + \cdots + \gamma_nE = 0 \quad ,$$

Multiplying this identity on the left by $(-1)^ng'F(t_1)F^{-1}(\tau)$ and on the right by b and using (55), we obtain a differential equation for the function

$$\psi(t) = g'F(t_1F)^{-1}(t)b \tag{57}$$

in the following form

$$\psi^{(n)}(t) - \gamma_1\psi^{(n-1)}(t) + \cdots + (-1)^{n-1}\gamma_{n-1}\psi(t) + (-1)^n\gamma_n\psi(t) = 0 \quad . \tag{58}$$

The initial conditions follow from (56) and (57):

$$\psi(t_1) = g'b,\ \dot{\psi}(t_1) = -g'Ab, \ldots, \psi^{(n-1)}(t_1) = (-1)^{n-1}g'A^{n-1}b \ .$$

Returning to the condition (54), we conclude that for it to be satisfied it is necessary and sufficient that

1) the vectors $b, Ab, \ldots, A^{n-1}b$ be linearly independent;

2) the characteristic equation for (58) have no roots with zero imaginary parts;

3) the difference $\tau = t_1 - t_0$ be sufficiently large.

Thus, we have proved the following theorem.

Theorem 14. The ordinary dynamical system with a single-input (52) may not be positively completely controllable.

Dynamical systems of odd order ($n = 2k+1$) are never positively controllable.

In a positively controllable system (52), there exists a number τ_0 such that for each $L > 0$, there is a state x_0 which cannot be brought to the origin in a time less that τ_0.

For positive controllability of the system (52), it is necessary and sufficient that it be controllable and that all characteristic roots of A have nonzero imaginary parts.

Remark. The number τ_0 is expressed through the smallest imaginary part ω_{min} of the characteristic roots $\lambda_i + \sqrt{-1}\,\omega_i : \tau_0 = \pi/\omega_{min}$.

2. Multi-input Systems

We study systems of the form

$$\frac{dx}{dt} = Ax + \sum_{\nu=1}^{r} b^{\nu} u_{\nu} \quad . \tag{59}$$

Introducing the notation

$$s(t,x) = F(t)F^{-1}(t_0)x, \; S(t,\tau,u) = \sum_{\nu=1}^{r} F(t)F^{-1}(\tau)b^{\nu}u_{\nu} \quad ,$$

we pass from the system (59) to (44) by means of (53). Let the set U(L) have the form

$$U(L) = \{u_1,\ldots,u_r: 0 \le u_{\nu} \le L \; , \; \nu = 1,\ldots,r\} \quad .$$

Then

$$\mu(L,g,t_1,\tau) = L \sum_{\nu=1}^{r} [g'F(t_1)F^{-1}(\tau)b^{\nu}]_- \quad .$$

In analogy with the single-input case, for each $\nu = 1,\ldots,r$, we obtain differential equations

$$\psi_{\nu}^{n} - \gamma_1\psi_{\nu}^{(n-1)} + \cdots + (-1)^{n-1}\gamma_{n-1}\dot{\psi}_{\nu} + (-1)^{n}\gamma_n\psi_{\nu} = 0 \quad ,$$

for the functions $\psi_{\nu}(t) = g'F(t_1)F^{-1}(t)b^{\nu}$. The initial conditions are

$$\psi_{\nu}(t_1) = g'b^{\nu} \; , \; \dot{\psi}_{\nu}(t_1) = -g'Ab^{\nu},\ldots,\psi^{(n-1)}(t_1)$$
$$= (-1)^{n-1}g'A^{n-1}b^{\nu}; \; \nu = 1,\ldots,r \quad .$$

The necessary and sufficient condition for positive controllability in the large for (59) assumes the form

$$\sum_{\nu=1}^{r} [g'F(t_1)F^{-1}(t)b^{\nu}]_- \not\equiv 0 \; , \quad t\varepsilon T \; , \text{ for all } g\varepsilon E_n \quad .$$

Hence, it follows that if the following conditions are satisfied, then the system is positively completely controllable:

1) rank $\{b^1,\ldots,b^r, Ab^1,\ldots,Ab^r,\ldots,A^{n-1}b^1,\ldots,A^{n-1}b^r\}=n$;

2) for each vector $\bar{b}$ from the linear hull of the vectors $b^1,\ldots,b^r$, it is possible to write $\bar{b} = -\sum_{i=1}^{r} \alpha_i b_i,\ \alpha_i > 0$ (the vectors $b^1,\ldots,b^r$ form a positive basis in the characteristic subspace).

The converse is also true: if the system is positively completely controllable, then the vectors $b^1,\ldots,b^r$ clearly form a non-negative basis (condition 2) in the characteristic subspace. From these remarks, it follows that any controllable system may be made positively completely controllable by adding one control. The new control u_{r+1} is added with the vector $b^{r+1} = \sum_{i=1}^{r} \alpha_i b^i$, $\alpha_i > 0$. These results reduce to the theorem

<u>Theorem 15.</u> <u>In order that the system (59) with r inputs be positively completely controllable, it is necessary and sufficient that</u>

$$\text{rank } [b^1, Ab^1,\ldots,A^{n-1}b^1,\ldots,b^r,Ab^r,\ldots,A^{n-1}b^r] = n \quad ,$$

<u>and the vectors $b^1,\ldots,b^r$ form a positive basis in the subspace spanned by them.</u>

Each controllable system may be made positively completely controllable by adding one more control u_{r+1} with vector $b^{r+1} = -\sum_{i=1}^{r} \alpha_i b^i$, $\alpha_i > 0$. The minimal number of inputs in a positively completely controllable system (59) is one more than the number of nontrivial invariant factors of the matrix A.

§10. Linear Control Systems with a Delay. The Defining Equation

Let the motion of the controlled object be described by the equation [15]

$$\frac{dx(t)}{dt} = Ax(t) + A_1x(t - h) + Bu(t) \quad , \qquad (60)$$

where $x = \{x_1,\ldots,x_n\}$ is an n-vector, $u = \{u_1,\ldots,u_r\}$ is an r-vector, h is a number characterizing the delay, and A, A_1, and B are constant matrices of appropriate sizes. In order that the motion of the system be defined for $t \geq 0$, we must give an initial condition

$$x_o(\cdot) = \{x(t) = \Phi(t) \quad , -h \leq t < 0 \quad , x(0) = x_o\} \quad , \qquad (61)$$

where $\Phi(t)$ is a continuous function, and x_o is an n-vector. The state of the object at any moment is not characterized by only a finite number of quantitites, but by a function

$$\{x(t + \theta) \quad , \quad -h \leq \theta < 0\} \quad ,$$

defined on an interval $[t - \theta, t)$. This peculiarity of delay systems essentially complicates solution of the control problem.

It is possible to distinguish two forms of controllability. If for each state (61), it is possible to find a time $t_1 > 0$ and a piecewise continuous control u(t), $0 \leq t \leq t_1$, for which $x(t_1) = 0$, then we will say that the state is <u>relatively controllable</u>. In many cases of such systems it is not sufficient to "cut off" control at time $t = t_1$ $(u(t) \equiv 0,\ t \geq t_1)$ since the system may depart from the equilibrium $x(t) \equiv 0$ due to the action of the delay. Thus, we introduce the concept of a <u>completely controllable state</u> of (61) when the trajectory goes to the origin and remains there under an admissible control.

In subsequent discussions, a large role will be played by the relations

$$Q_k(s) = AQ_{k-1}(s) + A_1Q_{k-1}(s - h) \quad , \ s \geq 0 \quad , \ k = 1,2,\ldots,$$

$$Q_0(0) = B \quad , \quad Q_0(s) \equiv 0 \quad , \quad s \neq 0 \quad , \qquad (62)$$

which are easily formed from the right-side of equation (60). We call (62) the defining equation of the system (60). It is clear that the solution of (62) is a sequence of matrices defined for $k = 1,2,\ldots$, $s = 0,h,2h,\ldots$, where for fixed k, $= Q_k(s) \equiv 0$ for $s = (k+1)h, (k+2)h,\ldots$. For each α, we denote Π_α as the set

$$\Pi_\alpha = \{Q_k(s),\ k = 0,1,\ldots,n-1;\ s \in [0,\alpha h]\}$$

the solution of the defining equation. The defining equation is called <u>α-nonsingular</u> if rank $\Pi_\alpha = n$. If this condition is satisfied for one $\alpha < \infty$, then we call the defining equation <u>nonsingular</u>.

§11. Relative Controllability of Linear, Stationary Systems with a Constant Delay

<u>Definition 10</u>. The initial state $x_o(\cdot)$ of the system (60) is called <u>relativelely T-controllable</u> if there exists a piecewise-continuous control $u(t)$, $t \in T = [0,t_1]$, such that the trajectory $x(t)$ starting at x_o, generated by the control $u(t)$, satisfies the condition $x(t_1) = 0$.

A system (60), all of whose initial states are relatively T-controllable, is called <u>relatively T-controllable</u>.

<u>Definition 11</u>. If the initial state $x_o(\cdot)$ (for the system (60)) is relatively T-controllable (T-controllable) for any t_1, then this state (system) is called <u>relatively controllable (controllable)</u>.

By direct substitution, we see that the solution of Eq.(60) is expressed by the formula

$$x(t) = s(t,x_o) + \int_0^t F(t,\tau)Bu(\tau)\, d\tau \quad ,$$

where

$$s(t,x_o) = F(t,0)x_o + \int_0^h F(t,\tau)A_1\Phi(\tau - h)\, d\tau \quad ,$$

where $F(t,\tau)$ is an nxn matrix function satisfying the equation

$$\frac{\partial F(t,\tau)}{\partial \tau} = -F(t,\tau)A - F(t,\tau + h)A_1 \quad , \quad \tau \le t \quad ,$$

$$F(t,t - 0) = E \quad , \quad F(t,\tau) = 0 \quad , \quad \tau \ge t + 0 \quad .$$

Setting $S(t,\tau,u) = F(t,\tau)Bu$, we arrive at (44) for $t_o = 0$. This means that, by virtue of the results of §8, a necessary and sufficient condition for relative T-controllability of the system (60) is the requirement

$$[g'F(t_1,t)B]_\nu = 0 \quad , \quad t\varepsilon T \quad , \quad ||g|| = 1 \quad ,$$

for some $\nu, \nu = 1,\ldots,r$. To simplify this condition, we use the function

$$\psi(t,\tau) = g'F(t_1,\tau)b = y'(\tau)b \quad . \qquad (63)$$

We consider the system of operator equations

$$\left.\begin{aligned} (pE + A' + A_1' e^{ph})\tilde{y}(\tau) &= v(\tau) \quad , \\ \tilde{\psi}(\tau) - \tilde{y}'(\tau)b &= 0 \quad . \end{aligned}\right\}$$

Here p is a differential operator, while e^{ph} is the translation operator

$$p^i e^{kph} z(\tau) = \frac{d^i z(\tau + kh)}{d\tau^i} \quad ,$$

and $v(\tau)$ is an arbitrary n-dimensional vector function. Solving the linear system (64) for $\tilde{\psi}(\tau)$, we obtain

$$\tilde{\psi}(\tau) = \frac{\begin{vmatrix} v(\tau) & pE + A' + A_1' e^{ph} \\ 0 & -b' \end{vmatrix}}{\det (pE + A' + A_1' e^{ph})} \quad . \tag{65}$$

The functions $y(\tau)$ and $\psi(g,\tau)$ satisfy the system (64) with $v(\tau) \equiv 0$. Substituting $v(\tau) = 0$ in (65), we obtain an equation for $\psi(g,\tau)$:

$$\det (pE + A' + A_1' e^{ph}) \psi(g,\tau) = 0 \quad ,$$

or, in differential form:

$$\sum_{i=0}^{n} \sum_{j=0}^{i} r_{ij} \frac{d^{n-i}\psi(g,\tau + jh)}{d\tau^{n-i}} = 0 \quad , \tau < t_1 \quad . \tag{66}$$

Here

$$r_{oo} = 1 \; , \; r_{10} = SpA = \sum_{i=1}^{n} a_{ii} \; , \; r_{11} = SpA_1 = \sum_{i=1}^{n} a_{1},_{ii} \quad ,$$

$$r_{no} = \det A \quad , \quad r_{nn} = \det A_1 \quad .$$

The remaining r_{ij} are also expressed by means of elements of the matrices A and A_1 (recurrence formulas for them are given, for example, in [39], p.77).

The initial conditions for (66) follow from (62), (63):

$$\psi^{(i)}(g,\tau) = \begin{cases} (-1)^i g' q_i(0) \ , & \tau = t_1 - 0 \ , \\ 0 \ , & \tau = t_1 + 0 \ , \end{cases} \tag{67}$$

$$\psi^{(i)}(g,\tau) \equiv 0 \quad , \ \tau > t_1 \quad , \quad i = 0,1,\ldots,n-1 \quad , \tag{68}$$

where $q_i(s)$ are computed using Eq.(62) ($Q_0(0) = q_0(0) = b$).

From (63) we obtain

$$\psi^{(k)}(g,\tau) = (-1)^k g' \sum_{i=0}^{k} F(t_1,\tau + ih) q_k(ih) , \ k = 0,1,\ldots, \tag{69}$$

Consequently, for k = 0 equality (69) coincides with (63). We assume that (69) holds for k = 0,1,...,p-1, and we prove that it also holds for k = p. Since

$$\psi^{(p)}(g,\tau) = \frac{d\psi^{(p-1)}(g,\tau)}{d\tau^{(p-1)}} = (-1)^{p-1} g' \sum_{i=0}^{p-1} \frac{dF(t_1,\tau + ih)}{d\tau} q_{p-1}(ih)$$

$$= (-1)^p g' \sum_{i=0}^{p-1} F(t_1,\tau + ih) A q_{p-1}(ih)$$

$$+ \sum_{i=1}^{p} F(t_1,\tau + ih) A_1 q_{p-1}((i-1)h) \quad ,$$

adding the zero terms $g'F(t_1,\tau + ph)Aq_{p-1}(ph)$ and $g'F(t_1,\tau)A_1q_{p-1}(-h)$ on the right, we obtain

$$\psi^{(p)}(g,\tau) = (-1)^p g' \sum_{i=0}^{p} F(t_1,\tau + ih)[Aq_{p-1}(ih)$$

$$+ A_1 q_{p-1}((i-1)h)] = (-1)^p g' \sum_{i=0}^{p} F(t_1,\tau + ih) q_p(ih) \ .$$

By virtue of the arbitrariness of the number p, the equality (60) holds for any k = 0,1,.... From (69), with regard to (62), we have

$$\left.\begin{aligned} \psi^{(i)}(g,t_1 - kh - 0) &= \psi^{(i)}(g,t_1 - kh + 0) , \\ &\qquad i = 0,1,\ldots,k-1 , \\ \psi^{(i)}(g,t_1 - kh - 0) &= \psi^{(i)}(g,t_1 - kh + 0) + (-1)^i g' q_i(kh) , \\ &\qquad i = k,k+1,\ldots; \quad k = 1,2,\ldots, \end{aligned}\right\} \quad (70)$$

Now it is possible to prove the following assertion.

Theorem 16. The system (60) is relatively T-controllable if and only if the defining equation of this system is α-nonsingular, $\alpha = [T/h]$.

Proof. 1. Single-input systems, $B = b$. Necessity.

Let the system (60) be relatively T-controllable. By virtue of (49),(63) this means that

$$\psi(g,\tau) \not\equiv 0 \quad \text{for all } \tau \in T \quad , \quad ||g|| = 1 \quad .$$

We show that the defining equation (62) is α-nonsingular. Assume the contrary: rank Π_α is less than n. Then there exists a vector g_o, $||g_o|| = 1$, for which

$$g_o' q_k(s) = 0 \; ; \; k = 0,1,\ldots,n-1 \; ; \; s \in [0,\alpha h] \quad . \qquad (71)$$

From (66),(67),(68), and (71) we have

$$\psi(g_o,\tau) \equiv 0 \quad , \quad t \in [t_1 - h, t_1] \quad .$$

Let

$$\psi(g_o,\tau) \equiv 0 \quad , \quad \tau\varepsilon[t_1 - kh, t_1] \quad , \tag{72}$$

where k is an arbitrary number from $[1,\alpha]$. We show that

$$\psi(g_o,\tau) \equiv 0 \quad , \quad \tau\varepsilon[t_1 - (k + 1)h, t_1 - kh] \ . \tag{73}$$

Since the function $\psi(g_o,\tau)$ satisfies the homogeneous equation (66) on $[t_1 - (k+1)h,\ t_1 - kh]$ with the zero initial function (72), the identity (73) is satisfied if

$$\psi^{(i)}(g_o, t_1 - kh - 0) = 0 \ , \ i = 0,1,\ldots,n-1 \quad .$$

But this is a corollary of the relations (70)-(72). Thus,

$$\psi(g_o,\tau) \equiv 0 \quad , \quad \tau\varepsilon[t_1 - (\alpha + 1)h, t_1] \quad ,$$

which contradicts (49). Necessity is proved.

<u>Sufficiency</u>. Assume (49) is not satisfied: there exists a vector g_o, $||g_o|| = 1$, such that

$$\psi(g_o,\tau) \equiv 0 \quad , \quad \tau\varepsilon T \quad . \tag{74}$$

Then

$$\psi^{(i)}(g_o,\tau) \equiv 0 \quad , \ \tau\varepsilon T \quad , \quad i = 1,2,\ldots, \tag{75}$$

i.e.

$$\psi^{(i)}(g_o,\tau + 0) = \psi^{(i)}(g_o,\tau - 0) = 0 \quad , \quad i = 0,1,\ldots, \tag{76}$$

at each interior point of the interval T. From (67), (74), and (75):

$$g_0' q_i(0) = 0 \quad , \quad i = 0,1,\ldots,n-1 \quad . \tag{77}$$

From (70) and (76) follows

$$g_0' q_i(kh) = 0 \quad , \quad i = k,\ldots,n-1 \; ; \quad k = 1,\ldots,\alpha \quad .$$

But, by the properties of the solutions $q_i(kh)$ of Eq.(62)

$$g_0' q_i(kh) = 0 \quad , \quad i = 0,\ldots,k-1 \; ; \quad k = 1,\ldots,\alpha \quad .$$

Combining these relations with (77), we are led to

$$g_0' q_i(kh) = 0 \quad , \quad i = 0,\ldots,n-1 \; ; \quad k = 0,\ldots,\alpha \quad . \tag{78}$$

But, the conditions (78) occur only for $g_0 \equiv 0$. This contradicts the earlier assumption about g_0. The theorem is proved for case 1.

2. Several Inputs. The proof is analogous.

Remark. In the case $t_1 \leq h$, the number $\alpha = 0$ and the condition for relative controllability of a system with delay (60) coincides with the condition for complete controllability of a system without delay:

$$\text{rank } \{B, AB, \ldots, A^{n-1}B\} = n \quad .$$

§12. Linear Systems with a Deviating Argument of Neutral Type. The Defining Equation.

Let there be given the equation

$$\frac{dx(t)}{dt} = Ax(t) + A_1x(t-h) + B\frac{dx(t-h)}{dt} + Cu(t) \quad , \qquad (79)$$

where $x = \{x_1,\ldots,x_n\}$, $u = \{u_1,\ldots,u_r\}$, h = constant, $h > 0$, and A, A_1, B, C are constant matrices of appropriate sizes. We assign the initial function

$$x_0(\cdot) = \{\Phi(\tau),\ \tau\varepsilon[-h,0)\ ;\ x(0) = x_0\} \quad , \qquad (80)$$

where $\Phi(\tau)$ is a continuously differentiable function and x_0 is an n-vector. Just as in the investigation of Eq.(60), here it is also possible to distinguish two forms of controllability: relative controllability and complete controllability. In this paragraph we study the relative controllability of the stationary system (79) in the sense of definitions 10 and 11 of §11.

We consider the system

$$\frac{dx(t)}{dt} = Ax(t) + A_1x(t-h) + B\frac{dx(t-h)}{dt} \quad . \qquad (81)$$

We introduce the nxn matrix function $Q_k(s)$, depending on the two arguments k,s, $k = 1,2,\ldots,;\ s = 0,h,2h,\ldots,$ and we establish the following relation between the functions $x(t)$ and $Q_k(s)$:

$$x(t) \to Q_{k-1}(s) \ , \quad x(t-h) \to Q_{k-1}(s-h) \ ,$$

$$\dot{x}(t) \to Q_k(s) \ , \quad x(t-h) \to Q_k(s-h) \quad .$$

Then for Eq.(81) we have

$$Q_k(s) = AQ_{k-1}(s) + A_1Q_{k-1}(s - h) + BQ_k(s - h) \tag{82}$$

$$(k = 0,1,2,\ldots,;\ s = 0,h,2h,\ldots,) \ .$$

For the unique solvability of (82), we require: $Q_o(0) = E$, $Q_o(s) \equiv 0$, if $s < 0$. We call (82) the defining equation for (79). For each α we set

$$\Pi_\alpha = \{Q_k(s)C,\ k = 0,1,\ldots,n-1;\ s\varepsilon[0,\alpha h]\}$$

and introduce the concepts of α-nonsingularity and nonsingularity of the defining equation (cf. §10).

Before passing to the study of relative controllability of the system (79), we obtain a formula for the solution $x(t)$ (cf. [15]). Let $F(t,\tau)$ be the unique function satisfying the equation

$$\frac{\partial F(t,\tau)}{\partial\tau} = -F(t,\tau)A - F(t,\tau + h)A_1 + \frac{\partial F(t,\tau + h)}{\partial\tau} B \ , \tag{83}$$

$$0 < \tau < t \ , \quad \tau \neq t - nh \ , \quad h = 0,1,\ldots,$$

$$F(t,t - 0) = E \ , \quad F(t,\tau) \equiv 0 \ , \quad \tau \geq t + 0 \ , \tag{84}$$

and

$$F(t,\tau) - F(t,\tau + h)B \tag{85}$$

continuous in τ, $\tau\varepsilon[0,t]$. We multiply both sides of (79) by $F(t,\tau)$ and integrate from 0 to t:

$$\int_0^t F(t,\tau)[\dot{x}(\tau) - B\dot{x}(\tau - h)]\,d\tau - \int_0^t F(t,\tau)Ax(\tau)\,d\tau$$
$$- \int_0^t F(t,\tau)A_1x(\tau - h)\,d\tau = \int_0^t F(t,\tau)Cu(\tau)\,d\tau \ .$$

From (84) and (85) we have

$$\int_0^t \frac{\partial}{\partial\tau}[F(t,\tau) - F(t, \quad + h)B]x(\tau)\, d\tau = x(t) - F(t,0)x_0$$

$$+ F(t,h)Bx_0 - \int_{-h}^{0} F(t,-\tau)B\dot{\Phi}(-\tau - h)\, d\tau$$

$$- \int_0^t F(t,\tau)[\dot{x}(\tau) - B\dot{x}(\tau - h)]\, d\tau \quad .$$

Since

$$\int_0^t F(t,\tau)A_1x(\tau - h)\, d\tau = \int_0^t F(t,\tau + h)A_1x(\tau)\, d\tau$$

$$+ \int_{-h}^{0} F(t,\tau +h)A_1\Phi(\tau)\, d\tau \quad ,$$

then taking into account that $F(t,\tau)$ is a solution of (83), we have

$$x(t) = \{F(t,0) - F(t,h)B\}\, x_0 + \int_{-h}^{0} \{F(t,\tau + h)A_1\Phi(\tau)$$

$$+ F(t,-\tau)B\dot{\Phi}(-\tau - h)\}\, d\tau + \int_0^t F(t,\tau)Cu(\tau)\, d\tau \quad . \quad (86)$$

§13. Relative Controllability of Stationary Linear Systems with a Deviating Argument of Neutral Type

We introduce the notation

$$s(t,x_0) = \{F(t,0) - F(t,h)B\}\, x_0 + \int_{-h}^{0} \{F(t,\tau + h)A_1\Phi(\tau)$$

$$+ F(t,-\tau)B\Phi(-\tau - h)\}\, d\tau \ , \quad S(t,\tau,u) = F(t,\tau)Cu(\tau).$$

We have obtained relation (44). Thus, by virtue of the results of §8, a necessary and sufficient condition for the relative T-controllability of (79) is the linear independence of the functions

$$[F(t_1,\tau)c^{\nu}]_i \quad , \; i = 1,\dots,n \; ; \; t\varepsilon T; \; T = [0,t_1] \quad ,$$

for some ν, $\nu = 1,\dots,r$ (c^{ν} is the νth column of C).

We confine our attention to the simplest case of a single-input system, C = c.

We introduce the functions $\psi(g,\tau) = g'F(t_1,\tau)c$, $y(\tau) = g'F(t_1,\tau)$ and we consider the system of operator equations

$$(pE + A + A_1e^{ph} + Bpe^{ph})\tilde{y}(\tau) = v(\tau) \quad , \; \tilde{\psi}(\tau) - \tilde{y}'(\tau)c = 0 \; .$$

We find $\tilde{\psi}(\tau)$ and set $v(\tau) \equiv 0$. Then we obtain the equation for $\psi(g,\tau)$:

$$\det\,(pE + A + A_1e^{ph} + Bpe^{ph})\psi(g,\tau) = 0 \quad , \quad \tau < t_1 \quad . \tag{87}$$

The differential form of this equation coincides with (66); the initial conditions are:

$$\psi^{(i)}(g,\tau) = \begin{cases} (-1)^i g'Q_i(0) \quad , & \tau = t_1-0 \quad , \\ 0 \quad , & \tau = t_1+0 \quad , \end{cases}$$
$$\psi^{(i)}(g,\tau) \equiv 0 \; , \quad \tau > t_1 \; , \; i = 0,1,\dots,n-1 \quad . \tag{88}$$

The functions $\psi^{(i)}(g,\tau)$ have discontinuities at the points $\tau = t_1 - kh$. We find the jumps of $\psi^{(i)}(g,\tau)$ at those points. For this, it suffices to find the jumps of the function $\dfrac{\partial^{(i)}F(t_1,\tau)}{\partial\tau^i}$. We have

$$\frac{\partial^{(i)} F(t_1,\tau)}{\partial \tau^i} = (-1)^i \sum_{j=0}^{k-1} F(t_1,\tau + jh)[AQ_{i-1}(jh) + A_1 Q_{i-1}([j - 1]h)] \quad , \qquad (89)$$

$$\tau \varepsilon [t_1 - kh, t_1 - (k - 1)h] \quad .$$

We prove this formula by induction. By virtue of (83), the formula (89) is valid for $k = 1$ and all i. From (83) it follows that (89) is valid for $\tau = 1$ on the interval $[t_1 - kh, t_1 - (k - 1)h]$. Let (89) hold for $k = \ell$ and for all i and for $i = p$ and all k. We establish the validity of (89) for $p + 1$ and $k = \ell + 1$. On the interval $[t_1 - (\ell + 1)h, t_1 - \ell h]$, the pth derivative of $F(t_1,\tau)$ is given by

$$\frac{\partial^{p+1} F(t_1,\tau)}{\partial \tau^{p+1}} = (-1)^p \sum_{j=0}^{\ell} \frac{\partial F(t_1,\tau + jh)}{\partial \tau} [AQ_{p-1}(jh) + A_1 Q_{p-1}([j - 1]h)] = (-1)^{p+1} \sum_{j=0}^{\ell} [F(t_1,\tau + jh)A + F(t_1,\tau + (j + 1)h)A_1 - \frac{\partial F(t_1,\tau + (j + 1)h)}{\partial \tau} B][AQ_{p-1}(jh) + A_1 Q_{p-1}([j - 1]h)$$

$$= (-1)^{p+1} \sum_{j=0}^{\ell} F(t_1,\tau + jh)[AQ_p(jh) + A_1 Q_p([j - 1]h)] \quad ,$$

i.e. one side of the discontinuity (89) is valid. Now we show that if (89) is true on $[t_1 - \ell h, t_1 - (\ell - 1)h]$ then it is true on $[t_1 - (\ell + 1)h, t_1 - \ell h]$. From (83)

$$\frac{\partial^{p+1}F(t_1,\tau)}{\partial\tau^{p+1}} = -\frac{\partial^{p}F(t_1,\tau)}{\partial\tau^{p}}A - \frac{\partial^{p}F(t_1,\tau+h)}{\partial\tau^{p}} + \frac{\partial^{p+1}F(t_1,\tau+h)}{\partial\tau^{p+1}}B \ .$$

This means

$$\frac{\partial^{p+1}F(t_1,\tau)}{\partial\tau^{p+1}} = (-1)^{p+1}\sum_{j=0}^{\ell} F(t_1,\tau+jh)[AQ_{p-1}(jh) + A_1Q_{p-1}([j-1]h)]A$$

$$+ (-1)^{p+1}\sum_{j=0}^{\ell-1} F(t_1,\tau+(j+1)h)[AQ_{p-1}(jh) + A_1Q_{p-1}([j-1]h)]A_1$$

$$+ (-1)^{p+1}\sum_{j=0}^{\ell-1} F(t_1,\tau+(j+1)h)[AQ_p(jh) + A_1Q_p([j-1]h)]B \ .$$

Using (82) we obtain

$$\frac{\partial^{p+1}F(t_1,\tau)}{\partial\tau^{p+1}} = (-1)^{p+1}\sum_{j=0}^{\ell} F(t_1,\tau+jh) \cdot [AQ_p(jh) + A_1Q_p([j-1]h)] \ ,$$

i.e. (89) is true for all i and k. We find the jumps of the function $\psi(g,\tau)$ at the points $\tau = t_1 - kh$. By virtue of the continuity of the function (85), we have

$$F(t_1,t_1 - kh + 0) = F(t_1,t_1 - kh - 0) - B^k \ ,$$

$$\psi^{(i)}(g,\tau) =$$

$$(-1)^i g' \sum_{j=0}^{k-1} F(t_1,\tau+jh)[AQ_{i-1}(jh) + A_1Q_{i-1}([j-1]h)] \ ,$$

$$\psi^{(i)}(g,t_1 - kh + 0) = (-1)^i g' \sum_{j=0}^{k-1} F(t_1,t_1 + (j - k)h + 0)$$

$$\cdot [AQ_{i-1}(jh) + A_1Q_{i-1}([j - 1]h)]$$

$$= (-1)^i g' \sum_{j=0}^{k-1} F(t_1,t_1 - (k - j)h - 0)$$

$$\cdot [AQ_{i-1}(jh) + A_1Q_{i-1}([j - 1]h)]$$

$$- (-1)^i g' \sum_{j=0}^{k-1} B^{k-j}[AQ_{i-1}(jh) + A_1Q_{i-1}([j - 1]h)] .$$

Now we compute $\psi^{(i)}(g,t_1 - kh - 0)$. Since

$$\psi^{(i)}(g,\tau)$$

$$= (-1)^i g' \sum_{j=0}^{k} F(t_1,\tau + jh) [AQ_{i-1}(jh) + A_1Q_{i-1}([j - 1]h)],$$

then

$$\psi^{(i)}(g,t_1 - kh - 0) = (-1)^i g' \sum_{j=0}^{k=1} F(t_1,t_1 - (k - j)h - 0)$$

$$\cdot [AQ_{i-1}(jh) + A_1Q_{i-1}([j - 1]h)]$$

$$+ (-1)^i g'[AQ_{i-1}(kh) + A_1Q_{i-1}([k - 1]h)] .$$

We have

$$\psi^{(i)}(g,t_1 - kh + 0) + (-1)^i g'[AQ_{i-1}(kh)$$

$$+ A_1Q_{i-1}([k - 1+h)] = \psi^{(i)}(g,t_1 - kh - 0)$$

$$- (-1)^i g' \sum_{j=0}^{k-1} B^{k-j}[AQ_{i-1}(jh) + A_1Q_{i-1}([j - 1]h)] ,$$

or

$$\psi^{(i)}(g,t_1 - kh + 0) + (-1)^i g' \sum_{j=0}^{k} B^{k-j}[AQ_{i-1}(jh) + A_1Q_{i-1}([j - 1)]h)] = \psi^{(i)}(g,t_1 - kh - 0) .$$

But

$$Q_i(kh) = AQ_{i-1}(kh) + Q_1Q_{i-1}([k - 1]h) + B[AQ_{i-1}([k - 1]h) + A_1Q_{i-1}([k - 2]h)] + \cdots + B^k[AQ_{i-1}(0) + A_1Q_{i-1}(-h)] .$$

Finally:

$$\left.\begin{aligned} \psi^{(i)}(g,t_1 - kh - 0) &= \psi^{(i)}(g,t_1 - kh + 0) \\ &\quad + (-1)^i g'Q_i(kh) , \\ i = 1,2,\ldots,;\quad & k = 0,1,\ldots, \end{aligned}\right\} \tag{90}$$

Now it is possible to prove the following result.

Theorem 17. The system (79) is relatively T-controllable if and only if the defining equation (82) is α-nonsingular, $\alpha = [T/h]$.

Proof. (Necessity) Let (79) be relatively controllable. This means that $\psi(g,\tau) \not\equiv 0$ for all $\tau \varepsilon T$, $||g|| = 1$.

We show that the defining equation (82) is α-nonsingular. Assume the contrary. Then there exists a vector g_o, $||g_o|| = 1$, for which

$$g_o'Q_k(s) = 0 \quad , \; k = 0,1,\ldots,n-1 \quad , \; s\varepsilon[0,\alpha h] \quad . \tag{91}$$

From (87),(88), and (91), we obtain

$$\psi(g_o,\tau) \equiv 0 \quad , \quad \tau\varepsilon[t_1 - h,t_1] \quad . \tag{92}$$

Let

$$\psi(g_o,\tau) \equiv 0 \quad , \quad \tau\varepsilon[t_1 - kh,t_1] \quad , \tag{93}$$

where k is an arbitrary number. We show that

$$\psi(g_o,\tau) \equiv 0 \quad , \quad \tau\varepsilon[t_1 - (k + 1)h, \; t_1 - kh] \quad . \tag{94}$$

On $[t_1 - (k + 1)h, \; t_1 - kh]$ the function $\psi(g_o,\tau)$ satisfies the homogeneous equation (87) with zero initial condition (93); consequently, (94) is valid if $\psi^{(i)}(g_o,t_1 - kh - 0) = 0$, $i = 0,1,\ldots,n-1$. But this is a consequence of (90)-(92). Thus, $\psi(g_o,\tau) \equiv 0$ for $\tau\varepsilon[t_1 - (\alpha + 1)h,t_1]$, which contradicts the linear independence of the functions $[F(t_1,\tau)c]_i$.

<u>Sufficiency</u>. Assume the contrary. Then there exists a vector g_o, $||g_o|| = 1$, such that $\psi(g_o,\tau) \equiv 0$, $\tau\varepsilon T$. Hence

$$\psi^{(i)}(g_o,\tau) \equiv 0 \quad , \; \tau\varepsilon T \quad , \; i = 1,2,\ldots, \tag{95}$$

$$\psi^{(i)}(g_o,\tau + 0) = \psi^{(i)}(g_o,\tau - 0) \quad , \; i = 0,1,\ldots, \tag{96}$$

at each interior point of the interval $[0,t_1]$. From (88), (95), we have $g_o'Q_i(0) = 0$, $i = 0,1,\ldots,n-1$. Conditions (90) and (96) give

$$g_o'Q_i(kh) = 0 \quad , \; i = 0,1,\ldots,n-1 \quad ; \; k = 1,\ldots,\alpha \quad .$$

Combining the last relations, we are led to equation

$$g_0' Q_i(kh) = 0 \quad , \quad i = 0,1,\ldots,n-1 \quad ; \quad k = 0,1,\ldots,\alpha \quad ,$$

which is impossible. The theorem is proved.

In the case of the vector control $u = \{u_1,\ldots,u_r\}$ the matrix C is written by its columns $c^1,\ldots,c^r$. Then Eq. (87) is given by the r equations

$$\sum_{i=0}^{n} \sum_{j=0}^{i} r_{ij} \frac{d^{n-i} \psi_s(g,\tau + jh)}{d\tau^{n-i}} = 0 \quad , \quad \tau < t_1$$

Here $\psi_s(g,\tau) = g'F(t_1,\tau)c^s$. Each of the conditions (88) gives r equations

$$\psi_s^{(i)}(g,\tau) = \begin{cases} (-1)^i g' Q_i^s(0) & , \quad \tau = t_1 - 0 \quad , \\ 0 \quad , & \quad \tau = t_1 + 0 \quad , \end{cases}$$

$$\psi_s^{(i)}(g,\tau) \equiv 0 \quad , \quad \tau \equiv t_1 \quad , \quad i = 0,1,\ldots,n-1 \quad ,$$

where $Q_i^s(0)$ is the sth column of the matrix $Q_i(0)$

Conditions (90) have the form

$$\psi_s^{(i)}(g,t_1 - kh - 0) = \psi_s^{(i)}(g,t_1 - kh + 0) + (-1)^i Q_i^s(kh) \quad ,$$

$$k = 0,1,\ldots, \quad ; \quad i = 1,2,\ldots, \quad ; \quad s = 1,\ldots,r \quad .$$

Now, using the scheme of proof of Theorem 17, it is not difficult to see that for relative T-controllability of the system (79), a necessary and sufficient condition is that the defining equation (82) be α-nonsingular for $\alpha = [T/h]$.

§14. Stationary Linear Systems with Several Delays

We consider the system

$$\left.\begin{aligned} \dot{x}(t) &= \sum_{i=0}^{m} A_i x(t - h_i) + Bu(t) \ , \quad t \geq 0 \ , \\ & \qquad h_o < h_1 < \cdots < h_m \ ; \\ x_o(\cdot) &= \{\Phi(\tau) \ , \ \tau \varepsilon [-h_m, 0) \ , \ x(0) = x_o\} \ , \ m \geq 2 \ , \end{aligned}\right\} \tag{97}$$

where A_i are nxn matrices, h_i are positive numbers, and $h_o = 0$. The defining equation for (97) has the form

$$\left.\begin{aligned} Q_k(s) &= \sum_{i=0}^{m} A_i Q_{k-1}(s - h_i) \ , \ 0 \leq s \ ; \ k = 1,2,\ldots, \\ Q_o(s) &= \begin{cases} B \ , & s = 0 \ , \\ 0 \ , & s = 0 \ . \end{cases} \end{aligned}\right\} \tag{98}$$

Equation (98) is called nonsingular if rank $\pi n-1 = n$

$$\pi_{n-1} = \{Q_k(s) \ , \ k = 0,\ldots,n-1 \ ; \ s\varepsilon[0,(n-1)h_m]\} \ .$$

We note that for each $k = 1,2,\ldots,$ the solution $Q_k(s)$ of Eq. (98) is identically zero everywhere except possibly at the points

$$s_{j_1,\ldots,j_k} = \sum_{\sigma=1}^{k} h_{j_\sigma} \ , \ j_p = 0,1,\ldots,j_{p-1} \ ; \ p \geq 1 \ ; \ j_o = m \ . \tag{99}$$

<u>Theorem 18. The system (97) is relatively controllable if and only if the defining equation (98) is nonsingular.</u>

We give a proof for the case $r = 1$.

Necessity. We consider the equation

$$\dot{x}(t) = \sum_{t=0}^{m} A_i x(t - h_i) + bu(t) \quad ,$$

$$x_o(\cdot) = \{x(\tau) = \Phi(\tau) \ , \ -h_m \le \tau < 0 \ ; \ x(0) = x_o\} \quad .$$

The solution $x(t)$ of this equation may be found by Cauchy's formula [15]. Let $\psi(g,\tau)$ denote the scalar function

$$\psi(g,\tau) = g'F(t_1,\tau)b \quad .$$

Here the function $F(t_1,\tau)$ satisfies the equation

$$\frac{dF(t_1,\tau)}{d\tau} = - \sum_{i=0}^{m} F(t_1,\tau + h_i)A_i \quad ,$$

$$F(t_1,\tau) = \begin{cases} E \ , & \tau = t_1 - 0 \ , \\ 0 \ , & \tau \ge t_1 \ . \end{cases}$$

The function $\psi(g,\tau)$ satisfies the operator equation

$$\det \ (pE + \sum_{i=0}^{m} A_i e^{ph_i}) \ \psi \ (g,\tau) = 0 \quad ,$$

which corresponds to the differential equation

$$\frac{d^n\psi(g,\tau)}{d\tau^n} + \sum_{i=1}^{n} \sum_{j_1=0}^{m} \sum_{j_2=0}^{j_i} \cdots \sum_{j_i=0}^{j_{i-1}} \frac{d^{n-i} \ (g,\tau + \sum_{s=1}^{i} h_{js})}{d\tau^{n-i}} = 0 \tag{100}$$

with the initial conditions

$$\psi^{(i)}(g,\tau) = \begin{cases} g'q_i(0)(-1)^i \ , & \tau = t_1 - 0 \ , \\ 0 \ , & \tau = t_1 + 0 \ , \end{cases} \tag{101}$$

$$\psi^{(i)}(g,\tau) \equiv 0 \quad , \quad \tau > t_1 \quad , \quad i = 0,\ldots,n-1 \quad , \tag{102}$$

where the coefficients $r_{j_1 j_2 \ldots j_i}$, $j_p = 0,1,\ldots,j_{p-1}$; $p = 1,\ldots,i$; $i = 1,\ldots,n$, $j_o = m$, in explicit form are expressed through the elements of the matrix A_i, while the vectors $q_i(0)$ are calculated by the equations

$$\left.\begin{aligned} q_k(s) &= \sum_{i=0}^{m} A_i q_{k-1}(s - h_i) \ , \ s \geq 0 \ , \ k = 1,2,\ldots, \\ q_o(s) &= \begin{cases} b \ , & s = 0 \ , \\ 0 \ , & s \neq 0 \ . \end{cases} \end{aligned}\right\} \tag{98a}$$

Let the original system be relatively controllable. By virtue of Theorem 16, there exists a $t_1, t_1 < \infty$, such that

$$\psi(g,\tau) \not\equiv 0 \quad , \quad \tau \varepsilon T \quad ,$$

for all g, $||g|| = 1$. From the numbers $s_{j_1 \ldots j_k}$, $k = 1,2,\ldots,$ of the defining relation (99), we form the strictly increasing sequence $\{s_i, i = 0,1,\ldots,\}$, assuming that coincident numbers enter into it only one time. Then the sequence $\{t_1 - s_i, i = 0,1,\ldots,\}$ will be strictly decreasing. We assume that rank π_{n-1}

$$\pi_{n-1} = \{q_k(s), \ k = 0,\ldots,n-1; \ s\varepsilon[0,(n-1)h_m]\} \quad ,$$

is less than n. Thus,

$$g_o' q_k(s) = 0 \ , \ s\varepsilon[0,kh_m] \ , \ k = 0,\ldots,n-1 \quad . \tag{103}$$

Then for each finite t_1 we have

$$\psi(g_0,\tau) \equiv 0 \quad , \quad \tau\varepsilon[t_1 - s_k, t_1] \quad ,$$

just as for the solution of the homogeneous equation (100) with the zero initial conditions (101) and (102). Generally, if

$$\psi(g_0,\tau) \equiv 0 \quad , \quad \tau\varepsilon[t_1 - s_k, t_1] \quad ,$$

where k is an arbitrary integer, $k \geq 1$, then

$$\psi(g_0,\tau) \equiv 0, \tau\varepsilon[t_1 - s_{k+1}, t_1 - s_k] \quad .$$

In order to see this, it is sufficient to demonstrate the condition

$$\psi^{(i)}(g_0, t_1 - s_k - 0) = 0 \ , \ i = 0,\ldots,n-1 \quad . \tag{104}$$

We show the validity of (104). By analogy with (69), we have the relation

$$\psi^{(k)}(g_0,\tau)$$
$$= (-1)^k g_0' \sum_{j_1=0}^{j_0} \sum_{j_2=0}^{j_1} \cdots \sum_{j_k=0}^{j_{k-1}} F(t_1, \tau + \sum_{s=1}^{k} h_{j_s}) q_k(\sum_{s=1}^{k} h_{j_s}),$$
$$k = 1,\ldots; \quad j_0 = m \quad ,$$

and, as a corollary,

$$\psi^{(i)}(g_0, t_1 - \sum_{s=1}^{k} h_{j_s} - 0)$$
$$= \psi^{(i)}(g_0, t_1 - \sum_{s=1}^{k} h_{j_s} + 0) \ , \ i = 0,\ldots,k-1 \quad , \tag{105}$$

$$\psi^{(i)}(g_o, t_1 - \sum_{s=1}^{k} h_{j_s} - 0) = \psi^{(i)}(g_o, t_1 - \sum_{s=1}^{k} h_{j_s} + 0)$$

$$+ (-1)^i g_o' q_i (\sum_{s=1}^{k} h_{j_s}) \ , \ i = k, k+1, \ldots;$$

$$j_p = 1, \ldots, j_{p-1}; \quad p = 1, \ldots, k; \quad j_o = m; \quad k = 1, 2, \ldots . \quad (106)$$

Equation (105) is obtained under the following assumptions: if for any fixed sequence of numbers j_s and j_p, $s = 1, \ldots, k_1$; $p = 1, \ldots k_1 + \ell$; $\ell \geq 1$, the points $s_{j_1} \ldots _{j_{k_1}}$ and $s_{j_1} \ldots _{j_{k+1}}$ coincide, i.e.

$$\sum_{s=1}^{k_1} h_{j_s} = \sum_{s=1}^{k_{1+1}} h_{j_s} \ ,$$

then the relations (105), for $k = k_1 + \ell$, are not considered at these points. The equation (106) is valid at every point $s_{j_k} \ldots$, $k = 1, 2, \ldots .$ From (105), (106) and (103), we obtain (104). By virtue of the arbitrariness of k, we see that

$$\psi(g, \tau) \equiv 0 \quad , \quad \tau \varepsilon T \quad ,$$

when $g = g_o$ for every terminal point t_1. Since the last identity contradicts (49), necessity is proved.

Sufficiency. We show that relative controllability of the system follows from the nonsingularity of the defining equation. Assume the contrary: the system is not relatively controllable. This means that there exists a vector g_o, $||g_o|| = 1$, such that

$$\psi(g_o, \tau) = g_o' F(t_1, \tau) b \equiv 0 \quad , \quad \tau \varepsilon T \quad , \quad (107)$$

for all t_1, $0 < t_1 < \infty$. From (107), we have the

equality

$$\psi^{(k)}(g_o, \tau - 0) = \psi^{(k)}(g_o, \tau + 0) \ , \ \tau \varepsilon T \ , \ k = 0, \ldots, n-1 \ . \quad (108)$$

Setting $t_1 > nh_m$ and considering (100) at the points $\tau = t_1 - s_\alpha$, $\alpha = 0, \ldots, N$; $s_N = nh_m$, we are led to the relations

$$g_o' q_k(s_\alpha) = 0 \ , \ k = 0, \ldots, n-1; \ \alpha = 0, \ldots, N \ ,$$

which, by virtue of the nonsingularity of the defining equation, can occur only if $g_o = 0$. This contradiction proves the theorem.

§15. Relative Controllability of Nonstationary Linear Systems

Let the motion of the controlled system be given by the equation

$$\left.\begin{aligned} \dot{x}(t) &= A(t)x(t) + A_1(t)x(t-h) + b(t)u(t) \ , \ t \geq 0 \ , \\ \dot{x}_o(\cdot) &= \{\Phi(\tau) \ , \ \tau \varepsilon [-h, 0) \ , \ x(0) = x_o\} \ , \end{aligned}\right\} \quad (109)$$

in which the components of $A(t)$, $A_1(t)$, and $b(t)$ are $(n-1)$-times continuously differentiable for $t \geq 0$. As above, we set the system (109) into correspondence with the equation

$$\begin{aligned} q_k(s,t) &= A(t)q_{k-1}(s,t) + A_1(t)q_{k-1}(s-h, t-h) \\ &\quad - \frac{dq_{k-1}(s,t)}{dt} \ , \ k \geq 1 \ , \ s \geq 0 \ , \ t \geq 0 \ , \\ q_o(s,t) &= \begin{cases} b(t) \ , & s = 0 \ , \\ 0 \ , & s \neq 0 \ , \end{cases} \end{aligned} \quad (110)$$

which we call the defining equation of the control system (109). We consider the set

$$\pi_\alpha(t) = \{q_k(s,t) \ , \ k = 0,\dots,n-1; \ s \in [0,\alpha h] \quad ,$$

formed from all the nonzero solutions of the defining equation (110) on the interval $[0,\alpha h]$.

We will call Eq. (110) nonsingular if for one $t_1 > 0$

$$\operatorname{rank} \pi_\alpha(t_1) = n \quad , \quad \alpha = \left[\frac{t_1}{n}\right] .$$

We prove the following result.

<u>Theorem 19</u>. <u>If the defining equation (110) is nonsingular, then the system (109) is relatively controllable.</u>

<u>Proof</u>. Let the number $t_1 > 0$ be such that $\operatorname{rank} \pi_\alpha(t_1) = n$. We show that for each initial state $x_0(\cdot)$ there exists a piecewise continuous control $u(t)$, $t \in T$, for which $x(t_1) = 0$. It is not difficult to show that the solution $x(t)$ of Eq. (109) may be represented at $t = t_1$ by the formula

$$x(t_1) = s(t_1,x_0) + \int_0^{t_1} F(t_1,\tau)b(\tau)u(\tau)\, d\tau \quad ,$$

where

$$s(t,x_0) = F(t,0)x_0 + \int_0^h F(t,\tau)A_1(\tau)\Phi(\tau - h)\, d\tau \quad ,$$

while $F(t,\tau)$ is a solution of the equation

$$\frac{dF(t,\tau)}{d\tau} = -F(t,\tau)A(\tau) - F(t,\tau + h)A_1(\tau + h) \ , \ \tau \le t \quad ,$$

$$F(t,t - 0) = E \ , \ F(t,\tau) \equiv 0 \ , \ t \ge t + 0 \quad .$$

Therefore, by virtue of the results of §10, we may assert that the system (109) is relatively controllable if and only if

$$\psi(g,\tau) = g'F(t_1,\tau)b(\tau) \not\equiv 0 \quad , \quad \tau\varepsilon T \quad ,$$

for all g, $||g|| = 1$. It is possible to prove that the function $\psi(g,\tau)$ satisfies the relation

$$\psi^{(k)}(g,\tau) = (-1)^k g' \sum_{i=0}^{k} F(t_1,\tau + kh)q_k(ih,\tau + ih) \quad ,$$

$$k = 0,1,\ldots,$$

from which follow the equalities

$$\left.\begin{aligned} \psi^{(i)}(g,t_1 - kh - 0) &= \psi^{(i)}(g,t_1 - kh + 0) \quad , \\ & i = 0,\ldots,k - 1 \quad , \\ \psi^{(i)}(g,t_1 - kh - 0) &= \psi^{(i)}(g,t_1 - kh + 0) \\ &+ (-1)^i g'q_i(kh,t_1) \quad , \\ i = k \ , \ k + 1,\ldots; \ k &= 1,2,\ldots \end{aligned}\right\} \qquad (111)$$

We assume that for some $g = g_0$, $||g_0|| = 1$, we have the identity

$$\psi(g_0,\tau) \equiv 0 \quad , \quad \tau\varepsilon T \quad . \qquad (112)$$

Then

$$\psi^{(i)}(g_0,t_1 - 0) = g_0'q_i(0,t_1) = 0 \ , \ i = 0,\ldots,n-1 \quad . \qquad (113)$$

In addition, from (112) we have

$$\psi^{(i)}(g_o, \tau - 0) = \psi^{(i)}(g_o, \tau + 0) = 0 \quad , \quad \tau \varepsilon T \quad , \qquad i = 0,\ldots,n-1 \quad . \tag{114}$$

Considering the relations (114) at the points $\psi = t_1 - kh$, $k = 1,\ldots,\alpha$, and taking into account (111), we obtain

$$g_o' q_i(kh, t_1) = 0 \quad , \quad i = k,\ldots,n-1 \quad ; \quad k = 1,\ldots,\alpha \quad . \tag{115}$$

Adjoining the relations

$$g_o' q_i(kh, t_1) = 0 \quad , \quad i = 0,\ldots,k-1 \quad ; \quad k = 1,\ldots,\alpha \quad ,$$

to Eqs. (113) and (115), we are led to the result

$$g_o' q_i(kh, t_1) = 0 \quad , \quad i = 0,\ldots,n-1 \quad ; \quad k = 0,\ldots, \quad .$$

Since, by assumption, $||g_o|| = 1$, the above relations imply the linear dependence of the vectors $\{q_i(kh, t_1), i = 0,\ldots,n-1 ; k = 1,\ldots,\alpha\}$ and, simultaneously, of the system $\pi_\alpha(t_1)$. This contradicts the condition of the theorem and the result is proved.

§16. On the Relative Controllability of Linearized Dynamical Systems with a Delay

Let there be given the equation of motion

$$\left.\begin{aligned} &\frac{dx(t)}{dt} = f(x(t), x(t - h), u(t)), \; f(0,0,0) = 0 \quad , \\ &x_o(\cdot) = \{x(\tau) = \Phi(\tau) \; , \; -h \leq \tau \leq 0 \; , \; x(0) = x_o\} \; , \end{aligned}\right\} \tag{116}$$

where $x = \{x_1,\ldots,x_n\}$, $u = \{u_1,\ldots,u_r\}$, and $f(x,y,u)$ is continuously differentiable in all arguments.

Definition 12. The nonlinear system is called *relatively controllable* if for any $t_1 > 0$, $L > 0$, there is a number $\alpha_0(t_1, L)$ such that each trajectory $x(t)$, corresponding to initial states from the domain,

$$||x_0(\cdot)|| \leq \alpha_0(t_1, L) \quad ,$$

may be transferred to the origin $x = 0$ by a piecewise continuous control $u(t)$, $||u(t)|| \leq L$.

For (116), we form the linear approximation

$$\frac{dx(t)}{dx} = Ax(t) + A_1 x(t - h) + Bu(t) \quad ,$$

where

$$A = \frac{\partial f(0,0,0)}{\partial x} \quad , \quad A_1 = \frac{\partial f(0,0,0)}{\partial y} \quad , \quad B = \frac{\partial f(0,0,0)}{\partial u} \quad .$$

We have the following theorem.

Theorem 20. *The dynamical system (116) is relatively controllable if its linear approximation is relatively controllable.*

This theorem may be proved by the scheme of §7. Necessity of the condition is completely obvious. For a similar goal, the theorem on the existence of roots [2] was applied in [74c]. Theorem 20, containing sufficient conditions for controllability of nonlinear systems, is analagous to theorems about the stability of linearized systems.

§17. Complete Controllability of Dynamical Systems with a Delay. Criteria for Complete Controllability.

We return to the linear system

$$\left.\begin{aligned} \frac{dx(t)}{dt} &= Ax(t) + A_1x(t - h) + Bu(t) \quad , \\ x_0(\cdot) &= x(\tau) = \Phi(\tau) \ , \ -h \le \tau < 0 \ , \ x(0) = x_o \quad . \end{aligned}\right\} \quad (117)$$

Definition 13. The state $x_0(\cdot)$ is called controllable if there exists a time $t_1 < \infty$, and a piecewise continuous control $u(t)$, $0 \le t \le t_1$, such that the trajectory $x(t)$ of (117) satisfies the condition

$$x(t) = 0 \quad , \quad t_1 - h \le t \le t_1 \quad .$$

Definition 14. A dynamical system with a delay is completely controllable if each of its initial states is controllable.

Criteria for complete controllability are obtained for a series of particular cases of system (117). We begin by isolating those systems for which these criteria coincide with the conditions of relative controllability.

1. Systems with n Inputs. Let the motion of the system be described by Eq. (117), where rank $B = n$. In these cases questions about complete controllability are particularly simple to solve. Clearly, under these assumptions $r \ge n$. Without loss of generality, it is possible to assume that the vectors $b^1, \ldots, b^n$ (the columns of B) are linearly independent. We introduce the new controls $v(t) = \{u_1(t), \ldots, u_n(t)\}$ and $w(t) = \{u_{n+1}, \ldots, u_r(t)\}$. Then Eq. (117) has the form

$$\dot{x}(t) = Ax(t) + A_1x(t - h) + B_1v(t) + B_2w(t) \quad ,$$

where B_1 is nonsingular. On the basis of Theorem 16, system (117) is relatively controllable. This means that for each initial state $x_o(\cdot)$, it is possible to find a number $t_1 < \infty$ and a function $u(\cdot)$, such that

$$x(t_1 - h) = 0 \quad .$$

We extend $u(t)$ to the interval $[t_1 - h, t_1]$ setting

$$v(t) = -B_1^{-1}A_1x(t - h) \quad , \quad w(t) \equiv 0 \quad .$$

Thus, we guarantee the condition

$$x(t) \equiv 0 \quad , \quad t_1 - h \leq t \leq t_1 \quad .$$

Consequently, the system (117) is completely controllable.

2. <u>nth Order Differential Equations with a Delayed Argument.</u>

$$\left. \begin{aligned} & x^{(n)}(t) + \sum_{i=1}^{n} a_i x^{(n-i)}(t) + \sum_{s=1}^{n} a_{1,s} x^{(n-s)}(t - h) \\ & \qquad = \sum_{j=0}^{m} b_j u^{(j)}(t) \ , \ t \geq 0 \ , \ m < n, \ \sum_{j=0}^{m} b_j^2 > 0 \ , \\ & x^{(k)}(\tau) = \Phi^{(k)}(\tau) \ , \ \tau \varepsilon [-h,0] \ , \ k = 0,\ldots,n-1 \quad . \end{aligned} \right\} \tag{118}$$

Equation (118) is completely controllable. Indeed, we write it in the form

$$\left. \begin{aligned} & \dot{y}(t) = Ay(t) + A_1y(t - h) + b\mu(t) \quad , \\ & y_o(\cdot) = \{\Phi(\tau) \quad , \quad \tau \varepsilon [-h,0]\} \quad , \end{aligned} \right\} \tag{119}$$

where $\mu(t) = \sum_{j=0}^{m} b_j u^{(j)}(t)$ is the new control,

$\Phi(\tau) = \{\Phi^o(\tau), \ldots, \Phi^{(n-1)}(\tau)$ is the initial function, and

$$A = \begin{Bmatrix} 0 & 1 & 0 & & \cdots & 0 \\ 0 & 0 & 1 & 0 & \cdots & 0 \\ \cdots & \cdots & \cdots & \cdots & \cdots & \cdots \\ 0 & \cdots & \cdots & \cdots & 0 & 1 \\ -a_n & \cdots & \cdots & \cdots & \cdots & -a_1 \end{Bmatrix},$$

$$A = \begin{Bmatrix} 0 & \cdots & 0 \\ \cdot & & \cdot \\ \cdot & & \cdot \\ \cdot & & \cdot \\ \cdot & & \cdot \\ 0 & \cdots & 0 \\ -a_{1,n} & \cdots & -a_{1,1} \end{Bmatrix}, \qquad b = \begin{Bmatrix} 0 \\ \cdot \\ \cdot \\ \cdot \\ \cdot \\ 0 \\ 1 \end{Bmatrix} .$$

It isn't difficult to prove that (119) is relatively controllable (in μ). This means that for each $y_o(\cdot)$, there exists a terminal time t_1 and a piecewise continuous function $\mu_1(t)$, $0 \leq t \leq t_1 - h$, such that

$$y(t_1 - h) = 0 \quad .$$

If we set

$$\mu_2(t) = \sum_{s=1}^{n} a_{1,s} y_{n-s+1}(t - h) \quad ,$$

for $t \varepsilon [t_1 - h, t_1]$, then the control

$$\mu(t) = \begin{cases} \mu_1(t) \quad , & t \varepsilon [0, t_1 - h] \quad , \\ \mu_2(t) \quad , & t \varepsilon [t_1 - h, t_1] \quad , \end{cases}$$

transfers the system (119) to the state $y(t) \equiv 0$, $t_1 - h \leq t \leq t_1$. Then the control

$$u(t) = \begin{cases} u_1(t) & , \quad t \varepsilon [0, t_1 - h] \quad , \\ u_2(t) & , \quad t \varepsilon [t_1 - h, t_1] \quad , \end{cases}$$

$$\sum_{s=0}^{m} b_x u_i^{(s)}(t) = \mu_i(t) \quad , \quad i = 1,2,$$

ensures that the trajectory of the system (118) satisfies the condition

$$x^{(k)}(t) \equiv 0 \ , \ t\varepsilon[t_1 - h, t_1] \ , \ k = 0,\ldots,n-1 \quad .$$

This means that the system is completely controllable.

3. Systems with $r(r < n)$ Inputs. Equation (118) is a particular case of the system (117) with the matrices

$$A_1 = \begin{Bmatrix} 0 & \cdots\cdots\cdots & 0 \\ \cdots & \cdots\cdots\cdots & \cdots \\ 0 & \cdots\cdots\cdots & 0 \\ a_{1;n-r+1,1} & \cdots & a_{1;n-r+1,n} \\ \cdots & \cdots\cdots\cdots & \cdots \\ a_{1;n,1} & \cdots\cdots & a_{1;n,n} \end{Bmatrix} = \begin{Bmatrix} 0 \\ \\ A_{1,1} \end{Bmatrix} ,$$

$$\begin{Bmatrix} 0 & \cdots\cdots & 0 \\ \cdots & \cdots\cdots & \cdots \\ 0 & \cdots\cdots & 0 \\ b_{n-r+1,1} & \cdots & b_{n-r+1,n} \\ \cdots & \cdots & \\ b_{n1} & \cdots\cdots & b_{nr} \end{Bmatrix} = \begin{Bmatrix} 0 \\ \\ B_1 \end{Bmatrix} , \quad \det B_1 \neq 0 \ . \ (120)$$

The following can be proved.

Theorem 21. In order that the system (117), with the matrices A_1 and B of the form (120), be completely controllable it is necessary and sufficient that it be relatively controllable.

The control

$$u(t) = \begin{cases} u_1(t) , & t\varepsilon[0,t_1 - h] , \\ u_2(t) = B_1^{-1}A_{1,1}x(t - h) , & t\varepsilon[t_1 - h,t_1] , \end{cases}$$

completely cancels out the initial perturbations of the system at time $t = t_1$ if $u_1(t)$ and t_1 are chosen by the condition $x(t_1 - h) = 0$. But, such a choice is possible for each $x_0(\cdot)$ by virtue of the relative controllability of the system. This proves the theorem.

4. Dynamical Systems with a Pure Delay. We consider the equation

$$\left.\begin{aligned} &\dot{x}(t) = A_1x(t - h) + Bu(t) , \\ &x_0(\cdot) = \{\Phi(\tau) , \tau\varepsilon[-h,0)\}, \; x(0) = x_o , \end{aligned}\right\} \qquad (121)$$

where $x = \{x_1,\ldots,x_n\}$, $u = \{u_1,\ldots,u_r\}$. We assume that rank $A_1 = n$, rank $B = r$.* For complete controllability of the system (121), it is necessary that it be relatively controllable. We note that (121) is relatively controllable if and only if rank $[B,A_1B,\ldots,A_1^{n-1}B] = n$ (see §13].

---

* This is not a restrictive assumption since if rank $B = r_1 < r$, the arguments remain valid if we reduce the dimension of the control to r_1 and set the components of the vector u, corresponding to the dependent columns of B, equal to zero.

We show that for a relatively controllable system (121), there exists a control $u(t)$, $t \varepsilon T$ (t_1 sufficiently large but finite) ensuring the condition

$$x(t) \equiv 0 \quad , \quad t \varepsilon [t_1 - h, t_1] \quad . \tag{122}$$

Along with (122), we consider the identity $x^{(n)}(t) \equiv 0$, $t \varepsilon [t_1 - h, t_1]$, which in explicit form is

$$A_1^n x(t - nh) + \sum_{i=0}^{n-1} A_1^i B u^{(n-i-1)}(t - ih) \equiv 0 \ , \ t \varepsilon [t_1 - h, t_1] \ . \tag{123}$$

We may consider (123) as a system of algebraic equations in the unknowns $u_j^{(n-i-1)}(t - ih)$, $i = 0,\ldots, n-1$; $j = 1,\ldots,r$. From the relative controllability of (121), it follows that the rank of the matrix of the system (123) is equal to n. Thus, the system (123) may be solved

$$v(t) = \{u_1^{(n-1)}(t) \ , \ u_2^{(n-1)}(t),\ldots,u_r^{(n-1)}(t),$$

$$u_{j_1}^{(n-i_1-1)}(t - i_1 h),\ldots,u_{j_{n-r}}^{(n-i_{n-r}-1)}(t - i_{n-r}h)\} \quad ,$$

setting

$$u_j^{(n-i-1)}(t - ih) \equiv 0 \ , \ i \neq i_s \ , \ j \neq j_s \ , \ s = 1,\ldots,n-r \ . \tag{124}$$

This solution has the form

$$v(t) = Gx(t - nh) \ , \ t \varepsilon [t_1 - h, t_1] \quad , \tag{125}$$

where G is a constant, nonsingular matrix. The relations (124), (125) determine the control $u(t)$ on the interval $[t_1 - nh, t_1]$.

From (123), we have the identity

$$x(t) \equiv \sum_{s=0}^{n} d_s \frac{t^{n-s}}{(n-s)!} \quad ,$$

$$A_1^i x(t - ih) + \sum_{j=0}^{i-1} A_1^j B u^{(i-j-1)}(t - jh) \tag{126}$$

$$\equiv \sum_{s=0}^{n-i} d_s \frac{t^{n-i-s}}{(n - i - s)!} \; , \; t \varepsilon [t_1 - h, t_1] \; , \; i = 1,\ldots,n \; .$$

Here $||d_0|| = 0$, while d_i, $i = 1,\ldots,n$, are arbitrary constant n-vectors. In order that (122) be satisfied, it is necessary and sufficient that $||d_i|| = 0$, $i = 1,\ldots,n$, in (126). Transforming (126), we are led to a system for the unknowns d_i, $i = 1,\ldots,n$:

$$\sum_{i=s}^{n-1} \left[\frac{t_1^{i-s-1}}{(i - s - 1)!} E - \frac{(t_1 - h)^{i-s} A_1}{(t - s)!} \right] d_{n-i} = Bu^{(s)}(t_1) .$$

The determinant of this system is different from zero since, by assumption, A_1 is nonsingular. Consequently, $d_i = 0$, $i = 1,\ldots,n$, if and only if

$$u^{(n-1)}(t) = 0 \; , \; s = 0,\ldots,n-1 \; . \tag{127}$$

From (125) we have

$$u^{(n-1)}(t) = G_1 x(t - nh) \; , \; t \varepsilon [t_1 - h, t_1] \quad , \tag{128}$$

where G_1 is a matrix composed from the first r rows of the matrix G. Differentiating (128) with respect to t, we obtain by virtue of (121)

$$u^{(n)}(t) = G_1A_1x(t - (n + 1)h) + G_1Bu(t - nh) \quad , \quad t\varepsilon[t_1 - h, t_1] \quad . \tag{129}$$

We may distinguish two cases:

1) rank $G_1B = r$. Then the system (129) with controls $w(t) = u(t - nh)$ is completely controllable. Consequently, there exists $w(t)$, $t\varepsilon[t_1 - h, t_1]$(or, what is the same, $u(t)$, $t\varepsilon[t_1 - (n + 1)h, t_1 - nh]$), ensuring (127). The control $u(t)$ may be arbitrary on the interval $[0, t_1 - (n + 1)h]$. Thus, if the system (121) is relatively controllable, condition (122) may be satisfied for all $t_1 \geq (n + 1)h$.

2) rank $G_1B < r$. Let the number k be such that rank $G_1A_1^iB$, $i = 1,\ldots,k-1$ is also less than r but the matrix $G_1A_1^kB$ has rank r. From the relative controllability of the system (121), it follows that such a k exists and is greater than n-1. Setting

$$u(t) \equiv 0 \ , \ t\varepsilon[t_1 - (n + k)h, t_1 - nh] \quad ,$$

from (128) and (121) we obtain

$$u^{(n+k)}(t) = G_1A_1^{k+1}x(t - (n + k + 1)h)$$
$$+ G_1A_1^kBu(t - (n + k)h) \ , \ t\varepsilon[t_1 - h, t_1] \ . \tag{130}$$

Equation (130) is completely controllable with respect to the perturbations $w(t) = u(t - (n + k)h)$. Consequently, the condition (127) may be guaranteed by choosing a suitable control $w(t)$, $t\varepsilon[t_1 - h, t_1]$($u(t)$, $t\varepsilon[t_1 - (n + k + 1)h, t_1 - (n + k)h]$), letting $u(t)$ be arbitrary on the interval $[0, t_1 - (n + k + 1)h]$.

Thus, we have proved the following theorem.

Theorem 22. The system (121) (A_1 a nonsingular matrix) is completely controllable if and only if it is relatively controllable.

5. Complete Controllability of a Single-Input System. For the system (121) with a single control ($r = 1$), the restriction on the matrix A_1 may be dropped. We consider this case in more detail.

Let the system

$$\left.\begin{aligned} \dot{x}(t) &= A_1 x(t-h) + bu(t) \quad , \\ x_0(\cdot) &= \{\Phi(\tau)\ ,\ \tau\varepsilon[-h,0),\ x(0) = x_o \quad , \end{aligned}\right\} \tag{131}$$

where u is a scalar, be relatively controllable. This means that the vectors $b, A_1 b, \ldots, A_1^{n-1} b$ are linearly independent. Choosing these vectors as new basis vectors, we transform the system (131) to the form [39]

$$\dot{y}(t) = A_{11} y(t-h) + b_1 u(t) \quad ,$$

$$y_0(\cdot) = \{\Phi_1(\tau)\ ,\ \tau\varepsilon[-h,0)\ ,\ y(0) = y_0 \quad ,$$

$$A_{11} = \begin{pmatrix} 0 & \cdots\cdots & 0 & \alpha_1 \\ 1 & 0 \ \cdots & 0 & \alpha_2 \\ \cdots & \cdots & \cdots & \cdots \\ 0 & \cdots\cdots 0 & 1 & \alpha_n \end{pmatrix} , \qquad b_1 = \begin{pmatrix} 1 \\ 0 \\ \vdots \\ 0 \end{pmatrix} . \tag{132}$$

We assume that α_{k+1} is the first nonzero number from the sequence $\alpha_1, \ldots, \alpha_n$. We shall separate the system (132) into two subsystems for the vectors $v = \{y_1, \ldots, y_k\}$ and $w = \{y_{k+1}, \ldots, y_n\}$. We have

$$\dot{v}(t) = D_1 v(t-h) + e_1 u(t) \ , \ \dot{w}(t) = D_2 w(t-h) + e_2 y_k(t-h) .$$

Here

$$D = \begin{pmatrix} 0 & \cdots\cdots\cdots & 0 \\ 1 & 0 \cdots\cdots & 0 \\ \cdots & \cdots\cdots\cdots & \cdots \\ 0 & \cdots\cdots 1 & 0 \end{pmatrix}, \quad e_1 = \begin{pmatrix} 1 \\ 0 \\ \vdots \\ 0 \end{pmatrix},$$

$$D_2 = \begin{pmatrix} 0 & \cdots\cdots & 0 & \alpha_{k+1} \\ 1 & 0 \cdots & 0 & \alpha_{k+2} \\ \cdots & \cdots & \cdots & \cdots \\ 0 & \cdots 0 & 1 & \alpha_n \end{pmatrix}, \quad e_2 = \begin{pmatrix} 1 \\ 0 \\ \vdots \\ 0 \end{pmatrix}.$$

In the case $k = 0$ ($\alpha_1 \neq 0$), the second subsystem is identical with the completely controllable system (132): $w = y$, $D_2 = A_{11}$, $e_2 = b_1$, $y_0(t - h) = u(t)$. We show that the condition $y(t) \equiv 0$, $t \in [t_1 - h, t_1]$ will be satisfied for (132) if we use the control

$$u(t) = \begin{cases} u_1(t) , & t \in [0, t_1 - nh] , \\ u_2(t) , & t \in [t_1 - nh, t_1 - kh], \\ 0 , & t \in [t_1 - kh, t_1] , \end{cases}$$

where $u_1(t)$ and t_1 are chosen such that

$$y_k^{(i)}(t_1 - h) = 0 \quad , \quad i = 0,\ldots,n-1 \quad , \tag{133}$$

while $u_2(t)$ satisfies

$$w^{(n)}(t) \equiv 0 \quad , \qquad t \in [t_1 - h, t_1] \quad ,$$

or, in explicit form,

$$Ly(t - nh) + \sum_{i=k}^{n-1} D_2^{i-k} e_2 u_2^{(n-i-1)}(t - ih) \equiv 0 \quad , \qquad t \in [t_1 - h, t_1] \; . \tag{134}$$

Here $D_2^0 = E$, and L is an $(n - k) \times n$ matrix of the form

$$L = \{l_{ij}\} = \{[D_2^{n-k+j-1} e_2]_i\} \quad .$$

We note that for each fixed t, $t \varepsilon [t_1 - h, t_1]$, the solution $\{u_2^{(n-k-1)}(t - kh),\ u_2^{(n-k-2)}(t - (k + 1)h), \ldots, u_2(t - (n - 1)h)\}$ of the system (134) exists by virtue of the linear independence of the vectors $e_2, D_2 e_2, \ldots, D_2^{n-k-1} e_2$ and is completely determined by the state $\{y(t),\ t\varepsilon[t_1 - (n - 1)h,\ t_1 - nh]\}$. If $u_2(t)$ is given by (134), then we have the identity

$$w(t) \equiv \sum_{s=0}^{n} d_s \frac{t^{n-s}}{(n - s)!} \quad ,$$

$$D_2^m w(t - mh) + \sum_{i=1}^{m} D_2^{i-1} e_2 y_k^{(m-i)}(t - ih) \tag{135}$$

$$\equiv \sum_{s=0}^{n-m} d_s \frac{t^{n-m-s}}{(n - m - s)!} \quad , \ m = 1,\ldots,n;\ t\varepsilon[t_1 - h, t_1] \ ,$$

where $||d_0|| = 0$, while d_i, $i = 1,\ldots,n$ are arbitrary constant vectors. Considering (135) at the moments $t = t_1 - h$ and $t = t_1$, we obtain for the d_i, $i = 1,\ldots,n$, a system of algebraic equations

$$\sum_{i=s}^{n-1} \left[\frac{t_1^{i-s-1}}{(i - s - 1)!} E - \frac{(t_1 - h)^{i-s}}{(i - s)!} D_2 \right] d_{n-i} = e_2 y_k^{(s)}(t_1 - h),$$

$$s = 0,\ldots,n-1 \quad , \tag{136}$$

with a nonzero determinant (the matrix D_2 is nonsingular). If the control $u_1(t)$, $t\varepsilon[0, t_1 - nh]$ is chosen according to condition (133), then the system (136) admits only the trivial solution $||d_i|| = 0$, $i = 1,\ldots,n$. This means that

$$w^{(k)}(t) \equiv 0 \quad , \quad t\varepsilon[t_1 - h, t_1] \quad , \quad k = 0,\ldots,n-1 \quad .$$

Thus, the control

$$u(t) = \begin{cases} u_1(t) & , \quad t\varepsilon[0, t_1 - nh] \quad , \\ u_2(t) & , \quad t\varepsilon[t_1 - nh, t_1 - kh] \quad , \end{cases}$$

under the condition (133) ensures the n-k coordinates $y_{k+1},\ldots,y_n$ are zero at $t = t_1$, Moreover, this same control "prepares" the trajectory of the system in order to damp out the perturbations in the remaining k coordinates by the cut-off control

$$u_3(t) \equiv 0 \quad , \quad t\varepsilon[t_1 - kh, t_1] \quad .$$

From (132), it follows that for $u = u_3(t)$

$$y_1(t) \equiv \text{const} \quad , \quad t\varepsilon[t_1 - kh, t_1] \quad .$$

In addition, by (132) and (133)

$$y_1(t_1 - kh) = \dot{y}_2(t_1 - (k - 1)h) = \cdots = y_k^{(k-1)}(t_1 - h) = 0 \quad .$$

Therefore,

$$y_1(t) \equiv 0 \quad , \quad t\varepsilon[t_1 - kh, t_1] \quad .$$

For arbitrary r, $1 \leq r \leq k-1$, let

$$y_r(t) \equiv 0 \quad , \quad t\varepsilon[t_1 - (k - r + 1)h, t_1] \quad .$$

We show that

$$y_{r+1}(t) \equiv 0 \quad , \quad t\varepsilon[t_1 - (k - r)h, t_1] \quad .$$

From Eq. (132) we have

$$u_{r+1}(t) \equiv \text{const} \quad , \quad t\varepsilon[t_1 - (k - r)h, t_1] \quad .$$

But, since

$$y_{r+1}(t_1 - (k - r)h) = y_{r+2}(t_1 - (k - r - 1)h)$$
$$= \cdots = y_k^{(k-r-1)}(t_1 - h) \quad ,$$

then

$$y_{r+1}(t) \equiv 0 \quad , \quad t\varepsilon[t_1 - (k - r)h, t_1] \quad .$$

Because of the arbitrariness of r, we conclude

$$y_{s+1}(t) \equiv 0 \ , \ t\varepsilon[t_1 - (k - s)h, t_1] \ , \ s = 0, \ldots, k-1 \quad .$$

To complete the proof, it remains to establish that a control $u_1(t)$ may be chosen according to (133). We solve Eq. (134) for $u_2^{(n-k-1)}(+ - kh)$:

$$u_2^{(n-k-1)}(t - kh = a'y(t - nh \quad t\varepsilon[t_1 - h, t_1] \quad . \tag{137}$$

Here $a = \{a_1, \ldots, a_n\}$ is some constant vector with $a_1 = -\alpha_{k+1}$. Using (132), we calculate the derivative $y_k^{(n)}(t-h)$ and, taking into account (137), we obtain

$$y_k^{(n)}(t - h) = a'A_{11}y(t - (n + 1)h) - \alpha_{k+1}u(t - nh) \ , \tag{138}$$
$$t\varepsilon[t_1 - h, t_1].$$

Since by assumption, $\alpha_{k+1} \neq 0$, Eq. (138), considered on the interval $[t_1 - h, t_1]$ as an equation without delay, is completely controllable. Consequently, condition (133) is satisfied for all $t_1 \geq (n + 1)h$ by choosing the control $u(t)$ only on the interval $[(n + 1)h, t_1]$, assuming it to be arbitrary at $t = (n + 1)h$ (for example, $u(t) \equiv 0$, $t \in [0, (n + 1)h]$). Thus, the system (132) is completely controllable on each interval T, $t_1 \geq (n + 1)h$. Since Eq. (132) was obtained from (131) using a nonsingular transformation, from the complete controllability of (132) we also obtain complete controllability for (131).

We formulate these results as a theorem.

Theorem 23. In order that the system (131) be completely controllable, it is necessary and sufficient that it be relatively controllable.

In sections 1-4 of this paragraph, we have shown that in some cases relative controllability of a system implies complete controllability. However, this is not true in general.

Example 10. The system

$$\dot{x}(t) = Ax(t) + A_1 x(t - h) + bu(t)$$

with matrices

$$A = \begin{pmatrix} 1 & 0 \\ -e & -1 \end{pmatrix}, \quad A_1 = \begin{pmatrix} 0 & 0 \\ 1 & 0 \end{pmatrix}, \quad b = \begin{pmatrix} 1 \\ 0 \end{pmatrix},$$

which is clearly relatively controllable for $h = 1$, is not completely controllable.

Example 11. The system (117) with matrices

$$A = \begin{pmatrix} 0 & 1 \\ 0 & 0 \end{pmatrix}, \qquad A_1 = \begin{pmatrix} 0 & -1 \\ 1 & 0 \end{pmatrix}, \qquad b = \begin{pmatrix} 0 \\ 1 \end{pmatrix}$$

is not completely controllable for any $h > 0$.

6. An Auxiliary Result. Below (sections 7,8) are considered two particular cases of system (117) for which it is possible to state sufficient conditions for complete controllability which do not coincide with conditions for relative controllability.

Beforehand, we formulate a result related to the relative controllability of systems with a delay, which will be useful for the analysis of these cases.

We assume that it is required to reduce the system (117):

$$\dot{x}(t) = Ax(t) + A_1 x(t - h) + Bu(t) \quad ,$$

$$x_o(\cdot) = \{\Phi(\tau), \tau \varepsilon [-h,0), \ x(0) = x_o\} \quad ,$$

to the position $x(t_1) = 0$ under the condition that the control $u(t)$ may be chosen only on the subinterval $[0,t_1 - h]$ of T, while on the remainder of the interval it is given in the form

$$u(t) = \sum_{i=0}^{s} V_i x(t - ih) \quad , \quad t \varepsilon [t_1 - h, t_1] \quad , \qquad (139)$$

where the V_i are some constant rxn matrices. Then on the interval $[t_1 - h, t_1]$, we have

$$x(t) = \sum_{i=0}^{s} D_i x(t - ih) \quad ,$$

where $D_o = A + BV_o$, $D_1 = A + BV_1$, $D_k = BV_k$, $k = 2,\ldots,s$.

The condition $x(t_1) = 0$ is written, using Cauchy's formula, in the form

$$\left.\begin{aligned} x(t_1 - h) + \int_{t_1-h}^{t_1} F_1(t_1 - h - \tau) \sum_{i=1}^{s} D_i x(\tau - ih)\,d\tau = 0 \ , \\ \dot{F}_1 = C_o F_1 \ , \ F_1(0) = E \ . \end{aligned}\right\} \quad (140)$$

Replacing $x(t_1 - h)$ and $x(\tau - ih)$ by Cauchy's formula, we reduce the problem to a problem of moments: to find a function

$$u(t) = \{u_1(t), \ldots, u_r(t)\} \ ,$$

generating a linear functional which assumes given values on given elements:

$$\int_0^{t_1-h} u'(\theta)[F_2(t_1 - h, \theta)B]_j d\theta = \beta_j \ , \ j = 1, \ldots, n \ .$$

Here

$$F_2(t_1 - h, \theta) = F(t_1 - h, \theta)$$
$$+ \int_{t_1-2}^{t_1-h} F_1(t_1 - 2h - \tau) \sum_{i=1}^{s} D_i F(\tau - (i - 1)h, \tau)\,d\tau \ , \qquad (141)$$

$[F_2(\tau, \theta)B]_j$ is a vector consisting of the elements of the jth row of the matrix $F_2(\tau, \theta)B$, while the quantities β_j are determined by the initial state $x_o(\cdot)$ of the system (117). We form the sequence of matrices $\Gamma_k(s, \sigma)$, $k = 1, 2, \ldots$, by the rule

$$\Gamma_o(s,\sigma) = Q_o(s) \quad ,$$

$$\Gamma_k(s,\sigma) = Q_k(s) + F_1(-\sigma) \sum_{i=1}^{s} D_i Q_{k-1}(s - (i - 1)h)$$

$$+ F_1(-\sigma + h) \sum_{i=1}^{s} D_i Q_{k-1}(s - ih) + \frac{d\Gamma_{k-1}(s,\sigma)}{d\sigma} \,, \tag{142}$$

$$k \geq 1 \ , \ s \geq 0 \ , \ \sigma \geq 0 \ ,$$

where $Q_k(s)$ is the solution of the defining equation (62). Then we have the following theorem.

Theorem 24. If

$$\text{rank } \{\Gamma_k(\alpha h,h) \ , \ \alpha = 0,\ldots,k; \ k = 0,\ldots,n-1\} = n \quad ,$$

then the system (117), with the control (139) given on $[t_1 - h, t_1]$, is relatively controllable.

Proof. If the condition of the theorem is satisfied and $t_1 > nh$, then it is possible to find a number k, $1 \leq k \leq r$, such that the relation

$$\psi_k(g,\theta) = g'[F_2(t_1 - h,\theta)B]^k \not\equiv 0 \ , \ \theta\varepsilon[0,t_1 - h] \ , \tag{143}$$

will be satisfied for any vector g, $||g|| = 1$. Here $[F_2(t_1 - h,\theta)B]^k$ is the kth column of the matrix $F_2(t_1 - h,\theta)B$. Assume the contrary. Then for some vector $g = g_o$, $||g_o|| = 1$, we have the equation

$$\left.\begin{aligned} \psi_k^{(m)}(g_o,t_1 - h - 0) = 0 \quad , \\ \psi_k^{(m)}(g_1,t_1 - ph - 0) - \psi_k^{(m)}(g_o,t_1 - ph + 0) = 0 \quad , \\ m = 0,\ldots,n-1 \ ; \ p = 2,\ldots,n \quad , \end{aligned}\right\} \tag{144}$$

where $\psi_k^m(g,\theta)$ is the mth derivative in θ of the function $\psi_k(g,\theta)$. Transforming (144) with the help of (140)-(143), we obtain

$$g_0'[\Gamma_m(\alpha h,h)]^k = 0 \quad ,$$

$$\alpha = 0,\ldots,m \ ; \ m = 0,\ldots,n-1 \ ; \ k = 1,\ldots,r \ .$$

But the last equation is possible only for $||g_0|| = 0$. This contradiction proves the theorem.

7. Systems with n-1 Inputs. Let the system (117) have the matrices

$$B = \begin{Bmatrix} b_{11} & \cdots\cdots & b_{1,n-1} \\ \cdots & \cdots & \cdots \\ b_{n-1,1} & \cdots & b_{n-1,n-1} \\ 0 & \cdots\cdots & 0 \end{Bmatrix} = \begin{Bmatrix} B_1 \\ \\ 0 \end{Bmatrix} \quad , \quad \det B_1 \neq 0 \quad . \tag{145}$$

Since for complete controllability it is necessary that the system be relatively controllable, first all we shall find conditions for the relative controllability of the system (117),(145).

Lemma 3. In order that the systems (117),(145) be relatively controllable, it is necessary and sufficient that

$$\text{rank}\ \{B,AB,A_1B\} = n \quad . \tag{146}$$

Proof. (Sufficiency) Under the condition (146), the defining equation (62) is nonsingular and, consequently, by Theorem 16 the system (117) is relatively controllable.

(Necessity). Assume that the system (117),(145) is relatively controllable but that rank $\{B, AB, A_1B\} < n$. Thus, taking into account the special form of the matrix B, we have

$$a_n'\bar{b}_j = 0 \ , \ a_{1,n}'\bar{b}_j = 0 \ , \ j = 1,\dots,n-1 \quad . \tag{147}$$

Here and below we use the notation: if $D = \{d_{ij}\}$ is a $k \times \ell$ matrix, then $d_s = \{d_{s1},\dots,d_{s\ell}\}$, $\bar{d}_p = \{d_{1p},\dots,d_{kp}\}$ $s = 1,\dots,k$; $p = 1,\dots,\ell$. From (147) it follows that $a_{nj} = 0$, $a_{1,nj} = 0$, $j = 1,\dots,n-1$, since by assumption $\det B_1 \neq 0$. But then in all the matrices $Q_k(s) = \{q_{ij}(k_1s)\}$, $k = 0,1,\dots,$ satisfying (62), the vectors $q_n(k,s)$ are zero for each fixed $s \in (-\infty,\infty)$. Consequently, the defining equation (62) is singular and then, by virtue of Theorem 16, the system (117),(145) is not relatively controllable, which is a contradiction. The lemma is proved.

In connection with Lemma 3, we consider two possibilities:

1) We assume that

$$\text{rank } \{B, A_1B\} = n \quad . \tag{148}$$

This means that $||B'a_{1,n}|| \neq 0$, i.e. it is possible to find a k, $1 \leq k \leq n-1$, such that $a_{1,n}'\bar{b}_k \neq 0$. If for each initial state $x_0(\cdot)$ there exists $t_1, t_1 < \infty$, and a $u(\cdot)$, such that

$$x(t_1 - h) = 0 \quad , \tag{149}$$

then subsequent control of the system by the law

$$u_i(\theta) = 0 \ , \ i \neq k \ , \ u_k(\theta) = -\frac{1}{a'_{1,n}\bar{b}_k}[a_{1,n}Ax(\theta) + a'_{1,n}A_1x(\theta - h)] , \ \theta\varepsilon[t_1 - 2h, t_1 - h] \ , \tag{150}$$

and

$$\left.\begin{aligned} u(\theta) &= -B_1^{-1}A_{11}x(\theta - h) \ , \ \theta\varepsilon[t_1 - h, t_1] \ , \\ A_{11} &= \begin{Bmatrix} a_{1,11} & \cdots\cdots & a_{1,1n} \\ \cdots & \cdots & \cdots \\ a_{1,n-11} & \cdots & a_{1,n-1n} \end{Bmatrix} , \end{aligned}\right\} \tag{151}$$

transfers it to the state $x(t) \equiv 0$, $t\varepsilon[t_1 - h, t_1]$. The equality (150) is obtained from the conditions $a'_{in}x(\theta) \equiv 0$, $\theta\varepsilon[t_1 - 2h, t_1 - h]$; hence, it follows that $a'_{in}x(\theta) \equiv$ constant on the interval $[t_1 - 2h, t_1 - h]$. But if the condition (149) is satisfied, then $a'_{in}x(\theta) \equiv 0$ $\theta\varepsilon[t_1 - 2h, t_1 - h]$. The last identity, together with (149) and (151), ensures the condition $x(\theta) \equiv 0$, $\theta\varepsilon[t_1 - h, t_1]$.

Thus, the system (117),(145) is completely controllable if there exists a control $u(t)$, $t\varepsilon[0, t_1 - 2h]$, solving problem (149) under the condition (150). Sufficient conditions for the solvability of problem (149), (150) follow from Theorem 24 under the assumptions that we substitute

$$\left.\begin{aligned} D_o &= \{d^o_{ij}\} = \left\{ a_{ij} - \frac{1}{a'_{1,n}\bar{b}_k} a'_j a_{1,n} b_k \right\} , \\ D_1 &= \{d^1_{ij}\} = \left\{ a_{1,ij} - \frac{b_{ik}}{a'_{1,n}\bar{b}_k} \bar{a}'_{1,j} a_{1,n} \right\} , \\ D_m &= \{0\} \ , \ m \geq 2 \end{aligned}\right\} \tag{152}$$

into the formulas (142).

2) Now let

$$\text{rank}\,\{B, A_1 B\} < n \quad , \tag{153}$$

but

$$\text{rank}\,\{B, AB\} = n \quad . \tag{154}$$

From (153) we have $a_{1,nj} = 0$, $j = 1,\dots,n-1$. We will assume $a_{1,nn} \neq 0$, since, if not, this case turns out to be covered by Theorem 21. By virtue of (154), the norm $||B'a_n|| \neq 0$. Consequently, it is possible to find a number k, $1 \leq k \leq n-1$, such that $a_n'\bar{b}_k \neq 0$. In order to ensure the condition $a_{1n}'\ddot{x}(\theta) \equiv 0$, $\theta \in [t_1 - 2h, t_1 - h]$, we set

$$\left.\begin{aligned} &u_i(\theta) \equiv 0 \quad , \quad i \neq k \quad , \quad u_k(\theta) = -\frac{1}{a_n'\bar{b}_k}\{a_n'Ax(\theta) \\ &+ a_n'A_1x(\theta - h) + a_{1,nn}[a_n'x(\theta - h) + a_{1,n}'x(\theta - 2h)]\}, \\ &\qquad\qquad\qquad\qquad \theta \in [t_1 - 2h, t_1 - h] \ . \end{aligned}\right\} \tag{155}$$

Then the trajectories of the system (117) will satisfy the identity

$$\left.\begin{aligned} &a_n'x(t - h) + a_{1,n}'x(t - 2h) \equiv d_1 \quad , \\ &a_{1,n}'x(t - h) \equiv d_1 t + d_2 \quad , \quad t \in [t_1 - h, t_1] \quad , \end{aligned}\right\} \tag{156}$$

where d_i, $i = 1,2$ are some constant quantities. From (156) we have the equation

$$(1 - t_1 + h)d_1 - d_2 = a_n' x(t_1 - h) \quad ,$$

$$t_1 d_1 + d_2 = a_{1,n}' x(t_1 - h) \quad ,$$

which, under condition (149), ensures that the constants d_i, $i = 1,2$, are identically zero. If the control is defined on the interval $[t_1 - h, t_1]$ by (151), then with regard for (149) and (156) (where $d_1 = d_2 = 0$), we obtain $x(t) \equiv 0$, $t_1 - h \leq t \leq t_1$. Consequently, in this case the system (117),(145) is completely controllable if it is relatively controllable on $[0, t_1 - h]$ with the control $u(t)$, $t \varepsilon [t_1 - 2h, t_1 - h]$ from (155). The latter occurs for the system (117),(145) if the conditions of Theorem 24 are satisfied with

$$D_0 = \{d_{ij}^0\} = \left\{ a_{ij} - \frac{b_{ik}}{a_n' \bar{b}_k} \bar{a}_j' a_n \right\} \quad ,$$

$$D_1 = \{d_{ij}^1\} = \left\{ a_{1,ij} - \frac{b_{ik}}{a_n' \bar{b}_k} [a_n' a_{i,j} + a_{nj}] \right\} \quad ,$$

$$D_2 = \{d_{ij}^2\} = - \frac{1}{a_n' \bar{b}_k} b_{ik} a_{1,nj} \quad ,$$

$$D_m = \{0\} \quad , \quad m \geq 3 \quad .$$

8. <u>Toward a Study of the General Case of Complete Controllability</u>. Generalizing the preceding case, we consider the system (117) with the matrices

$$A_1 = \begin{Bmatrix} a_{1,11} \cdots\cdots a_{1,1n} \\ \cdots\cdots\cdots\cdots\cdots\cdots \\ a_{1,r1} \cdots\cdots a_{1,rn} \\ \hline a_{1,r+11} \cdots a_{1,r+1n} \\ 0 \cdots\cdots\cdots 0 \\ \cdots\cdots\cdots\cdots\cdots\cdots \\ 0 \cdots\cdots\cdots 0 \end{Bmatrix} = \begin{Bmatrix} A_{11} \\ --- \\ A_{12} \end{Bmatrix},$$

$$B = \begin{Bmatrix} b_{11} \cdots b_{1r} \\ \cdots\cdots\cdots\cdots \\ b_{r1} \cdots b_{rr} \\ \hline 0 \cdots\cdots 0 \\ \cdots\cdots\cdots\cdots \\ 0 \cdots\cdots 0 \end{Bmatrix} = \begin{Bmatrix} B_1 \\ -- \\ 0 \end{Bmatrix}. \tag{157}$$

Two situations are possible:

1) rank $\{B\} \neq$ rank $\{B,A_1B\}$. In this case the investigation of complete controllability of the system (117), (157) proceeds according to the scheme of section 7 (the case rank $\{B,A_1B\} = n$). As a result, we are led to relations analagous to (150)-(152) with the difference that in (150),(152) the vector $a_{i,n}$ is replaced by the vector $a_{1,r+1}$, while in (151) A_{11} and B_1 are the matrices from (157).

2) Much more complex is the case rank B = rank $\{B,A_1B\}$. We will assume that $||a_{1,r+1}|| \neq 0$, since the case $||a_{1,r+1}|| = 0$ was already treated in section 3.

As a prelude, we prove an auxiliary proposition applying only in case 2).

Lemma 4. If

$$a'_{1,r+1}\bar{q}_j(p,0) = 0 \ , \ j = 1,\ldots,r \ ; \ p = 0,\ldots,k \ ,$$

then

$$a'_{1,r+1}\bar{q}_j(k + m,mh) = 0 \ , \ j = 1,\ldots r \ ; \ m = 1,2,\ldots,$$

Here $\bar{q}_j(p,s)$ is the jth column of the matrix $Q_p(s)$ from (62).

Proof. We proceed by double induction. First of all, set p = 0. Then from the relations

$$a'_{1,r+1}\bar{q}_j(0,0) = 0 \quad , \quad j = 1,\ldots,r \quad ,$$

it follows that $a_{1,r+1j} = 0$, $j = 1,\ldots,r$. Thus

$$a'_{1,r+1}\bar{q}_j(1,h) = a'_{1,r+1}A_1\bar{q}_j(0,0) = a_{1,r+1,r+1}a'_{1,r+1}\bar{q}_j(0,0) = 0.$$

Generally, if

$$a'_{1,r+1}\bar{q}_j(m,mh) = 0 \quad , \quad j = 1,\ldots,r \quad ,$$

then

$$a'_{1,r+1}\bar{q}_j(m + 1,(m + 1)h) = a'_{1,r+1}A_1\bar{q}_j(m,mh)$$

$$= a_{1,r+1,r+1}a'_{1,r+1}\bar{q}_j(m,mh) = 0 \quad .$$

We assume that the lemma is true for $p = 0,\ldots,k-1$. We show that it holds for p = k. Let

$$a'_{1,r+1}\bar{q}_j(p,0) = 0 \ , \ j = 1,\ldots,r \ ; \ p = 0,\ldots,k \quad .$$

This means that $[(A')^p a_{1,r+1}]_s = 0$, $s = 1,\dots,r$; $p = 0,\dots,k$. Then

$$a'_{1,r+1}\bar{q}_j(k + 1,h) = (A'a_{1,r+1})'\bar{q}_j(k,h) + a'_{1,r+1}A_1\bar{q}_j(k,0)$$

$$= a'_{1,r+1}A^2\bar{q}_j(k - 1,h) + a'_{1,r+1}AA_1\bar{q}_j(k - 1,0)$$

$$+ a'_{1,r+1}A_1\bar{q}_j(k,0) = \cdots = a'_{1,r+1}A^k\bar{q}_j(1,h)$$

$$+ \sum_{i=1}^{k} a'_{1,r+1}A^{k-i}A_1\bar{q}_j(i,0) = \sum_{i=0}^{k} a'_{1,r+1}A^{k-i}A_1\bar{q}_j(i,0)$$

$$= \sum_{i=0}^{k} [(A')^{k-i}a_{1,r+1}]_{r+1}a'_{1,r+1}\bar{q}_j(i,0) = 0 \ , \ j = 1,\dots,r \ .$$

Now we assume that

$$a'_{1,r+1}\bar{q}_j(k + m,mh) = 0 \quad , \quad j = 1,\dots,r \quad ,$$

and show that

$$a'_{1,r+1}\bar{q}_j(k + m + 1,(m + 1)h) = 0 \quad , \quad j = 1,\dots,r \quad .$$

Repeating the preceding argument, we obtain

$$a'_{1,r+1}\bar{q}_j(k + m + 1,(m + 1)h) = a'_{1,r+1}A\bar{q}_j(k + m,(m + 1)h)$$

$$+ a'_{1,r+1}A_1\bar{q}_j(k + m,mh) = a'_{1,r+1}A^2\bar{q}_j(k + m - 1,(m + 1)h)$$

$$+ a'_{1,r+1}AA_1\bar{q}_j(k + m - 1,mh) + a'_{1,r+1}A_1\bar{q}_j(k + m,mh)$$

$$= \cdots = a'_{1,r+1}A^k\bar{q}_j(m + 1,(m + 1)h)$$

$$+ \sum_{i=1}^{k} a'_{1,r+1}A^{k-i}A_1q_j(i + m,mh) = \sum_{i=0}^{k} a'_{1,r+1}A^{k-i}A_1\bar{q}_j(i + m,mh)$$

$$= \sum_{i=0}^{k} [(A')^{k-i} a_{1,r+1}]_{r+1} a'_{1,r+1} \bar{q}_j(i+m,mh) = 0 \quad ,$$

$$j = 1,\ldots,r \quad .$$

The lemma is proved.

Corollary. If the system (117),(157) is relatively controllable, then there exist numbers ℓ and p, $1 \le \ell \le n$, $1 \le p \le r$, such that

$$\left. \begin{array}{r} a'_{1,r+1}\bar{q}_j(k,0) = 0 \ , \ j = 1,\ldots,r \ ; \ k = 0,\ldots,\ell-1 \ , \\ a'_{1,r+1}\bar{q}_p(\ell,0) \neq 0 \ . \end{array} \right\} \quad (158)$$

Since by assumption, $||a_{1,r+1}|| = 0$, then by Lemma 4 we have the equality

$$a'_{1,r+1}\bar{q}_j(k,0) = 0 \ , \ j = 1,\ldots,r \ ; \ k = 0,\ldots,n-1 \quad ,$$

which means that

$$\text{rank } \{\bar{q}_j(k,0) \ , \ j = 1,\ldots,r \ ; \ k = 0,\ldots,n-1\} < n \quad .$$

This contradicts the relative controllabilty of (117), (157).

Now we pass directly to the question of the complete controllability of the system (117),(157). We consider the equation

$$a'_{1,r+1} x^{(\ell)}(t) = 0 \quad , \quad t \in [t_1 - 2h, t_1 - h] \quad ,$$

or, what is the same thing,

$$a'_{1,r+1} \sum_{i=0}^{\ell} P_\ell(ih) x(t - ih) + \sum_{j=1}^{r} a'_{1,r+1} q_j(\ell,0) u_j(t) = 0 \ . \quad (159)$$

Here, ℓ is the number from (158), while $P_k(s)$ is the solution of the defining equation (98) with the initial condition

$$P_o(s) = \begin{cases} E , & s = 0 \\ 0 , & s \neq 0 . \end{cases} \tag{160}$$

We set

$$u(t) = -B_1^{-1}A_{11}x(t - h) \quad , \; t\varepsilon[t_1 - h,t_1] \quad , \tag{161}$$

$$u_j(t) \equiv 0 \quad , \quad j \neq p \quad , \quad t\varepsilon[t_1 - 2h,t_1 - h] \quad . \tag{162}$$

Then, by virtue of (159)

$$u_p(t) = - \frac{1}{a_{1,r+1}'\bar{q}_p(\ell,0)} a_{1,r+1}' \sum_{i=0}^{\ell} P_\ell(ih)x(t - ih) \; , \tag{163}$$

$$t\varepsilon[t_1 - 2h,t_1 - h] \; .$$

Choosing such a control ensures that for the trajectories of the system we will have the identity

$$a_{1,r+1}' \sum_{i=0}^{s} P_s(ih)x(t - ih) \equiv \sum_{j=1}^{\ell-s} d_j \frac{(t + h)^{\ell-s-j}}{(\ell - s - j)!} \quad , \tag{164}$$

$$s = 0,\ldots,\ell-1 \; , \; t\varepsilon[t_1 - 2h,t_1 - h] \quad ,$$

where d_j, $j = 1,\ldots,\ell$, are arbitrary constants of integration. After a simple transformation of the relations (164), we are led to a system of algebraic equations for the d_j, $j = 1,\ldots,\ell$:

$$\sum_{j=1}^{\ell} d_j \frac{t_1^{\ell-j}}{(\ell-j)!} = a'_{1,r+1}x(t_1 - h) \quad ,$$

$$\sum_{j=1}^{\ell-k} d_j \frac{t_1^{-k-j}}{(\ell-k-j)!} - \sum_{i=0}^{k-1} [(A')^i a_{1,r+1}]_{r+1} \sum_{j=k}^{\ell+i} d_{j-k+1} \frac{(t_1-h)^{\ell+i-j}}{(\ell+i-j)!}$$

$$= a'_{1,r+1}A^k x(t_1 - h) \ , \ k = 1,\ldots,\ell-1 \quad ,$$

the determinant of which equals

$$\Delta(h) =$$

$$\begin{vmatrix} \frac{h^{\ell-1}}{(\ell-1)!} & \frac{h^{\ell-2}}{(\ell-2)!} & \cdots & h & 1 \\ \frac{h^{\ell-2}}{(\ell-2)!} & \frac{h^{\ell-3}}{(\ell-3)!} & \cdots & 1 & -[a_{1,r+1}]_{r+1} \\ \frac{h^{\ell-3}}{(\ell-3)!} & \frac{h^{\ell-4}}{(\ell-4)!} & \cdots & -[a_{1,r+1}]_{r+1} & -[A'a_{1,r+1}]_{r+1} \\ \cdots & \cdots & \cdots & \cdots & \cdots \\ 1 & -[a_{1,r+1}]_{r+1} & \cdots & -[(A')^{\ell-3}a_{1,r+1}]_{r+1} & -[(A')^{\ell-2}a_{1,r+1}]_{r+1} \end{vmatrix}$$

Let $t_1 < \infty$ and $u(\cdot)$ be such that (149) is satisfied. If for a given h the determinant $\Delta(h) \neq 0$, then $d_j = 0$, $j = 1,\ldots,\ell$ meaning

$$a'_{1,r+1}x(t - h) \equiv 0 \quad , \quad t\varepsilon[t_1 - h, t_1] \quad .$$

The last condition, in conjunction with (161) and (149), gives $x(t) \equiv 0$, $t_1 - h \le t \le t_1$. Thus, if h is not a root of the equation $\Delta(h) = 0$, then to prove complete controllability of (117),(157), it suffices to show the stability of the problem (149). It is possible to do this with the help of Theorem 24, assuming in (142) that

$$D_o = \{d_{kj}\} = \left\{ a_{ij} - \frac{b_{ip}}{a'_{1,r+1}\bar{q}_p(\ell,0)} a'_{1,r+1}\bar{p}_j(\ell,0) \right\} ,$$

$$D_1 = \{d^1_{ij}\} = \left\{ a_{1,ij} - \frac{b_{ip}}{a'_{1,r+1}\bar{q}_p(\ell,0)} a_{1,r+1}\bar{p}_j(\ell,h) \right\} ,$$

$$D_k = \{d^k_{ij}\} = -\left\{ \frac{b_{ip}}{a'_{1,r+1}\bar{q}_p(\ell,0)} a'_{1,r+1}\bar{p}_j(\ell,kh) \right\} ,$$

$$k = 2,\ldots,\ell \ ; \quad D_m = \{0\} \ , \quad m \geq \ell + 1 \ .$$

§18. A General Scheme for the Investigation of Complete Controllability

In the preceding paragraph, by particular examples we illustrated the possibility of reducing the functional problem of complete controllability to a finite-dimensional problem. Below we describe an approach to such a reduction in the general case.

We consider the system (117):

$$\dot{x}(t) = Ax(t) + A_1x(t - h) + bu(t) \quad ,$$

$$x_o(\cdot) = \{\Phi(\tau) \ , \ \tau\varepsilon[-h,0] \ , \ x(0) = x_o\} \quad ,$$

with arbitrary matrices A, A_1, and vector b. We assume that it is completely controllable, i.e. for any $x_o(\cdot)$, there exists a t_1 and $u(\cdot) = \{u(t), \ t\varepsilon[0,t_1]\}$ such that

$$x(t) \equiv 0 \quad , \quad t\varepsilon[t_1 - h, t_1] \quad . \tag{165}$$

Without loss of generality, it is possible to assume $t_1 \geq 2nh$, since if the system is completely controllable on $[0,t_2]$, $t_2 < 2nh$, then it is completely controllable on each interval $[0,t_1] \supset [0,t_2]$, We introduce several relations following from (165) which we use later for finding t_1 and $u(\cdot)$. Together with (165), the identity

$$\sum_{i=0}^{n} P_n(ih)x(t - ih) + \sum_{i=0}^{n-1} \sum_{j=i}^{n-1} q_j(ih)u^{(n-j-1)}(t - ih) \equiv 0 \quad , \tag{166}$$
$$t\varepsilon[t_1 - h, t_1] \; ,$$

is satisfied for the considered t_1 and $u(\cdot)$. Here $q_k(s)$ and $P_n(s)$ are the solution of the defining equation (62). From relations (166) follow the equality

$$x(t) \equiv \sum_{i=1}^{n} d_i \frac{t^{n-i}}{(n - i)!} \quad ,$$

$$\sum_{i=0}^{s} P_s(ih)x(t - ih) + \sum_{i=0}^{s-1} \sum_{j=i}^{s-1} q_j(ih)u^{(s-j-1)}(t - ih) \tag{167}$$
$$\equiv \sum_{i=1}^{n-s} d_i \frac{t^{n-s-i}}{(n - s - i)!} \quad , \; s = 1,\ldots,n-1 \; ; \; t\varepsilon[t_1 - h, t_1] \; ,$$

where the d_i, $i = 1,\ldots,n$, are arbitrary constant n-vectors. Considering (167) at the moments $t = t_1 - h$ and $t = t_1$, we obtain the system of algebraic equations for the d_i, $i = 1,\ldots,n$:

$$\sum_{i=1}^{n} \frac{t_1^{n-i}}{(n-i)!} d_i = x(t_1) \quad ,$$

$$\sum_{i=1}^{n} \left[\frac{t_1^{n-s-i}}{(n-s-i)!} E - \sum_{p=0}^{s=1} \frac{(t_1 - h)^{n-i-p}}{(n-i-p)!} A^{s-p-1} A_1 \right] d_i \tag{168}$$

$$= A^s x(t_1) + \sum_{i=1}^{s} A^{i-1} b u^{(s-i)}(t_1) \quad , \quad s = 1,\ldots,n-1 \quad .$$

The coefficient matrix of this system reduces, after some simple transformations, to the form

$$M(h) = \begin{vmatrix} \frac{h^{n-1}}{(n-1)!} E & \frac{h^{n-2}}{(n-2)!} E & \cdots & hE & E \\ \frac{h^{n-2}}{(n-2)!} E & \frac{h^{n-3}}{(n-3)!} E & \cdots & E & -A_1 \\ \frac{h^{n-3}}{(n-3)!} E & \frac{h^{n-4}}{(n-4)!} E & \cdots & -A_1 & -AA_1 \\ \cdots & \cdots & \cdots & \cdots & \cdots \\ E & -A_1 & \cdots & -A^{n-3} A_1 & -A^{n-2} A_1 \end{vmatrix} \quad . \tag{169}$$

We will assume that in Eq. (117) the delay h satisfies the condition det $M(h) \not\equiv 0$. This condition is satisfied for almost all h since det $M(h) \not\equiv 0$. By assumption, the control $u(\cdot)$ and the number t_1 are such that the trajectories of (117) satisfy (165). This means that in the relations (167) all $d_i = 0$, $i = 1,\ldots,n$. This is possible only if the system (168) is relatively controllable. Since by virtue of (165) $x(t_1) = 0$, for the homogeneous system (168) it is necessary and sufficient that the equations

$$\sum_{i=1}^{s} A^{i-1} b u^{(s-i)}(t_1) = 0 \ , \ s = 1,\ldots,n-1 \ ,$$

be satisfied and, as follows from them,

$$u^{(i)}(t_1) = 0 \ , \ i = 0,\ldots,n-2 \ . \tag{170}$$

Thus, if the control $u(t)$, $t \varepsilon [0,t_1]$, ensures condition (165) for the system (117), then it necessarily satisfies relations (166) and (170).

We transform (166). Introducing the notation $v(t) = \{u(t),\ldots,u(t - (n - 1)h)\}$, we have an integro-differential equation for the vector $v(t)$ [25,96,103]:

$$\sum_{i=0}^{n-1} Q_i v^{(n-i-1)}(t) + \int_{t_{1-h}}^{t_1} R(t,\tau)v(\tau)d\tau = f(t) \ , \ t\varepsilon[t_1 - h, t_1], \tag{171}$$

where

$$Q_i = \begin{Bmatrix} q_{i1}(0) & q_{i1}(h) & \cdots & q_{i1}((n-1)h) \\ q_{i2}(0) & q_{i2}(h) & \cdots & q_{i2}((n-1)h) \\ \cdots & \cdots & \cdots & \cdots \\ q_{in}(0) & q_{in}(h) & \cdots & q_{in}((n-1)h) \end{Bmatrix} ,$$

$$R(t,\tau) = \sum_{i=0}^{n-1} P_n(ih)S(t - ih,\tau) \ ,$$

$$S(t - kh,\tau) =$$

$$\begin{Bmatrix} [F(t - kh,\tau)b]_1 & \cdots & [F(t - kh,\tau - (n - 1)h)b]_1 \\ [F(t - kh,\tau)b]_2 & \cdots & [F(t - kh,\tau - (n - 1)h)b]_2 \\ \cdots & \cdots & \cdots \\ [F(t - kh,\tau)b]_n & \cdots & [F(t - kh,\tau - (n - 1)h)b]_n \end{Bmatrix}$$

$$f(t) = -\sum_{i=0}^{n} P_n(ih)s(t - ih) - \sum_{i=0}^{n} P_n(ih)\int_0^{t_1-nh} F(t - ih,\tau)bu(\tau)d\tau,$$

and $s(t)$, $F(t,\tau)$ are functions from the Cauchy formula On the basis of the preceding discussion, we find that if the system is completely controllable then there exists a solution of Eq. (171)

$$v(t) = Df(t) + \int_{t_1-h}^{t_1} \Gamma(t,\tau,h)f(\tau)d\tau \quad , \qquad t\varepsilon[t_1 - h,t_1) \quad , \tag{172}$$

satisfying the boundary condition (170): $v_1^{(i)}(t_1) = 0$, $i = 0,\ldots,n-2$. The relations (170) - (172) define the form of the control $u(t)$ on the interval $[t_1 - nh,t_1]$. We find the control $u(t)$ for $t\varepsilon[0,t_1 - nh]$, from the condition $x(t_1) = 0$, under the assumption that the matrix $F(t_1,t_1 - nh)$ is nonsingular. Transforming the condition $x(t_1) = 0$ with the help of (172), we obtain

$$\int_0^{t_1-nh} [F(t_1 - nh,\tau)b + K(\tau,h)b]_i u(\tau)d\tau = \gamma_i \ , \ i = 1,\ldots,n. \tag{173}$$

Here

$$K(\tau,h) = F^{-1}(t_1,t_1 - nh)\left[\int_{t_1-(n+1)h}^{t_1-nh} F(t_1,\theta + h)A_1F(\theta,\tau)d\theta\right.$$

$$- \int_0^{t_1} S(t_1,\theta)\, D \sum_{i=0}^{n} P_n(ih)F(\theta - ih,\tau)d\theta$$

$$\left. - \int_{t_1-h}^{t_1}\int_{t_1-h}^{t_1} S(t_1,\sigma)\Gamma(\sigma,\theta,h) \sum_{i=0}^{n} P_n(ih)F(\theta - ih,\tau)d\theta\, d\sigma\right] ,$$

$$\gamma = F^{-1}(t_1,t_1 - nh)\left[\int_{t_1-h}^{t_1} S(t_1,\tau)D \sum_{i=0}^{n} P_n(ih)s(\tau - ih)d\tau\right.$$

$$+ \int_{t_1-h}^{t_1}\int_{t_1-h}^{t_1} S(t_1,\tau)\Gamma(\tau,\theta,h) \sum_{i=0}^{n} P_n(ih)s(\theta - ih)d\theta\, d\tau$$

$$\left. - \int_{t_1-(n+1)h}^{t_1-nh} F(t_1,\tau + h)A_1s(\tau)d\tau\right] - s(t_1 - nh) .$$

As a result, we are led to the following conclusion: the control $u(t)$, $t\varepsilon[0,t_1]$, ensuring the condition (165) for the system (117) is the solution of the problem (173) on the interval $[0,t_1 - nh]$, while the integro-differential equation (171) and condition (170) are satisfied for $t\varepsilon[t_1 - nh,t_1]$.

The converse is also true: if for some t_1, $2nh < t_1 < \infty$, the problem (170), (171), (173) has a solution $u(t)$, $t\varepsilon[0,t_1]$, then the trajectories $x(t)$ of the system (117) corresponding to this control satisfy the condition (165).

Actually, if the control $u(t)$, $t \varepsilon T$, is found from (170),(171),(173), then the system (168) is homogeneous and under the condition $\det M(h) \neq 0$, it has only the trivial solution, hence (165) follows. Thus, in order that the system (117), with delay h satisfying the conditions $\det M(h) \neq 0$ and $\det F(t_1, t_1 - nh) \neq 0$, be completely controllable, it is necessary and sufficient that there exist a solution to the problem (170),(171),173).

Although formulating conditions of complete controllability in such a form is not sufficiently effective as a practical tool, describing the scheme gives practical methods for investigating controllability of concrete systems for almost all h. For this, it is not necessary to find the solution of the intego-differential equation (171) in closed form which poses a difficulty: it is sufficient to express the control $u(t)$ in the form of some functionals.

We illustrate the application of our scheme by some examples.

Example 12. Let the given system be of third-order

$$\left.\begin{aligned} &\dot{x}_1(t) = -x_1(t) + u(t) \ , \ \dot{x}_2(t) = -x_1(t - 1) \ , \\ &\dot{x}_3(t) = x_2(t - 1) \ . \end{aligned}\right\} \quad (174)$$

We want to know if this system is completely controllable.

According to the general scheme, we form the equations (166):

$$-x_1(t) + \ddot{u}(t) - \dot{u}(t) + u(t) = 0 \ ,$$

$$x_1(t - 1) + \dot{u}(t - 1) - u(t - 1) = 0 \ ,$$

$$-x_1(t - 2) + u(t - 2) = 0 \ .$$

The solution of the system (174), computed by Cauchy's formula, has the form

$$\begin{pmatrix} x_1(t) \\ x_2(t) \\ x_3(t) \end{pmatrix} = \begin{pmatrix} e^{-t} \\ 0 \\ 0 \end{pmatrix} x_{10} + \begin{pmatrix} 0 \\ 1 \\ 0 \end{pmatrix} x_{20} + \begin{pmatrix} 0 \\ 0 \\ 1 \end{pmatrix} x_{30}$$

$$+ \int_0^t \begin{pmatrix} 0 \\ 1 \\ 0 \end{pmatrix} \Phi_1(\tau - 1)d\tau + \int_0^t \begin{pmatrix} 0 \\ 0 \\ 1 \end{pmatrix} \Phi_2(\tau - 1)d\tau$$

$$+ \int_0^t \begin{pmatrix} e^{-t+\tau} \\ 0 \\ 0 \end{pmatrix} u(\tau)d\tau \quad ,$$

$$t\varepsilon[0,1] \quad ,$$

$$\begin{pmatrix} x_1(t) \\ x_2(t) \\ x_3(t) \end{pmatrix} = \begin{pmatrix} e^{-t} \\ -e^{-t+1} + 1 \\ 0 \end{pmatrix} x_{10} + \begin{pmatrix} 0 \\ 1 \\ t - 1 \end{pmatrix} x_{20} + \begin{pmatrix} 0 \\ 0 \\ 1 \end{pmatrix} x_{30}$$

$$+ \int_0^{t-1} \begin{pmatrix} 0 \\ 1 \\ t - \tau - 1 \end{pmatrix} \Phi_1(\tau - 1)d\tau + \int_{t-1}^1 \begin{pmatrix} 0 \\ 1 \\ 0 \end{pmatrix} \Phi_1(\tau - 1)d\tau$$

$$+ \int_0^1 \begin{pmatrix} 0 \\ 0 \\ 1 \end{pmatrix} \Phi_2(\tau - 1)d\tau + \int_0^{t-1} \begin{pmatrix} e^{-t+\tau} \\ -e^{-t+\tau+1} + 1 \\ 0 \end{pmatrix} u(\tau)d\tau$$

$$+ \int_{t-1}^t \begin{pmatrix} e^{-t+\tau} \\ 0 \\ 0 \end{pmatrix} u(\tau)d\tau \ , \ t\varepsilon[1,2] \quad ,$$

$$\begin{Bmatrix} x_1(t) \\ x_2(t) \\ x_3(t) \end{Bmatrix} = \begin{Bmatrix} e^{-t} \\ -e^{-t+1} + 1 \\ e^{t+2} + t - 3 \end{Bmatrix} x_{10} + \begin{Bmatrix} 0 \\ 1 \\ t - 1 \end{Bmatrix} x_{20} + \begin{Bmatrix} 0 \\ 0 \\ \; \end{Bmatrix}$$

$$+ \int_0^1 \begin{Bmatrix} 0 \\ 1 \\ t - \tau - t \end{Bmatrix} \Phi_1(t - 1)\,d\tau + \int_0^1 \begin{Bmatrix} 0 \\ 0 \\ 0 \end{Bmatrix} \Phi_2(\tau - 1)\,d\tau$$

$$+ \int_0^{t-2} \begin{Bmatrix} e^{-t+\tau} \\ -e^{-t+\tau+1} + 1 \\ e^{-t+\tau+2} + t - \tau - 3 \end{Bmatrix} u(\tau)\,d\tau$$

$$+ \int_{t-2}^{t-1} \begin{Bmatrix} e^{-t+\tau} \\ -e^{-t+\tau+1} + 1 \\ 0 \end{Bmatrix} u(\tau)\,d\tau + \int_{t-1}^{t} \begin{Bmatrix} e^{-t+\tau} \\ 0 \\ 0 \end{Bmatrix} u(\tau)\,d\tau \;,$$

$$t \geq 2 \quad .$$

Assuming $t_1 \geq 4$, we are led to the system of integro-differential equations (see (171))

$$\left.\begin{aligned} &-e^{-t}x_{10} - \int_0^t e^{-t+\tau}u(\tau)\,d\tau + \ddot{u}(t) - \dot{u}(t) + u(t) = 0 \quad , \\ &e^{-t+1}x_{10} + \int_0^{t-1} e^{-t+\tau+1}u(\tau)\,d\tau \\ &\qquad + \dot{u}(t - 1) - u(t - 1) = 0 \quad , \\ &-e^{-t+2}x_{10} + \int_0^{t-2} e^{-t+\tau+2}u(\tau)\,d\tau + u(t - 2) = 0 \quad , \\ &\qquad\qquad t \varepsilon [t_1 - 1, t_1] \quad . \end{aligned}\right\} \qquad (175)$$

It is clear that the function

$$u(t) \equiv 0 \quad , \quad t \varepsilon [t_1 - 3, t_1] \quad , \tag{176}$$

satisfies (175),(170), if

$$x_{10} + \int_0^{t_1-3} e^{\tau} u(\tau) d\tau = 0 \quad . \tag{177}$$

Taking into account the form of the solution of the system (174), it is not difficult to see that the condition $x_1(t_1) = 0$, under the assumption (176), coincides with (177), while the requirement $x_2(t_1) = x_3(t_1) = 0$ leads to the integral relations

$$\left. \begin{aligned} & x_{10} + x_{20} + \int_0^1 \Phi_1(\tau - 1) d\tau + \int_0^{t_1-3} u(\tau) d\tau = 0 \quad , \\ & -2x_{10} + x_{30} - \int_0^1 \tau\Phi_1(\tau - 1) d\tau + \int_0^t \Phi_2(\tau - 1) d\tau \\ & \qquad - \int_0^{t_1-3} (\tau + 2) u(\tau) d\tau = 0 \quad . \end{aligned} \right\} \tag{178}$$

The linear independence of the functions $e^{\tau}, 1, \tau + 2$ guarantees the solvability of the problem (177),(178) (in the general scheme (173)) for any initial state $x_o(\cdot)$, and the complete controllability of (174) if $\det M(1) \neq 0$. The validity of the last inequality follows from a simple calculation.

<u>Example 13</u>. We consider the system

$$\left. \begin{aligned} \dot{x}_1(t) &= x_1(t) + u(t) \quad , \\ \dot{x}_2(t) &= -ex_1(t) - x_2(t) + x_1(t - 1) \quad , \end{aligned} \right\} \tag{179}$$

which clearly is relatively controllable. We show that it is not completely controllable. If $t_1 \leq 2$, then the necessary conditions for maintaining the trajectory of (179) at the origin

$$u(t) \equiv 0 \quad , \quad x_1(t-1) \equiv 0 \quad , \quad t \varepsilon [t_1 - 1, t_1] \quad ,$$

cannot be satisfied for an arbitrary initial function $\Phi(\theta)$, $\theta \varepsilon [-1,0]$. Therefore, we set $t_1 > 2$. From the Cauchy formula we have

$$\begin{Bmatrix} x_1(t) \\ x_2(t) \end{Bmatrix} = \begin{Bmatrix} e^t \\ \frac{1}{2} e(e^{-t} - e^t) \end{Bmatrix} x_{10} + \begin{Bmatrix} 0 \\ e^{-t} \end{Bmatrix} x_{20}$$

$$+ \int_0^t \begin{Bmatrix} 0 \\ e^{-t+\tau} \end{Bmatrix} \Phi_1(\tau - 1) d\tau + \int_0^t \begin{Bmatrix} e^{t-\tau} \\ \frac{1}{2} e(e^{-t+\tau} - e^{t-\tau}) \end{Bmatrix} u(\tau) d\tau \; ,$$

$$t \varepsilon [0,1] \quad ,$$

$$\begin{Bmatrix} x_1(t) \\ x_2(t) \end{Bmatrix} = \begin{Bmatrix} e^t \\ \frac{1}{2} e^t(e^{-1} - e) \end{Bmatrix} x_{10} + \begin{Bmatrix} 0 \\ e^{-t} \end{Bmatrix} x_{20}$$

$$+ \int_0^1 \begin{Bmatrix} 0 \\ e^{-t+\tau} \end{Bmatrix} \Phi_1(\tau - 1) d\tau + \int_0^{t-1} \begin{Bmatrix} e^{t-\tau} \\ \frac{1}{2} e^{t-\tau}(e^{-1} - e) \end{Bmatrix} u(\tau) d\tau$$

$$+ \int_{t-1}^t \begin{Bmatrix} e^{t-\tau} \\ \frac{1}{2} e(e^{-t+\tau} - e^{t-\tau}) \end{Bmatrix} u(\tau) d\tau \quad , \quad t \geq 1 \quad .$$

We write the condition $x(t_1) = 0$:

$$
\left.\begin{aligned}
&x_{10} + \int_0^{t_1} e^{-\tau} u(\tau)\, d\tau = 0 \quad , \\
&x_{20} + \int_0^1 e^{\tau} \Phi_1(\tau - 1)\, d\tau \\
&\quad + \frac{e^{t_1}}{2} \int_{t_1 - 1}^{t_1} (e^{-t_1+\tau+1} - e^{t_1-\tau-1}) u(\tau)\, d\tau = 0
\end{aligned}\right\} \qquad (180)
$$

The problem (180) has a solution for any $x_o(\cdot)$ only in the case $u(\tau) \not\equiv 0$, $t \varepsilon [t_1 - 1, t_1]$. Now we form (166):

$$
\begin{aligned}
&x_1(t) + u(t) + \dot{u}(t) = 0 \quad , \\
&x_2(t) - eu(t) + u(t - 1) = 0 \ , \ t \varepsilon [t_1 - 1, t_1] \quad .
\end{aligned} \qquad (181)
$$

It is not difficult to see that the requirements (165),(170),(181) are uniquely satisfied by the function $u(t) \equiv 0$, $t \varepsilon [t_1 - 2, t_1]$. The inconsistency of the conditions (180),(181), and (170) means that the system (179) is not completely controllable. We note that the initial state

$$
\begin{aligned}
x_o(\cdot) = \{\Phi(\tau),\ \tau \varepsilon [-1,0)\ ,\ & x_{10} = \Phi_1(0) \quad , \\
& x_{20} = -\int_0^1 e^{\tau} \Phi_1(\tau - 1)\, d\tau\}
\end{aligned}
$$

of the system (179) is completely controllable.

§19. Sufficient Conditions for the Complete Controllability of Systems with a Small Delay

In the preceding paragraph, the problem about complete controllability of systems with a delay was reduced to several equivalent problems. Now we show that in the case of a small delay the conditions for complete controllability are expressed in explicit form by means of the parameters of the original system.

We return once again to the system (117). The corresponding integro-differential equation may be represented in the form

$$D(p)v(t) + \int_{t_1-h}^{t_1} R(t,\tau)v(\tau)d\tau = f(t) \quad . \tag{182}$$

The differential operator D(p) is defined by the formula

$$D(p) = \begin{Bmatrix} \sum_{i=0}^{n-1} q_{i1}(0)p^{n-i-1} & \sum_{i=1}^{n-1} q_{i1}(h)p^{n-i-1} & \cdots & q_{n-1,1}((n-1)h) \\ \sum_{i=0}^{n-1} q_{i2}(0)p^{n-i-1} & \sum_{i=1}^{n-1} q_{i2}(h)p^{n-i-1} & \cdots & q_{n-1,2}((n-1)h) \\ \cdots & \cdots & \cdots & \cdots \\ \sum_{i=0}^{n-1} q_{in}(0)p^{n-i-1} & \sum_{i=1}^{n-1} q_{in}(h)p^{n-i-1} & \cdots & q_{n-1,n}((n-1)h) \end{Bmatrix}$$

or, more compactly,

$$D(p) = \{d_{ij}\} = \sum_{k=j-1}^{n-1} q_{ki}((j-1)h)p^{n-k-1} \quad .$$

We set

$$D(p) = \{d_{ij}\} = \{d_{i,n-j+1}\}$$

It is known [90,126] that there exists a matrix L(p) with det L(p) = constant $\neq 0$, such that

$$D_1(p) = L(p)\bar{D}(p) = \begin{Bmatrix} m_{11}(p) & \cdots & & m_{1n}(p) \\ 0 & m_{22}(p) & \cdots & m_{2n}(p) \\ \cdots & \cdots & \cdots & \cdots \\ 0 & \cdots & 0 & m_{nn}(p) \end{Bmatrix}.$$

Here $m_{ik}(p)$ are some polynomials in p, in particular,

$$m_{nn}(p) = \sum_{s=1}^{\ell} d_s p^s \quad . \tag{183}$$

We note that the degrees of the polynomials $\Delta(p) = \det D(p)$ and $\Delta_1(p) = \det D_1(p)$ are equal and do not exceed $n(n-1)/2$.

We prove the following assertion.

<u>Theorem 25</u>. <u>If</u>

$$\text{rank } \{b,(A + A_1)b,\ldots,(A + A_1)^{n-1}b\} = n \quad , \tag{184}$$

$$\Delta(p) \not\equiv 0 \quad , \quad \sum_{i=n-2}^{\ell} \alpha_i^2 \neq 0 \quad , \tag{185}$$

<u>where ℓ and α_k, $k = n-2,\ldots,\ell$ are the numbers from (183), then for sufficiently small h the system (117) is completely controllable</u>.

From the results of §16, we find that to prove Theorem 25 it is sufficient to demonstrate solvability of the problem (170),(171),(173), under the given conditions. Thus, let h be a sufficiently small number. In this case, the condition $\Delta(p) \not\equiv 0$ ensures the existence of a solution v(t) of the system (182). The requirement $\sum_{i=n-2}^{\ell} \alpha_i^2 \neq 0$ guarantees that the components $v_1(t)$ of this solution depend on not less than n-2 arbitrary constants which, in turn, gives the possibility to satisfy condition (170). Since det $M(0) \neq 0$ and det $F(t_1, t_1) \neq 0$, for sufficiently small h, the matrices $M(h)$ and $F(t_1, t_1 - h)$ are nonsingular. Thus, the question is only one of choosing $u(t)$ on the interval $[0, t_1 - nh]$ in correspondence with (173).

Because of (184), the system (117) is completely controllable for $h = 0$. This means that

$$g'F(t_1, \tau)b \not\equiv 0 \quad , \qquad \tau \varepsilon [0, t_1] \quad ,$$

for each g, $||g|| = 1$. From the compactness of the sphere $||g|| = 1$, it follows that

$$\min_{||g||=1} \max_{\tau \varepsilon T} |g'F(t_1, \tau)b| = \alpha > 0 \quad .$$

Since $g'F(t_1 - nh, \tau)b$ is a continuous function of h on $[0, (t_1 - \tau)/n]$, it is possible to find numbers α_1 and h_1, $0 < \alpha_1 < \alpha$, $h_1 > 0$, such that

$$\min_{||g||=1} \max_{\tau \varepsilon [0, t_1 - nh]} |g'F(t_1 - nh, \tau)b| \geq \alpha_1(h_1) \tag{186}$$

for all $h \varepsilon [0, h_1]$. The number h_1 may be decreased in order to simultaneously have the inequality

$$\max_{\tau \varepsilon [0, t_{1-nh}]} ||K(\tau, h)b|| \leq \beta(h_1) \quad , \ h \varepsilon [0, h_1] \quad ,$$

$$0 < \beta(h_1) < \alpha_1(h_1) \quad .$$

with (186). Then, for arbitrary g, $||g|| = 1$, and $h \varepsilon [0, h_1]$, we have

$$\max_{\tau \varepsilon [0,t_1-nh]} |g'F(t_1 - nh,\tau)b + g'K(\tau,h)b|$$

$$\geq \max_{\tau \varepsilon [0,t_1-nh]} (|g'F(t_1 - nh,\tau)b| - |g'K(\tau,h)b|)$$

$$\geq \max_{\tau \varepsilon [0,t_1-nh]} |g'F(t_1 - nh,\tau)b| - \max_{\tau \varepsilon [0,t_1-nh]} |g'K(\tau,h)b|$$

$$\geq \min_{||g||=1} \max_{\tau \varepsilon [0,t_1-nh]} |g'F(t_1 - nh,\tau)b|$$

$$- \max_{\tau \varepsilon [0,t_1-nh]} ||K(\tau,h)b|| \geq \alpha_1(h_1) - \beta(h_1)$$

and

$$\max_{\tau \varepsilon [0,t_1-nh]} |g_o'F(t_1 - nh,\tau)b + g_o'K(\tau,h)b|$$

$$= \min_{||g||=1} \max_{\tau \varepsilon [0,t_1-nh]} |g'F(t_1 - nh,\tau)b + g'K(\tau,h)b|$$

$$\geq \alpha_1(h_1) - \beta(h_1) \quad .$$

Since $\alpha_1(h_1) - \beta(h_1) = \alpha_2(h_1) > 0$, then

$$g'F(t_1 - nh,\tau)b + g'K(\tau,h)b \not\equiv 0 \ , \ \tau \varepsilon [0,t_1 - nh] \ , \ h \varepsilon [0,h_1] \ .$$

Commentary on Chapter I

Concepts near to controllability have always played an important role for objects (systems, processes) capable of changing their behavior under directed perturbations. Interest in the controllability properties of dynamical systems strengthened after the beginning of the theory of optimal processes [19]. The first conditions for the controllability of stationary linear systems were discovered within the scope of this theory [38b]. The term "controllability" and the discovery of the true meaning of these conditions appeared later [58a].

Our definitions (§1) differ from the standard ones, but are quite near to many of them. As is evident from the contents of Chapters I,II, the term "controllability" is very overworked; to reflect new features of controlled objects, it is necessary to introduce new concepts which, being related to the concept of controllability, overload it.

The method of increments for studying controllability was introduced in [32i]. For linear systems it is equivalent to the method of undetermined multipliers and leads to one of the variants of Cauchy's formula. Theorem 3 on the minimal number of inputs to a control system was proved by other methods in [218].

Sufficient conditions for controllability of nonstationary linear systems have been repeatedly established by other techniques [74n]. The natural question is about the converse of this result. In the general case, the answer is negative [74n,212].

The critical case of controllability has still been studied very little. It is possible to note the work [59] for planar systems. However, the methods of this work are, perhaps, difficult to generalize to higher-dimensional systems.

In §11-16 we studied the controllability of dynamical systems with a delay. A new view on the conditions for controllability of ordinary systems was provided [66b] by the introduction of the concept of a defining equation for a control system. In questions of controllability, this equation plays the same role as the characteristic equation in problems of stability. Relative controllability was studied in [143]. Complete controllability of systems with a delay, a more difficult problem than relative controllability, was investigated in §11-16. This property was considered in [74n].

CHAPTER II

The Theory of Controllability in a Direction

§1. Statement of the Problem. Definitions. The Method of Investigation.

Let the dynamical system be described by the equation

$$\frac{dx}{dt} = f(x,u) \quad , \quad x(0) = 0 \quad , \quad t\varepsilon T = [0,t_1] \quad , \quad f(0,0) = 0 \quad , \qquad (1)$$

where x is a n-vector of states and u is an r-vector of controls. We assume that the controlling actions u(t), $t\varepsilon T$, are piecewise continuous functions. We will call such a control admissible. To each admissible control u(t), $t\varepsilon T$, there corresponds a unique trajectory x(t) of the system (1) emanating from the point x = 0.

Problem. To find conditions under which for each point x_1 there is an admissible control transferring the trajectory from x = 0 to $x = x_1$.

Definitions.

1. The origin is called TL-controllable in the direction p, $||p|| = 1$, if there exists a number $\alpha_o = \alpha_o(t_1,L,p) > 0$, such that for each α, $|\alpha| \leq \alpha_o$, there is an admissible control u(t), $||u(t)|| \leq L$, such that $p'x(t_1) = \alpha$.

2. A dynamical system is TL-controllable (at the origin) if the origin is TL-controllable in any direction, $\inf_{||p||=1} \alpha_o(t_1,L,p) > 0$.

3. The origin is called _controllable in the direction_ p, $||p|| = 1$, if it is TL-controllable in this direction for some $t_1 = t_1(p) < \infty$, $L = L(p) < \infty$, $(\alpha_o(t_1, L, p) > 0)$.

4. The dynamical system (1) is _controllable_ (at the origin) if the origin is controllable in any direction ($\inf\limits_{||p|| = 1} \alpha_o(t_1(p), L(p), p) > 0$).

5. The origin is called _completely controllable in the direction p_ if it is TL-controllable in this direction with any $t_1 > 0$, $L > 0$ ($\alpha_o(t_1, L, p) > 0$ for any $t_1 > 0$, $L > 0$).

6. A dynamical system is called _completely controllable_ (at the origin) if the origin is completely controllable in any direction ($\inf\limits_{||p|| = 1} \alpha_o(t_1, L, p) > 0$ for any $t_1 > 0$, $L > 0$).

7. The origin in the direction p is _controllable in the large_ if it is controllable in this direction and there can be found sequences $t_{1k}(p)$, $L_k(p)$ such

$$\alpha_o(t_{1k}(p), L_k(p), p) \to \infty \text{ for } k \to \infty .$$

8. A dynamical system is _controllable in the large_ (at the origin) if the origin is controllable in the large for any direction ($\inf\limits_{||p|| = 1} \lim\limits_{k \to \infty} \alpha_o(t_{1k}(p), L_k(p), p) = \infty$).

The method for investigating controllability follows in an almost obvious fashion from the preceding definitions. Along trajectories of the system (1) we introduce the functional

$$J(u) = p'x(t_1) . \tag{2}$$

Using formulas, the variations in the right side are expressed through the parameters of the system. Conditions under which the functional (2) depends upon the control are conditions for controllability.

§2. A Formula for Variations Relative to the Controls.

We compare the functional (2) on two controls: $u_1(t) \equiv 0$ and $u = u(t)$. Since the system (1) has the trivial solution $x \equiv 0$ under the first control, the increment $\Delta J(u) = J(u) - J(u_1)$ reduces to the expression (2). We introduce the n-vector function $\psi(t)$, $\psi(t_1) = p$. Using the formula for integration by parts, the right side of the expression (2) may be represented in the form

$$p'x(t_1) = \psi'(t_1)x(t_1) = \int_0^{t_1} \dot{\psi}'(t)\dot{x}(t)dt + \int_0^{t_1} \psi'(t)x(t)dt \quad . \tag{3}$$

We define the function $\psi(t)$ on the interval T by the equation

$$\dot{\psi}(t) = -A'\psi(t) \quad , \quad A = \frac{\partial f(0,0)}{\partial x} \quad .$$

We substitute this definition of $\dot{\psi}$, and also the value of x from (1), into (3) and make a simple transformation

$$p'x(t_1) = \int_0^{t_1} \psi'(t)f(0,u(t))dt$$

$$+ \int_0^{t_1} \psi'(t) \quad \frac{\partial f(0,u(t))}{\partial x} - \frac{\partial f(0,0)}{\partial x} \quad x(t)dt + \int_0^{t_1} o(||x(t)||)dt.$$

Finally, we write the formula in the following form:

$$J(u) = p'x(t_1) =$$

$$= \int_0^{t_1} \psi'(t)Bu(t)dt + \int_0^{t_1} \psi'(t) \sum_{\nu=1}^{r} C\nu x(t)u_\nu(t)dt$$

$$+ \int_0^{t_1} o(||x(t)||)dt + \int_0^{t_1} o(||u(t)||)dt \quad . \tag{4}$$

Here

$$B = \frac{\partial f(0,0)}{\partial u} \quad , \qquad C = \frac{\partial^2 f(0,0)}{\partial x \partial u_\nu} \quad .$$

§3. Controllability of Linear Systems

We confine our attention to single-input systems. We consider a particular case of (1):

$$\frac{dx}{dt} = Ax + bu \quad .$$

From (4) we obtain

$$J(u) = p'x(t_1) = \int_0^{t_1} \psi'(t)bu(t)dt \quad .$$

From this expression it is clear that if

$$\lambda(t) = \psi'(t)b \not\equiv 0 \quad , \tag{5}$$

then for any α_0 (outside of its dependence on p and t_1) it is possible to find an L such that along the admissible control* $u(t) = \beta L \text{ sign } \lambda(t)$, the functional J(u) assumes all values from $[-\alpha_0, \alpha_0]$ under variation of the parameter β in the interval [-1,1]. Conversely, if J(u)

---

* sign 0 = 0.

assumes values from some interval $[-\alpha_0, \alpha_0]$, then (5) is satisfied. The derivative of the function $\lambda(t)$ has the form

$$\dot{\lambda}(t) = -\psi'(t)AB, \ldots, \lambda^{(n)}(t) = (-1)^n \psi'(t)A^n b \quad . \tag{6}$$

On the other hand, every nxn matrix satisfies its characteristic equation (Cayley-Hamilton theorem [39])

$$A^n + \gamma_1 A^{n-1} + \cdots + \gamma_{n-1} A + \gamma_n E = 0 \quad .$$

Multiplying this equation on the left by $\psi'(t)$ and on the right by b, by virtue of (5),(6) we obtain the differential equation

$$\gamma^{(n)}(t) - \gamma_1 \lambda^{(n-1)}(t) + \cdots + \gamma_{n-1}(-1)^{n-1}\dot{\lambda}(t) + (-1)^n \gamma_n \lambda(t) = 0$$

with the initial condition

$$\lambda(t_1) = p'b \ , \ \dot{\lambda}(t_1) = -p'Ab, \ldots, \lambda^{(n-1)}(t_1) = (-1)^{n-1} p'A^{n-1} b \ .$$

Since a homogeneous linear differential equation has the solution zero if and only if all initial conditions are identically zero, from the condition $\lambda(t) \not\equiv 0$ we are led to the following assertion.

<u>Theorem 1. The origin of the dynamical system (1) is TL-controllable (controllable, completely controllable) in the large in the direction p if and only if each of the numbers</u>

$$p'b, \ p'Ab, \ldots, p'A^{n-1}b \quad .$$

<u>is different from zero.</u>

Corollary. For TL-controllability (controllability, complete controllability) in the large of the dynamical system (1), it is necessary and sufficient that

$$\text{rank } \{b, Ab, \ldots, A^{n-1}b\} = n \quad .$$

Remark. From the proof of the theorem, it follows that the function $\alpha_o(t_1, L, p)$ involved in Definitions 1-8 may be chosen independently of t_1 and p and with unbounded growth in L.

§4. Controllability of Linearized Systems.

For the dynamical system

$$\frac{dx}{dt} = f(x,u) \quad , \quad x(0) = 0 \quad , \tag{7}$$

let the linear model

$$\frac{dx}{dt} = Ax + bu \tag{8}$$

be controllable. We show that in this case the system (7) is completely controllable. We will choose the control u(t) in the form

$$u(t) = \mu\nu(t) \quad , \quad |\nu| \leq L \quad , \tag{9}$$

where μ is a small parameter. The solution x(t) of Eq. (7) under the control (9) is continuous in μ and has order of magnitude not lower than μ: $x(t) \sim \mu$. In this case, formula (4) may be written in the form

$$J(u) = \mu \int_0^{t_1} \psi'(t) b\nu(t) dt + o(\mu) \quad ,$$

where $o(\mu)$ is continuous in μ and has magnitute μ^2: $o(\mu) \sim \mu^2$. Controllability of the linear model (8) means that the quantity $\int_0^{t_1} \psi'(t)b\nu(t)dt$, under the control (9) may assume any value from the interval $[-M,M]$, where the quantity M depends on L, t_1, and the system parameters but does not depend on p and μ and where for any t_1, $M \to \infty$ if $L \to \infty$.

We show that for any t_1, L, there exists $\alpha_o(t_1, L)$ such that for any $\alpha \varepsilon [-\alpha_o, \alpha_o]$, there exists a control $u(t)$ for which $J(u) = \alpha$. In (9) we fix the value $\mu = \mu$, where μ_1 is such that $|o(\mu_1)| = 0.1\mu_1$. For definiteness, let $\alpha > 0$ and let $v(t)$ denote the control from (9) for which

$$\int_0^{t_1} \psi'(t)bv(t)dt = M \quad .$$

For the control $w(t) = \beta v(t)$, $0 \leq \beta \leq 1$, we have

$$\int_0^{t_1} \psi'(t)bw(t)dt = \beta M \quad .$$

Substituting the control $u(t) = \beta\mu_1 v(t)$ into (7), we obtain that the trajectory $x(t) = x(t,\beta)$ and the quantity $o(\mu_1) = 0(\mu_1,\beta)$ depend continuously on β and $x(t_1 0) = 0$, $o(\mu_1, 0) = 0$. This means that the functional $J(u) = J(u,\beta)$ is continuous in β and its values on the interval $0 \leq \beta \leq 1$ equal $J(u,0) = 0$, $J(u,1) = \mu_1 M + o(\mu_1) \geq 0.9\mu_1 M = \alpha_o$. Hence, it follows that there is a $\overline{\beta}$, $0 \leq \overline{\beta} \leq 1$, for which $J(u,\beta) = \alpha$. By construction, the function $\alpha_o(t_1, L, p)$ does not depend on p. Thus, we have proved the following theorem.

Theorem 2. The origin of the system (7) is completely controllable in the direction p if its linear model (8) *also possesses this property.*

Corollary. The dynamical system (7) is completely controllable if its linear approximation (8) is completely controllable.

§5. Linear Systems with a Nonlinear Input

1. Controllability in the Small.

We consider the problem of controllability for the system

$$\frac{dx}{dt} = Ax + \sum_{i=1}^{\ell} b^i u^i \quad . \tag{10}$$

From formula (4) we have

$$\begin{aligned} J(u) &= p'x(t_1) \\ &= \int_0^{t_1} \psi'(t) b^1 u(t)\,dt + \cdots + \int_0^{t_1} \psi'(t) b^\ell u^\ell(t)\,dt \quad . \end{aligned} \tag{11}$$

In order that the origin be completely controllable in the direction p, it is sufficient that

$$\psi'(t) b^1 \not\equiv 0 \quad , \quad t \in T \quad ,$$

which, by virtue of Theorem 1, is equivalent to the requirement that each of the numbers

$$p'b^1, p'Ab^1, \ldots, p'A^{n-1}b^1 \quad , \tag{12}$$

be different from zero. Let all the numbers in (12) equal zero. Then the first integral in (11) equals zero.

We consider the integral

$$\int_0^{t_1} \psi'(t)b^2u^2(t)dt \quad .$$

Since, for sufficiently small t_1, the function $\psi'(t)b^2$, $0 \leq t \leq t_1$, does not change its sign, it is possible to state: the origin is not completely controllable in the direction p with small controls if

$$p'A^ib^1 = 0 \ , \ i = 0,\ldots,n-1; \ \sum_{i=0}^{n-1} |p'A^ib^2| \neq 0 \quad ,$$

and t_1 is small. Extending this argument, we are led to the following assertion.

Theorem 3. The origin is completely controllable in the direction p if the first nonzero number from the sequence

$$p'A^ib^j \ , \ i = 0,\ldots,n-1; \ j = 1,\ldots,\ell \quad ,$$

has odd index j. When this index is even, then the origin is not completely controllable in the direction p.

Corollary. The dynamical system (10) is completely controllable if the first index j for which

$$\text{rank } \{b^j, Ab^j, \ldots, A^{n-1}b^j\} = n; \ j = 1,\ldots,\ell \quad .$$

is odd.

If the index is even, then the system (10) cannot be completely controllable.

2. Controllability in the Large

In the dynamical system (10), let the admissible control satisfy the condition: $|u(t)| \leq L$, where L is a sufficiently large number. In this case, it is natural to begin the investigation of complete controllability by considering the integral

$$\int_0^{t_1} \psi'(z) b^\ell u^\ell(t)\,dt \quad .$$

It is not difficult to see that for even ℓ and sufficiently small t_1, this integral assumes values of the same sign as $\psi'(t)b$, while for odd ℓ it assumes values of opposite signs. From similar considerations, it is easy to establish the following theorem.

Theorem 4. The origin is completely controllable in the large in the direction p if the first non-zero number in the series

$$p'A^i b^j \quad , \qquad i = 0,\ldots,n-1; \qquad j = \ell,\ldots,1 \quad ,$$

has odd index j.

Corollary. If the condition

$$\text{rank}\ \{b^j, Ab^j, \ldots, A^{n-1}b^j\} = n \ , \ j = \ell,\ldots,1 \quad ,$$

is first satisfied for an odd index j, then the system (10) is completely controllable in the large.

§6. On the Invariance and Autonomy of Dynamical Systems

Let us consider the system (8).

Definition. The origin is invariant to control in the direction p if for any admissible control we have the identity $p'x(t) \equiv 0$, $t \varepsilon T$.

Although the introduced concept is not the opposite of the concept of controllability, they have analagous methods of investigation. From Theorem 1, we have the result

Theorem 5. The origin is invariant to control in the direction p for the system

$$\frac{dx}{dt} = Ax + bu$$

if and only if

$$p'b = p'AB = \cdots = p'A^{n-1}b=0 \quad .$$

We pass to the dynamical system with several controls

$$\frac{dx}{dt} = Ax + \sum_{\nu=1}^{r} b^{\nu}u_{\nu}$$

and consider the directions p^{ν}, $\nu = 1,\dots,r$.

Definition 10. The origin is autonomously controllable relative to u_{ν} in the direction p^{ν} if it is controllable in this direction by u_{ν} and is invariant to controls u_j, $j \neq \nu$.

Conditions for autonomy are obtained from Theorems 1 and 5.

Theorem 6. The origin is autonomously controllable with respect to u_{ν} in the direction p^{ν} if and only if

$$\sum_{i=0}^{n-1} |(p^{\nu})'A^{i}b^{\nu}| \neq 0 \ , \quad \sum_{i=0}^{n-1} \sum_{\substack{j=1 \\ j\neq\nu}}^{r} |(p^{j})'A^{i}b^{j}| = 0 \quad .$$

Commentary on Chapter II

1. The concepts of controllability studied in this chapter are equivalent to the analogous notions from Chapter I only for systems whose accessible regions are convex. In particular, this occurs for linear systems (with linear or nonlinear inputs).

The foundation of the method given here for investigating the controllability of dynamical systems is analogous to that of relative controllability from §1.4. Under such an approach, the complex question of the controllability of a dynamical system is replaced by consideration of the simplest problem of controllability of a point with respect to a direction. Results of the solution of the latter problem may be directly used for the study of the problems of Chapter I in the case of linear systems. In the general case of nonlinear systems, passage from the problems of this chapter to the analagous problems of the preceding chapter may be effected by the concept of an autonomous control system. More concretely, if in a nonlinear system with r controls the origin is autonomously controllable with respect to r linearly independent directions p , then any point from some neighborhood of the origin lying in the linear hull of the vectors p^{ν}, $\nu = 1,\dots,r$, may be reached from the origin. The approach to the problems of the given chapter and the results obtained are most naturally applied in the theory of necessary conditions for controllability in the sense of Chapter I.

2. The next step in the study of controllability in the sense of the definitions of Chapter I is more complex than that considered in this chapter. It is the introduction of the concept of controllability in two, three,...,n, linearly independent directions. In the latter case, we have the problem of Chapter I. This question is not considered in this monograph.

3. In Chapter II, controllability is understood as the property of a system under which a trajectory from the origin may be transferred to another point. Earlier (Chapter I) we posed the problem of transferring the trajectory from the given point to the origin. This was done only to simplify notations. Later, in Definitions 1-8 of Chapter II, we brought in several controllability notions different from the analagous ones in Chapter I. This makes this chapter different from the preceding one in the essence of the considered questions.

4. The theory of invariance has a wide literature. Our approach is closest to the variational method of

L.I. Rozonoer [104b,114c]. In these works there may also be found some results for nonlinear systems. Along with complete invariance for nonlinear systems, invariance of some order is also of interest. The meaning of this last concept may be clarified by analogy with the concept of indifference introduced in §3.7 in connection with observability. Roughly speaking, a system is invariant of order k with respect to a direction p, if $p'x(t) = o(\beta^k)$ for all admissible controls constrained by the condition $|u(t)| \leq \beta$. In this new concept, the influence of the perturbations on the functional is taken into account more precisely than in ε-invariance [91].

The property of autonomy plays a large role in control theory. As is clear from §6, conditions for autonomy are obtained naturally from the theory of controllability and the theory of invariance.

CHAPTER III

The Observability of Dynamical Systems

§1. Statement of the Problem. Definitions. Method of Investigation

We will consider a dynamical system of the form

$$\frac{dx}{dt} = f(x) \quad , \quad x(0) = x_o \quad , \quad f(0) = 0 \quad , \tag{1}$$

where $x = \{x_1,\dots,x_n\}$ is a n-vector of states.

Let the information about the unknown state x_o be contained in the values of the m-vector z:

$$z = h(x) \quad , \quad h(0) = 0 \quad , \tag{2}$$

measured during the course of time $T = [0,t_1]$. We call the vector z the output of the system (1), the number m is the output dimension.

Problem. The functions f(x), h(x) are given. From the known output z(t), t T, of the system (1), find its initial state x_o.

We need the following definitions.

Definitions.

1. The direction p, $||p|| = 1$, is called T-observable at the origin) with respect to the measurements (2) on T, if there exists a number $\alpha_o = \alpha_o(t_1,p) > 0$, such that any initial state of the system (1) of the form

$$x_0 = \alpha p \quad , \quad |\alpha| \le \alpha_0 \quad , \tag{3}$$

may be reconstructed using the function $z(t)$, $t \in T$.

2. The dynamical system (1) is called <u>T-observable</u> (at the origin) if each direction p, $||p|| = 1$ is T-observable at the point $x = 0$.

3. If there exists an interval T for which the direction p, $||p|| = 1$, is T-observable (at $x = 0$), then this direction is called <u>observable</u> at $x = 0$ ($\alpha_0(t_1,p) > 0$ for some $t_1 = t_1(p)$).

4. A dynamical system is called <u>observable (at x = 0)</u> if each direction p, $||p|| = 1$, is observable at $x = 0$ ($\inf\limits_{||p|| = 1} \alpha_0(t_1(p),p) > 0$).

5. The direction p, $||p|| = 1$, is <u>completely observable</u> (at $x = 0$) if it is T-observable at $x = 0$ for any $t_1 > 0$ ($\alpha_0(t_1,p) > 0$ for all $t_1 > 0$).

6. The dynamical system (1) is <u>completely observable</u> (at $x = 0$) if each direction p, $||p|| = 1$, is completely observable ($\inf\limits_{||p|| = 1} \alpha_0(t_1,p) > 0$ for all $t_1 > 0$).

7. The direction p, $||p|| = 1$, is <u>observable in the large</u> (at $x = 0$) for the dynamical system (1) if it is observable and there exists a $t_1 = t_1(p)$ for which $\alpha_0(t_1(p),p) = \infty$.

8. A dynamical system observable in the large (at $x = 0$) at each direction p, $||p|| = 1$, is <u>observable in the large</u> at $x = 0$.

The method of investigating observability which is adopted in 3-6, consists of the following: We let $x(t,\alpha)$ denote the solution of the system (1) having an initial condition of the form (3). Substitution of $x(t,\alpha)$ into (2) generates a function* $z(t,\alpha) = h(x(t,\alpha))$. It is easy to see that the direction p, $||p|| = 1$, of the system (1) is T-observable if and only if there exists some time $t = \bar{t}$ such that the function $z = z(\bar{t},\alpha)$ is invertible in some neighborhood of the point $\alpha = 0$, i.e. $\alpha = \alpha(z)$. The condition for invertibility of a function is easily expressed by means of derivatives. It is known that a function $z = z(\bar{t},\alpha)$ is invertible in a neighborhood of $\alpha = 0$ if in the series of its derivatives

$$\left.\frac{\partial^i z(\bar{t},\alpha)}{\partial \alpha^i}\right|_{\alpha=1}, \quad i = 1,2,..., \tag{4}$$

The first nonzero derivative has odd index. If this index is even, then the function is not invertible in a neighborhood of $\alpha = 0$. For invertibility of the function $z = z(\bar{t},\alpha)$ for all α, it is necessary and sufficient that it be monotone (strictly) in the domain of definition.

One approach to the calculation of the derivatives (4) is to expand the function $z(\bar{t},\alpha)$ in powers of α. Since $z(\bar{t},0) \equiv 0$, $t \in T$, then the quantity $z(\bar{t},\alpha)$ may be treated as the increment $z(\bar{t},\alpha) - z(\bar{t},0)$. Thus, the method of increments for functionals along the trajectories of dynamical systems finds a natural application in the theory of observability.

§2. The Formula for Increments with Respect to Initial States

From the definition of the function** $z = h(x)$, we have

$$z(t,\alpha) = \frac{\partial h'(0)}{\partial x} x(t,\alpha) + o(\alpha) \quad . \tag{5}$$

---

* For simplicity, we consider single-output systems, i.e. here z is a scalar.

** Here, as above, z is a scalar.

On the other hand, for any differentiable function $\psi(t)$ and solution $x(t,\alpha)$ of Eq. (1) with the initial condition (3), we have the identity

$$\psi'(\bar{t})x(\bar{t},\alpha) = \psi'(0)p\alpha + \int_0^{\bar{t}} \frac{d\psi'(t)}{dt} x(t,\alpha)dt$$

$$+ \int_0^{\bar{t}} \psi'(t)\frac{dx(t,\alpha)}{dt} dt \quad . \tag{6}$$

We define the function $\psi(t) = \psi(t,\bar{t})$ by the equation

$$\frac{d\psi(t,\bar{t})}{dt} = -A'\psi(t,\bar{t}) \quad , \quad \psi(\bar{t},\bar{t}) = \frac{\partial h(0)}{\partial x} \quad , \quad A = \frac{\partial f(0)}{\partial x} \quad . \tag{7}$$

Substituting (6), (7), and (1) into (5), we obtain

$$z(\bar{t},\alpha) = \psi'(0,\bar{t})p\alpha - \int_0^{\bar{t}} \psi'(t,\bar{t})Ax(t,\alpha)dt$$

$$+ \int^{\bar{t}} \psi'(t,\bar{t})\left[f(0) + \frac{\partial f(0)}{\partial x} x(t,\alpha) + o(\|x(t,\alpha)\|)\right] dt \ .$$

Since $||x(t,\alpha)|| \leq k\alpha$, $t\varepsilon T$, k = constant, we finally have

$$z(\bar{t},\alpha) = p'\psi(0,\bar{t})\alpha + o(\alpha) \quad . \tag{8}$$

§3. Observability of Linear Systems

1. Single-Input Systems

Let the dynamical system (1) have the form

$$\frac{dx}{dt} = Ax \quad , \tag{9}$$

where A is an nxn matrix. We measure the quantity

$$z = c'x \quad ,$$

where c is an n-vector. In this case, formula (8) assumes the form

$$z(\bar{t},\alpha) = p'\psi(0,\bar{t})\alpha \quad ,$$

where $\psi(t,\bar{t})$ satisfies the equation

$$\frac{d\psi(t,\bar{t})}{dt} = -A'\psi(t,\bar{t}) \quad , \quad \psi(\bar{t},\bar{t}) = c \quad .$$

Hence, it is clear that the direction p in the system (9) is observable in the large if and only if

$$p'\psi(0,\bar{t}) \not\equiv 0 \quad , \quad \bar{t} > 0 \quad . \tag{10}$$

We express this condition through the given problem. The sequence of derivatives of the function

$$\lambda(t) = p'\psi(0,t) \tag{11}$$

have the form

$$\dot{\lambda}(t) = p'A'\psi(0,t),\ldots,\lambda^{(n)}(t) = p'(A^n)'\psi(0,t) \quad . \tag{12}$$

On the other hand, the matrix A' satisfies its own characteristic equation

$$\lambda^n + \gamma_1\lambda^{n-1} + \cdots + \gamma_{n-1}\lambda + \gamma_n = 0 \quad ,$$

that is

$$(A^n)' + \gamma_1(A^{n-1})' + \cdots + \gamma_{n-1}A' + \gamma_n E = 0 \quad .$$

Multiplying the last equation on the left by p' and on the right by $\psi(0,t)$, we obtain, by virtue of (11) and (12), differential equations for the function $\psi(t)$:

$$\lambda^{(n)}(t) + \gamma_1\lambda^{(n-1)} + \cdots + \gamma_{n-1}\dot{\lambda}(t) + \gamma_n\lambda(t) = 0 \quad , (13)$$

with the initial conditions

$$\lambda(0) = p'c \ , \ \dot{\lambda}(0) = p'A'c,\ldots,\lambda^{(n-1)}(0) = p'(A^{n-1})'c \ . \quad (14)$$

In order that the solution of the homogeneous equation (13) with the initial conditions (14) be nontrivial, it is necessary and sufficient that at least one of the numbers

$$p'c,\ p'A'c,\ldots,p'(A^{n-1})'c \qquad (15)$$

be nonzero. Combining (10) with (11), we have the following theorem.

Theorem 1. *The direction p of the linear system (9) is T-observable (observable, completely observable) in the large if and only if not all of the numbers (15) are zero.*

The linear dynamical system (9) is T-observable (observable, completely observable) in the large if and only if

$$\text{rank}\ \{c,A'c,\ldots,(A^{n-1})'c\} = n \quad .$$

2. Multi-Input Systems

In system (9), let the measurable quantities be

$$z_j = c_j'x \quad , \quad j = 1,\ldots,m \quad , \qquad (16)$$

where c_j are n-vectors. Repeating the arguments of section 1 for each output, it is not difficult to obtain the following result.

Theorem 2. The direction p for system (9) with the m outputs (16) is T-observable (observable, completely observable) in the large if and only if not all of the numbers

$$p'c_j \;,\; p'A'c_j,\ldots,p'(A^{n-1})'c_j \quad , \quad j = 1,\ldots,m \;,$$

equal zero.

For T-observability (observability, complete observability) in the large of the linear system (9) with outputs (16), it is necessary and sufficient that

$$\text{rank}\,\{C \;,\; A'C,\ldots,(A^{n-1})'C\} = n \quad .$$

Here C is an nxm matrix formed from the n-vectors c_j, $j = 1,\ldots,m$.

3. Observable Systems with a Minimal Number of Outputs.

Let the matrix A be given a system (9). It is required to find the minimal number of vectors $c_1,\ldots,c_m$ for which the system (9) with outputs (16) is observable. Based on the discussion of §1-3 connected with finding the minimal number of inputs, we obtain the followwing result.

Theorem 3. The minimal number of outputs for the observability of the system (9) equals the number of nontrivial invariant factors of the matrix A'.

§4. The Observability of a Dynamical System and Its Controllability With Respect to the Initial Conditions

We consider the dynamical system

$$\frac{dx}{dt} = Ax \quad , \quad x(0) = x_o \quad , \tag{17}$$

whose motion $x(t,x_o)$ is determined by the choice of the initial state x_o, $||x_o|| \neq 0$. We will say that the system has input d for perturbation in the initial conditions if the admissible initial conditions are representable in the form

$$x_o = dy \quad ,$$

where y is a scalar taking on all finite values, and d is a n-vector characterizing the input.

Definition 9. The origin is controllable in the direction p by initial perturbations if for any α we can find a number t_1 and a vector dy such that

$$p'x(t_1,dy) = \alpha \quad .$$

Definition 10. The dynamical system (17) is controllable by initial perturbations if the origin is controllable by initial perturbations in all directions.

For studying this new type of controllability, we use the method of increments. The increment of the functional

$$J(x_o) = p'x(t_1,x_o) \quad ,$$

is expressed (§1.2, §2.2) by the parameters of the

system (17) in the following form:

$$J(x_o) = \psi'(0,t_1)dy \quad , \tag{18}$$

where $\psi(t,t_1)$ is the solution of the equation

$$\frac{d\psi(t,t_1)}{dt} = -A'\psi(t,t_1) \quad , \quad \psi(t_1,t_1) = E \quad .$$

From formula (18), it is clear that the origin is controllable in the direction p by admissible vectors x_o if and only if

$$\lambda(t) = \psi'(0,t)d \not\equiv 0 \quad , \quad t > 0 \quad .$$

This condition is equivalent (cf. §3) to the following: at least one of the numbers

$$p'd \ , \ p'Ad,\dots,p'A^{n-1}d \tag{19}$$

is different from zero. Thus, we have proved the following theorem.

Theorem 4. In order that the origin be controllable in the large in the direction p, $||p|| = 1$, by initial perturbations, it is necessary and sufficient that not all of the numbers (19) equal zero.

A dynamical system is controllable in the large by initial perturbations if and only if

$$\text{rank}\ \{d \ , \ Ad,\dots,A^{n-1}d\} = n \quad . \tag{20}$$

Comparing this result with Theorem 1, we are led to the conclusion of the following theorem.

Theorem 5. *The dynamical system*

$$\frac{dx}{dt} = Ax$$

with input d for the initial perturbations, is controllable by those inputs if and only if the adjoint system

$$\frac{dx}{dt} = -A'x$$

is observable with respect to the measurements

$$z = d'x \quad .$$

By condition (20) and the corollary of Theorem 2.1, we are led to the result,

Theorem 6. *The dynamical system*

$$\frac{dx}{dt} = Ax$$

with input d for the initial perturbations, is controllable by these perturbations if the dynamical system

$$\frac{dx}{dt} = Ax + du \quad .$$

is controllable.

The converse is also true.

§5. Observability of Linearized Dynamical Systems

We consider the dynamical system

$$\frac{dx}{dt} = f(x) \quad , \quad f(0) = 0 \quad , \tag{21}$$

with output

$$z = h(x) \tag{22}$$

and the linear model

$$\frac{dx}{dt} = Ax \quad , \quad A = \frac{\partial f(0)}{\partial x} \quad , \tag{23}$$

$$z = c'x \quad , \quad c = \frac{\partial h(0)}{\partial x} \quad . \tag{24}$$

We prove the following assertion.

Theorem 7. _If the direction p, $||p|| = 1$, is observable for the linear model (23), (24), then it is completely observable for the system (21), (22)._

For the complete observability of the nonlinear systems (21), (22), it is sufficient that its linear model (23), (24) be observable.

Proof. It suffices to prove only the first part of the theorem. For the system (21), (22), from (8) we have the following formula

$$z(\bar{t},\alpha) = p'\psi(0,\bar{t})\alpha + o(\alpha) \quad .$$

By virtue of Theorem 1, from the observability of the direction p in the linear systems (23), (24), it follows that

$$p'\psi(0,t) \not\equiv 0 \quad , \quad 0 \le t \le t_1 \quad ,$$

for any $t_1 > 0$. Therefore, for any p we can find numbers $\bar{t}$, α_o, such that $p'\psi(0,t) \neq 0$ and the function $z(\bar{t},\alpha)$ is monotonic in α for $|\alpha| \le \alpha_o$. This means that the function

$z(\bar{t},\alpha)$ is invertible for $|\alpha| \leq \alpha_o$, which is equivalent to the observability of the system (21,), (22) in the direction p. The theorem is proved.

§6. Critical Cases of Observability

Definition 11. We call the direction p, $||p|| = 1$, critical for the dynamical system (21) with output (22), if it is not observable for the linear model (23), (24).

If the system (21), (22) has one critical direction, then we will say that in (21), (22) there occurs a critical case of observability.

For studying critical cases of observability, it is necessary to define formula (8) more precisely in order to explicitly describe the terms containing the higher powers of α. On the one hand, from (5) we have

$$z(\bar{t},\alpha) = \frac{\partial h(0)}{\partial x} x(\bar{t},\alpha) + \frac{1}{2} x'(\bar{t},\alpha) \frac{\partial^2 h(0)}{\partial x^2} x(\bar{t},\alpha) + o(||x||^2) \quad . \tag{25}$$

On the other hand, it is easy to verify the identity

$$x'(\bar{t},\alpha)\Psi(\bar{t},\bar{t})\, x(\bar{t},\alpha) + x'(0,\alpha)\Psi(0,\bar{t})x(0,\alpha)$$
$$= \int_0^{\bar{t}} [\dot{x}'(t,\alpha)\Psi(t,\bar{t})x(t,\alpha) + x'(t,\alpha)\Psi(t,\bar{t})\dot{x}(t,\alpha)]dt$$
$$+ \int_0^{\bar{t}} x'(t,\alpha)\dot{\Psi}(\bar{t},\bar{t})x(t,\alpha)dt \quad .$$

We define the matrix function $\Psi(t,\bar{t})$ by the equation

$$\frac{d\Psi(t,\bar{t})}{dt} = -A'\Psi(t,\bar{t}) - \Psi(t,\bar{t})A - \frac{1}{2}\psi'(t,\bar{t})D \tag{26}$$

with the initial condition

$$\Psi(\bar{t},\bar{t}) = \frac{1}{2}\frac{\partial^2 h(0)}{\partial x^2} \quad . \tag{27}$$

Here the symbol $\psi'D$ denotes the expression

$$[\psi'D]_{ij} = \sum_{k=1}^{n} \frac{\partial f_{k}(0)}{\partial x_i \partial x_j} \psi_k \quad .$$

After substituting (7), (23), (26), and (27) into (25), we obtain the desired formula

$$z(\bar{t},\alpha) = \alpha p'\psi(0,\bar{t}) + \alpha^2 p'\Psi(0,\bar{t})p + o(\alpha^2) \quad . \tag{28}$$

Now let p be a critical direction in the system (21), (22). Then, by virtue of Theorem 1, we have

$$p'\psi(0,\bar{t}) \equiv 0 \quad , \qquad \bar{t}\varepsilon T \quad ,$$

and formula (28) assumes the form

$$z(\bar{t},\alpha) = \alpha^2 p'\Psi(0,\bar{t})' + o(\alpha^2) \quad . \tag{29}$$

Hence, it follows that in the critical direction p, the parameter α may be reconstructed from the value $z(t,\alpha)$ only at those moments t when

$$p'\Psi(0,t)p = 0 \quad , \quad t\varepsilon T \quad .$$

The set of points t for which this equality is satisfied has zero measure. In order to investigate this singular case of observability, we introduce a definition.

Definition 12. The direction p, $||p|| = 1$, is called essentially observable (at x = 0) for a dynamical system, if it is observable with respect to values of z(t) defined on a set of positive measure. A dynamical system is called <u>essentially observable</u> if each direction p is essentially observable.

It is not difficult to note that in Theorems 1-7, words of the type "observability" may be replaced by the words "essential observability". By virtue of the analyticity of the function $\Psi(0,t)$, from (29) it follows that for the essential observability of a critical direction p it is necessary that

$$p'\Psi(0,t)p \equiv 0 \quad , \quad t\varepsilon T \quad .$$

Differentiating this identity and considering the expression so obtained at the time t = 0, we convince ourselves that the following theorem is valid.

<u>Theorem 8.</u> <u>In order that a critical direction p in the system (21), (22) be essentially observable, it is necessary that</u>

$$p'G_kp = 0 \quad , \quad k = 0,1,..., \tag{30}$$

<u>where the nxn matrices G_k are obtained from the recurrence relations</u>

$$G_0 = \frac{\partial^2 h(0)}{\partial x^2} \quad , \quad G_{k+1} = A'G_k + G_kA + \frac{1}{2}\, q_k'D \quad ,$$

$$q_0 = \frac{\partial h(0)}{\partial x} \quad , \quad q_{k+1} = Aq_k \quad .$$

If a dynamical system is essentially observable, then condition (30) is satisfied for each critical direction. In particular it follows from this theorem that the linear system

$$\frac{dx}{dt} = AX$$

with outputs

$$z = c'x + x'x$$

is not essentially observable if

$$\text{rank } \{c, A'c, \ldots, (A^{n-1})'c\} < n \quad .$$

For determination of sufficient conditions for observability in critical cases, we must make formula (29) more precise by separating the terms continuing α^3, $\alpha^4, \ldots$. The scheme obtained from this formula in the general case is described in §6.5. Theorem 8 may be used as a sufficient condition for observability if the observability is understood in the weak sense.

Definition 13. The direction p in (21), (22) is called one-sided T-observable if there exists an $\alpha_o > 0$ such that all intial states of the form

$$x_o = \alpha p \quad , \qquad |\alpha| \leq \alpha_o \quad ,$$

may be reconstructed from measurements z(t) on T and knowledge of sign α.

Theorem 9. A critical direction is one-sided observable if condition (30) is violated for one k, k = 0,1,....

§7. On Indifference and Autonomy in the Theory of Observability

We return to the linear system

$$\frac{dx}{dt} = Ax \tag{31}$$

with outputs

$$z = c'x \quad . \tag{32}$$

Definition 14. The direction p, $||p|| = 1$, for the dynamical system (31) is called T-indifferent to the observation (32) (at $x = 0$), if for all initial states of the form

$$x_o = \alpha p \quad , \quad -\infty < \alpha < \infty \quad ,$$

the output $z(\bar{t})$ does not depend on α whatever may be the number $\bar{t} \in T$.

The condition for indifference to an observation is a direct corollary of Theorem 1.

Theorem 10. The direction p, $||p|| = 1$, is T-indifferent to the observation (32) if and only if

$$p'c = p'A'c = \cdots = p'(A^{n-1})'c = 0 \quad .$$

We introduce the following definition.

Definition 15. The set of directions p_j, $||p_j|| = 1$, $j = 1,\ldots,m$, of the dynamical system (31), with outputs

$$z_j = c_j'x \quad , \quad j = 1,\ldots,m \quad , \tag{33}$$

are autonomously T-observable (at $x = 0$) if each direction p_j is T-observable with respect to the output z_j and is T-indifferent to observations in the other directions.

Combining the conditions of Theorems 1, 10, we are

led to the assertion

Theorem 11. The set of directions p_j, $||p_j|| = 1$ $j = 1,\ldots,m$, of the dynamical system (31) are autonomously observable with respect to the outputs (33) if and only if

$$\sum_{k=0}^{n-1} |p_j'(A^k)'c_j| > 0 \quad , \quad \sum_{k=0}^{n-1} |p_i'(A^k)'c_j) = 0$$

$$\text{for} \quad i \neq j \quad , \quad i,j = 1,\ldots,m \quad .$$

For nonlinear systems, the following defintion may turn out to be useful.

Definition 16. The direction p, $||p|| = 1$, of a dynamical system has order k T-indifference to the observations (22) if for each initial state of the form

$$x_o = \alpha p \quad ,$$

the output (22) has the form

$$z(\bar{t},\alpha) = o(\alpha^k) \quad , \quad \bar{t}\varepsilon T \quad .$$

The corresponding form of Definition 15 may also be modified. From these definitions, it follows that the properties of dynamical systems which they introduce may be investigated by the method of increments, the first results already having been obtained in Theorem 10. Each critical direction of a system has at least order one T-indifference.

§8. Methods of Functional Analysis in the Theory of Observability according to Kalman

1. Application of the Minimax Theorem

In this paragraph we use the definition of observability given by Kalman [58a].

Definition 17. The direction p of the linear system

$$\frac{dx}{dt} = Ax$$

is called observable with respect to the output

$$y = c'x \quad , \tag{35}$$

if there exists an interval $T = [0,t_1]$ such that from the function $y(t)$, $t \in T$, it is possible to reconstruct the value $p'x_0$ for each initial state x_0 of the system (34).

Definition 18. If each direction of the system (34) is observable, then the system is called completely observable.

In his proof of the theorem about observability, Kalman used linear operations of reconstruction. More precisely, the direction p of the system (34) is regarded as observable if there can be found a measurable function $\xi(t)$, $t \in T$, such that

$$p'x_0 = \int_0^{t_1} c'x(t,x_0)\xi(t)dt \quad , \tag{36}$$

for any x_0. Here $x(t,x_0)$ is the solution of the system corresponding to the initial value x_0. Since

$$x(t,x_0) = F(t)x_0 \quad ,$$

where F(t) is the fundamental matrix of the solution ($\dot{F} = AF$, $F(0) = E$), (36) assumes the form

$$p'x_o = \int_0^{t_1} c'F(t)\xi(t)dt\ x_o \quad . \tag{37}$$

We impose the constraint

$$|\xi(t)| \leq L \quad , \quad t\varepsilon T \quad ,$$

on the function $\xi(t)$, generating the reconstruction operator. This is done in order to seek the numbers L for which the observability problem has a solution. Clearly, equality occurs in (37) for all x_o if and only if for some L the direction p lies in the set

$$Q(L) = \{q:\ q = \int_0^{t_1} F'(t)c\xi(t)dt \ ,\ |\xi(t)| \leq L\} \ . \tag{38}$$

In other words, the problem of observability in the direction p has a solution if and only if the distance

$$\rho(L) = \min_{q\varepsilon Q(L)} ||p - q||$$

from the direction p to the set Q(L) equals zero for some L. Since, for each L, Q(L) is a closed, convex set, it is possible to use the minimax theorem. We have

$$\rho(L) = \min_{q\varepsilon Q(L)} \max_{||z|| \leq 1} z'(p-q) = \max_{||z|| \leq 1} \min_{q\varepsilon Q(L)} z'(p-q) \ .$$

Here we substitute the expression for q from (38), while the operation min over q is replaced by the operation min over ξ, $|\xi| \leq L$. Then

$$\rho(L) = \max_{||z|| \leq 1} (z'p - L \int_0^{t_1} |z'F'(t)c|dt) \quad .$$

In order that for each p there exist a number L for which $\rho(L) = 0$, it is necessary and sufficient that

$$z'F'(t)c \not\equiv 0 \quad , \quad t \in T \quad , \tag{39}$$

for all z, $||z|| \neq 0$.

Consequently, if we have $z'F'(t)c \equiv 0$ for some z, then the direction $p = \alpha z$ is unobservable: $\rho(L) = |\alpha| > 0$.

If (39) is satisfied, then assuming

$$L = \frac{||p||}{\min\limits_{||z|| < 1} \int_0^{t_1} |z'F'(t)c|dt} \quad ,$$

we obtain $\rho(L) \leq 0$, i.e. the system is observable. Forming the differential equation for the function $\lambda(t) = z'F'(t)c$ (cf. §1.3), we are led to the assertion:

<u>Theorem 12. The system (34), (35) is observable if and only if</u>

$$\text{rank } \{c, A'c, \ldots, (A^{n-1})'c\} = n \quad .$$

2. Use of the Theorem on the Separation of Convex Sets

Condition (39) for the solvability of the problem of observability is easily proved with the help of the theorem on the separation of convex sets. The <u>necessity</u> is obvious: if

$$p = \int_0^{t_1} F'(t)c\xi(t)dt \quad ,$$

then

$$0 = z'p - \int_0^{t_1} z'F'(t)c\xi(t)dt \leq z'p - L \int_0^{t_1} |z'F'(t)c| \, dt$$

for all z, $||z|| \leq 1$. These inequalities are possible for all p (i.e. there exists a value of L such that they are satisfied) only when (39) is satisfied. For proving sufficiency of (39), we will assume the contrary: Let (39) be satisfied but the system is unobservable, i.e. for any L the set Q(L) does not contain the vector p. Then, by the theorem on the separation of convex sets, we can find a hyperplane strictly separating Q(L) and p. This means that there exists a vector z, $||z|| = 1$, such that

$$z'p < \int_0^{t_1} z'F'(t)c\xi(t)dt$$

for all $|\xi(t)| \leq L$, $t \in T$. The last inequality is satisfied when the right side assumes its minimal value

$$z'p < - L \int_0^{t_1} |z'F'(t)c|dt \quad . \tag{40}$$

Let

$$\mu = \min_{||z|| \leq 1} \int_0^{t_1} |z'F'(t)c|dt \quad .$$

The minimum of the right side is assumed for $\mu > 0$ by virtue of (39). We set $L = |z'p|/\mu$. Then, from (40) have

$$z'p < -|z'p| \quad ,$$

which is impossible. The sufficiency of (39), and therefore Theorem 12, is proved with the help of the theorem on separating convex sets.

3. The Problem of Moments in the Theory of Observability

If we change the coordinate system, then the condition

$$p = \int_0^{t_1} F'(t)c\xi(t)dt \quad ,$$

for solvability of the problem of observability in the direction p may be treated as a problem of moments and we are again led to (39). We sharpen this result. For this, we successively integrate the right side of (39) n times by parts. Referring to the proof of Theorem 1.1 for details, we formulate the result:

Theorem 13. The observable directions of the system (34), and only those, have the representation

$$p = \sum_{i=0}^{n-1} \mu_i (A')^i c \; , \; -\infty < \mu_i < \infty \; , \; i = 0,\dots,n-1 \quad .$$

This result may be obtained from Theorem 1.1 as a corollary of the duality principle, the proof of which is found in Theorem 1.1; in order that the system

$$\dot{x} = Ax + bu$$

be controllable, it is necessary and sufficient that the system

$$\dot{x} = A'x \quad , \quad z = b'x$$

be observable.

§9. The Solvability of Two-Point Boundary Value Problems and Controllability with Respect to Initial Conditions

Let the dynamical system be described by the equation

$$x^{(n)} = f_1(x,\ldots,x^{(n-1)}) \quad , \quad t\varepsilon T = [0,t_1] \quad ,$$

$$f(0,\ldots,0) = 0 \quad ,$$

It is required to find conditions under which there exists a solution $x(t)$, $t\varepsilon T$, satisfying the conditions

$$x(0) = \cdots = x^{(n-2)}(0) = 0 \ , \ x^{(k)}(t_1) = \alpha \quad ,$$

where $0 \leq k \leq n$, α is a given number, and $f_1(x,\ldots,x^{(n-1)})$ is a scalar function.

An immediate generalization of this two-point boundary value problem has the form

$$\frac{dx}{dt} = f(x) \quad , \quad t\varepsilon T \quad , \quad f(0) = 0 \quad , \tag{41}$$

$$d_j'x(0) = 0 \ , \quad j = 1,\ldots,n-1 \quad , \tag{42}$$

$$c'x(t_1) = \alpha \quad , \tag{43}$$

where $x = \{x_1,\ldots,x_n\}$ is an n-vector of states, d_j, $j = 1,\ldots,n-1$, are a given set of linearly independent vectors, and α is a given number.

The formulated problem reduces to the problem of controllability relative to initial conditions in the following way. Let d be a unit vector orthogonal to the vectors d_j, $j = 1,\ldots,n-1$, i.e.

$$d'd_j = 0 \quad , \quad j = 1,\ldots,n-1 \quad .$$

Under the given assumptions, there exists such a vector and it is unique. We assume that the vector d characterizes the input of a system relative to initial conditions, i.e. we will assume that the admissible initial conditions of the system (41) have the form

$$x_0 = dy \quad , \quad -\infty < y < \infty \quad . \tag{44}$$

The boundary value problem (41)-(43) is equivalent to the following controllability problem: find a number y, $|y| < \infty$, for which the output

$$z = c'x(t_1) \quad ,$$

of the system (41), with initial condition (42), assumes a given value. We carry out the investigation of the last problem for the linear case:

$$\frac{dx}{dt} = Ax \quad , \quad x_0 = dy \quad . \tag{45}$$

It is not difficult to prove (§1.2,§2.2) that

$$c'x(t_1) = \psi'(0,t_1)dy \quad , \tag{46}$$

where $\psi(t,t_1)$ is the solution of the equation

$$\frac{d\psi(t,t_1)}{dt} = -A'\psi(t,t_1) \quad , \quad \psi(t_1,t_1) = c \quad .$$

From (46), we find that for any α the problem (45),(42), (43) has a solution if and only if

$$\psi'(0,t_1)d \neq 0 \quad .$$

Hence, it follows that if at least one of the numbers

$$c'd\ ,\ c'Ad, \ldots, c'A^{n-1}d\ , \tag{47}$$

is different from zero, then the problem (45), (43) has a solution for any α and for all t_1, with the exception of a finite number of points on each bounded interval. If all the numbers (47) are zero, then the problem (45), (42), (43) has no solution for any $\alpha \neq 0$, for all t_1.

For more flexibility in the application of the methods of control theory to two-point boundary value problems, it is advisable to change the original formulation (41)-(43). For given t_1, c, d_j, $j = 1, \ldots, n-1$, and any $\varepsilon > 0$, we wish to find conditions under which there exists a solution x(t) of Eq. (41) satisfying the initial condition (42) and the condition $c'x(\bar{t}) = \alpha$ for some $\bar{t}$ such that $|t_1 - \bar{t}| < \varepsilon$. This problem, in the linear case, has a solution if and only if there is at least one number in (47) different from zero.

Under the new formulation of the boundary value problem, we may investigate the problems of observability from §1-6 in a similar fashion. The results of the solution to problems of observability may also be immediately reformulated in terms of boundary value problems.

Commentary on Chapter III

The definition of observability from §1 differs from the original [58a] but is equivalent to the latter for linear systems. We have constructed our definition to be strictly dual to the definition of controllability in Chapter II. For this we have abandoned the notion of the co-state [58a] in favor of the concept of direction. A point (in the definition of controllability) and a direction (in the definition of observability) are dual. Observability in the sense of §1 is intuitively completely clear both for linear and nonlinear systems. The definitions themselves lead to natural methods of investigation. The effectiveness of such an approach is illustrated in

§3-6. Beginning in §3, the known conditions for observability of linear systems are obtained in an elementary fashion and the theorem on the minimal number of outputs for an observable system is easily proved. However, this latter result is a corollary of the duality principle [58a]. This principle is extended by new results in §4.

Observability in the sense of Kalman for linearized systems is proved in [3]. The critical case of observability has not been treated in the literature. Here emerge new qualitatively observable systems (essentially observable systems) which have not been noted earlier. Finally, Theorem 8 makes sense for systems observable in the sense of Kalman.

CHAPTER IV

The Theory of Identification of Dynamical Systems

§1. Statement of the Problem. Definitions. Method of Investigation.

Let the equation of perturbed motion of the dynamical system have the form

$$\frac{dx}{dt} = f(x,w) \ , \ x(0) = 0 \ , \ t\varepsilon T = [0,t_1] \ , \ f(0,0) = 0 \ , \quad (1)$$

where x is the n-vector of states of the system, while w is a q-vector of parameters. We assume that the function f(x,w) is known with accuracy up to the parameters $w\varepsilon W$. The variables admitting measurement (output variables) give the numbers $z_1,\ldots,z_m$, which are connected with the state x of (1) by the relations

$$z = h(x) \quad , \quad h(0) = 0 \quad . \qquad (2)$$

Problem. Find the value of $w\varepsilon W$ using the measurements z(t), $t\varepsilon T$.

Definitions.

1. The direction s(at the point w = 0 in the parameter space) is T-identifiable with respect to the measurements (2), if there exists a number $\beta_o = \beta_o(t_1,s) > 0$ such that each value of the parameter of the form

$$w = \beta s \quad , \quad |\beta| \leq \beta_o \quad , \qquad (3)$$

may be reconstructed by the measurements $z(t)$, $t \in T$.

2. The dynamical system (1) is called T-identifiable (at $2 = 0$) with respect to the measurements (2), if each directions s in the parameter space is T-identifiable relative to the measurements (2) ($\inf_{||s||=1} \beta_o(t_1, s) > 0$).

3. The direction s (at $w = 0$) is identifiable relative to (2), if there exists an interval $T = [0, t_1]$, $t_1 = t_1(s) < \infty$, on which this direction is T-identifiable ($\beta_o(t_1(s), s) > 0$).

4. A dynamical system is called identifiable (at $w = 0$) if each direction in its parameter space is identifiable ($\inf_{||s||=1} \beta_o(t_1(s), s) > 0$).

5. The direction s (at $2 = 0$) is completely identifiable relative to the measurements (2) if it is T-identifiable with respect to the measurements $z(t)$ for any T ($\beta_o(t_1, s) > 0$ for all $t_1 > 0$).

6. The dynamical system (1) is completely identifiable (at $2 = 0$) if each direction in its parameter space is completely identifiable ($\inf_{||s||=1} \beta_o(t_1, s) > 0$ for all $t_1 > 0$)

7. The direction s (at $w = 0$) is identifiable in the large with respect to the measurements (2), if it is identifiable and $\beta_o(t_1(s), s) = \infty$.

8. A dynamical system is called identifiable in the large (at $w = 0$) relative to the measurements (2), if each direction (at $w = 0$) is identifiable in the large relative to (2).

The idea of the method which is used in the sequel to investigate the identifiability of a dynamical system is explained on the case of a single-output system. To each value of the parameter β connected with the direction s, there corresponds a trajectory $x(t)$ of the system (1) which generates a function $z(t)$ from the measuring apparatus. Thus, the set of values $z(t)$ at any moment $t = \bar{t}$ is a function of β, $z = z(\bar{t},\beta)$. Clearly, the direction s is identifiable with respect to $z(\bar{t},\beta)$ if and only if the function $z = z(\bar{t},\beta)$ is invertible with respect to β. The conditions for invertibility, which is connected with the monotonicity of the function $z(\bar{t},\beta)$, are simply expressed through the derivatives of the function $z(t,\beta)$. For calculation of the derivatives of $z(\bar{t},\beta)$, the method of increments is very helpful. For this, it is sufficient to express $z(\bar{t},\beta) - z(\bar{t},0)$ in powers of β:

$$z(\bar{t},\beta) - z(\bar{t},0) = a_1\beta + \frac{a_2}{2!}\beta^2 + \cdots + \frac{a_n}{n!}\beta^n + o(|\beta|^n) \ . \quad (4)$$

If the coefficients $a_1,\ldots,a_n$ can be expressed through the given problem, then we obtain an explicit exression for the derivatives $\dot{z}(\bar{t},0), \ddot{z}(\bar{t},0),\ldots,z^{(n)}(\bar{t},0)$, $a_i = z^{(i)}(\bar{t},0)$. In (4), if the first nonzero coefficient has an odd index, then the function $z(\bar{t},\beta)$ is invertible in a neighborhood of $\beta = 0$ and, in the opposite case, it is not invertible.

§2. The Formula for Increments with Respect to a Parameter

We will consider system (1) with a single output ($m = 1$). If in the representation (3), the number $\beta = 0$, then to the value of the parameter $w = 0$, by virtue of (1), there corresponds only the trivial solution $x \equiv 0$ for which $z(t) = h(0) = 0$. To a nonzero value of β there corresponds a trajectory $x(t,\beta)$, which generates from (1) a function $z(t,\beta)$. The increment of this function at any moment $\bar{t}$ itself equals this function. We express $z(\bar{t},\beta)$ by the conditions of the problem. On the one hand, we have

$$z(\bar{t},\beta) = h(x(\bar{t},\beta)) = \frac{\partial h'(0)}{\partial x} x(\bar{t},\beta) + o(||x(\bar{t},\beta)||) \ . \quad (5)$$

We introduce the function $\psi(\bar{t},t)$ by the equation

$$\frac{d\psi(t,\bar{t})}{dt} = -\frac{\partial f'(0,0)}{\partial x}\psi(t,t) \quad , \quad \psi(t,t) = \frac{\partial h(0)}{\partial x} \quad . \tag{6}$$

Then, from the formula for integration by parts,

$$\psi'(t,\bar{t})x(t,\beta)\Big|_0^{\bar{t}} = \int_0^{\bar{t}} \frac{d\psi'(t,\bar{t})}{dt} x(t,\beta)dt + \int_0^{\bar{t}} \psi'(t,\bar{t})\frac{dx(t,\beta)}{dt} dt$$

with the help of (1), (5), (6) we have

$$z(\bar{t},\beta) = \int_0^{\bar{t}} \psi'(t,\bar{t})f(x(t,\beta),s\beta) + \int_0^{\bar{t}} \frac{d\psi'(t,\bar{t})}{dt} x(t,\beta)dt$$

$$+ o(||x(\bar{t},\beta)||)$$

$$= \int_0^{\bar{t}} \psi'(t,\bar{t}) \frac{\partial f(0,0)}{\partial x} x(t,\beta)dt + \int_0^{\bar{t}} \psi'(t,\bar{t}) \frac{\partial f(0,0)}{\partial w} s\beta\, dt$$

$$+ \int_0^{\bar{t}} \frac{d\psi'(t,\bar{t})}{dt} x(t,\beta)dt + \int_0^{\bar{t}} o(||x(t,\beta)|| + \beta)dt \quad .$$

Since $||x(t,\beta)|| \sim \beta$, finally we have

$$z(t,\beta) = \beta \int_0^{\bar{t}} \psi'(t,\bar{t}) \frac{\partial f(0,0)}{\partial w} s\, dt + o(\beta) \quad . \tag{7}$$

§3. Identification of Linear Systems

1. Single-Output Systems

Let the dynamical system be described by the equation

$$\frac{dx}{dt} = Ax + Bw \quad , \tag{8}$$

while the measured quantity is

$$z = c'x \quad . \tag{9}$$

For parameters w of the form $w = \beta s$, by virtue of (7) we have

$$z(\bar{t},\beta) = \beta \int_0^{\bar{t}} s'B'\psi(t,\bar{t})dt \quad ,$$

where $\psi(t,\bar{t})$ is the solution of the equation

$$\frac{d\psi(t,\bar{t})}{dt} = -A'\psi(t,\bar{t}) \quad , \qquad \psi(\bar{t},\bar{t}) = c \quad .$$

From formula (10), it follows that the direction s is T-identifiable in the large if and only if for at least one $\bar{t}$ T we have

$$\mu(\bar{t}) = \int_0^{\bar{t}} s'B'\psi(t,\bar{t})dt \not\equiv 0 \quad .$$

The function

$$\lambda(t) = \dot{\mu}(t) = s'B'c + \int_0^t s'B'A'\psi(\tau,t)d\tau$$

has the following derivatives

$$\dot{\lambda}(t) = s'B'A'c + \int_0^t s'B'(A')^2\psi(\tau,t)d\tau, \ldots,$$

$$\lambda^{(n)}(t) = s'B'(A^n)'c + \int_0^t s'B'(A')^{n+1}\psi(\tau,t)d\tau \quad .$$

But, since

$$A^n + \gamma_1 A^{n-1} + \cdots + \gamma_{n-1}A + \gamma_n E = 0 \quad ,$$

$\lambda(t)$ satisfies the differential equation

$$\lambda^{(n)}(t) + \gamma_1 \lambda^{(n-1)}(t) + \cdots + \gamma_{n-1}\dot{\lambda}(t) + \gamma_n \lambda(t) = 0 \quad , \quad (11)$$

with the initial conditions

$$\lambda(0) = s'B'c,\ \dot{\lambda}(0) = s'B'A'c,\ldots,\lambda^{(n-1)}(0) = s'B'(A^{n-1})'c \ . \qquad (12)$$

The differential equation (11), with the initial conditions (12), has a nontrivial solution if and only if there is a nonzero number in (12).

Theorem 1. The direction s in the parameter space is T-identifiable (identifiable, completely identifiable) in the large relative to the measurement (9) if and only if the numbers

$$s'B'c,\ s'B'A'c,\ldots,s'B'(A^{n-1})'c \quad ,$$

are not all zero. In order that the system (8) be identifiable in the large with respect to the measurements (9), it is necessary and sufficient that

$$\text{rank}\ \{B'c,\ B'A'c,\ldots,B'(A^{n-1})'c\} = q \quad .$$

2. Multi-output Systems

Let the measurable quantities for system (8) have the form

$$z_1 = c_1'x,\ldots,z_m = c_m'x \quad . \qquad (13)$$

Then, repeating the above arguments for each output z_j, we are led to the following result:

Theorem 2. The direction s in the parameter space is T-identifiable (identifiable, completely identifiable) in the large, relative to the measurements (13), if and only if not all of the numbers

$$s'B'c_j \ , \ s'B'A'c_j, \ldots, s'B'(A^{n-1})'c_j \ , \ j = 1,\ldots,m \ ,$$

are zero.

For T-identifiability of the system (8) with respect to the measurements (13), it is necessary and sufficient that

$$\text{rank}\,\{B'C \ , \ B'A'C, \ldots, B'(A^{n-1})'C\} = q \ . \qquad (14)$$

Here C is the nxm matrix consisting of the n-vectors c_j, $j = 1,\ldots,m$.

§4. The Relative Controllability of a Dynamical System and its Identifiability

If in (14) we set

$$B' = K \ , \ A' = A \ , \ C = B \ ,$$

then we are led to the condition (1.33) of Theorem 1.7. This means that the following theorem is valid.

Theorem 3. The dynamical system

$$\frac{dx}{dt} = Ax + Bu$$

is controllable relative to the subspace $K(Kx(t_1) = 0)$ if and only if the system

$$\frac{dx}{dt} = -A'x - K'w$$

is identifiable with respect to the measurements

$$z = B'x \quad .$$

Corollary. The r-input dynamical system

$$\frac{dx}{dt} = Ax + Bu$$

is controllable if and only if the dynamical system

$$\frac{dx}{dt} = -A'x - w$$

is identifiable by means of the r outputs

$$z = B'x \quad .$$

§5. The Identifiability of Linearized Systems

The nonlinear system

$$\frac{dx}{dt} = f(x,w) \quad , \quad x(0) = 0 \quad , \tag{15}$$

with outputs

$$z = h(x) \quad , \tag{16}$$

is linearized along w = 0. We obtain

$$\frac{dx}{dt} = Ax + Bw \quad , \quad z = c'x \quad . \tag{17}$$

From formula (7) applied to (15), we obtain

$$z(\bar{t},\beta) = \beta \int_0^{\bar{t}} \psi'(t,\bar{t})Bs\ dt + o(\beta) \quad . \tag{18}$$

If the direction s is identifiable for the linear model (17), then from Theorem 1 it follows that there exists a moment $\bar{t}$ such that

$$\int_0^{\bar{t}} \psi'(t,\bar{t})Bs\ dt \neq 0 \quad .$$

By virtue of (10), this means that the value $z(\bar{t},\beta)$ may be found for a number β in a neighborhood of zero.

Theorem 4. The nonlinear system (15) is identifiable relative to the measurements (16) if the linear model (17) is identifiable.

§6. Conditions for the Identifiability of Dynamical Systems in Critical Cases

We will say that there occurs a critical case in the problem of identification of the nonlinear system (15), (16) if the linear model (17) is not identifiable. The direction s in the parameter space is called critical for the dynamical system (15) if it is not identifiable for (17).

In critical cases of identification, there always exists at least one critical direction. Before seeking conditions for identifiability in critical cases, we obtain a formula for the increment of the function $z = h(x)$, improving upon the result (7). We have

$$z(\bar{t},\beta) = \frac{\partial h'(0)}{\partial x}\, x(\bar{t},\beta) + \frac{1}{2}\, x'(\bar{t},\beta)\frac{\partial^2 h(0)}{\partial x^2}\, x(\bar{t},\beta)$$
$$+ o(||x(\bar{t},\beta)||^2) \quad .$$

We introduce the n-vector function $\psi(t,\bar{t})$ and the nxn-matrix function $\Psi(t,\bar{t})$ by the equations

$$\frac{d\psi(t,\bar{t})}{dt} = -A'\psi(t,\bar{t}) \quad , \quad \psi(\bar{t},\bar{t}) = c \quad , \tag{19}$$

$$\frac{d\Psi(t,\bar{t})}{dt} = -\Psi(t,\bar{t})A - A'\Psi(t,\bar{t}) - \frac{1}{2}\psi'(t,\bar{t})D \ , \tag{20}$$

$$\Psi(\bar{t},\bar{t}) = \frac{1}{2}\,\frac{\partial^2 h(0)}{\partial x^2} \quad .$$

Here

$$c = \frac{\partial h(0)}{\partial x} \quad , \qquad [\psi D]_{ij} = \sum_{k=1}^{n} \frac{\partial f_k(0,0)}{\partial x_i\ \partial x_j}\,\psi_k \quad .$$

From the formula for integration by parts

$$\psi'(t)x(t)\Big|_0^{\bar{t}} = \int_0^{\bar{t}} \dot{\psi}'(t)x(t)dt + \int_0^{\bar{t}} \psi'(t)\dot{x}(t)dt \quad ,$$

$$x'(t)\Psi(t)x(t)\Big|_0^{t} = \int_0^{\bar{t}} \dot{x}'(t)\Psi(t)x(t)dt + \int_0^{\bar{t}} x'(t)\Psi(t)\dot{x}(t)dt$$
$$+ \int_0^{\bar{t}} x'(t)\dot{\Psi}(t)x(t)dt$$

We obtain

$$z(t,\beta) = \int_0^{\bar{t}} \psi'(t,\bar{t})\dot{x}(t,\beta)dt + \int_0^{\bar{t}} \dot{x}'(t,\beta)\Psi(t,\bar{t})x(t,\beta)dt$$
$$+ \int_0^{\bar{t}} x'(t,\beta)\Psi(t,\bar{t})\dot{x}(t,\beta)dt + \int_0^{\bar{t}} \dot{\psi}'(t,\bar{t})x(t,\beta)dt$$
$$+ \int_0^{\bar{t}} x'(t,\beta)\dot{\Psi}(t,\bar{t})x(t,\beta)dt + o(||x(\bar{t},\beta||^2) \quad .$$

Here we substitute the values $\dot{\psi}(t,\bar{t})$, $\dot{\Psi}(t,\bar{t})$ from Eqs. (19), (20) and the value $x(t,\beta)$, represented in the form

$$\dot{x} = Ax + Bw + Cwx + \frac{1}{2} Dxx + \frac{1}{2} Gww + o \quad .$$

The meaning of the new symbols A,B,C,D,G is made clear in the introduction. After obvious transformations we obtain

$$z(\bar{t},\beta) = \beta\int_0^{\bar{t}} \psi'(t,t)Bs\,dt + \beta\int_0^{\bar{t}} [\psi'(t,\bar{t})Csx(t,\beta)$$

$$+ s'B'\Psi(t,\bar{t})x(t,\beta) + x'(t,\beta)\Psi(t,\bar{t})Bs$$

$$+ \frac{1}{2} \beta\psi'(t,\bar{t})Gss]\,dt + o(\beta^2) \quad .$$

If s is a critical direction, then the first term on the right equals zero (Theorem 1); thus, we finally have

$$z(t,\beta) = \beta \int_0^{\bar{t}} [\psi'(t,\bar{t})Csx(t,\beta) + s'B'\Psi(t,\bar{t})x(t,\beta)$$

$$+ x'(t,\beta)\Psi(t,\bar{t})Bs + \frac{1}{2} \beta\psi'(t,t)Gss]dt + o(\beta^2) \,. \tag{21}$$

The leading term in β of the trajectory $x(t,\beta)$ may be described in the following form

$$x(t,\beta) = \beta \int_0^{t} F(t)F^{-1}(\tau)Bs\,d\tau + o(\beta) \quad .$$

We substitute this expression into (21):

$$z(t,\beta) = \beta^2 \int_0^{\bar{t}} \int_0^{t} [\psi'(t,\bar{t})CsF(t)F^{-1}(\tau)Bs$$

$$+ s'B'\Psi(t,\bar{t})F(t)F^{-1}(\tau)Bs$$

$$+ s'B'(F^{-1}(\tau))'F'(t)\Psi(t,\bar{t})Bs]d\tau\ dt$$

$$+ \frac{1}{2}\beta^2 \int_0^{\bar{t}} \psi'(t,\bar{t})Gss\ dt + o(\beta^2) \quad .$$

If the direction s is identifiable with respect to the value $z(\bar{t},\beta)$, then the coefficient of β^2 necessarily equals zero. By virtue of the analyticity of the inputs in this coefficient function, it is either identically zero in $\bar{t}$ or it may assume the value zero only on a set of measure zero. As it is singular, in order to eliminate the last case from consideration we introduce the following definition.

Definition 9. The direction s, $||s|| = 1$ (at $w = 0$), is called essentially identifiable if the values $z(t,\beta)$ at which it is identifiable are defined on a set of points t of positive measure. A dynamical system, all of whose directions are essentially identifiable, is called essentially identifiable.

If a critical direction s is essentially identifiable, then for all $t \in T$ we have

$$\int_0^{\bar{t}}\int_0^{\bar{t}} [\psi'(t,\bar{t})CsF(t)F^{-1}(\tau)Bs + s'B'\Psi(t,\bar{t})F(t)F^{-1}(\tau)Bs$$

$$+ s'B'(F^{-1}(\tau))'F'(t)\Psi(t,\bar{t})Bs]d\tau\,dt + \frac{1}{2}\int_0^{\bar{t}} \psi'(t,\bar{t})Gss\ dt \equiv 0 \ .$$

Differentiating this identity and considering the left side of the obtained expression at $\bar{t} = 0$, we obtain a sequence of equalities

$$c'Gss = 0, \quad \sum_{p=0}^{k-2} \lambda_p' CsA^{k-p-2}Bs + s'B' \sum_{p=0}^{k-2} (\Lambda_p A^{k-p-2} + (A')^{k-p-2}\Lambda p) Bs$$

$$+ \frac{1}{2} \lambda_{k+1}' Gss = 0 \ , \ k = 0,\ldots,n-1 \ , \qquad (22)$$

where

$$\lambda_o = c \ , \quad \lambda_{k+1} = A\lambda_k \ , \quad \Lambda_{k+1} = \Lambda_k A + A'\Lambda_k + \lambda_k D \ ,$$

$$\Lambda_o = \frac{\partial^2 h(0)}{\partial x^2} \ .$$

Theorem 5. *If the critical direction s, $||s|| = 1$, in the parameter space of the system (15) is essentially identifiable by the measurements (16), then it satisfies the relations (22). For the essential identifiability of the system (15) with respect to (16), it is necessary that each critical direction s satisfy (22).*

The conditions (22) may be used for the formulation of criteria for one-sided identifiability. The latter concept is completely analagous to the concept of one-sided observability (Definition 3.13).

Remark. In Theorems 1-4 "identifiability" may be replaced by "essential identifiability".

In this chapter we have not considered questions of indifference and autonomy in the theory of identification. These concepts are also introduced in identification theory just as in the theory of observability (Definitions 3.14-3.16). Since the corresponding criteria are obtained from the proofs of the theorems of this chapter without difficulty, the formulation of these notions is left to the reader.

§7. Methods of Functional Analysis in Identification Theory

To illustrate the possibilities for functional analysis in identification theory, it is simpler to use definitions which are equivalent, but different, to those introduced above (for linear systems).

Definition 10. The direction s in the parameter space of the dynamical system

$$\frac{dx}{dt} = Ax + Bw \quad , \quad x(0) = 0 \quad ,$$

is called identifiable relative to the measurements

$$y = c'x \quad ,$$

if there is a time t_1 such that the value $\gamma = s'w$ may be reconstructed from the measurements $y(t)$, $t \varepsilon T = [0, t_1]$ for all q-vectors w.

A dynamical system is called completely identifiable if each direction is identifiable.

Remark. In this definition, it is not explicitly assumed that the reconstruction operation is linear.

It follows directly from the definition that the direction s is identifiable if and only if there exists a measurable function $\xi(t)$, $t \varepsilon T$, such that

$$s'w = \int_0^{t_1} \xi(t) y(t) dt \quad \text{for all } w.$$

Here we substitute the value y(t) with the help of which

$$x(t) = \int_0^t F(t) F^{-1}(\tau) \; Bw \; d\tau \quad .$$

As a result, we obtain

$$s'w = \int_0^{t_1} \xi(t) \int_0^t c'F(t)F^{-1}(\tau)\ B\ d\tau\, w\, dt \quad .$$

The last equality occurs for all w if and only if

$$s' = \int_0^{t_1} \xi(t) \int_0^t c'F(t)F^{-1}(\tau)\ B\ d\tau\ dt \quad . \tag{23}$$

To solve the last problem, we may use different approaches from functional analysis.

1. Application of the Minimax Theorem

We introduce the set

$$Q(L) = \{q\colon\ q = \int_0^{t_1} \xi(t) \int_0^t B'F^{-1}(\tau))'F'(t)c\ d\tau\ dt\ , |\xi(t)| \le L\} .$$

We find the distance from the point s to the set Q(L):

$$\rho(L) = \min_{q\varepsilon Q(L)} ||s - q|| \quad .$$

Since the set Q(L) is closed and convex, it is possible to employ the minimax theorem. We have

$$\rho(L) = \min_{q\varepsilon Q(L)} \max_{||z|| \le 1} z'(s-q) = \max_{||z|| \le 1} \min_{q\varepsilon Q(L)} z'(s-q)$$

$$= \max_{||z|| \le 1} \left[z's - L \int_0^{t_1} \left| \int_0^t z'B'(F^{-1}(\tau))'F'(t)c\ d\tau \right| dt \right] . \tag{24}$$

The condition (23) is clearly equivalent to the following: for each s, there is a number L such that $\rho(L) = 0$. For this it is necessary and sufficient that

$$\left| \int_0^t z'B'(F^{-1}(\tau))'F'(t)c\ d\tau \right| \not\equiv 0 \quad , \quad t\varepsilon T \quad , \tag{25}$$

for all z, $||z|| \neq 0$. Actually, it is necessary because in the opposite case for $s = \beta z$, $\beta > 0$, we have $\rho(L) = \beta > 0$. Moreover, condition (25) is sufficient since setting

$$L = \frac{||s||}{\min\limits_{||z|| \le 1} \int_0^{t_1} \left| \int_0^t z'B'(F^{-1}(\tau))'F'(t)c\, d\tau \right| dt} ,$$

from (24) we obtain $\rho(L) \le ||s|| - ||s|| = 0$. The function

$$\lambda(t) = \int_0^t z'B'(F^{-1}(\tau))'F'(t)c\, d\tau \quad ,$$

is not identically zero if and only if its derivative

$$\dot{\lambda}(t) = z'Bc + \int_0^t z'B'(F^{-1}(\tau))'F'(t)A'c\, d\tau \quad ,$$

is not identically zero. Calculating n derivatives of the last function and using the relation $A^n c = -\gamma_1 A^{n-1} c - \cdots - \gamma_n c$, we obtain a homogeneous nth order differential equation for $\dot{\gamma}(t)$ with the initial conditions

$$\dot{\gamma}(0) = z'B'c \ , \ \ddot{\gamma}(0) = z'B'A'c, \ldots, \gamma^{(n)}(0) = z'B'(A^{n-1})'c \ . \tag{26}$$

Thus, condition (25) turns out to be equivalent to the nontriviality of the initial conditions (26). Since z is an arbitrary vector, from this follows Theorem 1 in which identifiability is understood in the sense of Definition 10.

2. Use of the Theorem on Separation of Convex Sets

The condition (25) may be obtained with the help of the theorem on the separation of convex sets, the scheme of application of which is completely analagous to that described in §3.8 for the study of observability. Details of the proof are left to the reader.

3. The Problem of Moments

The application to the solution of this problem of functional analysis to the considered problem may be completed without difficulty by the reader in analogy with §3.9. We note only that from (23), by the method from §1.3, it is easily obtained that all identifiable directions, and only those, have the form

$$s = \sum_{i=0}^{n-1} \mu_i B'(A^i)'c \quad , \quad -\infty < \mu_i < \infty \quad .$$

Commentary on Chapter IV

The concepts of identifiability of dynamical systems in the current chapter are introduced in analogy with the concepts of observability: here the parameters entering into the equations of motion of the system are reconstructed using measured output signals, while above it was necessary to find the parameters of the initial conditions. As follows from the definitions, in this chapter we considered identification under ideal conditions without noise. The general case of the problem (this is related to observability) is omitted from this book, although many of these problems may be formulated and solved using the theory of optimal processes with parameters (§6.7). In §4, we proved a duality principle between the identifiability of a system and its relative controllability. If we represent the process of relative controllability of the system

$$\dot{x} = Ax + Bu \quad , \quad Kx(t_1) = 0 \quad ,$$

on a block diagram with input block B, object block A, and output block K, then the process of identification of the parameters w in the system

$$\dot{x} = -A'x - K'w$$

relative to the output $z = B'x$, may be obtained from the constructed diagram very simply: 1) reverse time, 2) replace all matrices by their transposes in all blocks. To our thinking, this result is more symmetric than the usual duality principle of Kalman [58a]. Comparison with the duality principle of §4 shows that for each part figuring into the Kalman duality principle, we have a more symmetric dual concept and, thus, from one (not totally symmetric from our view) principle, two symmetries are obtained. Combining different parts of these principles, it is possible to obtain other equivalent (but not symmetric) duality principles.

CHAPTER V

The Problem of Existence of Optimal Controls

§1. The Theorem of Existence of Optimal Controls for Systems with a Delay

1. Statement of the Problem

Let the equation of controlled motion have the form

$$\frac{dx(t)}{dt} = f(x(t),x(t - h(x,u,t)),u(t,t)), \quad t \varepsilon T = [t_o,t_1], \quad (1)$$

and assume that the system trajectory belongs to some set $X \subset R^n$, while the control $u(t)$, $t\varepsilon T$, is a member of some set $U \subset R^r$. We are given the initial conditions

$$x(\tau) = \Phi(\tau) \quad , \quad \tau \varepsilon S_o \quad ,$$

$$S_o = \{\tau: \quad \tau = \tau(t) = t - h(x,u,t) \leq t_o, \; t\varepsilon T \; , \; x\varepsilon X, \; u\varepsilon U\} \; .$$

We let F denote the class of measurable vector functions $u(t)$, defined on T, which assume values in U:

$$u(t) \varepsilon U \quad , \quad t\varepsilon T \quad . \quad (2)$$

We will call elements $u(\cdot)\varepsilon F$ admissible controls of class F. We let D denote the subset of F consisting of piecewise-continuous functions, while D_1 is the set of all piecewise differentiable functions, $D_1 \subset D$.

For some $u(\cdot)\varepsilon F$, let there exist a solution $x(t)$ of Eq. (1) defined on T. We give the functional

$$J(u) = \phi_1(x(t_1)) \quad , \tag{3}$$

on the set of points $x(t_1)$, where $\phi_1(x)$ is a lower semi-continuous finite function, defined on the closure $\bar{X}$ of the set X. The problem of the existence of optimal controls consists in the establishment of conditions under which within the admissible controls of class F there is a control $u^o(t)$, $t \varepsilon T$, such that

$$J(u^o) \leq J(u) \quad , \quad \text{for all } u(\cdot) \varepsilon F$$

The control $u^o(t)$, $t \varepsilon T$, is called an optimal control and the trajectory $x^o(t)$ of (1) corresponding to it is an optimal trajectory.

To begin with, we introduce some auxiliary results.

2. The Existence of Admissible Solutions

Lemma 1. Let: 1) the function $f(x,y,u,t)$ be defined and continuous on $P = X \times X \times U \times T$, together with $\partial f/\partial x$ and $\partial f/\partial y$ and

$$||f(x",y",u,t) - f(x,y,u,t)|| \leq L_1(||x" - x||, ||y" - y||) ,$$

$$||f(x,y,u,t)|| \leq L_2 \quad ;$$

2) the functions $h(x,u,t)$, $\partial h/\partial x$, $\partial h/\partial u$, $\partial h/\partial t$ be defined and continuous on $Q = X \times U \times T$, where $h(x,u,t) \geq 0$ and

$$|h(x",u,t) - h(x,u,t)| \leq L_3||x" - x|| \; ;$$

3) the function $\phi(\tau)$ be continuous on S_o together with $d\phi/d\tau$ and satisfy the inequality

$$||\Phi(\tau'') - \Phi(\tau)|| \leq L_4|\tau'' - \tau| \quad ;$$

4) $t^* = t_o + \delta/L$, $\delta = \min_{x\varepsilon\partial\overline{X}} ||x - \Phi(t_o)||$,

where $\partial\overline{X}$ is the boundary of the set $\overline{X}$, $L = \max\{L_1,\ldots,L_4\}$. Then for $u(\cdot)\varepsilon D$, on the interval $[t_o,t^*]$ there exists a unique solution $x(t)$ of Eq. (1) having the given initial conditions.

This assertion is proved in [36b] by the method of successive approximations which are constructed according to the scheme

$$x_o(t) = \begin{cases} \Phi_1(t) \quad , & t_o \leq t \leq t^* \quad , \quad \Phi_1(t_1) = \Phi(t_o) \quad , \\ \Phi(t) \quad , & t\varepsilon S_o \quad , \end{cases}$$

$$x_{p+1}(t) = \Phi(t_1) + \int_{t_o}^{t} f(x_p(t),y_p(t),u(t),t)dt, \ t \geq t_o \ ,$$

$$x_p(t) = \Phi(t) \ , \ t\varepsilon S_o \ , \ y_p(t) = x_p(t-h(x_p(t),u(t),t).$$

Here $\Phi_1(t)\varepsilon X$ is an arbitrary continuous function. If $x(t^*) \notin \partial\overline{X}$, then the solution may be extended.

3. The Boundedness of Solutions

Lemma 2. Let: 1) the function $f(x,y,u,t)$ be defined on P and

$$x'f(x,y,u,t) \leq c(||x||^2 + ||y||^2 + 1) \quad ;$$

2) $h(x,u,t) > 0$, with the function $h(x,y,t)$ differentiable in all arguments; 3) the function $\Phi(t)$ be defined and bounded on S_o: 4) the function $u(t)$ be defined and differentiable on T. Then the solution $x(t)$ of Eq. (1), $t\varepsilon T$, with the given initial conditions, is bounded on T.

Additional properties of the solutions of (1) (integral continuity relative to $u(t)$, $\Phi(t)$), are important for actual calculation of optimal controls and are proved in [36b].

4. Theorem on the Existence of Optimal Controls in Systems with a Delay

Let $T_t = [t_o, t]$, $S_t = S_o \cup T_t$. Let $X_t(\cdot)$ denote the set of curves $x(\theta)$ defined on S_t satisfying the conditions

$$||x(\theta'') - x(\theta)|| \leq L|\theta'' - \theta|, \quad \theta, \theta'' \varepsilon S_t \quad ,$$

$$x(\tau) = \Phi(\tau) \quad , \quad \tau \varepsilon S_o \quad .$$

On $X_t(\cdot)$ we define the point set

$$R(x(\cdot),t) = \{z: z = f(x(t), x(t - h(x(t),u,t)),u,t), \; u \varepsilon U\} \quad .$$

Theorem 1. Let: 1) the conditions of Lemma 1 be satisfied; 2) the solution $x(t)$ of Eq. (1) be defined and uniformly (with respect to $u(\cdot)\varepsilon F$) bounded; 3) for each $x(\cdot)\varepsilon X_t(\cdot)$, $t\varepsilon T$, the set $R(x(\cdot),t)$ be convex and closed. Then there exists an optimal control for the problem (1)-(3).

Proof. Let $u(t)$, $t\varepsilon T$, be some admissible control generating through (1) a trajectory $x(t)$, $t\varepsilon T$, on which the functional (3) assumes a finite value a. The set of solutions satisfying the condition $\Phi_1(x(t_1)) \leq a$ is nonempty. If it is finite, then the theorem is proved. Let this set be infinite. Let Q' denote a sequence of $\{x(t), t\varepsilon T\}$ for which the sequence $\{\Phi_1(x(t_1))\}$ is minimized. By virtue of the lower semicontinuity of the function $\Phi_1(x)$ and the boundedness of the set $\{\Phi_1(x(t_1))\}$, there exists a number $\bar{a}$ such that $\bar{a} = \inf \{\Phi_1(x(t_1))\} = \Phi_1(x(t_1))$. Now we show that from Q' it is possible to extract a subsequence converging uniformly to an optimal trajectory.

By condition 2), the set Q' consists of uniformly bounded functions. Moreover, the elements of Q' are uniformly continuous since, for all $x(\cdot)\varepsilon Q'$, by virtue of condition 1) the derivative dx/dt is bounded: $||dx/dt|| \leq L$. Therefore, from Q' it is possible to choose a subsequence $\{x_p(t)\}$ converging uniformly to a function $\bar{x}(t)$. It remains to show that $\bar{x}(t)$, $t\varepsilon T$, is generated through Eq. (1) by some admissible control u(t), $t\varepsilon T$. First, we show that for almost all $t^*\varepsilon T$ there exists a derivative $dx(t^*)/dt$ and there is a vector $u^* = u(t^*)$ such that

$$\frac{d\bar{x}(t^*)}{dt} = f(\bar{x}(t^*),\ \bar{x}(t^* - h(\bar{x}(t^*),u^*,t^*)),u^*,t^*) \ . \tag{4}$$

The function $x_p(t)$, $t\varepsilon T$, is absolutely continuous and almost everywhere we have $||dx_p(t)/dt|| \leq L$. Therefore, the limit function $\bar{x}(t)$ is absolutely continuous and the almost everywhere defined function g(t) = $d\bar{x}(t)/dt$ satisfies the inequality $||q(t)|| \leq L$. Let the vector q(t) be defined at the point $t^*\varepsilon T$. Then for an arbitrarily small $\varepsilon > 0$, there is a $\delta > 0$ such that

$$\frac{\bar{x}(t) - \bar{x}(t^*)}{t - t^*} \varepsilon [T(\bar{x}(\cdot),t^*]_\varepsilon \ , \ \bar{x}(\cdot)\varepsilon X_{t^*}(\cdot) \quad , \tag{5}$$

for $|t - t^*| \leq \delta$ (here $[\ ]_\varepsilon$ is a closed ε-neighborhood of the set [ ]). Actually, by construction of the function $\bar{x}(t)$

$$\frac{\bar{x}(t) - \bar{x}(t^*)}{t - t^*} = \lim_{p\to\infty} \frac{x_p(t) - x_p(t^*)}{t - t^*} \ .$$

Further,

$$x_p(t) = x_p(t^*) + \int_{t^*}^{t} q_p(t)dt \ , \quad q_p(t) = \frac{dx_p(t)}{dt} \quad ,$$

and, therefore,

$$\frac{\bar{x}(t) - \bar{x}(t^*)}{t - t^*} = \frac{1}{t - t^*} \lim_{p\to\infty} \int_{t^*}^{t} q_p(\tau)d\tau \quad ,$$

$$= \lim_{p\to\infty} \int_{0}^{t} q_p(t^* + \sigma(t - t^*))d\tau \quad . \tag{6}$$

The function $q_p(t)$ satisfies the condition : $q_p(t) \varepsilon R(x_p(\cdot),t)$, $x_p(\cdot)\varepsilon X_t(\cdot)$ and for sufficiently large p and sufficiently small δ, we have

$$q_p(t^* + \sigma(t - t^*))\varepsilon[R(\bar{x}(\cdot),t^*)]_\varepsilon \quad ,$$

$$\bar{x}(\cdot)\varepsilon X_{t^*}(\cdot) \quad , \ 0 \le \sigma \le 1 \ , \ |t - t^*| \le \delta \quad . \tag{7}$$

The last property comes about from the definition of the function $\bar{x}(t)$ and the upper semicontinuity relative to the set $R(x(\cdot),t)$, $x(\cdot)\varepsilon X_t(\cdot)$ (cf. Lemma 3). According to condition 3), the set $R(x(\cdot),t)$, $x(\cdot)\varepsilon X_t(\cdot)$ is convex; thus, from (7) it follows that

$$\int_0^1 q_p(t^* + \sigma(t - t^*))d\sigma \ \varepsilon[R(x(\cdot),t^*)]_\varepsilon \ ,$$

$$\bar{x}(\cdot)\varepsilon X_{t^*}(\cdot) \quad .$$

But, by virtue of (6), this is equivalent to (5).

We note that

$$\left\|\frac{d\bar{x}(t^*)}{dt} - \frac{\bar{x}(t) - x(t^*)}{t - t^*}\right\| \le \varepsilon, \quad \text{if } |t - t^*| \le \delta \quad ,$$

for sufficiently small δ. Therefore, from (5) we have

$$\frac{d\bar{x}(t^*)}{dt} \varepsilon [R(\bar{x}(\cdot),t^*)]_{2\varepsilon} \quad , \quad \bar{x}(\cdot)\varepsilon X_{t^*}(\cdot) \quad .$$

Since ε is arbitrary and $R(\bar{x}(\cdot),t^*)$ is closed (condition 3), we have

$$\frac{d\bar{x}(t^*)}{dt} \varepsilon R(\bar{x}(\cdot),t^*) \quad , \quad \bar{x}(\cdot)\varepsilon X_{t^*}(\cdot) \quad ,$$

which is equivelent to (4). To complete the proof of the theorem, it suffices to use Lemma 4 (see below) by virtue of which the values $u(t^*)$ may be chosen such that the function $u(t^*)$, $t^*\varepsilon T$, is measurable.

In conclusion, we prove two assertions used in the proof of the theorem. The set $R(x(\cdot),t)$ is called <u>upper semicontinuous relative to inclusion</u>, if for any $\varepsilon > 0$ there exists $\delta > 0$ such that for

$$||t'' - t|| < \delta \ ,$$

$$||x''(\cdot) - x(\cdot)|| = \max_{\theta\varepsilon[t_o,t'']\cap[t_o,t]} ||x''(\cdot) - x(\cdot)|| < \delta,$$

$$x(\cdot) \ \varepsilon \ X_t(\cdot) \quad , \quad x''(\cdot) \ \varepsilon \ X_{t''}(\cdot)$$

we have the inclusion

$$R(x''(\cdot),t'') \subseteq [R(x(\cdot),t]_\varepsilon \quad .$$

<u>Lemma 3. Let $f(x,y,u,t)$ be defined and continuous on the set $X \times X \times U \times T$, $U(x,u,t)$ defined and continuous on $X \times U \times T$, where</u>

$$|h(x'',y,t'') - h(x,y,t)| \leq L(||x'' - x|| + |t'' - t|)$$

and $h(x,u,t) > 0$ at each point $\{x,y,t\} \varepsilon X \times U \times T$. Then the set $R(x(\cdot),t)$ is upper semicontinous relative to inclusion.

Proof. Assume the contrary: for given t and $x(\cdot)$, it is possible to find $\varepsilon > 0$ such that for $\delta > 0$ sufficiently small the set $R(x''(\cdot),t'') \bar{\subset} [R(x(\cdot),t)]_{\varepsilon}$ for $||x''(\cdot) - x(\cdot)|| < \delta$ and $|t'' - t| < \delta$. This means that there exists a $\bar{u} \varepsilon U$ such that

$$||f(x''(t''),x''(t'' - h(x''(t''),u,t'')),u,t'')$$

$$- f(x(t),x(t - h(x(t),\bar{u},t)),\bar{u},t)|| > \varepsilon$$

for $||x''(\cdot) - x(\cdot)|| < \delta$, $|t'' - t| < \delta$. For definiteness, we set $t'' > t$. We have

$$||x''(t'') - x(t)|| \leq ||x''(t'') - x''(t)||$$

$$+ ||x''(t) - x(t)|| < L(t'' - t) + \delta < (L + 1)\delta \quad .$$

Analogously,

$$||x''(t'' - h(x''(t''),\bar{u},t'')) - x(t - h(x(t),\bar{u},t))||$$

$$= ||x''(\tau'') - x(\tau)|| \leq ||x''(\tau'') - x''(\tau)||$$

$$+ ||x''(\tau) - x(\tau)|| \quad .$$

Since

$$||x''(\tau'') - x''(\tau)|| \leq L|\tau'' - \tau|$$

$$\leq L[t'' - t + |h(x''(t''),\bar{u},t) - h(x(t),\bar{u},t)|]$$

$$< L(\delta + L[||x''(t'') - x(t)|| + t'' - t])$$

$$< L[1 + L(L + 2)]\delta ,$$

then

$$||x\ (\tau'') - x(\tau)|| < [L + L^2(L + 2) + 1]\delta \ .$$

Thus, for $u = \bar{u}$ and $\varepsilon > 0$, the inequality

$$||f(x'',u'',\bar{u},t'') - f(x,y,\bar{u},t),|| > \varepsilon$$

occurs if

$$||x'' - x|| < (L + 1)\delta \ , \quad t'' - t < \delta \ ,$$

$$||y'' - y|| < [L + L^2(L + 2) + 1]\delta \ ,$$

where $\delta > 0$ is sufficiently small. But this contradicts the continuity of $f(x,y,u,t)$ on the set $X \times X \times U \times T$. Lemma 3 is proved.

Lemma 4 (A.F. Fillipov [125]). Let: 1) $f(t,u)$ be a continuous n-vector function of the argument $t \varepsilon T$, $u = \{u_1,...,u_r\}$; 2) $Q(t)$ be a closed, bounded set, upper semicontinuous relative to inclusion in t; 3) $f(t,q(t)) = R(t)$; 4) $y(t) \varepsilon R(t)$ be a measurable n-vector function. Then there exists a measurable r-vector function $u(t) = \{u_1,...,u_r\} \varepsilon Q(t)$, such that $f(t,u(t)) = u(t)$ almost everywhere on T.

Proof. First let $r = 1$. The function $u(t)$, $t \varepsilon T$, is defined in the following way. For each $t \varepsilon T$, we set $u(t) = u$, where u is the smallest number satisfying the equality $f(t,u) = y(t)$. Clearly, such a number exists since the set $\{u: f(t,u) = y(t)\}$ is closed by virtue of the continuity of $f(t,u)$. By definition of a measurable function, for any $\varepsilon > 0$ there is a set $\sigma \subseteq T$, meas $\sigma > t_1 - \varepsilon - t_0$, on which the function $y(t)$ is continuous. We show that for any a, the set $\{t: u(t) \leq a, t \varepsilon \sigma\}$ is closed In the opposite case there exists a sequence $t_k \varepsilon \sigma$ such that

$$t_k \to \tilde{t} \quad , \quad u(t_k) \leq u(\tilde{t}) - \varepsilon_1 \quad , \quad \varepsilon_1 > 0 \quad . \tag{8}$$

Since $|u| \leq$ constant, from $\{t_k\}$ it is possible to extract a subsequence on which $u(t_k) \to u$. From the fact that $u(t_k) \varepsilon Q(t)$ and $Q(t)$ is a closed upper semicontinuous set relative to inclusion, it follows that $\tilde{u} \varepsilon Q(t)$. From (8) we have

$$\tilde{u} \leq u(\tilde{t}) - \varepsilon_1 \quad , \quad \varepsilon_1 > 0 \quad . \tag{9}$$

In the identity $f(t_k,u(t_k)) = y(t_k)$, we pass to the limit with the chosen sequence $\{t_k\}$. Using the continuity of $f(t,u)$, we obtain

$$f(\tilde{t},\tilde{u}) = y(\tilde{t}) \quad . \tag{10}$$

From (9) and (10) it follows that $u(\tilde{t})$ is not the smallest number satisfying the equation

$$f(\tilde{t},u) = y(\tilde{t}) \quad .$$

This contradicts the definition of $u(t)$. Thus, the set $\{t: u(t) \leq a, t\varepsilon\sigma\}$ is closed for all a, which is equivalent to the measurability of $u(t)$ on σ and, by virtue of the arbitrariness of ε, this function is measurable on all of T.

In the general case $(r > 1)$, the function $u(t) = \{u_1(t),\ldots,u_r(t)\}$ is constructed in the following way. From the solution to the equation $f(t,u) = u(t)$, we choose that whose first coordinate u_1 is smallest; if there are several such solutions, then from them we choose the one with the smallest second coordinate u_2, and so on. For each $t\varepsilon T$, we assume $u(t) = u$. We show the measurability of the function so obtained by induction. We have already proved the measurability of $u(t)$ for $r = 1$; we assume that $u(t)$ is measurable for $r = s-1$. We show its measurability for $r = s$. For any $\varepsilon > 0$, we can find $\sigma\varepsilon T$, meas $\sigma > t_1 - t_o - \varepsilon$, such that the functions $y(t)$, $u_1(t),\ldots,u_{s-1}(t)$ are continuous in σ. The proof of the closure of the set $\{t: u_s(t) \leq a\}$ for any a follows the proof given above for $r = 1$. Thus, $u_s(t)$ is a measurable function and the lemma is proved.

§2. The Existence of Optimal Controls in the Theory of Optimal Chattering Regimes

Following the theory of optimal chattering regimes [38e,221], we replace the problem (1)-(3), $h = h(x,t)$ by the problem of minimizing the functional

$$J(v,w) = \phi_1(z(t_1)) \tag{11}$$

along trajectories of the system

$$\frac{dz}{dt} = \sum_{\gamma=1}^{n+1} w_\gamma(t) f(z(t,z(t-h(z(t),t)), \; v_\gamma(t,t)) \;, \tag{12}$$

using measurable functions $w_\gamma(t)$, $v_\gamma(t)$, $t\varepsilon T$, satisfying the conditions

$$w_\gamma(t) \geq 0 \ , \ \sum_{\gamma=1}^{n+1} w_\gamma(t) = 1 \ , \ v_\gamma(t)\varepsilon U \ , \ \gamma = 1,\ldots,n+1 \ . \tag{13}$$

By virtue of Theorem 1, the problem (11)-(13) has a solution: the optimal controls $w^o(t)$, $v^o(t)$, $t\varepsilon T$, as will be shown in Chapter VI, satisfy the maximum condition

$$\sum_{\gamma=1}^{n+1} \xi'(t)w_\gamma^o(t)f(z^o(t),z^o(t - h(z^o(t),t)),v_\gamma^o(t),t)$$

$$= \max_{w_\gamma \geq 0,\sum w_\gamma=1} \ \max_{v_\gamma \varepsilon U} \sum_{\gamma=1}^{n+1} \xi'(t)w_\gamma f(z^o(t),z^o(t-h(z^o(t),t),v_\gamma,t) \tag{14}$$

where $\xi(t)$ is the solution of the system

$$\frac{d\xi(t)}{dt} = \frac{\delta\pi(z^o,\xi,v^o,w^o)}{\delta z(t)} \quad , \ \xi(t_1) = -\frac{\partial\phi_1(z^o(t_1))}{\partial z} \ . \tag{15}$$

Here $\frac{\delta\pi(z,\xi,v,w)}{\delta z(t)}$ is the variational derivative at the point t of the functional

$$\pi(z,\xi,v,w)$$

$$= \int_{t_o}^{t_1} \sum_{\gamma=1}^{n+1} \xi'(t)w_\gamma(t)f(z(t),z(t-h(z(t),t)),v_\gamma(t),t)dt \ .$$

relative to $z(t)$.

From (14) it follows that each vector function $v_\gamma^o(t)$, $t\varepsilon T$, satisfies the condition

$$\xi'(t)f(z \ (t),z^o(t - h(z^o(t),t)),v_\gamma^o(t),t)$$

$$= \max_{u\varepsilon U} \xi'(t)f(z^o(t),z^o(t - h(z^o(t),t)),u,t) \quad , \tag{16}$$

and, therefore, an optimal chattering regime (the solution of problem (11)-(13) with $0 < w^o(t) < 1$) does not arise, while the solution of the problem (1)-(3) will exist if the maximum value in u of the expression

$$\xi'(t)f(z^o(t) \; , \; z^o(t - h(z^o(t),t)),u,t) \; , \qquad (17)$$

is assumed at a unique point for almost all $t \in T$. We use this fact to establish criteria for the existence of optimal controls in the problem of minimizing the functional

$$J(u) = \phi_1(x(t_1)) + \int_{t_o}^{t_1} f_{n+1}(x(t),x(t-h(x(t),t)),u(t),t)dt \; , \qquad (18)$$

along trajectories of the system (1) with $h = h(x,t)$. The problem (1),(2),(18) reduces to the problem (1)-(3) if we introduce the variable

$$x_{n+1}(t) = \int_{t_o}^{t} f_{n+1}(x(\tau),x(\tau - h(x(\tau),\tau)),u(\tau),\tau)d\tau \; .$$

Because of the special structure of the obtained system of equations, the adjoint system (15) has a solution in which $\xi_{n+1}(t) \equiv -1$ and the function (17) assumes the form

$$\xi'(t)f(z^o(t),z^o(t - h(z^o(t),t)),u,t) - f_{n+1}(z^o(t),z^o(t - h(z^o(t),t)),u,t) \; .$$

Hence, the theorem:

Theorem 2. In addition to the conditions of Lemma 1, let: 1) $f(x,y,u,t) = f^1(x,y,t) + Bu$; 2) the function $f_{n+1}(x,y,u,t)$ be strictly convex in u for all $x,y \in X$, $t \in T$.

Then there exists an optimal control in the problem (1), (2),(18).

Extension of Theorem 2 to the general case of problem (1),(2),(18) is accomplished in [36c].

§3. Proof of the Existence of Optimal Controls by the Method of Increments

1. Systems Linear in the State

At first, we illustrate the idea of this paragraph by the following problem.

It is required to minimize the functional

$$J(u) = c'x(t_1) \quad , \tag{19}$$

along trajectories of the system

$$\frac{dx}{dt} = A(t)x + b(u,t) \;, \; u\varepsilon\overline{U}, \quad x(t_o) = x_o \quad , \tag{20}$$

where A(t), b(u,t) are continuous functions. We introduce the auxiliary system

$$\begin{aligned} &\frac{dz}{dt} = A(t)z + \sum_{\gamma=1}^{n+1} w_\gamma(t) b(v_\gamma(t),t), \; z(t_o) = z_o \;, \\ &w_\gamma(t) \geq 0 \quad , \quad \sum_{\gamma=1}^{n+1} w_\gamma(t) = 1 \quad , \quad v_\gamma(t)\varepsilon U \quad , \end{aligned} \tag{21}$$

along which we find the minimum of the functional

$$J(v,w) = c'z(t_1) \quad . \tag{22}$$

By virtue of Theorem 1, optimal controls $v_\gamma^o(t)$, $w_\gamma^o(t)$ exist for the problem (21),(22). The system (20) is a particular case of the system (21) and, therefore,

$$J(v^o,w^o) \leq \inf J(u) \quad . \tag{23}$$

We consider the control $u^*(t) = v_1^o(t)$ for the system (20) and the control $v_\gamma^*(t) = v_\gamma^o(t)$, $w^*(t) = \{w_1 \equiv 1, w_2 \equiv 0, \ldots, w_{n+1} \equiv 0\}$ for the system (21). Under these controls, the trajectories of the systems (20),(21) coincide and

$$J(u^*) = J(v^*,w^*) \quad . \tag{24}$$

But the increment $\Delta J(v^o,w^o) = J(v^*,w^*) - J(v^o,w^o)$ equals

$$\Delta J(v^o,w^o)$$

$$= - \int_{t_o}^{t_1} \xi'(t)b(v_1^o(t),t) - \sum_{\gamma=1}^{n+1} \xi'(t)w_\gamma^o(t)b(v_\gamma^o(t),t)\ dt$$

Since, by virtue of property (16), the integrand is identically zero, $J(v^*,w^*) = J(v^o,w^o)$. Comparing this equality with (23),(24), we conclude that the following theorem is true.

Theorem 3. Optimal controls exist for the problem (19),(20).

2. Systems with a Delay

We consider the system

$$\frac{dx(t)}{dt} = A(t)x(t) + A_1(t)x(t-h) + b(u(t),t), \; u(t)\varepsilon\bar{U} \quad ,$$

$$x(\tau) = \Phi(\tau) \quad , \quad \tau\varepsilon S_o \quad , \tag{25}$$

and the functional

$$J(u) = \phi_1(x(t_1)) + \int_{t_0}^{t_1} f_{n+1}(x(t), x(t-h), u(t), t)\,dt , \quad (26)$$

where $\phi_1(x)$, $f_{n+1}(x,y,u,t)$ are continuously differentiable and concave in $\{x,y\}$ for each $t \in T$; $A(t)$, $A_1(t)$, $b(u,t)$ are continuous matrix functions. We introduce the variable x_{n+1} by the equation

$$\frac{dx_{n+1}(t)}{dt} = f_{n+1}(x(t), x(t-h), u(t), t) \quad , \quad x_{n+1}(t_0) = 0 \ . \quad (27)$$

Then the problem (25),(26) leads to the problem of minimization of the function

$$\phi_1(x(t_1)) + x_{n+1}(t_1) \quad ,$$

along trajectories of the equations (25),(27). Repeating the basic arguments preceding Theorem 3 (the necessary formula for the increment is obtained in [36c]), it is not difficult to be convinced of the following assertion.

Theorem 4. The problem (25),(26) has a solution if

$$f_{n+1}(x,y,u,t) = f^1_{n+1}(x,y) + f^2_{n+1}(u,t) \quad .$$

§4. The Existence of Optimal Controls in Continuous Systems and the Maximum Principle for Discrete Systems

Up to now (§2,3), the proof of existence theorems for optimal controls have drawn on the theory of chattering regimes, supplemented (§3) by the increment of the criterion function. Below, we present a new method for establishing existence theorems for optimal controls. It is based on a difference approximation to the differential equations.

Let it be required to minimize the functional

$$J(u) = c'x(t_1) \quad ,$$

over all measurable functions $u(t)$, $t\varepsilon T$, $u(t)\varepsilon U = \{u: |u| \leq 1\}$, where the equations of motion are

$$\left.\begin{aligned} &\frac{dx}{dt} = f(x) + bu \quad , \quad x(t_0) = x_o \quad , \\ &f(x)\varepsilon C^{n-2} \quad , \qquad b = \text{const.} \end{aligned}\right\} \tag{28}$$

Instead of (28), we consider the discrete system

$$\left.\begin{aligned} &\frac{y(\tau_{n+1}^N) - y(\tau_h^N)}{\Delta t} = f(y(\tau_h^N)) + bv(\tau_h^N) \quad , \\ &h = 0,\ldots,N \quad , \quad y(\tau_0^N) = x_o \quad , \\ &\tau_h^N = t_0 + h\Delta t \quad , \quad \Delta t = \frac{t_1 - t_0}{N} \quad , \end{aligned}\right\} \tag{29}$$

along which we will minimize the function

$$J^N(v) = c'y(\tau_N^N) \quad , \tag{30}$$

with respect to the finite number of variables $v(\tau_h^N)$, $v(\tau_h^N)\varepsilon U$. Clearly, the problem (29),(30) always has a solution $v^o(\tau_h^N)$. We let $V^N(t)$ denote the piecewise continuous function constructed from $v^o(\tau_h^N)$ as follows:

$$V^N(t) = v^o(\tau_h^N) \quad \text{for} \quad \tau_h^N \leq t < \tau_{h+1}^N \quad . \tag{31}$$

Let $Y^N(t)$ be the polygon constructed along the trajectory $y^o(\tau_h^N)$. By virtue of (29), it corresponds to the sequence $v^o(\tau_h^N)$:

$$Y^N(t) = y^o(\tau_h^N) + (t - \tau_h^N)\frac{y^o(\tau_{h+1}^N) - y^o(\tau_h^N)}{\Delta t} ,$$

$$\tau_h^N \leq t < \tau_{h+1}^N ;$$

We let $K^N(t)$ denote the polygon

$$K^N(t) = \xi(\tau_h^N) + (t - \tau_h^N)\frac{\xi(\tau_{h+1}^N) - \xi(\tau_h^N)}{\Delta t} , \quad (32)$$

$$\tau_h^N \leq t < \tau_{h+1}^N ,$$

constructed along the solutions of the system

$$\frac{\xi(\tau_{h-1}^N) - \xi(\tau_h^N)}{\Delta t} = \frac{\partial f'(y^o(\tau_h^N))}{\partial y} \xi(\tau_h^N) , \quad \xi(\tau_h^N) = -c .$$

The family of functions $Y^N(t)$, $K^N(t)$, $N \geq 1$, are clearly compact and, thus, contain uniformly convergent subsequences which, in order not to complicate the notation, we again denote by $Y^N(t)$, $K^N(t)$. The optimal control $v^o(\tau_h^N)$ satisfies [110,160b] the maximum principle

$$\xi'(\tau_h^N) bv^o(\tau_h^N) = \max_{v \varepsilon U} \xi'(\tau_h^N) bv . \quad (33)$$

From (32) we have

$$b'K^N(\tau_h^N)V^N(\tau_h^N) \geq b'K^N(\tau_h^N)V^{N+1}(\tau_{h_1}^{N+1}) ,$$

$$b'K^{N+1}(\tau_{h_1}^{N+1})V^{N+1}(\tau_{h_1}^{N+1}) \geq b'K^{N+1}(\tau_{h_1}^{N+1})V^N(\tau_h^N) .$$

Hence

$$b'K^{N+1}(\tau_{h_1}^{N+1})V^N(\tau_h^N) - b'K^N(\tau_h^N)V^N(\tau_h^N)$$

$$\leq b'K^{N+1}(\tau_{h_1}^{N+1})V^{N+1}(\tau_{h_1}^{N+1}) - b'K^N(\tau_h^N)V^N(\tau_h^N)$$

$$\leq b'K^{N+1}(\tau_{h_1}^{N+1})V^{N+1}(\tau_{h_1}^{N+1}) - b'K^N(\tau_h^N)V^{N+1}(\tau_{h_1}^{N+1}) \ . \quad (34)$$

Let $t \varepsilon T$ be an arbitrary point, h, h_1 such that

$$t_o + h\,\frac{t_1 - t_o}{N} \leq t \leq t_o + (h+1)\,\frac{t_1 - t_o}{N} \ ,$$

$$t_o + h_1\,\frac{t_1 - t_o}{N+1} \leq t \leq t_o + (h_1+1)\,\frac{t_1 - t_o}{N+1} \ .$$

From (32) it follows that

$$K^N(t) = \xi(\tau_h^N) + o(\tfrac{1}{N}) \ , \ K^{N+1}(t) = \xi(\tau_{h_1}^{N+1}) + o\left(\frac{1}{N+1}\right) .$$

From (34) we obtain: $H^N(t) = b'K^N(t)V^N(t) \to H(t)$ for $N \to \infty$. Since

$$K(t) = \lim_{N\to\infty} K^N(t) \ ,$$

then

$$V^N(t) \to V(t) \quad \text{for} \quad N \to \infty \ ,$$

at all points t where $b'K(t) \neq 0$. We note that the function $K(t)$ is the solution of the equation

$$\frac{dK(t)}{dt} = -\frac{\partial f'(Y(t))}{\partial y}\,K(t) \ , \quad K(t_1) = -c \ , \quad (36)$$

$$(Y(t) = \lim_{N\to\infty} Y^N(t)) \ .$$

Let the meas $\{t: b'K(t) = 0\} = 0$. Then the convergence of (35) occurs almost everywhere on T and, by Lebesque's theorem, the function $V(t)$, $t\varepsilon T$, is measurable. From (29), it is not difficult to obtain that

$$\frac{dY}{dt} = f(Y) + bV \quad .$$

The control $V(t)$, $t\varepsilon T$, is admissible since $V^N(t)\varepsilon U$ and U is a closed set. Finally, in order to prove that $V(t)$, $t\varepsilon T$, is a solution of the problem (19),(28) it suffices to note that

$$\lim_{N\to\infty} J^N(v^o) \leq \inf J(u)$$

and

$$c'y^o(\tau_N^N) \to c'Y(t_1) \quad .$$

We formulate the result of this paragraph in the form of a theorem.

Theorem 5. Let $V^N(t)$, $N > 1$, $t\varepsilon T$, be a sequence of controls constructed along the solutions of the problem (29), (30), such that $Y^N(t) \to Y(t)$ for $N \to \infty$ and the conditions of Theorem 1.9 are satisfied along $Y(t)$ for almost all $t\varepsilon T$.

Then $V^N(t) \to V(t)$ almost everywhere, where $b'K(t) \not\equiv 0$, $K(t)$ a solution of the system (36), and $V(t)$, $t\varepsilon T$, is the optimal control for the problem (19),(28).

§5. Proof of the Existence Theorem for Optimal Controls Using a Difference Approximation to the Continuous System

Let us assume that we wish to minimize the functional

$$J(u) = \phi_1(x(t_1)) \tag{37}$$

along trajectories of the system

$$\frac{dx}{dt} = f(x,u) \quad , \quad x(t_0) = x_0 \quad , \quad t\varepsilon T \quad . \tag{38}$$

We assume that the admissible controls $u(t)$, $t\varepsilon T$, are measurable functions having values in a closed, bounded set U. We make the following assumptions about the functions $\phi_1(x)$, $f(x,u)$: 1) $\phi_1(x)$ is a finite lower semicontinuous function, 2) the function $f(x,u)$ is continuous in both its arguments and satisfies a Lipschitz condition in x. If, in addition, the trajectories $x(t)$, $t\varepsilon T$, of Eq. (38), corresponding to all admissible controls, are uniformly bounded, while the set

$$R(x) = \{z: \quad z = f(x,y), \; u\varepsilon U\} \quad ,$$

is convex for each x, then the problem (3),(2),(38) has a solution in the class of measurable controls.

To prove this assertion, we replace the system (38) by the equations

$$\frac{y(\tau_{h+1}) - y(\tau_h)}{\Delta t} = f(y(\tau_h), v(\tau_h)) \quad ,$$
$$y(\tau_0) = x_0 \quad , \quad h = 0, \ldots, N \quad , \tag{39}$$

where $\tau_h = t_0 + h\Delta t$, $\Delta t = (t_1 - t_0)/N$. From the closure of the set U, the continuity of $f(x,y)$, and the lower semicontinuity of $\phi_1(x)$, it follows that the problem of minimization of the function $\phi_1(y(\tau_N))$ over the finite number of variables $v(\tau_N)\varepsilon U$, always has a solution $v^o(\tau_h)$ to which, by virtue of (39), corresponds a sequence $y^o(\tau_h)$.

We let $Y_N(t)$, $t \varepsilon T$, denote the polygon formed by the points τ_h, $y^o(\tau_h)$. The sequence of polygons $\{Y^N(t)\}$, $N \geq 1$, forms a family of uniformly bounded functions (by virtue of the analagous property of the system (38) and the convergence of the solution of the system (39) to the solution of (38) as $N \to \infty$) satisfying the Lipschitz condition

$$||Y^N(t + \Delta t) - Y^N(t)|| \leq L|\Delta t| \ , \ L = \text{const.} \quad (40)$$

Therefore, it contains a subsequence of polygons, uniformly convergent to a function $\overline{Y}(t)$, $t \varepsilon T$, which also satisfies condition (40). From the last property, it follows that the derivative $dY(t)/dt$ is defined almost everywhere on T and is a measurable function. Let the point $t^* \varepsilon T$ be such that the derivative $d\overline{Y}(t^*)/dt$ exists. We introduce the numbers $\overline{h}$, θ, $0 \leq \theta \leq 1$: $t^* = \tau_h + \theta \Delta t$. Clearly, we have the equality

$$\frac{Y^N(t^* + \Delta t) - Y^N(t^*)}{\Delta t} = \frac{Y^N(t^* + \Delta t) - Y^N(\tau_{\overline{h}+1})}{t^* + \Delta t - \tau_{\overline{h}+1}}$$

$$+ (1 - \theta)\frac{Y^N(\tau_{\overline{h}+1}) - Y^N(t^*)}{\tau_{\overline{h}+1} - t^*} = \theta f(Y^N(\tau_{\overline{h}+1}), v^o(\tau_{\overline{h}+1}))$$

$$+ (1 - \theta) f(Y^N(\tau_{\overline{h}}), v^o(\tau_{\overline{h}})) \quad . \quad (41)$$

The set $R(Y(t^*))$ is upper semicontinuous relative to inclusion, i.e. for any $\varepsilon > 0$, it is possible to find a $\delta > 0$ such that from

$$||\overline{Y}(t^*) - Y|| \leq \delta \quad (42)$$

follows

$$R(Y) \subset [R(\overline{Y}(t^*))]_\varepsilon \quad . \quad (43)$$

By virtue of the definition of the function $\overline{Y}(t)$, there exists a number $N_o \geq 2L(t_1 - t_o)/\delta$ such that for $N \geq N_o$ we have

$$||\overline{Y}(t) - Y^N(t)|| \leq \frac{\delta}{2} .$$

The last inequality, together with (40), allows us to write

$$||\overline{Y}(t^* - Y^N(\tau_{\overline{h}})|| \leq \delta , \quad ||Y(t^*) - Y^N(\tau_{\overline{h}+1})|| \leq \delta .$$

Hence, by virtue of (42),(43), we have

$$f(Y^N(\tau_{\overline{h}}) , v^o(\tau_{\overline{h}}))\varepsilon R(Y^N(\tau_{\overline{h}}))\subset[R(\overline{Y}(t^*))]_\varepsilon , \tag{44}$$

$$f(Y^N(\tau_{\overline{h}+1}) , v^o(\tau_{\overline{h}+1}))\varepsilon R(Y^N(\tau_{\overline{h}+1}))\subset[R(\overline{Y}(t^*))]_\varepsilon . \tag{45}$$

The set $[R(Y)]_\varepsilon$ is convex together with $R(Y)$. Thus, from (42),(44),(45) for sufficiently large N we have

$$\frac{Y^N(t^* + \Delta t) - Y^N(t^*)}{\Delta t} \varepsilon [R(\overline{Y}(t^*))]_\varepsilon .$$

Let $N \to \infty$. Then $\Delta t \to 0$, $\varepsilon \to 0$ and, by virtue of the closure of the set $R(\overline{Y}(t^*))$ and the definition of the number t^*, we obtain $d\overline{Y}(t^*)/dt \; \varepsilon \; R(\overline{Y}(t^*))$. From this condition and Lemma 4, there follows the existence of a measurable function $\overline{u}(t)\varepsilon U$, $t\varepsilon T$, such that

$$\frac{d\overline{Y}(t)}{dt} = f(\overline{Y}(t),\overline{u}(t)) .$$

It is not difficult to show that

$$\lim_{N\to\infty} \min \phi_1(y(\tau_N)) \leq \inf J(u) ,$$

$$\phi_1(\overline{Y}(t_1)) = \lim_{N\to\infty} \phi_1(Y^N(t_1)) .$$

Here the first fact is a corollary of the fact that (39) approximates (38), while the second is a consequence of the lower semicontinuity of the function $\phi_1(x)$. From these two facts we obtain $J(\bar{u}) = \inf J(u)$, i.e. $\bar{u}(t)$, $t \in T$, is an optimal control.

§6. The Solution of the Problem of Existence of Optimal Controls by Methods of Functional Analysis

In §1.8 it was shown that by use of the methods of functional analysis, the problem of finding controls under which the trajectories of the system satisfy given boundary conditions leads to a condition of invertibility at the origin of some function. If, in addition, we impose an extremum condition (for definiteness minimization) on the control, then the problem of the existence of an optimal control reduces, in many cases, to an equivalent question about the left continuity of the mentioned function. Detailed proofs connected with this circle of problems are carried out in §6.9-6.14, where we solve not only the problem of the existence of optimal controls but also we find the form of the optimal control.

Commentary on Chapter V

1. If the delay $h(x,u,t)$ is independent of u, then in Theorem 1 instead of the set $R(x(\cdot),t)$ it suffices to consider the set

$$R(x,y,t) = \{z: \quad z = f(x,y,u,t) \ , \ u \in U\} \ ,$$

defined on the vectors x,y and the number t.

2. Theorem 1 remains true if in problem (1)-(3) we impose the condition

$$F(x(t),t) \leq 0 \quad , \quad t \in T \quad , \qquad (46)$$

where $G(x,t)$ is a continuous function. In connection with this, it is interesting to note that Theorem 3 is, generally speaking, false with the additional condition (46).

Example 1.

$$\dot{x}_1 = u \quad , \quad \dot{x}_2 = x_1 - u^2 \quad , \quad x_1(0) = x_2(0) = 0 \ ,$$

$$U = \{u: \ |u| \leq 1\} \ , \ T = [0,1] \ , \ c_1 = 0 \ , \ G(x,t) = -x_1 \quad .$$

In this example, the sequence of controls $u^N(t) = \text{sign} \sin Nt$, $N \geq 1$, is minimizing with $J(u^N) \to -1 = \inf J(u)$; however, there does not exist a measurable function $\bar{u}(t)$, $t \in T$, for which $J(\bar{u}) = -1$.

3. Although the conditions of Theorem 1 appear to be excessively strong (especially the requirement of convexity of the set $R(x(\cdot),t)$), generally speaking, it is impossible to weaken them.

Example 2.

$$\dot{x}_1 = u \quad , \quad \dot{x}_2 = x_1^2 \quad , \quad x_1(0) = x_2(0) = 0 \quad ,$$

$$U = \{u: \ |u| \leq 1\} \quad , \quad T = [0,1] \quad , \quad \phi_1(x) = x_2 \quad .$$

It is not difficult to see that $\inf J(u) = 0$. However, there does not exist a measurable function $\bar{u}(t)$, $t \in T$, minimizing the functional $J(u) = x_2$.

4. In general, the conditions of Theorems 2, and 4 may not be weakened.

Example 3.

$$\dot{x}_1 = u \quad , \quad f_{n+1} = x_1^2 - u^2 \quad , \quad x_1(0) = 0 \quad ,$$

$$U = \{u: \ |u| \leq 1\} \quad , \quad T = [0,1] \quad , \quad \phi_1(x) = x_2 \quad .$$

It is easily seen that $\inf J(u) = -1$, but the infimum over all measurable functions is not assumed.

5. The problem of the existence of optimal controls currently has an extensive literature. The first results on the existence of optimal controls were obtained in [38b,125]. Although it is impossible, in general, to weaken the conditions of A.F. Fillipov's theorem, special classes of problems for which these conditions may be weakened are of interest. In this direction the works [145,205,221a]. The results of §1.5 differ

from the known results both by the method of study and by the content. A good perspective, to our mind, is given by the idea of the proof of Theorem 5. This theorem and many numerical examples (see, in particular, Chapters V and IX) lead to the idea that the problem of the existence of optimal controls for continuous systems and the possibility of transferring the Pontryagin maximum principle to discrete systems are intimately connected. We formulate this thought in the form of the following proposition.

<u>Hypothesis</u>: In order that the optimal control problem for a continuous system have a solution in the class of measurable, bounded controls, it is necessary and sufficient that for sufficiently small quantization intervals (Δt small), the optimal controls of the discrete system approximating the continuous system satisfy the Pontryagin maximum principle.

The proof of this conjecture would be of great interest both for continuous and discrete systems. On the one hand, it shows the universality of the Pontryagin maximum principle, on the other, it is the basis of the application of this principle for the computation of optimal controls on digital computers. This last circumstance is one of the most important motivations for investigating the conjecture.

It is known that dynamic programming is the most general procedure for studying discrete processes. However, application of this method for the computation of optimal processes, even for comparatively simple systems, meets with serious difficulties connected with the limited high-speed storage of computers. Thus [129,131,177] there have been undertaken many attempts to extend the maximum principle to discrete systems. This question will be extensively investigated in Chapter IX. Although, generally speaking, there is no maximum principle for discrete systems, we will formulate valid hypotheses which do not destroy the immense value of the maximum principle for a large class of discrete systems obtained as approximations of continuous systems.

If the above conjecture turns out to be true, then there appears a new scheme of proving the theorem of existence of optimal controls. This scheme (partially used for the proof of Theorem 5), applied to the problem of the existence of optimal controls, reduces to a proof of the maximum principle for discrete systems. The latter problem is simply solved in a number of cases (cf. Chapter IX).

6. The existence theorem for optimal controls in the problem (1)-(3) is proved in [36b] by a scheme presented in [56] for ordinary dynamical systems.

PART II: THE MAXIMUM PRINCIPLE IN THE THEORY OF OPTIMAL PROCESSES

CHAPTER VI

Necessary Conditions for Optimal Controls

§1. The Method of Increments

1. Statement of the Problem. Formulas for the Increment of Scalar Functions.

In the class D of admissible controls (piecewise continuous functions $u(t)$, $t\varepsilon T$, with values in a bounded set U), we consider the solution $x(t), t\varepsilon T$, of the equation

$$\frac{dx}{dt} = f(x,y(t),t) \quad , \; x(t_o) = x_o \quad , \; t\varepsilon T = [t_o,t_1] \; ,$$
$$x = \{x_1,\ldots,x_n\} \quad , \; f = \{f_1,\ldots,f_n\} \quad , \; u = \{u_1,\ldots,u_r\}, \qquad (1)$$

where the $f_i(x,y,t)$ are continuous together with $\partial f_i(x,u,t)/\partial x_j, i,j = 1,\ldots,n$. In what follows, we will always assume that at points τ of discontinuity (of the first kind) of the function $u(t)$, that $u(\tau) = \lim u(t)$, $t \to \tau$, $t > \tau$. To the control $u(t)$, let there correspond some value of the functional

$$J(u) = \phi(x(t_1)) \quad . \qquad (2)$$

The objective of this paragraph is to find conditions (necessary conditions) of optimality which optimal controls $u^o(t)$, $t\varepsilon T$ (assuming they exist) satisfy in the problem (1),(2). For proof of the basic conditions, we will rely on the formula for the increment of a scalar function

$$\Delta\phi(x(t_1)) = \phi(x(t_1) + \Delta x(t_1)) - \phi(x(t_1)) \quad ,$$

defined along trajectories of the system (1). We let x(t), x(t) + Δx(t) denote the trajectories of Eq. (1) corresponding to the controls u(t), u(t) + Δu(t), respectively. The following formulas are valid

$$\Delta\phi(x(t_1)) = -\int_{t_0}^{t_1} [H(x(t),\psi(t),u(t) + \Delta u(t),t) - H(x(t),\psi(t),u(t),t)]dt - \int_{t_0}^{t_1} \frac{\partial[H'(x,\psi,u + \Delta u,t) - H'(x,\psi,u,t)]}{\partial x}\Delta x(t)dt - \int_{t_0}^{t_1} o_1(||\Delta x||)dt + o_4 \quad , \tag{3}$$

$$\Delta\phi(x(t_1)) = -\int_{t_0}^{t_1} [H(x + \Delta x,\psi,u + \Delta u,t) - H(x + \Delta x,\psi,u,t)]dt - \int_{t_0}^{t_1} o_2(||\Delta x||)dt + o_4 \quad , \tag{4}$$

$$\Delta\phi(x(t_1)) = -\int_{t_0}^{t_1} [H(x,\psi + \Delta\psi,u + \Delta u,t) - H(x,\psi + \Delta\psi,u,t)]dt - \int_{t_0}^{t_1} o_3(||\Delta x||)dt + o_4 \quad . \tag{5}$$

Here

$$H(x,\psi,u,t) = \psi' f(x,u,t) \quad ,$$

$$\dot{\psi} = \frac{H(x,\psi,u,t)}{\partial x} \quad , \qquad \psi(t_1) = -\frac{\partial\phi(x(t_1))}{\partial x} \quad ,$$

$$\Delta\dot{\psi} = -\frac{\partial H(x+\Delta x,\psi+\Delta\psi,u+\Delta u,t)}{\partial x} + \frac{H(x,\psi,u,t)}{\partial x} \quad ,$$

$$\Delta\psi(t_1) = 0 \quad ;$$

The quantities $o_1 - o_4$ are defined by the expansions

$$H(x + \Delta x,\psi,u + \Delta u,t) - H(x,\psi,u + \Delta u,t)$$

$$= \frac{\partial H'(x,\psi,u + \Delta u,t)}{\partial x} \Delta x + o_1(||\Delta x||) \quad ,$$

$$H(x + \Delta x,\psi,u,t) - H(x,\psi,u,t)$$

$$= \frac{\partial H'(x,\psi,u,t)}{\partial x} \Delta x + o_2(||\Delta x||) \quad ,$$

$$H(x + \Delta x,\psi + \Delta\psi,u + \Delta u,t) - H(x,\psi + \Delta\psi,u + \Delta u,t)$$

$$= \frac{\partial H'(x + \Delta x,\psi + \Delta\psi,u + \Delta u,t)}{\partial x} \Delta x + o_3(||\Delta x||) \quad ,$$

$$\phi(x + \Delta x) - \phi(x) = \frac{\partial\phi'(x)}{\partial x} \Delta x + o_4(||\Delta x||) \quad .$$

Proof of the formulas (3)-(5). The increments $\Delta x(t)$, $\Delta\psi(t)$ of the trajectories $x(t)$, $\psi(t)$ satisfy the equations

$$\frac{d\Delta x(t)}{dt} = f(x + \Delta x, u + \Delta u, t) - f(x,u,t) \ , \ \Delta x(t_o) = 0, \ t \varepsilon T \ ,$$

$$\frac{d\Delta\psi(t)}{dt} = - \frac{\partial H(x + \Delta x, \psi + \Delta\psi, u + \Delta u, t)}{\partial x} + \frac{\partial H(x,\psi,u,t)}{\partial x} \quad ,$$

$$\Delta\psi(t_1) = 0 \quad .$$

Integrating by parts, we obtain

$$\int_{t_o}^{t_1} \psi'(t) d\Delta x(t) = \psi'(t_1)\Delta x(t_1) - \psi'(t_o)\Delta x(t_o) - \int_{t_o}^{t_1} \Delta x'(t) d\psi(t)$$

After substitution of the values $\psi(t_1)$, $\Delta x(t_o)$, we obtain

$$J(u) = - \int_{t_o}^{t_1} \psi'(t) d\Delta x(t) - \int_{t_o}^{t_1} \Delta\, x'(y) d\psi(t) + o_4(||\Delta x(t_1)||) \ .$$

We replace $d\Delta x(t)$ and $d\psi(t)$ by their expressions. Then we have

$$\Delta J(u) = - \int_{t_o}^{t_1} [H(x + \Delta x, \psi, u + \Delta u, t) - H(x,\psi,u,t)]dt$$
$$+ \int_{t_o}^{t_1} \Delta x' \frac{\partial H(x,\psi,u,t)}{\partial x} dt + o_4 \quad .$$

In the right-hand side, we add the expression $H(x,\psi,u+\Delta u,t) - H(x,\psi,u+\Delta u,t)$. Again grouping terms we are led to (3). If in the right-side we add $H(x+\Delta x,\psi,u,t) - H(x+\Delta x,\psi,u,t)$ and group terms, then we obtain formula (4). To obtain (5), we integrate the expression

$$\int_{t_o}^{t_1} [\psi'(t) + \Delta\psi'(t)]d\Delta x(t) = [\psi'(t_1) + \Delta\psi'(t_1)]\Delta x(t_1)$$
$$- [\psi'(t_o) + \Delta\psi'(t_o)]\Delta x(t_o) - \int_{t_o}^{t_1} \Delta x'(t)[d\psi(t) + d\Delta\psi(t)] \ ,$$

by parts. We take into account the values $\psi(t_1)$, $\Delta\psi(t_1)$,

$\Delta x(t_0)$ and use the expressions for $d\Delta x(t)$, $d\psi(t)$, $d\Delta\psi(t)$. As a result we obtain

$$\Delta J(u) = - \int_{t_0}^{t_1} [H(x + \Delta x, \psi + \Delta\psi, u + \Delta u, t)$$

$$- H(x, \psi + \Delta\psi, u, t)]dt$$

$$+ \int_{t_0}^{t_1} \Delta x' \frac{\partial H(x + \Delta x, \psi + \Delta\psi, u + \Delta u, t)}{\partial x} dt + o_4 \quad .$$

In the right side we add the expression $H(x, \psi + \Delta\psi, u + \Delta u, t) - H(x, \psi + \Delta\psi, u + \Delta u, t)$ and, regrouping terms, we obtain formula (5).

2. A Special Increment of the Control

Let

$$\Delta_{\varepsilon\theta} u(t) = \begin{cases} 0 \quad , & t_0 \le t < \theta, \ \theta + \varepsilon \le t \le t_1 , \\ u^* - u(t) \quad , & \theta \le t \le \theta + \varepsilon \quad , \end{cases} \tag{6}$$

where θ, ε are arbitrary numbers, $\theta \varepsilon T$, $\theta + \varepsilon \ \varepsilon T$, $\varepsilon > 0$, u^* an arbitrary point of U. We substitute the increment $\Delta_{\varepsilon\theta} u(t)$ into (3). Then we obtain

$$\Delta_{\varepsilon\theta} J(u) = - \int_{\theta}^{\theta+\varepsilon} [H(x(t), \psi(t), u^*, t) - H(x(t), \psi(t), u(t), t)dt$$

$$- \int_{\theta}^{\theta+\varepsilon} \Delta_{\varepsilon\theta} x'(t) \frac{\partial [H(x(t), \psi(t), u^*, t) - H(x(t), \psi(t), u(t), t)]}{\partial x} dt$$

$$- \int_{\theta}^{t_1} o_1(||\Delta_{\varepsilon\theta} x(t)||dt + o_4(||\Delta_{\varepsilon\theta} x(t_1)||) \quad .$$

Here the symbol $\Delta_{\varepsilon\theta} x(t)$ denotes the increment of the trajectory $x(t)$, generated by the increments (6).

3. The L.S. Pontryagin Maximum Principle

First we mention three simple assertions.

Lemma 1. For any numbers θ, ε, all functions $h(t,u)$ which are continuous in t, and boundedly piecewise-continuous in the controls $u(t)$, satisfy the equality

$$\int_{\theta}^{\theta+\varepsilon} h(t,u(t))dt = \varepsilon h(\theta,u(\theta)) + o(\varepsilon) ,$$

for sufficiently small ε.

To prove sufficiency, we apply Taylor's formula

$$\alpha(\varepsilon) = \alpha(0) + \frac{\partial\alpha(0)}{\partial\varepsilon}\varepsilon + o(\varepsilon) .$$

Lemma 2. If the function $\alpha(t)$ is nonnegative for $t \geq t_o$, then for $c \geq 0$, $L \geq 0$, from the inequality

$$\alpha(t) \leq c + L \int_{t_o}^{t} \alpha(s)ds , \quad t \geq t_o ,$$

it follows that

$$\alpha(t) \leq c \exp L(t - t_o) , \quad t \geq t_o .$$

Actually, under the conditions of the Lemma, we have the sequence of inequalities

$$\alpha(t) \leq c + L \int_{t_o}^{t} \left[c + L \int_{t_o}^{s} \alpha(\sigma)d\sigma \right] ds \leq c + cL(t - t_o)$$

$$+ L^2 \int_{t_o}^{t} (t-s)\alpha(s)ds \leq c \left[1 + L(t-t_o) + \frac{L^2(t - t_o)}{2!} + \cdots + \right]$$

$$= c \exp L(t - t_o) .$$

Lemma 3. In the expression $\beta d + \gamma(\beta)$, let the quantity d be constant, let $\beta \to 0$ through positive values, and let $\gamma(\beta) \to 0$, where $\gamma(\beta)/\beta \to 0$ for $\beta \to 0$.

If $\beta d + \gamma(\beta) > 0$ for all sufficiently small β, then $d \geq 0$.

Actually, if $d < 0$ then $d + \gamma(\beta)/\beta < 0$ for sufficiently small β. Then $\beta d + \gamma(\beta) = \beta(d + \gamma(\beta)/\beta) < 0$ which contradicts the condition, proving Lemma 3.

We estimate the quantity $||\Delta_{\varepsilon\theta}x(t)||$, $t\varepsilon T$. It is clear that $\Delta_{\varepsilon\theta}x(t) \equiv 0$, $t_o \leq t \leq \theta$. For all $t, \theta \leq t < \theta + \varepsilon$, we have

$$\frac{d\Delta_{\varepsilon\theta}x(t)}{dt} = f(x + \Delta_{\varepsilon\theta}x, u^*, t) - f(x,u,t) \quad , \quad \Delta_{\varepsilon\theta}x(\theta) = 0 \quad .$$

Hence, passing to the integral form, we obtain

$$||\Delta_{\varepsilon\theta}x(t)|| \leq L \int_{\theta}^{t} ||\Delta_{\varepsilon\theta}x(\tau)||d\tau + \int_{\theta}^{\theta+\varepsilon} ||\Delta_{u*}f(x,u,t)||dt$$

$$(\Delta_{u*}f(x,u,t) = f(x,u^*,t) - f(x,u,t)) \quad .$$

We apply Lemma 1 to the second integral and use Lemma 2. Then we obtain

$$||\Delta_{\theta\varepsilon}x(t)|| \leq O(\varepsilon) \exp L\varepsilon = O_1(\varepsilon) \quad .$$

For $\theta + \varepsilon \leq t < t_1$, the equations for $\Delta_{\varepsilon\theta}x(t)$ have the form

$$\frac{d\Delta_{\varepsilon\theta}x}{dt} = f(x + \Delta_{\varepsilon\theta}x, u, t) - f(x,u,t) \quad , \quad ||\Delta_{\varepsilon\theta}x(\theta + \varepsilon)|| \leq O_1(\varepsilon).$$

Passing to the integral form and applying Lemma 2, we obtain

$$||\Delta_{\varepsilon\theta}x(t)|| \le O_1(\varepsilon) \exp L(t - \theta - \varepsilon) = O_2(\varepsilon) \quad ,$$

$$\theta + \varepsilon \le t \le t_1 \quad .$$

Thus, over the whole interval T the numbers $\Delta_{\varepsilon\theta}x(t)$ have order of magnitute less than ε:

$$||\Delta_{\varepsilon\theta}x(t)|| \le O(\varepsilon) \quad , \quad t\varepsilon T \quad .$$

Thus, using Lemma 1 the expression for $\Delta_{\varepsilon\theta}J(u)$ may be written in the form

$$\Delta_{\varepsilon\theta}J(u) = -\varepsilon[H(x(\theta),\psi(\theta),u^*,\theta)$$

$$- H(x(\theta),\psi(\theta),u(\theta),\theta)] + o(\varepsilon) \quad , \tag{7}$$

for sufficiently small ε.

Now let $u(t) = u^o(t)$ be the optimal control. Then for any $\varepsilon > 0$, we have $\Delta_{\varepsilon\theta}J(u^o) \ge 0$, i.e. we may apply Lemma 3 to (7), by virtue of which

$$H(x^o(\theta),\psi(\theta),u^o(\theta),\theta) \ge H(x^o(\theta),\psi(\theta),u^*(\theta),\theta) \quad ,$$

where $u^*\varepsilon U$, $\theta\varepsilon T$ are arbitrary numbers. This result is known as the <u>maximum principle</u>, a more precise formulation of which is now given.

<u>Theorem 1. Let $u^o(t)$, $x^o(t)$, $t\varepsilon T$, be the optimal control and trajectory in problem (1),(2); $\psi(t)$, $t\varepsilon T$ the solution of the equation</u>

$$\frac{d\psi(t)}{dt} = - \frac{\partial f'(x^o(t),u^o(t),t)}{\partial x}\psi(t) \quad , \quad \psi(t_1) = -\frac{\partial\phi(x^o(t_1))}{\partial x} \quad . \tag{8}$$

Then at each moment $t\varepsilon T$, the optimal control satisfies the maximum condition

$$\psi'(t)f(x^o(t),u^o(t),t) \geq \psi'(t)f(x^o(t),u,t) \text{ for all } u\varepsilon U . \tag{9}$$

4. Convex Functionals

The assumption of convexity of the function $\phi(x)$ allows us to remove the constraint on its smoothness. Let $\phi(x)$ be a continuous convex function. If the vector $x^o(t_1)$ is not a minimum point of the function $\phi(x)$ in n-dimensional space, then, using the theorem on the separation of convex sets and the variational equation for (1), we are led to the existence of a vector $c(||c|| \neq 0)$ such that

$$c'\Delta_{\varepsilon\theta}x^o(t_1) \geq o(\varepsilon) \quad , \tag{10}$$

$$c'\omega \leq 0 \quad . \tag{11}$$

Here ε is a sufficiently small positive numbers, $\theta\varepsilon T$ $(\theta + \varepsilon \varepsilon T)$ is an arbitrary point, and ω is any vector directed from the point $x^o(t_1)$ into the interior of the set $\{x: \phi(x) \leq \phi(x^o(t_1))\}$. Using techniques developed in [18d,109], it is not difficult to show that the vector c may be chosen independent of θ,ε. Below (§2) for the investigation of optimization problems with a delay, it is proved that property (10) of optimal controls leads to the following assertion.

Theorem 2. Along trajectories of (1) let the functions $u^o_\nu(t)$, $\nu = 1,\ldots,r$, minimize the functional (2) with convex function $\phi(x)$, where $\phi(x^o(t_1)) \neq \inf \phi(x)$, $x\varepsilon E_n$. Then there exists a vector $c(||c|| \neq 0)$ satisfying

(11), such that the maximum condition (9) is satisfied on the optimal controls with the function $\psi(t)$ being a solution of the equation

$$\frac{d\psi(t)}{dt} = - \frac{\partial f(x^o(t),u^o(t),t)}{\partial x}\psi(t) , \quad \psi(t_1) = -c .$$

§2. The Maximum Principle in Systems with a Delay

1. A Basic Lemma

We return to the system with a delay

$$\frac{dx(t)}{dt} = f(x(t),x(t - h(x(t),u(t),t)),u(t),t) , \quad t\varepsilon T ,$$
$$x(\tau) = \Phi(\tau), \ \tau\varepsilon S_o, \ h(x,u,t) > 0 , \ u(\cdot)\varepsilon D . \tag{12}$$

We consider the control $u(t)$, $t\varepsilon T$, for which

$$\frac{dh(x(t),u(t),t)}{dt} \le 1 - \alpha_1 , \quad \alpha_1 > 0 .$$

Let $\tau(t)$ denote the expression

$$\tau(t) = t - h(x(t),u(t),t) ,$$

while $t = r(\tau)$ is the solution of the equation $\tau = \tau(t)$. Let

$$H(x,y(x,u,t),\psi,u,t) = \psi' f(x,y(x,u,t),u,t) ,$$

$$y(x,u,t) = x(t - h(x,u,t)) .$$

We will say that the control $u(t)$, $t\varepsilon T$, satisfies the maximum condition with the function $\psi(t)$, $t\varepsilon T$, if

$$\left.\begin{array}{l} H(x(t),y(x(t),u(t),t),\psi(t),u(t),t) \\ \quad \geq H(x(t),y(x(t),u(t),t),\psi(t),u,t) \\ \quad \text{for all } u \varepsilon U \quad , \quad t \varepsilon T \quad . \end{array}\right\} \tag{13}$$

Lemma 4 (Basic). Let the functions $f(x,y,u,t)$, $\Phi(t)$, $\partial f(x,y,u,t)/\partial x$, $\partial f(x,y,u,t)/\partial y$, $h(x,u,t)$, $\partial h(x,u,t)/\partial x$, $\partial h(x,u,t)/\partial u$, $\partial h(x,u,t)/\partial t$, $d\Phi(t)/dt$ be defined and continuous where

1) $||f(\bar{x},\bar{y},u,t) - f(x,y,u,t)|| \leq L_1(||\bar{x} - x|| + ||\bar{y} - y||)$,

$$||f(x,y,u,t)|| \leq L_2 \ , \quad \left|\left|\frac{d\Phi}{dt}\right|\right| \leq L_2 \quad ,$$

$$|h(\bar{x},u,t) - h(x,u,t)| \leq L_3||\bar{x} - x|| \quad ,$$

$$||\Delta_{\bar{u}}f(x,y(x,u,t),u,t)|| \leq G||\bar{u} - u|| \quad ;$$

2) if θ_1 is a point of continuity of the function $u(t)$, then

$$h(x(\theta_1),u(\theta_1 - 0),\theta_1) > h(x(\theta_1),u(\theta_1 + 0),\theta_1) \quad ;$$

3) $L_3G \exp [L_1(1 + L_2L_3)(t_1 - t_0)] \leq 1$.

Then each piecewise continuosly differentiable function $u(t) = \{u_1(t),...,u_r(t)\}$, for which there exists a vector c ($||c|| \neq 0$) such that for sufficiently small $\varepsilon > 0$ and any $\theta \varepsilon T$, the inequality

$$c'\Delta_{\varepsilon\theta}x(t_1) \geq o(\varepsilon) \quad , \tag{14}$$

is satisfied, satisfies the maximum condition (13) with the function $\psi(t)$:

$$\left.\begin{aligned}
\frac{d\psi(t)}{dt} &= \frac{\partial H(x(t),y(x(t),u(t),t),\psi(t),u(t),t)}{\partial x} \\
&- \left.\frac{\partial H(x(\bar{t}),y(x(\bar{t}),u(\bar{t})\bar{t}),\psi(\bar{t}),u(\bar{t}),\bar{t})}{\partial y}\right|_{\bar{t}=r(t)} \frac{dr(t)}{dt} \\
&+ \frac{\partial H'(x(t),y(x(t),u(t),t),\psi(t),u(t),t)}{\partial y} \left.\frac{dx(s)}{ds}\right|_{s=\tau(t)} \\
&\cdot \frac{\partial h(x(t),u(t),t)}{\partial x} \quad , \\
& t_0 \leq t \leq t' = t_1 - h(x(t_1),u(t_1),t_1) \quad ; \\
\frac{d\psi(t)}{dt} &= - \frac{\partial H(x(t),y(x(t),u(t),t),\psi(t),u(t),t)}{\partial x} \\
&+ \frac{\partial H'(x(t),y(x(t),u(t),t),\psi(t),u(t),t)}{\partial y} \left.\frac{dx(s)}{ds}\right|_{s=\tau(t)} \\
&\cdot \frac{\partial h(x(t),u(t),t)}{\partial x} \\
& t' \leq t \leq t_1 \quad , \qquad \psi(t_1) = -c \quad .
\end{aligned}\right\} \quad (15)$$

We only mention the scheme of proof referring to [36c] for details. By the scheme of §1, but with more tedious calculations, it is possible to show that for sufficiently small ε there exists a constant β such that $||\Delta_{\varepsilon\theta}x(t)|| \leq \beta_{\varepsilon}$. Further, beginning with the identity

$$\begin{aligned}
c'\Delta x(t_1) &= -\psi'(t_1)\Delta x(t_1) \\
&= - \int_{t_0}^{t_1} \dot{\psi}'(t)\Delta x(t)dt - \int_{t_0}^{t_1} \dot{\psi}'(t)\Delta\dot{x}(t)dt \quad ,
\end{aligned}$$

By the scheme of §2.2, we see that

$$c'\Delta x(t_1) = -\int_{t_0}^{t_1} \Delta_{\tilde{u}} H(x(t),y(x(t),u(t),t),\psi(t),u(t),t)dt$$

$$-\int_{t_0}^{t_1} \Delta x'(t)\frac{\partial}{\partial x}\Delta_{\tilde{u}} H(x(t),y(x(t),u(t),t),\psi(t),u(t),t)dt$$

$$-\int_{t_0}^{t_1} \Delta x'[\tau(\tilde{x}(t),\tilde{u}(t),t)]$$

$$\cdot \frac{\partial}{\partial y} H(x(t),x[\tau(x(t),\tilde{u}(t),t)],\psi(t),\tilde{u}(t),t)dt$$

$$+\int_{t_0}^{t_1} \Delta x'[\tau(x(t),u(t),t)]$$

$$\cdot \frac{\partial}{\partial y} H(x(t),y(x(t),u(t),t),\psi(t),u(t),t)dt$$

$$+\int_{t_0}^{t_1} \frac{\partial}{\partial y}\{H'(x(t),x[\tau(x(t),\tilde{u}(t),t)],\psi(t),\tilde{u}(t),t)\}$$

$$\cdot \left.\frac{dx(s)}{ds}\right|_{s=\tau(x(t),\tilde{u}(t),t)} \Delta x(t)\frac{\partial h(x(t),\tilde{u}(t),t)}{\partial x}\, dt$$

$$-\int_{t_0}^{t_1} \frac{\partial}{\partial y}\{H'(x(t),y(x(t),u(t),t),\psi(t),u(t),t)\}$$

$$\cdot \left.\frac{dx(s)}{ds}\right|_{s=\tau(x(t),u(t),t)} \frac{\partial h'(x(t),u(t),t)}{\partial x}\, \Delta x(t)dt$$

$$-\int_{t_0}^{t_1} o(||\Delta x(t)||)dt \quad .$$

Here $\tilde{u}(t) = u(t) + \Delta u(t)$, $\tilde{x}(t) = x(t) + \Delta x(t)$.

The proof is completed by isolating the principal terms in ε from the last expression and applying Lemma 3.

Remarks.

1. The function $r(t)$ is defined on the interval $[\tau(\theta_1 - 0), \tau(\theta_1 + 0)]$ by the condition $r(t) \equiv r(\theta_1 - 0)$.

2. If the delay $h(x,u,t)$ does not depend explicitly upon u, then the assertion of the lemma is true for measurable functions u.

2. Free Endpoint Problems

Among all piecewise-continuously differentiable functions $u(t)\varepsilon U$, $t\varepsilon T$, we seek a control $u^o(t)$, $t\varepsilon T$, minimizing the functional

$$J(u) = \phi(x(t_1)) \quad ,$$

along (12), with $\phi(x)$ being differentiable. If $u^o(t)$, $t\varepsilon T$, is the optimal control, then for any $\varepsilon > 0$, $\theta\varepsilon T$, $u^*\varepsilon U$, we have

$$\Delta_{\varepsilon\theta}J(u^o) = \phi(x^o(t_1 + \Delta_{\varepsilon\theta}x(t_1)) - \phi(x^o(t_1))$$

$$= \frac{\partial\phi'(x^o(t_1))}{\partial x}\Delta_{\varepsilon\theta}x(t_1) + o(||\Delta_{\varepsilon\theta}x(t_1)||) = c'\Delta_{\varepsilon\theta}x(t_1) + o(\varepsilon) \geq 0$$

i.e. the basic condition (14) of Lemma 4 is satisfied. Thus, we have the following assertion.

Theorem 3. For functions $f(x,y,u,t)$, $\Phi(t)$, $h(x,u,t)$, let the conditions of Lemma 4 be satisfied. Then the piecewise-continuously differentiable optimal control $u^o(t)$, $t\varepsilon T$, in problem (12),(2), satisfies the maximum condition (13) with functions from (15), where

$$c = \frac{\partial\phi(x^o(t_1))}{\partial x}$$

3. Free Time Problems

Along trajectories of the system (12), let the functional

$$J(u,t_1) = \phi(x(t_1),t_1) \quad , \tag{16}$$

be minimized where the function $\phi(x,t)$ is continuous, along with $\partial\phi(x,t)/\partial x$, $\partial\phi(x,t)/\partial t$, and t_1 is a fixed number. If t_1^o, $u^o(t)$, $t\varepsilon[t_o,t_1^o]$ are the optimal time and control, then clearly the function $u^o(t)$ must satisfy the maximum condition from Theorem 3 on $[t_o,t_1^o]$. In order to obtain conditions of optimality for t_1^o, it suffices to consider processes of different duration and to set

$$\tilde{u}(t) = \begin{cases} u^o(t) \quad , & t_o \le t < t_1^o \quad , \\ u^o(t_1^o) \quad , & t_1^o \le t < t_1^o + \varepsilon \end{cases}$$

Then from the inequality

$$\Delta J(u^o,t_1^o) = \phi(\tilde{x}(t_1^o+\varepsilon),t_1^o+\varepsilon) - \phi(x^o(t_1^o),t_1^o) \ge 0$$

it follows that

$$\frac{\partial\phi(x^o(t_1^o),t_1^o)}{\partial t}\varepsilon + \frac{\partial\phi'(x^o(t_1^o),t_1^o)}{\partial x}$$

$$\cdot\ f(x^o(t_1^o),y^o(x^o(t_1^o),u^o(t_1^o),t_1^o),u^o(t_1^o),t_1^o)\varepsilon + o(\varepsilon) \ge 0 \tag{17}$$

for all sufficiently small $\varepsilon > 0$. Applying Lemma 3 to (17), we obtain

$$\frac{\partial\phi_1(x^o(t_1^o,t_1^o))}{\partial t} \ge H(x^o(t_1^o),y^o(x^o(t_1^o),u^o(t_1^o),t_1^o),\psi(t_1^o),u^o(t^o),t_1^o).$$

An inequality in the opposite sense is obtained if we compare the optimal t_1^o, $u^o(t)$ with $t_1^o - \varepsilon$, $\varepsilon > 0$, and $\tilde{u}(t) = u^o(t)$, $t_o \le t \le t_1^o - \varepsilon$.

Theorem 4. Let the conditions of Theorem 3 be satisfied and let t_1^o, $u^o(t)$, $t_o \le t \le t_1^o$ be the solution of problem (12), (16). Then, in addition to the conclusions of Theorem 3 ($c = \partial\phi(x^o(t_1^o),t_1^o)/\partial x$), we have the equality

$$\frac{\partial\phi(x^o(t_1^o),t_1^o)}{\partial t} = H(x^o(t_1^o),y^o(x^o(t_1^o),u^o(t_1^o),t_1^o),\psi(t_1^o),u^o(t_1^o),t_1^o) .$$

4. Problems with a Moving Endpoint

We assume that at the right end of the trajectory $x(t)$, $t\varepsilon T$, Eq.(12) is subject to the constraint

$$g(x(t_1)) \le 0 \quad , \tag{18}$$

where the scalar function $g(x)$ is defined and continuous together with $\partial g(x)/\partial x$.

Theorem 5. For the functions $f(x,y,u,t)$, $h(x,u,t)$, $\Phi(t)$, let the conditions of Lemma 4 be satisfied and let the piecewise differentiable control $u^o(t)$, $t\varepsilon T$, in problem (12),(18), be such that the vectors $\partial\phi(x^o(t_1))/\partial x$ and $\partial g(x^o(t_1))/\partial x$ are not collinear. Then there are numbers $\lambda,\mu \ge 0$, $\lambda + \mu = 1$ ($\mu = 0$ if $g(x^o(t_1)) < 0$) such that the function $u^o(t)$, $t\varepsilon T$, satisfies the maximum condition (13) with $\psi(t)$ (the solution of system (15)) for

$$c = \lambda\partial\phi(x^o(t_1))/\partial x + \mu\partial g(x^o(t_1))/\partial x \quad .$$

Proof. We show that if the conditions of Theorem 5 are satisfied then there exists a nonzero vector c satisfying the conditions of the basic lemma, i.e. $c'\Delta_{\varepsilon\theta}x(t_1) \geq o(\varepsilon)$, $o(\varepsilon) \sim \varepsilon^2$, for any $u^*\varepsilon U$, $\theta\varepsilon T$, if ε is sufficiently small. Since all remaining conditions of this lemma are assumed to be satisfied, this will then prove the assertions of Theorem 5.

We replace the equation

$$\frac{d\Delta_{\varepsilon\theta}x(t)}{dt} = f(\tilde{x}(t),\tilde{y}(t),u(t),t) - f(x(t),y(t),u(t),t), \qquad t \geq \theta + \varepsilon \ ,$$

by the corresponding variational equation

$$\left.\begin{aligned} &\frac{d\delta_{\varepsilon\theta}x(t)}{dt} = A(t)\delta_{\varepsilon\theta}x(t) + A_1(t)\delta_{\varepsilon\theta}x[\tau(t)] \ , \ t \geq \theta + \varepsilon \ , \\ &\quad x(t) \equiv 0 \ , \quad t < \theta + \varepsilon \ , \quad \delta_{\theta\varepsilon}x(\theta + \varepsilon) \\ &\qquad = \Delta_{u*}f(x(\theta + \varepsilon),y(\theta + \varepsilon),u(\theta + \varepsilon),\theta + \varepsilon)\varepsilon \ . \end{aligned}\right\} \quad (19)$$

Here

$$A(t) = \begin{Bmatrix} \dfrac{\partial f_1}{\partial x} - \dfrac{\partial f_1'}{\partial y} \cdot \left.\dfrac{dx(s)}{ds}\right|_{s=\tau(t)} \dfrac{\partial h}{\partial x} \\ \cdots\cdots\cdots\cdots\cdots\cdots\cdots \\ \dfrac{\partial f_n}{x} - \dfrac{\partial f_n'}{\partial y} \quad \left.\dfrac{dx(s)}{ds}\right|_{s=\tau(t)} \dfrac{\partial h}{\partial x} \end{Bmatrix}, \quad A_1(t) = \begin{Bmatrix} \dfrac{\partial f_1}{\partial y} \\ \vdots \\ \dfrac{\partial f_n}{\partial y} \end{Bmatrix},$$

where the vectors $\partial f_i/\partial x$, $\partial f_i/\partial y$, $i = 1,\ldots,n$, $\partial h/\partial x$ and the scalar function $\tau(t)$ are calculated along the trajectory $x(t)$ and the control $u(t)$. Taking into account that $\tilde{u}(t) \equiv u(t)$ for $t\varepsilon[\theta + \varepsilon,t_1]$, we have

$$\Delta_{\varepsilon\theta}x(t) - \delta_{\varepsilon\theta}x(t) = \Delta_{\varepsilon\theta}x(\theta + \varepsilon) - \delta_{\varepsilon\theta}x(\theta + \varepsilon)$$

$$+ \int_{\theta+\varepsilon}^{t} [f(\tilde{x}(s),\tilde{x}[\tau(\tilde{x},u,s),u,s) - f(x(s),x[\tau(x,u,s)],u,s)]ds$$

$$- \int_{\theta+\varepsilon}^{t} (A(s)\delta_{\varepsilon\theta}x(s) + A_1(s)\delta_{\varepsilon\theta}x[\tau(s)])ds$$

$$= o(\varepsilon) + \int_{\theta+\varepsilon}^{t} \frac{\partial f}{\partial x}\Delta_{\varepsilon\theta}x(s) - \frac{\partial f}{\partial y}\Delta_{\varepsilon\theta}x[\tau(\tilde{x},u,s)]$$

$$- \frac{\partial f}{\partial y}\left.\frac{dx(\sigma)}{d\sigma}\right|_{\sigma=\tau(s)}\frac{\partial h'}{\partial x}\Delta_{\varepsilon\theta}x(s)\ dx + \int_{\theta+\varepsilon}^{t} o(||\Delta_{\varepsilon\theta}x(s)||$$

$$+ ||\Delta_{\varepsilon\theta}x[\tau(\tilde{x},u,s)]||)ds - \int_{\theta+\varepsilon}^{t} (A(s)\delta_{\varepsilon\theta}x(s)$$

$$+ A_1(s)\delta_{\varepsilon\theta}x[\tau(s)])ds = \int_{\theta+\varepsilon}^{t} A(s)[\Delta_{\varepsilon\theta}x(s) - \delta_{\varepsilon\theta}x(s)]ds$$

$$+ \int_{\theta+\varepsilon}^{t} A_1(s)[\Delta_{\varepsilon\theta}x[\tau(\tilde{x},u,s)] - \Delta_{\varepsilon\theta}x[\tau(x,u,s)]]ds$$

$$+ \int_{\theta+\varepsilon}^{t} A_1(s)[\Delta_{\varepsilon\theta}x[\tau(s)] - \delta_{\varepsilon\theta}x[\tau(s)]]ds + o(\varepsilon) \quad .$$

In the proof of the basic lemma it was shown that

$$\int_{\theta+\varepsilon}^{t} A_1(s)[\Delta_{\varepsilon\theta}x[\tau(\tilde{x},u,x)] - \Delta_{\varepsilon\theta}x[\tau(x,y,s]]ds = o(\varepsilon) \quad .$$

It is not difficult to see that the following inequality is true:

$$\left|\left| \int_{\theta+\varepsilon}^{t} A_1(s)[\Delta_{\varepsilon\theta}x[\tau(s)] - \delta_{\varepsilon\theta}x[\tau(s)]]ds\right|\right|$$

$$= \left|\left|\int_{\tau(\theta+\varepsilon)}^{\tau(t)} A_1(r(s))[\Delta_{\varepsilon\theta}x(s) - \delta_{\varepsilon\theta}x(s)]r'(s)ds\right|\right|$$

$$\leq \int_{\tau(\theta+\varepsilon}^{\theta+\varepsilon} \|A_1(r(s))\| \, \|\Delta_{\varepsilon\theta}x(s)\| \, |r'(s)|ds$$

$$+ \int_{\theta+\varepsilon}^{t} \|A_1(r(s))\| \, \|\Delta_{\varepsilon\theta}x(s) - \delta_{\varepsilon\theta}x(s)\| \, |r'(s)|ds$$

$$\leq \int_{\theta}^{\theta+\varepsilon} \|A_1(r(s))\| \, \|\Delta_{\varepsilon\theta}x(s)\| \, |r'(s)|ds$$

$$+ \int_{\theta+\varepsilon}^{t} \|A_1(r(s))\| \, \|\Delta_{\varepsilon\theta}x(s) - \delta_{\varepsilon\theta}x(s)\| \, |r'(s)|ds$$

$$\leq L_1 \frac{\beta}{\alpha_1} \varepsilon^2 + L_1 \frac{1}{\alpha_1} \int_{\theta+\varepsilon}^{t} \|\Delta_{\varepsilon\theta}x(s) - \delta_{\varepsilon\theta}x(s)\|ds$$

$$= o(\varepsilon) + \frac{L_1}{\alpha_1} \int_{\theta+\varepsilon}^{t} \|\Delta_{\varepsilon\theta}x(s) - \delta_{\varepsilon\theta}x(s)\|ds \quad .$$

Taking this into account, we obtain

$$\|\Delta_{\varepsilon\theta}x(t) - \delta_{\varepsilon\theta}x(t)\| \leq \int_{\theta+\varepsilon}^{t} \|A(s)\| \, \|\Delta_{\varepsilon\theta}x(s) - \delta_{\varepsilon\theta}x(s)\|ds$$

$$+ \frac{L_1}{\alpha_1} \int_{\theta+\varepsilon}^{t} \|\Delta_{\varepsilon\theta}x(s) - \delta_{\varepsilon\theta}x(s)\|ds + o(\varepsilon)$$

$$\leq L_1\left(\frac{\alpha_1 + 1}{\alpha_1} + L_2L_3\right) \int_{\theta+\varepsilon}^{t} \|\Delta_{\varepsilon\theta}x(s) - \delta_{\varepsilon\theta}x(s)\|ds + o(\varepsilon)$$

Hence,

$$||\Delta_{\varepsilon\theta}x(t) - \delta_{\varepsilon\theta}x(t)||$$

$$\leq o(\varepsilon) \exp\left(L_1\frac{\alpha_1 + 1}{\alpha_1} + L_2L_3\right)(t_1 - \theta - \varepsilon), \qquad (20)$$

where $t \varepsilon [\theta + \varepsilon, t_1]$ and all $u^* \varepsilon U$. We describe the solution of the linear equation (19) at time t_1 by Cauchy's formula

$$\delta_{\varepsilon\theta}x(t_1) = F(t_1,\theta + \varepsilon)\Delta_{u*}f(x(\theta + \varepsilon),y(\theta + \varepsilon), u(\theta + \varepsilon),\theta + \varepsilon)\varepsilon \quad . \quad (21)$$

From (21), it is clear that for fixed θ,u*, and different $\varepsilon \geq 0$, the set $\{\delta_{\varepsilon\theta}x(t_1)\}$ is represented by some ray $\Pi_{\theta u*}$, emanating from the point $x(t_1)$.

We let $a = \partial\phi(x(t_1))/\partial x$, $b = \partial g(x(t_1))/\partial x$ and consider the two cones: $K_1 = \{\delta x\colon a'\delta x \leq 0$ and $K_2 = \{\delta x\colon b'\delta x \leq 0\}$. Let $K_3 = K_1 \cap K_2$. From the noncollinearity of the vectors a and b, it follows that the cone K_3 has a nonempty interior. We show that the ray $\Pi_{\theta u*}$ does not contain an interior point of the set K_3. Let the $\delta_{\varepsilon\theta}x(t_1) \varepsilon \Pi_{\theta u*}$ be an interior point for K_3. Then all points $\delta_{\varepsilon\theta}x(t_1)$, $0 < \varepsilon \leq \bar{\varepsilon}$, will be interior to K_3. By virtue of (20), for sufficiently small ε the points $\Delta_{\varepsilon\theta}x(t_1)$ will also be interior to K_3, which contradicts the optimality of u(t).

Thus, the sets $\Pi_{\theta u*}$ and K_3 may be separated, i.e. there exists a vector $c(u*,\theta)$ such that $c'\Delta_{\varepsilon\theta}x(t_1) \geq 0$, where $c(u*,\theta)$ is the support vector to K_3 at the point $x(t_1)$.

Following [109], it is possible to construct a vector $c(u*,\theta)$, independent of u* and θ, $u*\varepsilon U$, $\theta\varepsilon T$. In other words, there exists a nonzero vector c, supporting K_3 at the point $x(t_1)$, such that $c'\delta_{\varepsilon\theta}x(t_1) \geq 0$ for all $u*\varepsilon U$, $\theta\varepsilon T$.

Turning to (20), we obtain

$$c'\Delta_{\varepsilon\theta}x(t_1) \geq o(\varepsilon).$$

We express c through a and b. By construction $c'\delta x \leq 0$, $\delta x\varepsilon K_3$, i.e. $c\varepsilon K_3^-$, where K_3^- is the cone of nonpositive linear forms over K_3. It is known that K_3^- $= (K_1 \cap K_2)^- = \overline{K_1^- + K_2^-}$, i.e. each element $c \in K_3^-$ is representable in the form

$$c = \lambda a + \mu b \quad , \quad \lambda + \mu > 0 \quad , \quad \lambda, \mu \geq 0 \quad .$$

The theorem is now completely proved.

§3. A New Form of the Necessary Conditions for Optimality

1. Variational Derivatives of the First and Second Type

For description of the maximum principle in optimization problems for systems described by the ordinary differential equations (1), it is very convenient to use the function

$$H(x,\psi,u,t) = \psi' f(x,u,t) \quad , \tag{22}$$

with the help of which the equations for the basic and adjoint variables may be represented in the compact, easily remembered form:

$$\frac{dx}{dt} = \frac{\partial H(x,\psi,u,t)}{\partial \psi} \quad , \tag{23}$$

$$\frac{d\psi}{dt} = - \frac{\partial H(x,\psi,u,t)}{\partial x} \quad , \tag{24}$$

Using the function (22), the maximum principle has the form

$$H(x^o(t),\psi(t),u^o(t),t) = \max_{u \in U} H(x^o(t),\psi(t),u,t) \quad .$$

After the appearance of the maximum principle, different investigations were undertaken with a view towards generalizing this result to systems differing from (1) At present, a series of works are known in this direction [21d,30,52a,53a,104a,127]. In all these works, the form of the described conditions of optimality is based upon functions of the type (22). The equations for the variables x, ψ and maximum conditions for systems essentially different from (1), have a nonregular character and are difficult to remember. For confirmation of this feeling, it suffices to turn to Theorem 3. Equation (15) for the adjoint variables ψ is especially complex and difficult to remember. For other systems, the maximum condition will appear unusual if it is expressed in terms of the function (22).

Below, we propose a new form for describing the necessary conditions for optimality. As a prelude, we introduce two definitions. Let the functional $J(x)$ be given on the curve $x(t) \varepsilon C$, $t \varepsilon T$. If there exists an integrable function $g(t)$, $t \varepsilon T$, for which

$$J(x + \Delta x) - J(x) = \int_{t_0}^{t_1} g'(t) \Delta x(t) dt + o(\Delta x(\cdot)) ,$$

and $o(\Delta x(\cdot))/||\Delta x(\cdot)|| \to 0$ for $||\Delta x(\cdot)|| \to 0$, then we call $g(t)$ the <u>variational derivative of the first order</u> at the moment t of the functional $J(x)$ relative to the function $x(t)$, $t \varepsilon T$, and we denote it by the symbol $\delta J(x)/\delta x(t)$.

If the functional $J(x)$ depends on $x(t)$ and its derivatives $x^{(\ell)}(t)$, $t \varepsilon T$, then we retain the definition of the variational derivative of the first order, adding one condition: $o(\Delta x(\cdot))/||\Delta x(\cdot)|| \to 0$ for $||\Delta x(\cdot)|| \to 0$ in the metric space of continuous functions $x(t)$ having continuous derivatives $x^{(\ell)}(t)$, $\ell = 1,\ldots,k$:

$$||x(\cdot)|| = \max_{t\varepsilon T} \sum_{\ell=0}^{k} ||x^{(\ell)}(t)|| \quad .$$

We let the symbol $\delta_y J(x)/\delta x(t)$ denote the variational derivative of second order at the moment t of the functional J(x), relative to the function x(t), $t\varepsilon T$, understanding by this that there exists a function $\bar{g}(t)$ satisfying the condition

$$J(x + \Delta_{\varepsilon\theta}x) - J(x) = \varepsilon\bar{g}(\theta) + o(\varepsilon) \quad ,$$

where

$$\Delta_{\varepsilon\theta}x(t) = \begin{cases} 0 \ , & t_0 \leq t < \theta \ , \ \theta + \varepsilon \leq t < t_1 \quad , \\ y - x(t) \ , & \theta \leq t < \theta + \varepsilon \quad , \end{cases}$$

and $o(\varepsilon)/\varepsilon \to 0$ as $\varepsilon \to 0$. We illustrate the new form of the conditions of optimality on a series of systems beginning with the simplest type (1).

2. Ordinary Dynamical Systems

On the continuous curves x(t), $\psi(t)$, $t\varepsilon T$, and on the piecewise-continuous function u(t), $t\varepsilon T$, we define the functional

$$\pi(x,\psi,u) = \int_{t_0}^{t_1} \psi'(t)f(x,u,t)dt \quad .$$

Starting with the definitions introduced above, it is not difficult to show that Eqs. (1),(8) and condition (9) take on the forms

$$\frac{dx(t)}{dt} = \frac{\delta\pi(x,\psi,u)}{\delta\pi(t)} \quad , \tag{25}$$

$$-\frac{d\psi(t)}{dt} = \frac{\delta\pi(x,\psi,u)}{\delta\psi(t)} \quad , \tag{26}$$

$$\frac{\delta_{u*}\pi(x,\psi,u)}{\delta\pi(t)} \leq 0 \quad \text{for all } t\varepsilon T, \quad u*\varepsilon U \quad . \tag{27}$$

In the given case, the new f rm of writing the optimality conditions is not as complex as the traditional form [109].

3. Differential Equations with a Delay

For the system (12), we introduce the functional

$$\pi(x,\psi,u)$$

$$= \int_{t_o}^{t_1} \psi'(t)f(x(t),x(t-h(x(t),u(t),t)),u(t),t)dt \ ,$$

defined on continuous $x(t)$, $\psi(t)$ and piecewise-continuous $u(t)$, where $x(t) = \Phi(t)$, $t\varepsilon S_o$ and $\psi(t) \equiv 0$, $t > t_1$. Then Eqs. (12),(15), and condition (13) may be written in the form (25)-(27).

4. A Generalization of Ordinary Dynamical Systems

The differential equation (1) represents a particular case of the equation

$$R^N(p,t)x(t) = f(x,u,t) \quad , \tag{28}$$

where

$$R^N(p,t) = A^o(t)p^N + A^1(t)p^{N-1} + \cdots + A^N(t) \quad ,$$

$$A^i(t)\varepsilon C^{N-i} \quad , \quad i = 0,\ldots,N \quad , \quad p \equiv \frac{d}{dt} \quad .$$

If $A^o(t)$, $t\varepsilon T$, is a nonsingular matrix (the general case is considered below) then in the problem of minimizing the functional (2) along trajectories of Eq. (28)

(here we do not dwell on questions about the initial conditions), the maximum condition is satisfied for the optimal control $u^o(t)$, $t \in T$. We introduce the functional

$$\pi(x,\psi,u) = \int_{t_o}^{t_1} \psi'(t) f(x(t),u(t),t)dt \quad ,$$

and the expression

$$R_N'(p,t)\psi(t) = \sum_{i=0}^{N} (-1)^i \frac{d^i}{dt^i} [(A^{N-i}(t))'\psi(t)] \quad .$$

In the new notation, the equations for the primary and the adjoint variables, and also the maximum conditions, take on the form

$$R^N(p,t)x(t) = \frac{\delta\pi(x,\psi,u)}{\delta\psi(t)} \quad , \tag{29}$$

$$R_N'(p,t)\psi(t) = \frac{\delta\pi(x,\psi,u)}{\delta x(t)} \quad , \tag{30}$$

$$\frac{\delta_{u*}\pi(x,\psi,u)}{\delta u(t)} \leq 0 \quad \text{for all } t \in T \text{ , } u* \in U \quad . \tag{31}$$

The latter form of the maximum principle is more symmetric than (25)-(27). Finally, if the question is only about ordinary dynamical systems, then together with (29)-(31), it is possible to use the following description:

$$R^N(p,t)x(t) = \frac{\partial H(x,\psi,u,t)}{\partial x}$$

$$R_N'(p,t)\psi(t) = \frac{\partial H(x,\psi,u,t)}{\partial\psi}$$

$$H(x(t),\psi(t),u(t),t) = \max_{u \in U} H(x(t),\psi(t),u,t) \quad ,$$

which generalizes (22)-(24) and is more symmetric.

5. Implicit Differential Equations

We consider a control system described by the equation

$$R^N(p,t)x(t) = f(x(t),\dot{x}(t),\ldots,x^{(k)}(t),u(t),t) \ , \qquad (32)$$

and we assume that at $t = t_o$, an initial condition is given for this equation such that to each choice of the piecewise-continuous controls $u_\nu(t)$, $t\varepsilon T$, $\nu = 1,\ldots,r$, there corresponds a unique trajectory $x(t)$. Along this trajectory we define the functional

$$J(u) = \phi(x(t_1),\dot{x}(t_1),\ldots,x^{(m)}(t_1)) \ . \qquad (33)$$

We will assume that the functions $f(x,y^1,\ldots,y^k,u,t)$, $\Phi(x,y^1,\ldots,y^k)$ are continuous, together with their derivatives $\partial f(x,y^1,\ldots,y^k,u,t)/\partial x$, $\partial\phi(x,y^1,\ldots,y^m)/\partial x$, $\partial f(x,y^1,\ldots,y^k,u,t)/\partial y\rho$, $\partial\phi(x,y^1,\ldots,y^m)/\partial y^i$, $\ell = 1,\ldots,k$; $i = 1,\ldots,m$. We pose the problem of minimizing the functional (33) over controls from class D. We recall that D is the set of piecewise-continuous functions $u(t)$ having values in a bounded set U. This problem is solved by the scheme of §1. For formulation of the result and expressing the necessary conditions of optimality, we introduce the functional

$$\pi(x,\psi,u) = \int_{t_o}^{t_1} \psi'(t)f(x(t),\dot{x}(t),\ldots,x^{(k)}(t),u(t),t)dt \ ,$$

defined on the functions $x(t)\varepsilon C^k$, $\psi(t)\varepsilon C$, $u(t)\varepsilon D$. We write the equation for $\psi(t)$:

$$R_N'(p,t)\psi(t) = \frac{\delta\pi(x,\psi,u)}{\delta x(t)} \ , \qquad (34)$$

and the relation for $\phi(t_1)$:

$$\alpha\frac{\partial\phi(x(t_1),\dot{x}(t_1),\ldots,x^{(m)}(t_1))}{\partial x_q} = -\beta R'_{N-q-1}(p,t_1)\psi(t_1)$$

$$+ \gamma \sum_{j=q+1}^{k} (-1)^j \frac{d^{j-q-1}}{dt^{j-q-1}}$$

$$\cdot \left[\frac{\partial f'(x(t_1),\dot{x}(t_1),\ldots,x^{(k)}(t_1),u(t_1 - 0),t_1)}{\partial x^j} \psi(t_1)\right] , \quad (35)$$

$$q = 0,\ldots,m \quad .$$

Here the numbers α,β,γ depend upon N,k,m in the following way:

1) $N > k$, $m < k$: $\alpha = 1$, $\beta = 1$, $\gamma = 1$, if $q = 0,\ldots,m$;
   $\alpha = 0$, $\beta = 1$, $\gamma = 1$, if $q = m + 1,\ldots,k-1$;
   $\alpha = 0$, $\beta = 1$, $\gamma = 0$, if $q = k,\ldots,N-1$.

2) $N > k$, $m \geq k$: $\alpha = 1$, $\beta = 1$, $\gamma = 1$, if $q = 0,\ldots,k-1$
   $\alpha = 1$, $\beta = 1$, $\gamma = 0$, if $q = k,\ldots,m$;
   $\alpha = 0$, $\beta = 1$, $\gamma = 0$, if $q = m + 1,\ldots,N-1$.

3) $N > m > 0$, $k = 0$: $\alpha = 1$, $\beta = 1$, $\gamma = 0$, if $q = 0,\ldots,m$;
   $\alpha = 0$, $\beta = 1$, $\gamma = 0$, if $q = m + 1,\ldots,N-1$.

4) $N < k$, $m < N$; $\alpha = 1$, $\beta = 1$, $\gamma = 1$, if $q = 0,\ldots,m$;
   $\alpha = 0$, $\beta = 1$, $\gamma = 1$, if $q = m + 1,\ldots,N-1$;
   $\alpha = 0$, $\beta = 0$, $\gamma = 1$, if $q = N,\ldots,k-1$.

5) $N < m$, $k \geq N$: $\alpha = 1$, $\beta = 1$, $\gamma = 1$, if $q = 0,\ldots,N-1$;
   $\alpha = 1$, $\beta = 0$, $\gamma = 1$, if $q = N,\ldots,m$;
   $\alpha = 0$, $\beta = 0$, $\gamma = 1$, if $q = m + 1,\ldots,k-1$.

6) $N = 0$, $k > 0$, $k > m \geq 0$: $\alpha = 1$, $\beta = 0$, $\gamma = 1$, if $q = 0,\ldots,m$;
$\alpha = 0$, $\beta = 0$, $\alpha = 1$, if $q = m + 1,\ldots,k-1$.

7) $N = k$, $m < N$: $\alpha = 1$, $\beta = 1$, $\gamma = 1$, if $q = 0,\ldots,m$;
$\alpha = 0$, $\beta = 1$, $\gamma = 1$, if $q = m + 1,\ldots,N-1$.

We will say that the dynamical system (32) satisfies condition A along the control $u(t)$, $t \varepsilon T$, in problems (32),(33) if for all sufficiently small $\varepsilon > 0$, the following conditions are satisfied:

1) $||\Delta_{\varepsilon\theta} x_{(j)}(t)|| \leq \beta\varepsilon$, $\beta = \text{const}$, $j = 1,\ldots,\max\{k,m\}$;

2) the system (34),(35) has a solution along $u(t)$, $t \varepsilon T$.

It is not difficult to assert that the dynamical system (1),(12), under minimization of the functional (2), possesses property A.

<u>Theorem 6. Let $u^o(t)$, $t \varepsilon T$, be the optimal control minimizing the functional (33) and assume that along this control the system (32) possesses property A. Then the function $u^o(t)$ is such that it satisfies conditions (29)-(31),(35).</u>

6. Systems with an Aftereffect

Let the motion of a dynamical system be described by the equation

$$\frac{dx(t)}{dt} = X(x(\cdot),u(\cdot),t)$$

$$\equiv \int_\alpha^\beta d\tau \int_\gamma^\delta f'(x(t),x(t-\tau),u(t),u(t-\sigma),t,\tau,\sigma)\,d\sigma$$

$$+ \int_\alpha^\beta f^2(x(t),x(t-\tau),u(t),u(t-\sigma_1),t,\tau)\,d\tau$$

$$+ \int_{\delta}^{\gamma} f^3(x(t), x(t-\tau_1), u(t), u(t-\sigma), t, \sigma) d\sigma$$

$$+ f^4(x(t), x(t-\tau_2), u(t), u(t-\sigma_2), t), \quad t \varepsilon T, \ \beta > \alpha \geq 0,$$

$$\delta > \gamma \geq 0 \quad , \qquad (36)$$

under the initial conditions

$$u(t) = \Phi^1(t) \quad , \quad t_o - \max\{\delta, \sigma_1, \sigma_2\} \leq t \leq t_o \quad ,$$

$$x(t) = \Phi(t) \quad , \quad t_o - \max\{\beta, \tau_1, \tau_2\} \leq t \leq t_o \quad .$$

We consider the problem of minimizing the functional (2) using controls $u(t)$, $t\varepsilon T$, from class D.

Assuming that the functions $\phi(x)$, $f'(x,y,u,v,t,\tau,\sigma)$, $f^2(x,y,u,v,t,\tau)$, $f^3(x,y,u,v,t,\sigma)$, $f^4(x,y,u,v,t)$, $\Phi(t)$, $\Phi^1(t)$ are continuous together with the functions $\partial\phi/\partial x$, $\partial f'/\partial x$, $\partial f'/\partial y$, $\partial f^2/\partial x$, $\partial f^2/\partial y$, $\partial f^3/\partial x$, $\partial f^3/\partial y$, $\partial f^4/\partial x$, $\partial f^4/\partial y$, it is not difficult to establish necessary conditions for optimality in this problem following §1.

Theorem 7. The optimal control $u^o(t)$, $t\varepsilon T$, for the problem (36),(2) satisfies conditions (29)-(31), where

$$\pi(x,\psi,u) = \int_{t_o}^{t_1} \psi'(t) X(x(\cdot), u(\cdot), t) dt \quad ,$$

$$\psi(t_1) = -\frac{\partial\phi(x^o(t_1))}{\partial x} \quad , \quad \psi(t) \equiv 0 \quad , \quad t > t_1 \ .$$

7. Integral Equations

We consider the integral equation

$$x(t) = \int_{t_o}^{t_1} f(x(\tau), u(\tau), \tau, t) d\tau \quad , \quad t\varepsilon T \quad , \qquad (37)$$

which depends upon the control function $u(t)$, $u(\cdot)\varepsilon D$, $t\varepsilon T$. If Eq. (37) has a continuous solution $x(t)$ for even one function $u(t)$, $t\varepsilon T$, $u(\cdot)\varepsilon D$, then there arises the problem of minimzing the functional

$$J(u) = \phi\left(\int_{t_0}^{t_1} x(\tau)d\tau\right) , \tag{38}$$

over such solutions.

We assume that $f(x,u,\tau,t)$, $\phi(x)$ are defined and continuous, together with $\partial f(x,u,\tau,t)/\partial x$, $\partial\phi/\partial x$ and, moreover, we will assume that one of the following properties is satisfied:

1) $f(x,u,\tau,t) = 0$, if $\tau \geq t$;
2) $L(t_1 - t_0) < 1$, where $L = \max_{u\varepsilon U, t\varepsilon T, \tau\varepsilon T} ||f(x,u,\tau,t)||$

in a neighborhood of the optimal trajectory $x^o(t)$, $t\varepsilon T$.

To formulate the necessary conditions of optimality, we introduce the functional

$$\pi(x,\psi,u) = \int_{t_0}^{t_1} \psi'(t)dt \int_{t_0}^{t_1} f(x(\tau),u(\tau),\tau,t)dt - \phi\left(\int_{t_0}^{t_1} x(\tau)d\tau\right) . \tag{39}$$

Theorem 8. The optimal control $u^o(t)$, $t\varepsilon T$, in problem (37),(38) satisfies conditions (29)-(31) for the functional (39).

8. Partial Differential Equations

The scheme of §1 may be naturally generalized to problems of optimizing systems having distributed parameters. Not dwelling upon proofs, we introduce several results in new forms, greatly facilitating their review.

In the domain $\Gamma = \{t,s\colon\ t_0 \le t \le t_1, s_0 \le s \le s_1\}$, let the state $x(t,s)$ of the control system be described by the equation

$$\frac{\partial^2 x(t,s)}{\partial t\,\partial s} = f\left(x(t,s),\ \frac{\partial x(t,s)}{\partial t},\ \frac{\partial x(t,s)}{\partial s},\ u(t,s), t, s\right) \qquad (40)$$

with the boundary conditions

$$x(t_0,s) = \Phi(s),\ x(t,s_0) = \Phi^1(t),\ \Phi(s_0) = \Phi^1(t_1) \quad . \qquad (41)$$

We will assume that the functions

$$f(x,y,z,u,t,s),\ \partial f/\partial x,\ \partial f/\partial y,\ \partial f/\partial z,\ \Phi(s),\ \Phi^1(t),\ d\Phi/ds,\ d\Phi^1/dt\ ,$$

are defined and continuous.

As the class of admissible controls $u(t,s)$, we choose the family of piecewise-continuous functions having values in a given bounded set U in R^r. The criterion of the process $x(t,s)$ is measured by the functional

$$J(u) = \phi(x(t_1,s_1))\ , \qquad (42)$$

and we consider the problem of finding the optimal $u^o(t,s)$ in the admissible set such that

$$J(u^o) \le J(u) \quad .$$

To formulate the results about the necessary conditions for optimality, we make the concepts introduced above more precise. The variational derivative of a functional defined on functions of two variables is introduced in a manner analagous to the case of a single variable. The change consists of the replacement (in the definition of the variational derivative of the first type) of the integral by a double integral and by the introduction (in the definition of the variational derivative of the second type) of the square $[\theta_1 \leq t \leq \theta_1 + \varepsilon, \theta_2 \leq s \leq \theta_2 + \varepsilon]$ instead of the interval $[\theta, \theta + \varepsilon]$.

We define the functional

$$\pi(x,\psi,u) = \int_{t_0}^{t_1} dt \int_{s_0}^{s_1} \psi'(t,s) \cdot \left(f(x(t,s), \frac{\partial x(t,s)}{\partial t}, \frac{\partial x(t,s)}{\partial s}, u(t,s), s, t) \right) ds .$$

on the functions $x(t,s)$, $\psi(t,s)$, $u(t,s)$.

Theorem 9. Let $u^o(t,s)$, $\{t,s\} \varepsilon \Gamma$ be the optimal control in the problem (40),(42). It is generated by the twice continuously differentiable solutions $x^o(t,s)$, $\psi(t,s)$ to the equations

$$\frac{\partial^2 x^o(t,s)}{\partial t \, \partial s} = \frac{\delta \pi(x^o,\psi,u^o)}{\delta \psi(t,s)} , \tag{43}$$

$$\frac{\partial^2 \psi(t,s)}{\partial t \, \partial s} = \frac{\delta \pi(x^o,\psi,u^o)}{\delta x(t,s)} \tag{44}$$

with the boundary conditions (41) and

$$\psi(t_1,s_1) = - \frac{\partial \phi(x^o(t_1,s_1))}{\partial x} ,$$

$$\frac{\partial \psi(t,s_1)}{\partial t} = - \frac{\delta \pi(x^o,\psi,u^o)}{\delta x_s(t,s_1)} , \quad \frac{\partial \psi(t_1,s)}{\partial s} = - \frac{\delta \pi(x^o,\psi,u^o)}{\delta x_t(t_1,s)} ,$$

$$x_s(t,s) \equiv \frac{\partial x(t,s)}{\partial s} , \quad x_t(t,s) \equiv \frac{\partial x(t,s)}{\partial t} .$$

Then the control $u^o(t,s)$ satisfies the maximum condition

$$\frac{\delta_{u^*}\pi(x^o,\psi,u^o)}{\delta u(t,s)} \leq 0 \quad , \quad \{t,s\}\varepsilon\Gamma, u^*\varepsilon U \quad . \tag{45}$$

§4. An Identity for the Trajectories of Dynamical Systems

In the theory of the existence of optimal controls [56], during the study of necessary and sufficient conditions for optimality [105] it is often convenient to have a relation between the solutions of systems (1) and (8).

Let the function $f(x,u,t)$ be such that

$$f(\lambda x,u,t) = f^o(\lambda x,u,t) + \sum_{i=1}^{k} \lambda^{m_i} f^i(x,u,t) \quad .$$

Then the function $H(x,\psi,u,t)$ has the property

$$H(\lambda x,\psi,u,t) = H^o(\lambda x,\psi,u,t) + \sum_{i=1}^{k} \lambda^{m_i} H^i(x,\psi,u,t) \quad ,$$

$$H(x,\lambda\psi,u,t) = \lambda H(x,\psi,u,t) \quad ,$$

where $H^i(x,\psi,u,t) = \psi' f^i(x,u,t)$, $H^o(x,\psi,u,t) = \psi' f^o(x,u,t)$.

Theorem 10. The identity

$$\psi'(t)x(t) = \psi'(t_1)x(t_1) + \int_t^{t_1} \left[x'(\tau)\frac{\partial H^o}{\partial x} - H^o(x(\tau),\tau(\tau),u(\tau),\tau) \right] d\tau + \sum_{i=1}^{k} (m_i - 1) \int_t^{t_1} H^i(x,\psi,u,\tau)d\tau \ ,$$

is satisfied along the trajectories $x(t)$, $\psi(t)$, $t\varepsilon T$, of Eqs.(1) and (8).

Corollary. If the system of equations (1) is homogeneous in $x(f^o(x,u,t) \equiv 0)$, then

$$\psi'(t)x(t) = \psi'(t_1)x(t_1) + \sum_{i=1}^{k} (m_i - 1) \int_t^{t_1} H^i(x(\tau),\psi(\tau),u(\tau),\tau)d\tau \quad .$$

Proof. By Euler's formula for homogeneous functions, $x' \frac{\partial g(x)}{\partial x} = mg(x)$, if $g(\lambda x) = \lambda^m g(x)$. Thus, using Eqs. (1) (8), we have

$$\int_t^{t_1} \sum_{i=0}^{k} H^i(x,\psi,u,\tau)d\tau = \int_t^{t_1} H(x,\psi,u,\tau)d\tau$$

$$= \int_t^{t_1} \psi' \frac{\partial H}{\partial \psi} d\tau = \int_t^{t_1} \psi' \frac{dx}{dt} d\tau = \psi' x \Big|_t^{t_1} - \int_t^{t_1} x' \frac{d\psi}{dt} d\tau$$

$$= \psi' x \Big|_t^{t_1} + \int_t^{t_1} x' \frac{\partial H}{\partial x} d\tau$$

$$= \psi' x \Big|_t^{t_1} + \int_t^{t_1} x' \frac{\partial H^o}{\partial x} d\tau + \sum_{i=1}^{k} m_i \int_t^{t_1} H^i(x,\psi,u,\tau)d\tau .$$

Equating the last terms of this expression, we obtain the stated assertion.

§5. Singular Controls in Optimization Problems

In this paragraph, because of the high dimensionality of the matrices we apply the coordinate form of description where it is always assumed that summation takes place over repeated indices. The range of values of index is determined when it is introduced.

1. Simple Necessary Conditions of Optimality for Singular Controls

a) First Approach. Theorem 1 is ineffective for the study of optimal controls if the optimal trajedtory $x^o(t)$, $t\varepsilon T$, of system (1) is such that

$$\frac{\partial\phi(x^o(t_1))}{\partial x_i} = 0 \quad , \quad i = 1,\ldots,n \quad . \tag{46}$$

In this case, the adjoint variables $\psi_i(t)$, computed by (8), are identically zero on T and the maximum condition (9) reduces to a trivial equality, giving no information about the function $u^o(t)$, $t\varepsilon T$. In this paragraph, the singular case (40) is studied with the help of special techniques.

At first we obtain a special formula for the increment of the functional (2) along trajectories $x(t)$, $t\varepsilon T$, satisfying the condition (46).

Let $\phi(x)\varepsilon C^2$. Then

$$\Delta J(u) = \frac{1}{2}\,\frac{\partial^2\phi(x(t_1))}{\partial x_i\ \partial x_j}\,\Delta x_i(t_1)\Delta x_j(t_1) + o(||\Delta x(t_1)||^2) \quad ,$$

$$j = 1,\ldots,n \quad . \tag{47}$$

On the other hand, for continuous functions $\Delta x_i(t)$, $m_{ij}(t,s)$ with piecewise-continuous derivatives $\Delta\dot{x}_i(t)$, $\partial m_{ij}(t,s)/\partial t$, $\partial m_{ij}(t,s)/\partial s$, $\partial^2 m_{ij}(t,s)/\partial t\partial x$, we have the identity

$$m_{ij}(t_1,t_1)\Delta x_i(t_1)\Delta x_j(t_1) = \int_{t_o}^{t_1} \frac{\partial m_{ij}(t_1,s)}{\partial s}\,\Delta x_j(s)ds\Delta x_i(t_1)$$

$$+ \int_{t_o}^{t_1} \frac{\partial m_{ij}(t,t_1)}{\partial t}\,\Delta x_i(t)dt\Delta x_j(t_1)$$

$$+ \int_{t_o}^{t_1} \int_{t_o}^{t_1} m_{ij}(t,s)\Delta\dot{x}_1(t)\Delta\dot{x}_j(s)dt\,ds$$

$$- \int_{t_o}^{t_1} \int_{t_o}^{t_1} \frac{\partial^2 m_{ij}(t,s)}{\partial t\,\partial s} \Delta x_i(t)\Delta x_j(s)ds\,dt \quad . \tag{48}$$

We give the functions $m_{ij}(t,s)$ by the equations

$$\frac{\partial^2 m_{ij}(t,s)}{\partial t\,\partial s} = m_{j_1 j_2}(t,s)\frac{\partial f_{j_1}(x(t),u(t),t)}{\partial x_i} \cdot \frac{\partial f_{j_2}(x(s),u(s),s)}{\partial x_i} ,$$

$$j_1, j_2 = 1,\ldots,n \quad , \tag{49}$$

with the boundary conditions

$$m_{ij}(t_1,t_1) = -\frac{1}{2}\frac{\partial^2 \phi(x(t_1))}{\partial x_i\,\partial x_j} \quad ,$$

$$\frac{\partial m_{ij}(t,t_1)}{\partial t} = -\frac{\partial f_k(x,u,t)}{\partial x_i} m_{kj}(t,t_1) \;,\; t_o \le t \le t_1 \;, \tag{50}$$

$$\frac{\partial m_{ij}(t_1,s)}{\partial s} = -\, m_{ik}(t_1,s)\frac{\partial f_k(x,u,s)}{\partial x_j} \;,\; t_o \le s \le t_1 \;.$$

Substituting these conditions into (47),(48), after some elementary transformations (§1) we obtain

$$\Delta J(u) = -\int_{t_o}^{t_1} \int_{t_o}^{t_1} m_{ij}(t,s)\Delta_{\tilde{u}} f_i(x,u,t)\Delta_{\tilde{u}} f_j(x,u,s)ds\,dt + \eta, \tag{51}$$

where

$$\Delta_{\tilde{u}} f_i(x,u,t) = f_i(x,\tilde{u},t) - f_i(x,u,t) \quad ,$$

$$\eta = \eta_1 + \eta_2 + \eta_3 \quad , \quad \eta_1 = o(||\Delta x(t_1)||^2) \quad .$$

Here η_2, η_3 are remainder terms whose explicit forms are omitted due to their awkwardness. For what follows, it is important only that these terms on the special control increments have magnitute (in ε) higher than the terms which do appear explicitly in (51).

We make the special increment $\Delta_{\varepsilon\theta}u(t)$ defined in (10) on the control $u(t)$. From the estimate $||\Delta_{\varepsilon\theta}x(t)|| \leq \beta\varepsilon$, for the special increment $\Delta_{\varepsilon\theta}x(t)$ and from the definitions of the quantities η_1, η_2, η_3, it follows that $\eta \sim \eta^3$ on the special control increment. Thus, Eq. (51) with the help of Lemma 1, leads to

$$\Delta_{\varepsilon\theta}J(u) = -\varepsilon^2 m_{ij}(\theta,\theta)\Delta_{u*}f_i(x,u,\theta)\Delta_{u*}f_j(x,u,\theta) + o(\varepsilon^2) \quad .$$

We assume that $u(t) = u^o(t)$ is the optimal control. Then $\Delta_{\varepsilon\theta}J(u^o) \geq 0$ for any $\varepsilon > 0$, $\theta\varepsilon T$, and by Lemma 3 we are led to the following assertion:

<u>Theorem 11. Let $\phi(x)\varepsilon C^2$ and let the control $u^o(t)$, $t\varepsilon T$, minimize the functional (2), where along the optimal trajectory $x^o(t)$, $t\varepsilon T$, the system (1) satisfies condition (46). Then at each point $t\varepsilon T$, for any $u^*\varepsilon U$, we have</u>

$$m_{ij}(t,t)\Delta_{u*}f_i(x^o(t),u^o(t),t)\Delta_{u*}f_j(x^o(t),u^o(t),t) \leq 0 \quad , \tag{52}$$

<u>where $m_{ij}(t,s)\varepsilon C^2$ is the solution of Eq. (49) with the condition (50).</u>

We illustrate the application of Theorem 11 with a simple example.

<u>Example 1.</u> $\dot{x}_1 = u$, $x_1(0) = 2$, $|u| \leq 1$, $T = [0,1]$, $\phi(x) = \sin(\pi x/2)$, x,u scalars. We check two controls: 1) $u(t) \equiv 1$; 2) $u(t) \equiv -1$ which, as it is easy to see, satisfy the necessary conditions of optimality (Theorem 1)

by virtue of condition (46) being satisfied by both controls.

1) The trajectory $x(t) = 2 + t$, $x(1) = 3$, $\partial^2\phi(3)/\partial x^2 = \pi^2/4$. The equation for $m(t,s)$:

$$\frac{\partial^2 m(t,s)}{\partial t\, \partial s} \equiv 0 \ , \ m(1,1) = -\frac{\pi^2}{8} \ , \ \frac{\partial m(t,1)}{\partial t} = 0, \ \frac{\partial m(1,s)}{\partial s} = 0 \ .$$

Therefore, $m(t,s) = -\pi^2/8$, $\Delta_{u*}f(x,u,t) = u* - u$. Condition (52) has the form

$$-\frac{\pi^2}{8}(u* - 1)^2 \leq 0 \ ,$$

i.e. the control $u(t) \equiv 1$ is a candidate for optimality.

2) The trajectory $x(t) = 2 - t$, $x(1) = 1$, $\partial^2\phi(1)/\partial x^2 = -\pi^2/4$. $m(t,s) = \pi^2/8$. Condition (52) has the form

$$\frac{\pi^2}{8}(u* + 1)^2 \leq 0 \ ,$$

which shows the nonoptimality of the control $u(t) \equiv -1$.

b) A Second Approach. The necessary conditions for optimality expressed in Theorem 11 are formulated using an nxn-matrix function depending on two arguments. In the work [32j] there is described a scheme for obtaining necessary conditions in the singular case (40), which uses an auxiliary nxn-matrix function of only a single argument. A more general scheme is developed in paragraph 3, so we now present the results of [32j] for the case (46) without proof.

Theorem 12. Let $\phi(x)\varepsilon C^2$ and let condition (46) be satisfied along the optimal control $u^o(t)$, $t\varepsilon T$, for problem (1),(2). Then at each moment $t\varepsilon T$

$$\eta_{ij}(t)\Delta_{u*}f_i(x^o(t),u^o(t),t)\Delta_{u*}f_j(x^o(t),u^o(t)) \leq 0$$

for all $u^* \varepsilon U$, where the function $n_{ij}(t)$ is the solution of the equation

$$\frac{dn_{ij}(t)}{dt}$$

$$= -\frac{\partial f_{jl}(x^o(t),u^o(t),t)}{\partial x_i} n_{jlj}(t) - \frac{\partial f_{jl}(x^o(t),u^o(t),t)}{\partial x_j} n_{ijl}(t)$$

$$n_{ij}(t_1) = -\frac{1}{2}\frac{\partial^2 \phi(x^o(t_1))}{\partial x_i\ \partial x_j} \quad .$$

2. The Formula for kth Order Increments

Let two piecewise-continuous functions $u(t) = \{u_1(t),\ldots,u_r(t)\}$, $\tilde{u}(t) = \{\tilde{u}_1(t),\ldots,\tilde{u}_r(t)\}$, given on the interval $T = [t_o,t_1]$, correspond to the solutions $x(t) = \{x_1(t),\ldots,x_n(t)\}$, $\tilde{x}(t) = x(t) + \Delta x(t)$ of Eq.(1) on T. Assuming that the functions $f(x,u,t)$, $\phi(x)$ are sufficiently smooth (a precise assumption is introduced in the statement of the result), we find a formula for the increment

$$\Delta\phi(x(t_1)) = \phi(\tilde{x}(t_1)) - \phi(x(t_1)) \quad . \tag{53}$$

On the one hand, from (2) for a given k we have

$$\Delta\phi(x(t_1)) = \frac{\partial\phi(x(t_1))}{\partial x_{j_1}}\Delta x_{j_1}(t_1)$$
$$+ \frac{1}{2}\frac{\partial^2\phi(x(t_1))}{\partial x_{j_1}\ \partial x_{j_1}}\Delta x_{j_1}(t_1)\Delta x_{j_2}(t_1)$$
$$+ \cdots + \frac{1}{k!}\frac{\partial^k\phi(x(t_1))}{\partial x_{j_1}\cdots\partial x_{j_k}}\Delta x_j(t_1)\cdots\Delta x_{j_k}(t_1) + o(\|\Delta x(t_1)\|^k). \tag{54}$$

Here the index j_s assumes the values $1,\ldots,n$, for each $s = 1,\ldots,k$. On the other hand, integrating the expression

$$\int_{t_o}^{t_1} \psi_{j_1 \cdots j_\ell}(t)\Delta x_{j_1}(t) \cdots \Delta x_{j_{\ell-1}}(t)\Delta \dot{x}_{j_\ell}(t)dt \quad ,$$

$$\ell = 1,\ldots,k \quad ,$$

by parts, we obtain ($\Delta x_i(t_o) = 0$) the identity

$$\psi_{j_1 \cdots j_\ell}(t_1)\Delta x_{j_1}(t_1) \cdots \Delta x_{j_\ell}(t_1) = \int_{t_o}^{t_1} \dot{\psi}_{j_1 \cdots j_\ell}(t)\Delta x_{j_1}(t)$$

$$\cdots \Delta x_{j_\ell}(t)dt + \int_{t_o}^{t} \sum_{m=1}^{\ell} \psi_{j_1 \cdots j_\ell}(t)\Delta x_{j_1}(t)$$

$$\cdots \Delta x_{j_{m-1}}(t)\Delta \dot{x}_{j_m}(t)\Delta x_{j_{m+1}}(t) \cdots \Delta x_{j_\ell}(t)dt \quad . \qquad (55)$$

We set

$$\psi_{j_1 \cdots j_\ell}(t_1) = -\frac{1}{\ell!}\frac{\partial^\ell \phi(x(t_1))}{\partial x_{j_1} \cdots \partial x_{j_\ell}} \quad . \qquad (56)$$

Then (53), with the help of (54),(55), assumes the form

$$\Delta\phi(x(t_1)) = -\int_{t_o}^{t_1} \sum_{\ell=1}^{k} \sum_{m=1}^{\ell} \psi_{j_1 \cdots j_\ell}(t)\Delta x_{j_1}$$

$$\cdots \Delta x_{j_{m-1}} \Delta \dot{x}_{j_m} \Delta x_{j_{m+1}} \cdots \Delta x_{j_\ell} dt$$

$$-\int_{t_o}^{t_1} \sum_{\ell=1}^{k} \dot{\psi}_{j_1 \cdots j_\ell}(t)\Delta x_{j_1} \cdots \Delta x_{j_\ell} dt + o(\|\Delta x(t_1)\|^k) \quad .$$

We transform the expression

$$\int_{t_o}^{t_1} \sum_{\ell=1}^{k} \sum_{m=1}^{\ell} \psi_{j_1 \cdots j_\ell}(t)\Delta x_{j_1} \cdots \Delta x_{j_{m-1}} \Delta \dot{x}_{j_m} x_{j_{m+1}} \cdots \Delta x_{j_\ell} dt,$$

taking into account the equation

$$\Delta \dot{x}_i = f_i(x(t) + \Delta x \ , \ u(t) + \Delta u,t) - f_i(x(t),u(t),t)$$

and the variation

$$\Delta_{\tilde{u}} f_i(x,u,t) = f_i(x,\tilde{u},t) - f_i(x,u,t) \quad .$$

We have

$$\int_{t_0}^{t_1} \sum_{\ell=1}^{h} \sum_{m=1}^{\ell} \psi_{j_1 \cdots j_\ell}(t) \Delta x_{j_1} \cdots \Delta x_{j_{m-1}} \Delta \dot{x}_{j_m} x_{j_{m+1}} \cdots \Delta x_{j_\ell}(t) dt$$

$$= \int_{t_0}^{t_1} \psi_j \Delta_{\tilde{u}} f_j(x,u,t) dt$$

$$+ \int_{t_0}^{t_1} \sum_{q=1}^{k-1} \sum_{m=1}^{q+1} \psi_{j_1 \cdots j_{m-1} j_m \cdots j_q}(t) \Delta_{\tilde{u}} f_j(x,u,t) \Delta x_{j_1} \cdots \Delta x_{j_q} dt$$

$$+ \int_{t_0}^{t_1} \sum_{q=1}^{k} \sum_{\ell=1}^{q} \sum_{m=1}^{\ell} \frac{1}{(q-\ell+1)!} \psi_{j_1 \cdots j_{m-1} j_m \cdots j_{\ell-1}}(t)$$

$$\cdot \frac{\partial^{q-\ell+1} f_j(x,u,t)}{\partial x_{j_\ell} \cdots \partial x_{j_q}} \Delta x_{j_1} \cdots \Delta x_{j_q} dt$$

$$+ \int_{t_0}^{t_1} \sum_{q=1}^{k} \sum_{\ell=1}^{q} \sum_{m=1}^{\ell} \frac{1}{(q-\ell+1)!} \psi_{j_1 \cdots j_{m-1} j_m \cdots j_{\ell-1}}(t)$$

$$\cdot \frac{\partial^{q-\ell+1} \Delta_{\tilde{u}} f(x,u,t)}{\partial x_{j_\ell} \cdots \partial x_{j_q}} \Delta x_{j_1} \cdots \Delta x_{j_q} dt$$

$$+ \int_{t_0}^{t_1} \sum_{q=1}^{k} \sum_{m=1}^{q} \psi_{j_1 \cdots j_q}(t) \Delta x_{j_1} \cdots$$

$$\cdots \Delta x_{j_{m-1}} o_{j_m}(||\Delta x||^{k-q+1}) \Delta x_{j_{m+1}} \cdots \Delta x_{j_q} dt \; .$$

We define the functions $\psi_{j_1 \cdots j_q}(t)$ by the equations

$$\dot{\psi}_{j_1 \cdots j_q}(t)$$

$$= - \sum_{\ell=1}^{q} \frac{1}{(q - \ell + 1)!} \sum_{m=1}^{\ell} \psi_{j_1 \cdots j_{m-1} j_{q+m-\ell+1} \cdots j_q}(t) \cdot \frac{\partial^{q- \quad +1} f_j(x,u,t)}{\partial x_{j_m} \cdots \partial x_{j_{q+m-\ell}}} \tag{57}$$

and the initial conditions (56). Then (53) assumes the form

$$\Delta\phi(x(t_1)) = - \int_{t_0}^{t_1} \psi_j \Delta_{\tilde{u}} f_j(x,u,t)\,dt$$

$$- \int_{t_0}^{t_1} \sum_{q=1}^{k-1} \sum_{m=1}^{q+1} \psi_{j_1 \cdots j_{m-1}\ j_m \cdots j_q}(t) \Delta_{\tilde{u}} f_j(x,u,t) \Delta x_{j_1} \cdots \Delta x_{j_q}\,dt$$

$$- \int_{t_0}^{t_1} \sum_{q=1}^{k} \sum_{\ell=1}^{q} \frac{1}{(q - \ell + 1)!} \sum_{m=1}^{\ell} \psi_{j_1 \cdots j_{m-1}\ j_m \cdots j_{\ell-1}}(t) \cdot \frac{\partial^{q+\ell-1} \Delta_{\tilde{u}} f_j(x,u,t)}{\partial x_{j_\ell} \cdots \partial x_{j_q}} \Delta x_{j_1} \cdots \Delta x_{j_q}\,dt$$

$$- \int_{t_0}^{t_1} \sum_{q=1}^{k} \sum_{m=1}^{q} \psi_{j_1 \cdots j_q}(t) \Delta x_{j_1} \cdots \Delta x_{j_{m-1}}\, o_{j_m}(||\Delta x||^{k-q+1})\ r_{j_{m+1}} \cdots \Delta x_{j_q}\,dt + o(||\Delta x(t_1)||^k) \quad .$$

We introduce the notation

$$Q_{j_1 \cdots j_q}(x,\psi,u,\tilde{u},t) = \sum_{m=1}^{q+1} \psi_{j_1 \cdots j_{m-1}\ j_m \cdots j_q}(t) \Delta_{\tilde{u}} f_j(x,u,t)$$

$$+ \sum_{\ell=1}^{q} \frac{1}{(q - \ell + 1)!} \sum_{m=1}^{\ell} \psi_{j_1 \cdots j_{m-1}\ j_m \cdots j_{\ell-1}}(t) \frac{\partial^{q-\ell+1} \Delta_{\tilde{u}} f_j(x,u,t)}{\partial x_{j_\ell} \cdots \partial x_{j_q}}$$

As a result, we obtain a formula for the increment of the function $\phi(x(t_1))$:

$$\Delta\phi(x(t_1)) = - \int_{t_0}^{t_1} \psi_j(t)\Delta_u f_j(x,u,t)dt$$

$$- \int_{t_0}^{t_1} \psi_j(t) \frac{\partial \Delta_{\tilde{u}} f_j(x,u,t)}{\partial x_{j_1}} \Delta x_{j_1} dt$$

$$- \int_{t_0}^{t_1} \psi_j(t) o_j(||\Delta x||)dt + o(||\Delta x(t_1)||), \quad k = 1;$$

$$\Delta\phi(x(t_1)) = - \int_{t_0}^{t_1} \psi_j(t)\Delta_{\tilde{u}} f_j(x,u,t)dt$$

$$- \int_{t_0}^{t_1} \sum_{q=1}^{k-1} Q_{j_1 \cdots j_q}(x,\psi,u,\tilde{u},t)\Delta x_{j_1} \cdots \Delta x_{j_q} dt$$

$$- \int_{t_0}^{t_1} \sum_{q=1}^{k} \sum_{m=1}^{q} \psi_{j_1 \cdots j_q}(t)\Delta x_{j_1}$$

$$\cdots \Delta x_{j_{m-1}} o_{j_m}(||\Delta x||^{k-q+1})\Delta x_{j_{m+1}} \cdots \Delta x_{j_q} dt + o(||\Delta x(t_1)||^k),$$

$$k \geq 2 .$$

3. First Order Necessary Conditions for the Optimality of Singular Controls

The control $u(t) = \{u_\nu(t)\}$, $u(t)\varepsilon U$, $t\varepsilon T$, $\nu = 1,\ldots,r$, is called a singular control of the first order if

$$\psi_i(t)\Delta_{u*} f_i(x,u,t) = 0 \quad , \tag{59}$$

is an identity in $t\varepsilon T$, $u*\varepsilon U$, along trajectories $x(t)$, $\psi(t)$, $t\varepsilon T$, of Eqs.(1),(8), corresponding to the control $u(t)$.

In paragraph 1, we have considered first-order singular controls for the particular case when (59) is satisfied because $\psi_i(t) \equiv 0$, $t\varepsilon T$. To study the general case of first-order singular controls, we use formula (58) for $k = 2$. We form the control variation $\tilde{u}_\nu(t)$, $t\varepsilon T$, using the special increments $\Delta_{\varepsilon\theta}u_\nu(t)$, $t\varepsilon T$, defined in (6). In §1 it was shown that the increment $\Delta_{\varepsilon\theta}x(t)$, $t\varepsilon T$, of the trajectory $x(t)$, corresponding to $\Delta_{\varepsilon\theta}u(t)$, satisfies the bound

$$||\Delta_{\varepsilon\theta}x(t)|| \leq \beta\varepsilon \quad , \qquad \beta = \text{const.}$$

Applying the mean value theorem (cf. Lemma 1), it is not difficult to show that

$$\Delta_{\varepsilon\theta}x_i(t) = (t - \theta)\Delta_{u*}f_i(x(\theta),u(\theta),\theta) + o_i(t - \theta)$$

for $0 \leq t \leq \theta + \varepsilon$. Taking these facts into account, by a direct calculation we isolate the principal terms in ε on the right side of (58):

$$\Delta_{\varepsilon\theta}\phi(x(t_1)) = -\varepsilon^2 Q_i(x,\psi,u,u^*,\theta)\Delta_{u*}f_i(x,u,\theta) + o(\varepsilon^2) \quad , \tag{60}$$

(the coefficient of ε equals zero by virtue of (59)). We consider the problem of minimizing the function (2) along trajectories of Eq.(1) and we assume that $u^o(t)$, $t\varepsilon T$, is the optimal first-order singular control. Then for any $\theta\varepsilon T$, $\varepsilon > 0$, $\theta + \varepsilon \;\varepsilon\; T$ and $u^*\varepsilon U$, the quantity $\Delta_{\varepsilon\theta}J(u^o)$ is nonnegative. Thus, by virtue of Lemma 3 we may draw the following conclusion from (60).

Theorem 13. Let the functions $f(x,u,t)$, $\phi(x)$ be continuous and defined together with $\partial f_i(x,u,t)/\partial x_j$, $\partial^2 f_i(x,u,t)/\partial x_{j_1}\partial x_{j_2}$, $\partial\phi(x)/\partial x_i$, $\partial^2\phi(x)/\partial x_i\partial x_j$. Then, in order that the first order singular control $u^o(t)$, $t\varepsilon T$, be optimal for the problem (1),(2), the condition

$$[\psi_{ij}(t) + \psi_{ji}(t)]\Delta_{u*}f_i(x^o(t),u^o(t),t)\Delta_{u*}f_j(x^o(t),u^o(t),t)$$

$$+ \psi_{j_1}(t)\frac{\partial \Delta_{u*} f_{j_1}(x^o(t),u^o(t),t)}{\partial x_j} \Delta_{u*}f_j(x^o(t),u^o(t),t) \leq 0 \quad (61)$$

must be satisfied for all $t\varepsilon T$, $u*\varepsilon U$. Here $x^o(t)$, $t\varepsilon T$, is the trajectory of the system (1) corresponding to the control $u^o(t)$, $t\varepsilon T$; $\psi_i(t)$, $\psi_{ij}(t)$, $t\varepsilon T$, are the solutions of the equations

$$\dot{\psi}_i(t) = -\frac{\partial f_{j_1}(x^o(t),u^o(t),t)}{\partial x_i}\psi_i(t) \quad , \quad \psi_i(t_1) = -\frac{\partial \phi(x^o(t_1))}{\partial x_i};$$

$$\dot{\psi}_{ij}(t) = -\frac{\partial f_{ji}(x^o(t),u^o(t),t)}{\partial x_i}\psi_{j_1 j}(t)$$

$$-\frac{\partial f_{j_1}(x^o(t),u^o(t),t)}{\partial x_j}\psi_{ij_1}(t) - \frac{1}{2}\frac{\partial^2 f_{j_1}(x^o(t),u^o(t),t)}{\partial x_i\,\partial x_j}\psi_{j_1}(t)$$

$$\psi_{ij}(t_1) = -\frac{1}{2}\frac{\partial^2 \phi(x^o(t_1))}{\partial x_i\,\partial x_j} \quad .$$

We illustrate the applicatoin of Theorem 13 with two simple examples.

Example 2. $\dot{x}_1 = u$, $\dot{x}_2 = -x_1^2$, $x_1(o) = x_2(o) = 0$, $T = [0,1]$, $U = \{u: |u| \leq 1\}$, $\phi(x) = x_2$.

The control $u(t) \equiv 0$ is singular of the first order, since along it we have $x_1(t) = x_2(t) \equiv 0$, $\psi_1(t) \equiv 0$ and the function $H(x,\psi,u,t) = \psi_1 u - \psi_2 x_1^2$ does not depend on u. The equations for $\psi_i(t)$, $\psi_{ij}(t)$ have the form

$$\dot{\psi}_1 = 2\psi_2 x \;,\quad \dot{\psi}_2 = 0 \;,\quad \psi_1(1) = 0 \;,\quad \psi_2(1) = -1 \;;$$

$$\begin{Bmatrix} \dot{\psi}_{11} & \dot{\psi}_{12} \\ \dot{\psi}_{21} & \dot{\psi}_{22} \end{Bmatrix} = - \begin{Bmatrix} 0 & -2x_1 \\ 0 & 0 \end{Bmatrix} \begin{Bmatrix} \psi_{11} & \psi_{12} \\ \psi_{21} & \psi_{22} \end{Bmatrix} - \begin{Bmatrix} \psi_{11} & \psi_{12} \\ \psi_{21} & \psi_{22} \end{Bmatrix} \begin{Bmatrix} 0 & 0 \\ -2x_1 & 0 \end{Bmatrix}$$

$$- \begin{Bmatrix} 1 & 0 \\ 0 & 0 \end{Bmatrix} , \quad \begin{Bmatrix} \psi_{11}(1) & \psi_{12}(1) \\ \psi_{21}(1) & \psi_{22}(1) \end{Bmatrix} = \begin{Bmatrix} 0 & 0 \\ 0 & 0 \end{Bmatrix} .$$

Hence

$$\psi_{12}(t) = \psi_{21}(t) = \psi_{22}(t) \equiv 0 \;,\quad \psi_{11}(t) = 1 - t \;.$$

The necessary condition for optimality (61) leads to the inequality $(1-t)(u^* - u(t))^2 \leq 0$, which is violated if $t < 1$. Thus, the control $u(t) \equiv 0$ is not optimal.

Example 3. The equations for vertical rocket flight in a homogeneous gravitational field under constant resistance have the form

$$\dot{x}_1 = x_2 \;,\quad \dot{x}_2 = \frac{u - 4}{x_3} - g \;,\quad \dot{x}_3 = -\frac{1}{2} u \;,$$

$$x_1(0) = x_2(0) = 0 \;,\quad x_3(0) = 2 \;,$$

where x_1 is the relative height, x_2 is the velocity of the rocket, and x_3 is the rocket's mass. Let it be required to find a program of thrust application u, $|u| \leq 10$, so that the quantity $x_1(1) + x_3(1)$ is maximized at $t = 1$.

Setting $\phi(x) = -x_3 - x_1$, we are led to the problem (1),(2). It is not difficult to check that the control $u(t) \equiv 4$ is a singular first-order control. However, it is not optimal since the necessary condition (61) is not satisfied here (along $u(t) \equiv 4$, the left side of (61) equals

$$-\frac{(t-1)}{2x_3^2}(u^* - 4)^2 \quad ,$$

which is positive for $t < 1$).

4. Second-Order Singular Controls

The control $u(t) = \{u_\nu(t)\}$, $t\varepsilon T$, is called a <u>second-order singular control</u> if along it we have the following identity satisfied for $t\varepsilon T$, $u^*\varepsilon U$:

$$\psi_i(t)\Delta_{u^*}f_i(x(t),u(t),t) = 0 \quad ,$$

$$Q_i(x(t),\psi(t),u(t),u^*,t)\Delta_{u^*}f_i(x(t),u(t),t) = 0 \quad .$$

We introduce the set $V(u^*)$ of r-dimensional vectors; the vector $v^* = \{v_\nu^*\}\varepsilon V(u^*)$ if there exists a differentiable function $u_\nu^*(t)$ such that $u_\nu^*(0) = u^*$, $\dot{u}_\nu^*(0) = v_\nu^*$ and $\{u_\nu^*(t),t\} \varepsilon \{U,t\}$, $0 \leq t \leq \varepsilon$, for ε sufficently small. If $u \varepsilon \text{ int } U$, then the set $V(u)$ contains a sphere of arbitrarily large radius.

To obtain necessary conditions for the optimality of second-order singular controls, we use special control increments of the form $\Delta_{\varepsilon\theta}u(t)$

$$= \begin{Bmatrix} u(t) \\ u^*(t-\theta) - u(t), & \theta \leq t < \theta+\varepsilon, \; u^*\varepsilon U \quad , \quad v^*\varepsilon V(u^*) \;, \\ 0 \quad , & t_o \leq t < \theta, \quad \theta+\varepsilon \leq t \leq t_1 \quad . \end{Bmatrix} \qquad (62)$$

Let $u(t)$, $t\varepsilon T$ be a piecewise-continuous control with a piecewise-continuous derivative. In the increment (58), computed along the control $u(t)$, and in the special increments (62), we separate the leading terms in ε.

We have (θ is a point of smoothness of the functions $u_\nu(t)$, $t\varepsilon T$)

$$\Delta_{\varepsilon\theta}J(u) = \Delta_{\varepsilon\theta}\phi(x(t_1))$$

$$= -\int_\theta^{\theta+\varepsilon} Q_i(x(t),\psi(t),u(t),u^*(t-\theta),t)\Delta_{\varepsilon\theta}x_i(t)dt$$

$$- \int_\theta^{\theta+\varepsilon} Q_{ij}(x(t),\psi(t),u(t),u^*(t-\theta),t)\Delta_{\varepsilon\theta}x_i(t)\Delta_{\varepsilon\theta}\dot{x}_j(t)dt$$

$$+ o(\varepsilon^3) \quad .$$

Taking into account that $\Delta_{\varepsilon\theta}x_i(\theta) = 0$, we obtain

$$\Delta_{\varepsilon\theta}J(u) = -\frac{1}{6}\left[Q_i(x(\theta),\psi(\theta),u(\theta),u^*,\theta)\frac{d^2\Delta_{\varepsilon\theta}x_i(\theta + 0)}{dt^2}\right.$$

$$\left.+ 2\left.\frac{dQ_i(x(t),\psi(t),u(t),u^*(t-\theta),t)}{dt}\right|_{t=\theta+0}\Delta_{u^*}f_i(x(\theta),u(\theta),\theta)\right]$$

$$\cdot\ \varepsilon^3 - \frac{1}{3}Q_{ij}(x(\theta),\psi(\theta),u(\theta),u^*,\theta)\Delta_{u^*}f_i(x(\theta),u(\theta),\theta)$$

$$\cdot\ \Delta_{u^*}f_j(x(\theta),u(\theta),\theta)\varepsilon^3 + o(\varepsilon^3) \quad .$$

We now substitute for the values of $d^2\Delta_{\varepsilon\theta}x_i(\theta+0)/dt^2$, dQ_i/dt:

$$\frac{d^2\Delta_{\varepsilon\theta}x_i(\theta + 0)}{dt^2} = \frac{\partial f_i(x(\theta),u^*\theta)}{\partial x_j} f_j(x(\theta),u^*,\theta)$$

$$- \frac{\partial f_i(x(\theta),u(\theta),\theta)}{\partial x_j} f_j(x(\theta),u(\theta),\theta)$$

$$+ \frac{\partial f_i(x(\theta),u^*,\theta)}{\partial u_\nu} v^*_\nu - \frac{\partial f_i(x(\theta),u(\theta),\theta)}{\partial u_\nu} \dot{u}_\nu(\theta)$$

$$+ \frac{\partial \Delta_{u^*} f_i(x(\theta),u(\theta),\theta)}{\partial t} \; ; \; \left.\frac{dQ_i(x(t),\psi(t),u(t),u^*(t-\theta),t)}{dt}\right|_{t=\theta+0}$$

$$= \left\{\frac{d}{dt}\left[\psi_{ij}(\theta) + \psi_{ji}(\theta) + \psi_j(\theta)\frac{\partial}{\partial x_j}\right]\right\}\Delta_{u^*} f_j(x(\theta),u(\theta),\theta)$$

$$+ \left[\psi_{ij}(\theta) + \psi_{ji}(\theta) + \psi_j(\theta)\frac{\partial}{\partial x_i}\right]\left[\frac{\partial f_j(x(\theta),u^*,\theta)}{\partial u_\nu} v^*_\nu\right.$$

$$\left. - \frac{\partial f_j(x(\theta),u(\theta),\theta)}{\partial u_\nu} \dot{u}_\nu(\theta)\right]$$

Finally, the increment $\Delta_{\varepsilon\theta}J(u)$ may be written in the following form:

$$\Delta_{\varepsilon\theta}J(u) = -\varepsilon^3 R(x,\psi,u,\dot{u},u^*,\theta)$$

$$-\varepsilon^3 P_\nu(x,\psi,u,u^*,\theta)v^*_\nu + o(\varepsilon^3) \quad , \tag{63}$$

where

$$R(x,\psi,u,\dot{u},u^*,\theta)$$

$$= \frac{1}{6} Q_i(x,\psi,u,u^*,\theta)\left[\frac{\partial f_i(x,u^*,\theta)}{\partial x_j} f_j(x,u^*,\theta)\right.$$

$$- \frac{\partial f_i(x,u,\theta)}{\partial x_j} f_j(x,u,\theta)\Bigg] + \frac{1}{3} \Delta_{u*} f_i(x,u,\theta)$$

$$\cdot \left\{ \frac{d}{dt} \left[\psi_{ij}(\theta) + \psi_{ji}(\theta) + \psi_j(\theta) \frac{\partial}{\partial x_i} \right] \right\} \Delta_{u*} f_j(x,u,\theta)$$

$$+ \frac{1}{6} Q_i(x,\psi,u,u*,\theta) \frac{\partial f_i(x,u,\theta)}{\partial u_\nu} \dot{u}_\nu(\theta)$$

$$+ \frac{1}{3} \Delta_{u*} f_i(x,u,\theta) \left[\psi_{ij}(\theta) + \psi_{ji}(\theta) + \psi_j(\theta) \frac{\partial}{\partial x_i} \right]$$

$$\cdot \frac{\partial f_i(x,u,\theta)}{\partial u_\nu} \dot{u}_\nu(\theta) + \frac{1}{6} Q_i(x,\psi,u,u*,\theta) \frac{\partial \Delta_{u*} f_i(x,u,\theta)}{\partial t}$$

$$+ \frac{1}{3} Q_{ij}(x,\psi,u,u*,\theta) \Delta_{u*} f_i(x,u,\theta) \Delta_{u*} f_j(x,u,\theta) \quad ,$$

$$P_\nu(x,\psi,u,u* \theta) = \frac{1}{6} Q_i(x,\psi,u,u*,\theta) \frac{\partial f_i(x,u*,\theta)}{\partial u_\nu}$$

$$+ \frac{1}{3} \Delta_{u*} f_i(x,u,\theta) \left[\psi_{ij}(\theta) + \psi_{ji}(\theta) + \psi_j(\theta) \frac{\partial}{\partial x_i} \right] \frac{\partial f_j(x,u*,\theta)}{\partial u_\nu}$$

Let $u^o(t)$ be the optimal second order singular control. Then for all $\theta \varepsilon T$, $\varepsilon > 0$, $\theta + \varepsilon \, \varepsilon T$, $u^* \varepsilon U$, $v^* \varepsilon V(u^*)$, we have the inequality $\Delta_{\varepsilon\theta} J(u^o) \geq 0$ and, from (63), by virtue of Lemma 3, we obtain the following assertion:

Theorem 14. Let the functions $f_i(x,u,t)$, $\phi(x)$ be defined and continuous, together with the functions $\partial f_i(x,u,t)/\partial x_j$, $\partial \phi(x)/\partial x_i$, $\partial f_i(x,u,t)/\partial u$, $\partial f_i(x,u,t)/\partial t$, $\partial^2 f_i(x,u,t)/\partial x_{j_1} \partial x_{j_2}$, $\partial^2 f_i(x,u,t)/\partial x_j \partial t$,

$\partial^2 f_i(x,u,t)/\partial x_j\, \partial u_\nu,\ \partial^2\phi(x)/\partial x_i\, \partial x_j,$
$\partial^3 f_i(x,u,t)/\partial x_{j_1}\, \partial x_{j_2}\, \partial x_{j_3},\ \partial^3 f_i(x,u,t)/\partial x_{j_1}\, \partial x_{j_2}\, \partial u_\nu,$
$\partial^3 f_i(x,u,t)/\partial x_{j_1}\, \partial x_{j_2}\, \partial t,\ \partial^3\phi(x)/\partial x_{j_1}\, \partial x_{j_2}\, \partial x_{j_3}.$

In order that the piecewise-continuous control $u^o(t)$, $t\varepsilon T$, having the piecewise-continuous derivative $u^o(t)$, be an optimal second-order singular control in the problem (1),(2), it is necessary that for all $t\varepsilon T$, $u^*\varepsilon U$, $v^*\varepsilon V(u^*)$, we have the inequality

$$R(x^o(t),\psi(t),u^o(t),u^o(t),u^*,t) + P_\nu(x^o(t),\psi(t), u^o(t),u^*,t)v_\nu^* \leq 0 \quad .$$

If U contains interior points, then in addition

$$P_\nu(x^o(t),\psi(t),u^o(t),u^*,t) = 0$$

for all $t\varepsilon T$, $u^*\ \varepsilon$ int U.

5. Toward Necessary Conditions for Optimality of kth-Order Singular Controls

Further development of a scheme for studying singular controls in the general case runs across difficulties connected with describing the optimality conditions in a compact form. Below we study a particular case of singular controls of arbitrary order when the necessary conditions for optimality may be simply formulated.

Let the optimal control $u^o(t)$, $t\varepsilon T$, in problem (1), (2) be such that we have the equalities

$$\left.\begin{aligned} &\psi_i(t)\Delta_{u^*} f_i(x^o(t),u^o(t),t) = 0 \quad , \\ &Q_{j_1}\cdots_{j_\ell}(x^o(t),\psi(t),u^o(t),u^*,t) = 0 \ , \quad \ell = 1,\ldots,k-1 \end{aligned}\right\} \quad (64)$$

satisfied for all $t \in T$, $u^* \in U$. Then the principal term in the increment $\Delta_{\varepsilon\theta} J(u^o)$, computed along $u^o(t)$ and $\Delta_{\varepsilon\theta} u(t)$ from (6), has order ε^{k+1} and may be written in the form

$$\Delta_{\varepsilon\theta} J(u^o) = -\varepsilon^{k+1} Q_{j_1 \cdots j_k}(x^o(t), \psi(t), u^o(t), u^*, t)$$

$$\cdot \Delta_{u^*} f_{j_1}(x^o(t), u^o(t), t) \cdots \Delta_{u^*} f_{j_k}(x^o(t), u^o(t), t) + o(\varepsilon^{k+1}) .$$

Thus, for the problem (1),(2) we may state a theorem.

Theorem 15. Let the functions $f_i(x,u,t)$, $\phi(x)$ be defined and continuous, together with the functions $\partial^\alpha f_i(x,u,t)/\partial x_{j_1} \cdots \partial x_{j_\alpha}$, $\partial^\alpha \phi(x)/\partial x_{j_1} \cdots \partial x_{j_\alpha}$, $\alpha = 1, \ldots, k+1$.

For the optimality of the piecewise-continous control $u^o(t)$, $t \in T$, satisfying condition (64), it is necessary that we have the inequality

$$Q_{j_1 \cdots j_k}(x^o(t), \psi(t), u^o(t), u^*, t) \Delta_{u^*} f_{j_1}(x^o(t), u^o(t), t)$$

$$\cdots \Delta_{u^*} f_{j_k}(x^o(t), u^o(t), t) \leq 0 ,$$

for all $t \in T$, $u^* \in U$ (the functions $Q_{j_1 \cdots j_k}(x, \psi, u, u^*, t)$ were defined in paragraph 2).

§6. The Maximum Principle for Pontryagin Extremals

A fundamental result in the theory of optimal processes--the Pontryagin maximum principle--is a powerful tool for isolating a subset of the set of all admissible controls, within which the optimal controls are located, if they exist. The set of extremals may contain either a finite or infinite number of elements. Therefore, a natural desire is to define additional conditions on the set of extremals so that the even smaller subset of optimal controls is obtained. It is desirable to continue this process further in order that at each step we single out an optimal control.

In this paragraph, we undertake an attempt to realize this program.

1. Pontryagin Extremals

Let there be defined a class of admissible controls on $T = [t_0, t_1]$--the set of r-dimensional piecewise-continuous vector functions $u(t) = \{u^1(t), \ldots, u^r(t)\}$ with values in a given set U:

$$u(t) \in U \quad , \quad t \in T \quad . \tag{65}$$

We consider the optimization problem of minimizing the functional

$$J(u) = c'x(t_1) \quad , \tag{66}$$

over all admissible controls, where J is defined on the trajectories $x(t) = \{x_1(t), \ldots, x_n(t)\}$ of the system

$$\frac{dx}{dt} = f(x,u) \quad , \quad x(t_0) = x_0 \quad , \quad t \in T \quad . \tag{67}$$

Here c, x_0 are constant n-vectors, while $f(x,u)$ is a function which is continuous, together with $\partial f(x,u)/\partial x$. We call an admissible control $u(t)$, $t \in T$, a Pontryagin extremal for the problem (65)-(67) if, together with the corresponding functions $x(t)$, $\psi(t) = \{\psi_1(t), \ldots, \psi_n(t)\}$, it satisfies the conditions

$$H(x(t), \psi(t), u(t)) = \max_{u \in U} H(x(t), \psi(t), u) \quad ,$$

$$\frac{dx}{dt} = \frac{\partial H(x(t), \psi(t), u(t))}{\partial \psi} \quad , \quad x(t_0) = x_0 \quad ,$$

$$\frac{d\psi}{dt} = -\frac{\partial H(x(t),\psi(t),u(t))}{\partial x} \quad , \quad \psi(t_1) = -c \quad ,$$

$$H(x,\psi,u) = \psi' f(x,u) \quad .$$

According to the maximum principle, the optimal control in the problem (65)-(67) must be a Pontryagin extremal. If the extremal is unique, then the optimization problem is finished when we find it. The situation is more complicated in those problems which admit several extremals. Even in simple cases the number of extremals may turn out to be infinite.

Example 4. $\dot{x}_1 = u$, $\dot{x}_2 = -x_1^2$, $x_1(o) = x_2(o) = 0$, $T = [0,\varepsilon]$, $J(u) = -x_2(\varepsilon)$, $U = \{u: \ |u| \leq 1\}$.

It is not difficult to see that for each $\varepsilon \geq 0$, the controls

$$u^p(t) = \pm \text{ sign } \cos \frac{2p+1}{2\varepsilon} \pi t \quad , \quad p = 0,1,\ldots.$$

are among the extremals for this problem.

2. The Maximum Principle for Pontryagin Extremals

Lemma 5. The value of the functional

$$J(v) = c'y(t_1) \tag{68}$$

along trajectories of the system

$$\frac{dy}{dt} = f(y,v) \quad , \quad y(t') = 0 \quad , \quad t' \leq t \leq t_1 \quad ,$$

is given by the formula

$$J(v) = - \int_{t'}^{t_1} H(y(t),\psi(t),v(t))dt + \int_{t'}^{t_1} \frac{\partial H'(y(t),\psi(t),v(t))}{\partial x} y(t)dt \quad .$$

Lemma 6. The value of the functional (68) equals

$$J(v) = - \int_{t'}^{t_1} H(y(t),\psi(t),v(t))dt - \int_{t'}^{t_1} y'(t) \cdot \left\{ [\Psi(t) + \Psi'(t)]f(y(t),v(t)) - \frac{\partial H(y(t),\psi(t),v(t))}{\partial x} \right\} dt$$
$$+ \int_{t'}^{t_1} y'(t) \left[\Psi(t) \frac{\partial f(y,v)}{\partial x} + \frac{\partial f'(y,v)}{\partial x} \Psi(t) + \frac{1}{2} \frac{\partial^2 H(y,\psi,v)}{\partial x^2} \right] \cdot y(t)dt \quad .$$

Here $\Psi(t)$ is the solution of the system

$$\frac{d\Psi(t)}{dt} = - \Psi(t)\frac{\partial f(y,v)}{\partial x} - \frac{\partial f'(y,v)}{\partial x} \Psi(t) - \frac{1}{2} \frac{\partial^2 H(y,\psi,v)}{\partial x^2} \quad ,$$
$$\Psi(t_1) = 0 \quad .$$

Lemma 7. If the functional J(v) admits the expansion

$$J(v) = \theta a_1 + \theta^2 a_2 + \cdots + \varepsilon^{k-1} a_{k-1} + \varepsilon^k a_k + o(\varepsilon^k), \tag{69}$$

for all sufficiently small ε and all controls v(t), $t' \le t \le t_1$ from some set $U(\cdot)$, where the numbers $a_1,\ldots,a_k$ are independent of ε, then from the condition

$$J(v^o) = \min_{v \varepsilon U(\cdot)} J(v)$$

follows

$$a_k(v^o) = \min_{v \varepsilon U(\cdot)} a_k(v) \quad .$$

Proof. For $v^o \varepsilon U(\cdot)$ and $v \varepsilon U(\cdot)$, $v \neq v^o$, we have

$$J(v) - J(v^o) \geq 0 \quad ,$$

which, by virtue of the expansion (69) and the properties of the numbers $a_1, \ldots, a_{k-1}$, implies the inequality

$$\varepsilon^k (a_k(v) - a_k(v^o)) + o(\varepsilon) \geq 0 \quad .$$

Lemma 7 has been reduced to Lemma 3.

We denote the set of Pontryagin extremals for the problem (65)-(67) by Ω. It is known that the function

$$H_u(t) = H(x(t), \psi(t), u(t)) \quad ,$$

calculated along Pontryagin extremals and the corresponding functions $x(t)$, $\psi(t)$, equals a constant:

$$H_u(t) \equiv \text{const} \quad , \quad t \varepsilon T \quad . \tag{70}$$

We will say that the Pontryagin extremal $u^*(t)$, $t \varepsilon T$, satisfies the maximum condition if

$$H_{u^*}(t_1) = \max_{u(\cdot) \varepsilon \Omega} H_u(t_1) \quad . \tag{71}$$

By virtue of the property (70), condition (71) may be verified at every point $t \varepsilon T$.

Theorem 16 (The Maximum Principle for Pontryagin Extremals). The optimal control $u^o(t)$, $t \varepsilon T$, in problem (65) - (67) for sufficiently small $t_1 - t_o$, satisfies the maximum condition for a Pontryagin extremal.

Proof. From the principle of optimality, it follows that if $u^o(t)$, $t \varepsilon T$, is the optimal control in the problem (65)-(67), then for any t', $t_o \leq t' \leq t_1$, the control $v^o(t) = u^o(t)$, $t' \leq t \leq t_1$, will be optimal for the problem of minimizing the functional

$$J(v) = c'z(t_1) \quad ,$$

along trajectories of the system

$$\frac{dz}{dt} = f(z,v) \quad , \quad z(t') = x^o(t') \quad , \quad t' \leq t \leq t_1 \quad , \quad v(t) \varepsilon U \ .$$

We introduce the variable

$$y(t) = z(t) - x^o(t') \quad , \quad t' \leq t \leq t_1 \quad .$$

Then the control $v^o(t)$ will minimize the functional

$$J(v) = c'y(t_1) \quad ,$$

along trajectories of the system

$$\frac{dy}{dt} = f(y + x^o(t'),v) \quad , \quad y(t') = 0 \quad .$$

Let $t_1 - t' = \varepsilon$, and let $v(t)$, $t' \leq t \leq t_1$, be a Pontryagin extremal. Then from Lemma 5, property (70), and the estimate

$$||y(t)|| \leq L\varepsilon \quad , \quad t' \leq t \leq t_1 \quad ,$$

we obtain

$$J(v) = -H_v(t_1)\varepsilon + o(\varepsilon) \quad ,$$

which, by virtue of Lemma 7, gives

$$H_{v^o}(t_1 = \max H_v(t_1) \quad .$$

This property is equivalent to the assertion of the theorem if $t_1 - t_o$ is sufficiently small.

Remark. Theorem 16 is true for problems of the type (65)-(67) in which the time t_1 is not fixed, or when constraints $x(t_1)\varepsilon G$ are imposed upon the point $x(t_1)$, where G is some set.

Application of Theorem 16 to Example 4 shows that the controls $u^p(t)$, $p \leq 1$, will not be optimal if ε is sufficiently small.

3. Nonstationary Systems

In the problem (65)-(67), let Eq.(67) be replaced by the nonstationary version

$$\frac{dx}{dt} = f(x,y,t) \quad , \quad x(t_o) = x_o \quad . \tag{72}$$

Introducing the additional variable $x_{n+1}(t) = t$, we reduce the problem (65),(66),(72) to the problem (65)-(67) with $dx_{n+1}(t)/dt = 1$.

Theorem 17. If the interval length $t_1 - t_o$ in problem (65),(66) is sufficiently small, then the optimal control $u^o(t)$, $t\varepsilon T$, satisfies the maximum condition (71) for Pontryagin extremals.

Remark. In contrast to the stationary case, the condition (71) is, generally speaking, true only for $t = t_1$.

4. Necessary Conditions of Optimality for First-Order Extremals

Being only a necessary condition for optimality, property (71) does not isolate only optimal controls. In other words, there may be nonoptimal controls among the extremals satisfying the condition (71).

Example 5. $\dot{x}_1 = u$, $\dot{x}_2 = \frac{\pi}{2} u \cos \frac{\pi x_1}{2}$, $x_1(0) = x_2(0) = 0$, $T = [0,1]$, $J(u) = x_2(1)$, $U = \{u: |u| \leq 1\}$.

Here the controls $u^1(t) \equiv -1$, $u^2(t) \equiv +1$ are extremals and

$$H_{u^1}(1) = H_{u^2}(1) = \max_{u(\cdot)\varepsilon\Omega} H_u(1) = 0 \quad .$$

However, the control $u^2(t) \equiv +1$ is not optimal.

We call admissible controls satisfying condition (71) and which are extremals for the problem (65)-(67), first-order extremals.

We denote the set of first-order extremals for the problem (65)-(67) by Ω_1.

We will say that the maximum condition is satisfied for first-order extremals if

$$\Xi_u(t_1) = \max_{u(\cdot)\varepsilon\Omega_1} \Xi_u(t_1) \quad , \tag{73}$$

where

$$\Xi_u(t_1) = - \frac{\partial H'(x(t_1),\psi(t_1),u(t_1))}{\partial x} f(x(t_1),u(t_1)) \quad .$$

Theorem 18 (The Maximum Principle for First-Order Extremals). Let the interval $t_1 - t_0$ be sufficiently small in the problem (65)-(67). Then an optimal control $u^0(t)$, $t\varepsilon T$, from the class of piecewise-continuous functions $u(t)$, $u(t)\varepsilon U$, satisfies the maximum condition (73) for first-order extremals.

The proof of this theorem proceeds in the same way as the proof of Theorem 16, using Lemma 6 (instead of Lemma 5).

Application of Theorem 18 to Example 5 shows that the control $u^2(t) \equiv +1$ cannot be optimal.

§7. Optimization Problems with Parameters

1. Necessary Conditions for Optimality

Up to now, optimal controls have been obtained from Pontryagin extremals by the introduction of new necessary conditions on the set Ω. It is possible to move in another direction to attain these goals. Each extremal is characterized by some initial contion, $\psi(t_0) = \psi_0$, on the adjoint variable. Thus, seeking $u(t)$ from Eq.(67) by using the maximum principle

$$u(t) = u(x(t), \psi(t)) ,$$

we are led to the following optimization problem with parameters: find the minimum of the function $I(\psi_0) = c'x(t_1)$ along trajectories of the equation

$$\frac{dx}{dt} = \bar{f}(x, \psi) \quad , \quad x(t_0) = x_0 \quad ,$$

$$\frac{dx}{dt} = -A'(x, \psi)\psi \quad , \quad \psi(t_0) = \psi_0 \quad , \quad \psi(t_1) = -c \quad .$$

Here

$$\overline{f}(x,\psi) = f(x,u(x,\psi)) \ , \quad A(x,\psi) = \frac{\partial f(x,u(x,\psi))}{\partial x} \quad .$$

In this paragraph we consider only free-endpoint problems. The techniques needed to pass to problems with constrained or fixed terminal conditions are described in [114b].

Thus, let there be given two sets V and W, p- and q-dimensional vector spaces, respectively. Among, the elements $v \varepsilon V$, $w \varepsilon W$ of these sets, it is required to find those optimal parameters v^o, w^o such that

$$I(v^o,u^o) = \min_{v\varepsilon V, w\varepsilon W} I(v,w) \quad , \tag{74}$$

where $I(v,w) = c'x(t_1)$ is a function defined on the trajectories of the system

$$\frac{dx}{dt} = f(x,w) \quad , \quad t\varepsilon T = [t_o,t_1] \quad , \quad w\varepsilon W \quad , \tag{75}$$

with the initial condition

$$x(t_o) = g(v) \quad , \quad v\varepsilon V \quad . \tag{76}$$

Without further statement, we will always assume that the functions $f(x,w)$, $g(v)$ are continuous, together with the function $\partial f(x,w)/\partial x$. We use the scheme of §9.1,9.4 to study the problem of optimization with parameters in this paragraph. Below, we produce a formulation of the basic results.

We let $\sigma(z,Z)$ denote the star-shaped neighborhood of the point z relative to the set Z: $y\varepsilon\sigma(z,Z)$ if there is a sequence of numbers ε_i, $i = 1,2,\ldots,$; $\varepsilon_1 = 1$, $\varepsilon_i \to 0$, $i \to \infty$, such that $(1 - \varepsilon_i)z + \varepsilon_i y\varepsilon Z$ for all $i \geq 1$. We introduce the function

$$H(x,\psi,w) = \psi' f(x,w) \ , \ h(\psi,v) = \psi' g(v) \ .$$

We will let $x^o(t)$ and $\psi(t)$ denote solutions to the basic system (75) and to the adjoint system

$$\frac{d\psi}{dt} = - \frac{\partial H(x,\psi,w)}{\partial x} \ , \qquad \psi(t_1) = -c \ ,$$

corresponding to the optimal parameters v^o, w^o. An expression of the type $f(x,W)$ denotes the set

$$\{z: \ z = f(x,w), w \varepsilon W\} \ .$$

<u>Theorem 19. For the problem (74)-(76) we have the following assertions:</u>

1) α) $H(x^o(t),\psi(t),w^o) \geq \frac{1}{t_1 - t_o} \int_{t_o}^{t_1} H(x^o(t),\psi(t),w)dt$,

<u>for all w such that for $t\varepsilon T$,</u>

$$f(x^o(t),w)\varepsilon\sigma(f(x^o(t),w^o),f(x^o(t),W));$$

β) $h(\psi(t_o),v^o) \leq h(\psi(t_o),v)$

<u>for all v such $g(v)\varepsilon\sigma(g(v^o),vg(V))$</u>.

2. <u>The Maximum Principle. If for each x the sets $f(x,W)$, $g(V)$ are convex, then</u>

α) $H(x^o(t),\psi(t),w^o)$

$$= \max_{w\varepsilon W} \frac{1}{t_1 - t_o} \int_{t_o}^{t_1} H(x^o(t),\psi(t),w)dt \ ; \qquad (77)$$

$$\beta) \quad h(\psi(t_o),v^o) = \max_{v \in V} h(\psi(t_o),v) \quad . \tag{78}$$

3. If the functions $f(x,w)$, $g(v)$ are differentiable in v,w, then

$$\alpha) \quad \int_{t_o}^{t_1} \frac{\partial H'(x^o(t),\psi(t),w^o)}{\partial w} dt\, w^o \geq \int_{t_o}^{t_1} \frac{\partial H'(x^o(t),\psi(t),w^o}{\partial w} dt\, w$$

for all $w \in \sigma(w^o,W)$;

$$\beta) \quad \frac{\partial h'(\psi(t_o),v^o)}{\partial v} v^o \geq \frac{\partial h'(\psi(t_o),v^o)}{\partial v} v$$

4. If the functions $f(x,w)$, $g(v)$ are differentiable in w,v and the sets W,V are convex, then

$$\alpha) \quad \int_{t_o}^{t_1} \frac{\partial H'(x^o(t),\psi(t),w^o)}{\partial w} dt\, w^o$$

$$= \max_{w \in W} \int_{t_o}^{t_1} \frac{H'(x^o(t),\psi(t),w_o)}{\partial w} dt\, w; \tag{79}$$

$$\beta) \quad \frac{\partial h'(\psi t_o),v^o)}{\partial v} v^o = \max_{v \in V} \frac{\partial h'(\psi(t_o),v^o)}{\partial v} v \quad . \tag{80}$$

5. If, in addition to the conditions of assertion 4, the functions $H(x,\psi,w)$, $h(\psi,v)$ are concave in w,v, then for w^o,v^o conditions (77),(78) are satisfied.

The conditions of this assertion are realized, for example, for the system (75) in which the functions $f_n(x,w)$ do not depend on x_n, the functions $f_1(x,w),\ldots,$ $f_{n-1}(x,w),g_1(v),\ldots,g_{n-1}(v)$ are linear in w,v, the functions $f_n(x,w)$, $g_n(v)$ are convex in w,v, and the constants c_i equal zero except the last one $c_n - 1$.

6. If the functions $f(x,w)$, $g(v)$ are differentiable in w,v and the sets W,V are open, then

$$\alpha)\quad \int_{t_o}^{t_1} \frac{\partial H(x^o(t),\psi(t),w^o)}{\partial w}\, dt = 0; \tag{81}$$

$$\beta)\quad \frac{\partial h(\psi(t_o),v^o)}{\partial v} = 0 \quad . \tag{82}$$

7. If the functions $f(x,w)$, $g(v)$ are differentiable in w,v, then the gradient of the function $I(v,w)$ at the points v^1,w^1 equals

$$\alpha)\quad \frac{\partial I(v^1,w^1)}{\partial w} = - \int_{t_o}^{t_1} \frac{H(x^1(t),\psi(t),w^1)}{\partial w}\, dt \ ;$$

$$\beta)\quad \frac{\partial (v^1,w^1)}{\partial v} = - \frac{\partial h(\psi(t_o),v^1)}{\partial v} \quad ,$$

8. If $f(x,w) = Ax + b(w)$, then the conditions (77), (78) are necessary and sufficient for the optimality of v^o, w^o.

9. The S-Maximum Principle:

$$\alpha)\quad H(x^o(t),\psi(t),w^o)$$

$$\geq \frac{1}{t_1 - t_o} \int_{t_o}^{t_1} H(x^o(t),\psi(t),w)dt - \varepsilon_1 \ , \ \varepsilon_1 \geq 0 \ , \tag{83}$$

for all $w \varepsilon W$;

$$\beta)\quad h(\psi(t_o),v^o) \geq h(\psi(t_o),v) - \varepsilon_2 \ , \ \varepsilon_2 \geq 0 \ , \tag{84}$$

for all $v \varepsilon V$. For bounded sets V,W, for any $\varepsilon_1 > 0$, $\varepsilon_2 > 0$, it is possible to find a number τ such that conditions (83),(84) are satisfied in the problem (74)-(76) with $t_1 - t_o \leq \tau$.

To clarify the essence of the basic conditions in the preceding assertions, we consider

Example 6. $\dot{x}_1 = w_1$, $\dot{x}_2 = w_2$, $x_3 = x_1^2 + x_2^2$, $x_1(0) = v_1$, $x_2(0) = v_2$, $x_3(0) = 0$, $0 \leq t \leq \tau$, $c_1 = c_2 = 0$, $c_3 = 1$,

$$V,W = \{u_1,u_2\colon (u_1 - 1)^2 + (u_2+1)^2 \geq 8 \quad ,$$

$$(u_1 - 2)^2 + (u_2 + 2)^2 \geq 18\} \ .$$

At first, we consider the problem of optimization only with respect to the parameters w_1, w_2, setting $v_1 = v_2 = 0$. We have $x_1(t) = w_1 t$, $x_2(t) = w_2 t$, $x_3(\tau) = (w_1^2 + w_2^2)\tau^3/2$. Hence $w_1^o = -1$, $w_2^o = 1$. Further, $H(x,\psi,w) = \psi_1 w_1 + \psi_2 w_2 + \psi_3(x_1^2 + x_2^2)$, $\dot{\psi}_1 = -2x_1\psi_3$, $\dot{\psi}_2 = -2x_2\psi_3$, $\dot{\psi}_3 = 0$, $\psi_1(\tau) = \psi_2(\tau) = 0$, $\psi_3(\tau) = -1$, $x_1^o(t) = -t$, $x_2^o(t) = t$, $\psi_3(t) = -1$, $\psi_1(t) = \tau^2 - t^2$, $\psi_2(t) = t^2 - \tau^2$, $\int_0^\tau H(x^o(t),\psi(t),w)\,dt = 2/3\tau^3(w_1 - w_2) - 2\tau^3/3$.

At the point $w_1 = -1$, $w_2 = 1$, the last function admits an absolute minimum over W. Thus, condition (77) does not appear in this example. For the same reason, property (79) is violated. Also, property (81) is not satisfied in this example.

We pass to the optimization over the parameters v_1, v_2 under the conditions that $w_1 = w_2 = 0$. We have $x_1(t) = v_1$, $x_2(t) = v_2$, $x_3(\tau) = \tau(v_1^2 + v_2^2)$. Hence, $v_1^o = -1$, $v_2^o = 1$.

We check the necessary conditions for optimality:

$$h(\psi,v) = \psi_1 v_1 + \psi_2 v_2 \quad ,$$

$$\dot{\psi}_1 = -2x_1\psi_3 \quad , \quad \dot{\psi}_2 = -2x_2\psi_3 \quad , \quad \dot{\psi}_3 = 0 \quad ,$$

$$\psi_1(\tau) = \psi_2(\tau) = 0 \quad , \quad \psi_3(\tau) = -1 \quad ,$$

$$x_1^o(t) - -1 \ , \ \psi x_2^o(t) = 1 \ , \ \psi_3(t) = -1 \ , \ \psi_1(t) = 2(\tau - 1) \ ,$$

$$\psi_2(t) = 2(t - \tau) \ , \ h(\psi(0),v) = 2\tau(v_1 - v_2) \quad .$$

The last function does not satisfy any of the properties (78),(80),(82). From the fact that properties (77), (78) are violated for any $\tau > 0$, it follows that assertion 9 must be improved in the general case.

2. The Observation and Identification of Dynamical Systems as a Problem of Optimization with Respect to Parameters

A sufficiently general problem about observation and identification may be formulated in the following form. At the output of some object, we measure the quantity

$$y(t) \quad , \quad t\varepsilon T = [t_o,t_1] \quad .$$

Relative to the object, we know:

1) the differential equation of motion

$$\frac{dx}{dt} = f(x,w) \ , \tag{85}$$

up to a q-vector of parameters $w\varepsilon W$;

2) the initial condition

$$x(t_o) = g(v) \quad , \tag{86}$$

up to a p-vector of parameters $v\varepsilon V$;

3) the output mechanism

$$y = h(x) \quad .$$

Knowing the sets W,V, find the vectors $v^o \varepsilon V$, $w^o \varepsilon W$ such that

$$I(v^o,w^o) = \min_{v\varepsilon V,\ w\varepsilon W} I(v,w) \quad ,$$

where

$$I(v,w) = \int_{t_o}^{t_1} f_o(y(t),h(x(t)))dt \quad , \tag{87}$$

y(t) is the known measurement function, h(t) = h(x(t)) is a function of the trajectory x(t) of the system (85), corresponding to the values v,w; $f_o(y,h)$ is some measure of closeness of the vectors y,h.

Introducing the additional variable

$$x_o(t) = \int_{t_o}^{t} f_o(y(t),h(t))dt \quad ,$$

we reduce the formulated problem to a problem of optimization over parameters.

Theorem 19 allows us to not only check the optimality of concrete values of v,w, but also to construct an iterative process to refine initial estimates given by the method of descent (knowing the gradient).

Remarks. 1) Under the above assumptions on the optimal values v^o,w^o, the functional (87) equals zero. If the same problem is posed in a different setting, then $I(v^o,w^o) \geq 0$. Concretely, if we assume that the equations of motion are unknown, while Eq.(85) takes a desired form for the equation of motion (for example, from the point of view of convenient realization), then

$I(v^o, w^o) > 0$ in those cases when the real motion $y(t)$ does not lie in the set $h(t)$ generated by the system (85),(86) for different v, w. The last interpretation of the problem is very natural and is widely used.

2) To account for errors in the measured values does not pose serious difficulties. In this case, instead of (87) we take some average (for example, the mean) value.

3. The Problem of Reducing the Order of the Differential Equation

The study of complex objects whose motion is described by high-order differential equations is filled with great difficulties. Therefore, for preliminary investigations we frequently try to find an object simpler than the original, whose dynamical characteristics reflect the essential interesting properties of the original object. In many cases, this comes down to the replacement of the original differential equations by simpler equations. To accomplish this goal, different principles are used (the location of the roots of the characteristic equation, the isolation of slow trajectories, etc). Being sufficiently simple, these principles do not always lead to the goal because of their lack of rigorous formulation. Among the different principles, the statement of similar problems in the form of optimal control problems allows us to give them a precise form and opens the way for their solution by powerful methods.

In general, the problem of reducing the order of a differential equation consists of the following: We are given an n-dimensional equation of motion

$$\frac{dx}{dt} = f(x) \quad , \quad x(t_o) = x_o \quad .$$

It is required to find parameters v^o, w^o for which the motion of the m-dimensional system

$$\frac{dy}{dt} = g(y,w) \quad , \quad y(t_o) = v \quad ,$$

most completely mirrors the original motion x(t) in the sense of minimizing a given functional

$$J = J(v,w) \quad ,$$

defined on the motions x(t), y(t). Different problems arise depending upon the concrete form of the criteria. A linear variant of the formulated problem is usually given in the following way.

We are given the equation

$$x^{(n)} + a_1 x^{(n-1)} + \cdots + a_{n-1}\dot{x} + a_n x = 0 \quad ,$$

$$x^i(t_o) = x_{io} \quad , \qquad i = 0,\ldots,n-1 \quad .$$

It is required to find coefficients and initial conditions for the equation

$$y^{(m)} + b_1 y^{(m-1)} + \cdots + b_{m-1}\dot{y} + b_m y = 0 \quad ,$$

$$y^{(j)}(t_o) = y_{jo} \quad , \quad u = 0,\ldots,m-1 \quad ,$$

such that, for example, the mean-square deviation between x(t) and y(t) is minimized:

$$\int_0^\infty [x(t) - y(t)]^2 dt = \min_{b,y_o} \quad .$$

For the solution of similar problems, it is very natural to use the theory of optimal processes (in particular, the results of paragraph 1) for optimization problems.

§8. Methods of Functional Analysis in the Theory of the Maximimum Principle

The theorems and facts of functional analysis are widely used in all modern theories of necessary conditions of optimality. However, historically by the functional approach to the theory of optimal processes has been understood the collection of all theorems and schemes for their application which allow us (for linear systems) to carry through the solution for the optimization problem to the end. For this, we must surmount a basic difficulty of the maximum principle--the solution of boundary-value problems. For the functional approach, the variational problem is reduced to operations with functions of a finite number of variables. Usually, these operations are of the type associated with computing the minimum of a convex function of a finite number of variables. It is possible to assume that the last problem is sufficiently well developed, both in theory and in practice.

In the succeeding paragraphs, we illustrate the application to optimal processes of the following facts from functional analysis:

a) the minimax theorem;
b) the theorem on the separation of convex sets;
c) the L-problem of moments;
d) the theorem on the existence of a supporting hyperplane;
e) the theorem on the imbeddability of convex sets;
f) the Neyman-Pearson lemma.

As will be clear from our subsequent presentation, only in cases c) and f) are we forced to consider operations in functional spaces. In the remaining cases, the operations are carried out in finite-dimensional spaces. We note that the results under c) and f) follow from a) and b).

We introduce several considerations from the theory of convex sets and functions. The set $X \subset E_n$ is called convex if along with any two points $x_1 \varepsilon X$, $x_2 \varepsilon X$, the line joining these two points is also contained in X, i.e.

$$\lambda x_1 + (1 - \lambda) x_2 \varepsilon X \quad ,$$

for any λ, $0 \leq \lambda \leq 1$. Examples of convex sets in the plane are squares and circles. Moreover, the last set is strictly convex, i.e. for any $x_1 \varepsilon X$, $x_2 \varepsilon X$, $0 < \lambda < 1$, any point $\lambda x_1 + (1 - \lambda) x_2$ is an interior point of the set X. By boundary points we understand all points of X for which any neighborhood contains points not in X. Points of X which are not boundary points, are called interior points. A point $x \varepsilon X$ is called an extreme point if there do not exist points x_1, x_2 such that $x = \lambda x_1 + (1 - \lambda) x_2$, $0 < \lambda < 1$. For example, the boundary points of the square and the circle are just the boundaries of the closed polygon and circle, respectively. The extreme points are the vertices of the square and the boundary of the circle.

The smallest convex set containing a given X is called the convex hull of X: conv X. It is known that

$$\text{conv } X = \{x\colon x = \sum_{i=1}^{n+1} \alpha_i x_i, x_i \varepsilon X, \alpha_i \geq 0, \sum_{i=1}^{n+1} \alpha_i = 1\} \quad .$$

The set X is called a convex cone if from x, $y \varepsilon X$ it follows that $\alpha x + \beta y \varepsilon X$ for all $\alpha, \beta \geq 0$.

The function f(x), defined on the convex set X, is called <u>convex</u> if

$$f(\lambda x_1 + (1 - \lambda)x_2) \leq \lambda f(x_1) + (1 - \lambda)f(x_2) ,$$

for all x_1, $x_2 \varepsilon X$, $0 \leq \lambda \leq 1$. It is called <u>strictly convex</u> if strict inequality holds for all $0 < \lambda < 1$. The function f(x) is called <u>concave</u> if -f(x) is a convex function.

A convex function f(x) is continuous at interior points of X and possesses a derivative in any direction g at these points, i.e. the limit

$$\lim_{\substack{\alpha\to 0\\ \alpha>0}} \frac{f(x + \alpha g)}{\alpha} = \frac{\partial f}{\partial g} ,$$

exists for all $x \varepsilon \operatorname{int} X$, $g \varepsilon X$. If f(x,y), $y \varepsilon Y$ is a set of convex functions, then f(x) = sup f(x,y), $y \varepsilon Y$, is a convex function on int X. A local minimum of a convex function is also a global minimum. Moreover, it occurs at a unique point if the function is strictly convex. In the last case, the function y(x), $x \varepsilon \operatorname{int} X$, is continuous from the definition

$$f(x,y(x)) = \sup_{y \varepsilon Y} f(x,y) .$$

The set

$$X = \{x: f(x) \leq c\} ,$$

is convex if f(x) is a convex function. If for any c this set is convex, then f(x) is called a <u>quasiconvex</u> function. The set X is closed if f(x) is continuous or lower semicontinuous. The latter means that for $\varepsilon > 0$, there exists a $\delta = \delta(x,\varepsilon)$ such that

$$f(\bar{x}) - f(x) \leq -\varepsilon ,$$

for any $\bar{x}$ contained in the set $\{\bar{x}: ||\bar{x} - x|| \leq \delta\}$. A lower semicontinuous function achieves its minimum on compact (closed and bounded) sets.

§9. Application of the Minimax Theorem to the Determination of Optimal Controls

1. The Minimax Theorem. Let X and Y be convex sets in E_n, where X is bounded, Y compact.

If for each $y\varepsilon Y$, the function $f(x,y)$ is defined, continuous, and convex on the closure $\bar{X}$ of X, while for each $x\varepsilon\bar{X}$ it is defined, continuous, and concave on Y, then

$$\inf_{x\varepsilon X} \max_{y\varepsilon Y} f(x,y) = \max_{y\varepsilon Y} \inf_{x\varepsilon X} f(x,y) \quad . \tag{88}$$

Proof. We first assume that $f(x,y)$ is strictly convex in x for each y and strictly concave in y for each x. We let $x(y)$ denote the point of $\bar{X}$ for which

$$f(x(y),y) = \inf_{x\varepsilon X} f(x,y) = m(y) \quad . \tag{89}$$

By virtue of the continuity and strict convexity of $f(x,y)$ and the convexity of the set X, the functions $x(y)$, $m(y)$ are continuous with $m(y)$ concave. Let the point y^* be such that

$$m(y^*) = \max_{y\varepsilon Y} m(y) = \max_{y\varepsilon Y} \inf_{x\varepsilon X} f(x,y) \quad . \tag{90}$$

The strict concavity in y of the function $f(x,y)$ means that for any $y\varepsilon Y$ and any t, $0 < t < 1$, the following inequality is satisfied for an element $\tilde{y} = (1-t)y^* + ty$:

$$\tilde{f}(x,\tilde{y}) > (1 - t)f(x,y^*) + tf(x,y)$$

$$\geq (1 - t)m(y^*) + tf(x,y) \quad .$$

Hence, for $\tilde{x} = x(\tilde{y})$ we have

$$m(\tilde{y}) = f(\tilde{x},\tilde{y}) > (1 - t)m(y^*) + tf(\tilde{x},y) \quad .$$

But, from (90) it follows that $m(y^*) \geq m(\tilde{y})$. Therefore,

$$f(\tilde{x},y) < \frac{1}{t} m(\tilde{y}) - \frac{1 - t}{t} m(y^*) \leq m(y^*) = f(x(y^*),y^*) \quad .$$

Let $t \to 0$. Then $\tilde{y} \to y^*$, $\tilde{x} \to x(y^*) = x^*$ and $f(x^*,y) \leq f(x^*,y^*)$ which, together with (89), means that

$$f(x^*,y) \leq f(x^*,y^*) \leq f(x,y^*) \quad , \tag{91}$$

for all $x \varepsilon X$, $y \varepsilon Y$. By the conditions of the theorem, for any $\varepsilon > 0$ and $\tilde{x}^* \varepsilon X$, there is an element $x \varepsilon X$ such that

$$|f(x^*,y) - f(\tilde{x},y)| \leq \frac{\varepsilon}{2}$$

for all $y \varepsilon Y$. Therefore, from (91) it follows that for each $\varepsilon > 0$ there exist points $x \varepsilon X$, $y^* \varepsilon Y$ such that

$$f(\tilde{x},y) \leq f(\tilde{x},y^*) + \varepsilon \ , \quad f(x,y^*) \geq f(\tilde{x},y^*) - \varepsilon \quad , \tag{92}$$

for all $x \varepsilon X$, $y \varepsilon Y$. From the first inequality in (92) we obtain

$$\max_{y \varepsilon Y} f(\tilde{x},y) \leq f(\tilde{x},y^*) + \varepsilon \ , \ \inf_{x \varepsilon X} \max_{y \varepsilon Y} f(x,y) \leq f(\tilde{x},y^*) + \varepsilon . \tag{93}$$

Analogously, from the second inequality we have

$$\inf_{x \varepsilon X} f(x,y^*) \geq f(\tilde{x},y^*) - \varepsilon, \ \max_{y \varepsilon Y} \inf_{x \varepsilon X} f(x,y) \geq f(\tilde{x},y^*) - \varepsilon . \tag{94}$$

By virtue of the arbitrariness of $\varepsilon > 0$, from (93), (94) assertion (88) of the theorem follows (for a strictly convex-concave function f(x,y)).

In general, instead of f(x,y) we consider the function

$$f_\delta(x,y) = f(x,y) + \delta x'x - \delta y'y \quad .$$

For each $\delta > 0$, the function $f_\delta(x,y)$ satisfies the conditions under which (88) was proven. Letting $\delta \to 0$, we isolate a subsequence for which $y^*_\delta \to y^*$ which, by virtue of the compactness of the set Y, is always possible. The remainder of the proof requires no explanation.

Remarks. 1) if the set X is closed, then the operation inf in (88) is replaced by min.

2) The theorem remains true without the boundedness of the set X.

3) Under the conditions of the theorem, there exists a situation of ε-equilibrium (92). If, in addition, X is a closed set, then the function f(x,y) possesses saddle points (91). The last assertion may be obtained directly from (88) if X,Y are compact sets, while the function f(x,y) is continuous in each argument.

4. From the proof of the theorem (cf. (91) and what follows), it is clear that the presence of a saddle point $\{x^*,y^*\}$, $x^* \varepsilon X$, $y^* \varepsilon Y$ implies (88), where inf may be replaced by min. However, in general, the existence of a saddle point does not follow from the satisfaction of (88).

5) If one part of (88) is defined, then so is the other and they are equal.

2. Minimization of the Norm of the Terminal State

Let the n-dimensional process x(t), $t \varepsilon T = [t_0,t_1]$, be given by the expression

$$x(t) = s(t,x_0) + \int_{t_0}^{t} S(t,\tau)u(\tau)d\tau \quad , \tag{95}$$

where $s(t,x)$, $S(t,\tau)$ are continuous functions, while the control $u(t)$, $t\varepsilon T$, is chosen from the convex family of r-dimensional functions $U(\cdot)$.

Problem. Among all admissible controls, find that control $u^o(\cdot) = \{u^o(t), t\varepsilon T\}$, for which the functional

$$J(u) = ||x(t_1)|| \quad ,$$

is minimized for a given $t = t_1$.

This problem is simple to solve using the minimax theorem. We let R denote the accessible set for the system (95):

$$R = \{x: \ x = x(t_1) = s(t_1,x_0) + \int_{t_0}^{t_1} S(t_1,\tau)u(\tau)d\tau, u(\cdot)\varepsilon U(\cdot)\} \quad .$$

From the convexity of the family $U(\cdot)$, it follows that R is also convex. In the language of sets, we seek a vector $x\varepsilon R$ whose norm is minimal:

$$\inf_{u(\cdot)\varepsilon U(\cdot)} J(u) = \inf ||x|| \quad .$$

But, since $||x|| = \max_{||g||\leq 1} g'x$, using the minimax theorem (it is clear that in this case all the conditions are satisfied), we obtain

$$\delta^o = \inf J(u) = \inf_{x\varepsilon R} \max_{||g||\leq 1} g'x = \max_{||g||\leq 1} \inf_{x\varepsilon R} g'x \quad ,$$

$$= \max_{||g||\leq 1} \inf_{u(\cdot)\varepsilon U(\cdot)} \left[g's(t_1,x_0) + \int_{t_0}^{t_1} g'S(t_1,t)u(t)dt\right],$$

$$= \max_{||g|| \leq 1} \left[g's(t_1, x_0) + \inf_{u(\cdot)\varepsilon U(\cdot)} \int_{t_0}^{t_1} g'S(t_1,t)u(t)dt \right] .$$

The problem has been reduced to the study of the expression

$$\mu(g) = \inf_{u(\cdot)\varepsilon U(\cdot)} \int_{t_0}^{t_1} g'S(t_1,t)u(t)dt \quad . \tag{97}$$

In many problems, the family $U(\cdot)$ is such that computation of (97) presents no difficulty. If the lower bound is assumed on some element from $U(\cdot)$, then the original problem has a solution; in the opposite case, it is possible to construct a sequence of functions $\{u^k(g,t)\}\varepsilon U(\cdot)$, such that

$$\lim_{k\to\infty} \int_{t_0}^{t_1} g'S(t_1,t)u^k(g,t)dt = \mu(g) \quad . \tag{98}$$

We call the sequence $u^k(t)$ a generalized optimal control if $\lim_{k\to\infty} J(u^k) = \delta^o$.

The problem (95),(96) has a solution on $U(\cdot)$ if for any g, $||g|| \neq 0$, the linear functional (97) assumes its lower bound on $U(\cdot)$. The optimal control $u^o(t)$ satisfies the maximum principle

$$- \int_{t_0}^{t_1} g_o'S(t_1,t)u^o(t)dt = \max_{u(\cdot)\varepsilon U(\cdot)} \int_{t_0}^{t_1} [-g_o'S(t_1,t)u(t)]dt \tag{100}$$

where g_o is a vector for which the right side of (99) assumes its maximum. The generalized optimal control is such that

$$\lim_{k\to\infty} \int_{t_0}^{t_1} g_o'S(t_1,t)u^k(t)dt = \mu(g_o) \quad .$$

3. Standard Classes of Admissible Controls

In order to clarify the value of the obtained results, we consider extensions of the type of constraints on the control functions. Let the class $U(\cdot)$ consist of r-dimensional piecewise-continuous vector functions $u(t)$, $t\varepsilon T$, which are bounded in norm

$$||u(t)|| \leq 1 \quad , \quad t\varepsilon T \quad .$$

Then $\mu(g)$ (from (97)) clearly equals

$$\mu(g) = -\int_{t_0}^{t_1} ||g'S(t_1,t)||dt \quad ,$$

(the lower bound is assumed for all g). Further, let $U(\cdot)$ consist of r-dimensional measurable functions $u(t)$, $t\varepsilon T$, satisfying the condition

$$\int_{t_0}^{t_1} ||u(t)||^p dt \leq 1 \quad , \quad 1 < p < \infty \quad .$$

Then it is not difficult to see that the lower bound in (97) is assumed and equals

$$\mu(g) = -\left[\int_{t_0}^{t_1} ||g'S(t_1,t)||^q dt\right]^{1/q} \quad , \quad \frac{1}{p} + \frac{1}{q} = 1 \quad .$$

Finally, we consider the family $U(\cdot)$ of measurable functions $u(t)$, $t\varepsilon T$, satisfying the constraint

$$\int_{t_0}^{t_1} ||u(t)||dt \leq 1 \quad .$$

In this case, there exist no elements in $U(\cdot)$ on which the lower bound

$$-\mu(g) = \max_{t\varepsilon T} ||g'S(t_1,t)||$$

is assumed for the functional. One of the sequences $u^k(t)$ possessing property (98) has the form

$$u^k(t) = \begin{cases} \frac{1}{k} , & \bar{t} - \frac{1}{2^k} \leq t \leq \bar{t} + \frac{1}{2^k} , \\ 0 , & \text{at other points} . \end{cases}$$

Here $\bar{t}$ is a point at which the function $||g'S(t_1,t)||$ assumes a maximum (for simplicity, we assume that $\bar{t}$ is a unique point).

We summarize the results of points 2 and 3.

1) The variational problem (95),(96) is reduced by means of the minimax theorem, to a problem of minimizing a concave function over a sphere in a finite-dimensional space.

2) The question of the existence of an optimal control reduces to a question about the existence of the minimum of the linear functional (97) over a given class of admissible controls. The solution of this question is completely determined by the given class of controls.

3) The optimal control satisfies the maximum principle (100) in which the vector g_o, playing here the role of the initial condition $\psi(t_o)$, may be easily computed.

4. Optimization of Systems with a Nonlinear Input

The minimax theorem may be directly applied for the solution of the following problem: minimize the functional (96), defined on trajectories of the system

$$x(t) = s(t,x_o) + \int_{t_o}^{t_1} S(t,\tau,u(\tau))d\tau , \qquad (101)$$

over all r-dimensional measurable functions $u(t)$, $t\varepsilon T$, taking values in a given compact set U.

The system (101) is nonlinear in u, but the basic fact which allows us to apply the minimax theorem is the following: since the accessible set

$$R = \{x: \quad x = s(t_1,x_0) + \int_{t_0}^{t_1} S(t_1,t,u(t))dt, u(t)\varepsilon U\} \quad ,$$

is closed and convex (cf. §1.6),

$$\delta^o = \min_{u(\cdot)\varepsilon U(\cdot)} ||x(t_1)|| = \max_{||g||\leq 1} \{g's(t_1,x_0) + \mu(g)\}$$

$$= g_o's(t_1,x_0) + \mu(g_o) \quad , \qquad (102)$$

where

$$\mu(g) = \int_{t_0}^{t_1} \min_{u\varepsilon U} g'S(t_1,t,u)dt \quad .$$

Theorem 21. The optimization problem (101),(96) reduces to the finite-dimensional problem (102). The optimal control $u^o(t)$ exists and satisfies the maximum principle

$$-g_o'S(t_1,t,u^o(t)) = \max_{u\varepsilon U} [-g_o'S(t_1,t,u)] \quad .$$

5. Generalizations

Application of the maximum principle is not restricted to functionals of the type (96). We may consider the functional

$$I(u) = \phi(x(t_1)) \quad ,$$

where $\phi(x)$ is a quasiconvex, lower semicontinuous function. In this case, instead of the accessible set R,

$$Q(\delta) = \{x: \ \phi(x) \leq \delta\}$$

and we define the function $||z||$, $z \varepsilon P(\delta)$, on the set $P(\delta) = R^n - Q(\delta)$. The minimal δ for which $||z|| = 0$, clearly equals inf $I(u)$. Therefore, first we find $\lambda(\delta) = \inf ||z||$, reducing this problem to a finite dimensional one:

$$\lambda(\delta) = \inf_{z\varepsilon P} ||z|| = \inf_{z\varepsilon P} \max_{||g|| \leq 1} g'z = \max_{||g|| \leq 1} \inf_{z\varepsilon P} g'z .$$

The remaining calculations are clear.

Finally, we can effectively use the scheme presented not only for the minimization of functionals of the terminal state. It is not difficult to see (details will be given in succeeding paragraphs) that we may use this method to solve minimal time problems with fixed or moving endpoints, problems of minimizing integral criteria, and others. Moreover, the last reduction of the variational problem to a finite-dimensional one may be further extended to study qualitative questions in optimal control (uniqueness, dependence or parameters, etc.). These questions are also considered in succeeding paragraphs.

6. Conclusion

Starting with the examination of examples, we may formulate the basic steps in the reduction of concrete optimization problems to the minimax theorem. 1) Give the problem a geometric formulation. 2) If we succeed in representing the original problem as a problem concerning the existence of a common point of two convex sets, then introduce the distance $||z||$ between the sets. 3) Use the representation $||z|| = \max_{||g|| \leq 1} g'z$ and apply the minimax theorem.

Each of these steps has a different realization in concrete problems. To acquire practice in the use of these methods, we recommend the independent solution of the problems in §10,11 by reducing them to the minimax theorem. It is also interesting to consider variations of these problems.

§10. The Theorem on the Separation of Convex Sets and its Extension to Optimal Control Problems

In this paragraph, we first give an elementary proof of the theorem on the separation of convex sets. Next we illustrate the effectiveness of the approach to control problems based on this theorem by the solution of a series of optimization problems using dynamical systems.

1. The Theorem on the Separation of Convex Sets

Let X,Y be convex sets, one of which is bounded and assume the closures X,Y have no common point. Then there exists a vector g, $||g|| = 1$, such that for any $x \varepsilon X$, $y \varepsilon Y$, we have the inequality

$$g'x > g'y \quad . \tag{103}$$

Proof. We define the function $\rho(x,y) = (x - y)'(x - y)$ on the sets X,Y. Let the bounded set be X. Then on X the function

$$\rho(x) = \min_{y \varepsilon \overline{Y}} \rho(x,y) = \rho(x,y(x)) > 0 \quad ,$$

is bounded and continuous. This function assumes a minimum:

$$\rho^o = \min_{x \varepsilon \overline{X}} \rho(x,y(x)) = \rho(x_o,y(x_o)) > 0 \quad .$$

We set $z = x - y$, $z_o = x_o - y_o$, $y_o = y(x_o)$. For any $z \varepsilon Z = X-Y$ and all t, $0 \leq t \leq 1$, the point $z_t = z_o + t(z - z_o)$ satisfies the inequality $z_t' z_t \geq z_o' z_o$. Hence

$$2tz_o'(z - z_o) + t^2(z - z_o)'(z - z_o) \geq 0 \quad .$$

From the fact that the last inequality must be satisfied for all t, $0 \leq t \leq 1$, it follows that $z_o'(z - z_o) \geq 0$, or $z_o'z \geq z_o'z_o > 0$. We set $g = -z_o/||z_o||$. Then

$$g'z = g'(x - y) < 0 \quad ,$$

which is equivalent to (103). The theorem is proved.

The scheme of reducing optimal control problems to the theorem on the separation of convex sets always contains the basic step of treating the original problem as a problem of the existence of a common point of two convex sets. The remaining details are sufficiently routine that they may be seen from the examples given below.

Remark. If both the sets X and Y are bounded, then the theorem on separation of convex sets follows from the minimax theorem. We define a function $f(z,g) = g'z$ on the sets $Z, G = \{g: \ ||g|| \leq 1\}$. This function satisfies all conditions of the minimax theorem. Therefore,

$$\inf_{z \varepsilon Z} \max_{||g|| \leq 1} g'z = \max_{||g|| \leq 1} \inf_{z \varepsilon Z} g'z \quad . \tag{104}$$

From the compactness of G follows the existence of a g_o such that

$$\inf_{z \varepsilon Z} (-g_o'z) = \max_{||g|| \leq 1} \inf_{z \varepsilon Z} g'z \quad .$$

Since the left part of (104) is positive (Z does not contain the origin), $g_o \neq 0$. From (104) we obtain $-g_o'z > 0$

for all $z \in Z$ or

$$g_o'x > g_o'y \quad \text{for all } x \in X,\ y \in Y \quad , \tag{105}$$

which is equivalent to (103) (if we divide both parts of (105) by $||g_o||$).

We may also obtain the minimax theorem from the theorem on the separation of convex sets. This is an extension of the method of the last proof.

2. Conditions for the Solvability of One Functional Problem

We let $U(\cdot)$ denote some class of functions $u(t)$, $t \in T$. Let the transformation S assign to each element $u(\cdot)$ from $U(\cdot)$, a corresponding vector x from the finite-dimensional space E_n. The set

$$R = \{x: \ x = Su(\cdot), u(\cdot) \in U(\cdot)\} \quad ,$$

is bounded and convex. Further, we assign linear transformations from $E_n \to E_n$, Q_1, Q_2, lower semicontinuous functions $\phi_1(x)$, $\phi_2(x)$, an element $c \in E_n$, and nonnegative numbers δ_1, δ_2.

<u>Problem</u>. To find conditions under which

$$Su(\cdot) + Q_1 v + Q_2 w + c = 0 \ , \quad \phi_1(v) \leq \delta_1 \ , \quad \phi_2(w) \leq \delta_2 \ ,$$

$$u(\cdot) \in U(\cdot) \quad , \quad v, w \in E_n \quad . \tag{106}$$

Theorem 22. Problem (106) has a generalized solution if and only if

$$\max_{||g||=1} \{g'c + \inf_{u(\cdot)\varepsilon U(\cdot)} g'Su(\cdot) + \min_{\phi_1(v)\leq\delta_1} g'Q_1b + \min_{\phi_2(w)\leq\delta_2} g'Q_2w\} \leq 0 \quad . \tag{107}$$

Proof. Let $u(\cdot)$, v,w be a solution of the problem (106). Then

$$g'[Q_1v + Q_2w + c] = -g'Su(\cdot) \leq - \inf_{u(\cdot)\varepsilon U(\cdot)} g'Su(\cdot)$$

for any $g\varepsilon E_n$. On the other hand,

$$g'[Q_1v + Q_2w + c] \geq g'c + \min_{\phi_1(v)\leq\delta_1} g'Q_1v + \min_{\phi_2(w)\leq\delta_2} g'Q_2w \quad .$$

This means

$$g'c + \inf_{u(\cdot)\varepsilon U(\cdot)} g'Su(\cdot) + \min_{\phi_1(v)\leq\delta_1} g'Q_1v + \min_{\phi_2(w)\leq\delta_2} g'Q_2w \leq 0 \tag{108}$$

for any $g\varepsilon E_n$. Therefore, relation (107) is satisfied.

We assume that condition (107) is satisfied. Then for any g, $||g|| = 1$, inequality (108) is valid. We consider the set

$$R = \{x: \ x = Su(\cdot) + c, u(\cdot)\varepsilon U(\cdot)\} \quad ,$$

$$Q = \{x: \ x = -Q_1v - Q_2w, \phi_1(v) \leq \delta_1, \phi_2(w) \leq \delta_2\} \quad .$$

The set Q is clearly convex. The set R is bounded and convex (by assumption). If the problem (106) has no generalized solution, then the set $\overline{R} \cap Q$ is empty and, according to the theorem on the separation of convex sets, there exists a nonzero vector $\bar{g}$ such that

$$\min_{\phi_1(v)\leq\delta_1}\ \min_{\phi_2(w)<\delta_2}\ \bar{g}'[Q_1v+Q_2w] > \sup_{u(\cdot)\varepsilon U(\cdot)}\ \bar{g}'[-Su(\cdot)-c] ,$$

i.e

$$\min_{\phi_1(v)\leq\delta_1}\ \bar{g}'Q_1v + \min_{\phi_2(w)\leq\delta_2}\ \bar{g}'Q_2w + \inf_{u(\cdot)\varepsilon U(\cdot)}\ \bar{g}'Su(\cdot)+g'c<0. \tag{109}$$

The inequality (109) contradicts (108) which proves the theorem.

We assume that the transformation S is linear:

$$Su(\cdot) = \int_{t_0}^{t_1} S(t_1,\tau)u(\tau)d\tau \quad .$$

We set

$$\left.\begin{array}{l} U(\cdot) \equiv U_p(\cdot) = u(\cdot): \int_{t_0}^{t_1} ||u(t)||^p dt \leq 1 \quad , p > 1 , \\ \phi_1(x) = ||x - c^1|| \quad , \quad \phi_2(x) = ||x - c^2|| \quad , \end{array}\right\} \tag{110}$$

where c^1,c^2 are given vectors in E_n.

<u>Problem</u>. Find conditions under which

$$\left.\begin{array}{l} Su(\cdot) + Q_1v + Q_w w + c = 0 \ , \quad u(\cdot)\varepsilon U_p(\cdot) \quad , \\ ||v - c^1|| \leq \delta_1 \ , \quad ||w - c^2|| \leq \delta_2 \quad . \end{array}\right\} \tag{111}$$

<u>Theorem 23. The problem (111) has a solution if and only if</u>

$$\max_{||g||=1} \{g'[c + Q_1c^1 + Q_2c^2] - ||g'S|| - \delta_1||g'Q_1||-\delta_2||g'Q_2||\} \leq 0 \quad . \tag{112}$$

The assertion is proved analagously to Theorem 22.

Remark. Let $U(\cdot)$ be a class of vector functions $u(t)$, $t_0 \le t \le t_1$, such that the set $SU(\cdot)$ is closed and convex. Then the conditions for the existence of a solution to the problem (111), where $u(\cdot)\varepsilon U(\cdot)$, have the form

$$\max_{||g||=1} \{g'[c + Q_1c^1 + Q_2c^2] + \min_{u(\cdot)\varepsilon U(\cdot)} g'Su(\cdot)$$
$$- \delta_1||g'Q_1||-\delta_2||g'Q_2||\} \le 0 \quad .$$

We assume that $||g'S|| > 0$ for all g, $||g|| \neq 0$. Condition (112) is equivalant to the inequality $\lambda \le 1$,

$$\lambda = \max_{||g'S||=1} \{g'[c+Q_1c^1+Q_2c^2] - \delta_1||g'Q_1||-\delta_2||g'Q_2||\} \quad . \tag{113}$$

Theorem 24. In order that the element $g = g_o$ be a solution to the problem (113), it is necessary and sufficient that for any solution $u(\cdot)$, $||u(\cdot)|| = \lambda$, the v,w, of problem (111) satisfy the equations

$$g_o'Su(\cdot) = -||g_o'S|| \; ||u(\cdot)|| = -\lambda \quad , \tag{114}$$

$$g_o'Q_1[v - c^1] = -\delta_1||g_o'Q_1|| \quad , \tag{115}$$

$$g_o'Q_2[w - c^2] = -\delta_2||g_o'Q_2|| \quad . \tag{116}$$

Proof. (Necessity) Let g_o be a solution of problem (113) and let the functions $u(\cdot)$ and the vectors v,w satisfy the relations (111), where $||u(\cdot)|| = \lambda$. Clearly, we have the inequality

$$\lambda = -g_o'[Su(\cdot) + Q_1(v - c^1) + Q_2(w - c^2)] -\delta_1||g_o'Q_1||$$
$$- \delta_2||g_o'Q_2|| \le ||g_o'S|| \; ||u(\cdot)|| - g_o'Q_1(v - c^1)$$

$$- g_0'Q_2(w - c^2) - \delta_1||g_0'Q_1|| - \delta_2||g_0'Q_2|| \quad . \quad (117)$$

Hence, we obtain

$$g_0'Q_1(v - c^1) + g_0'Q_2(w - c^2) \leq -\delta_1||g_0'Q_1|| - \delta_2||g_0'Q_2|| \quad .$$

But

$$|g_0'[Q_1(v - c^1) + Q_2(w - c^2]| \leq \delta_1||g_0'Q_1|| + \delta_2||g_0'Q_2|| \quad .$$

Therefore, Eqs.(115)-(116) are satisfied. Now, relation (114) follows from (117).

(Sufficiency) Let $u(\cdot)$, v,w be a solution to problem (11) with $||u(\cdot)|| = \lambda$. If for some $\bar{g}$, $||\bar{g}|| = 1$, conditions (114)-(116) are satisfied, then instead of (117) we have

$$\bar{g}'(c + Q_1c^1 + Q_2c^2) - \delta_1||\bar{g}'Q_1|| - \delta_2||\bar{g}'Q_2|| = \lambda||\bar{g}'S|| \quad ,$$

i.e. the element $\bar{g}$ is a solution of problem (113) which proves the theorem.

3. The Problem of Minimizing a Quasiconvex Function of the Terminal State

We consider the system

$$\frac{dx}{dt} = A(t)x + b(u,t) \quad , \quad (118)$$

where $A(t)$ is a continuous nxn matrix function, $b(u,t)$ is a continuous n-vector function, and $u = \{u_1,\ldots,u_r\}$ is the control. We let $U(\cdot)$ denote the class of measurable r-dimensional functions $u(t)$, $t\varepsilon T$, with values in some given set U. Let there be given numbers $t_1 > t_1$, $\delta_1 > 0$ and quasiconvex, lower semicontinuous functions

$\phi_1(x)$, $\phi_2(x)$. Let the sets $\{x: \phi_1(x) \leq \alpha_1\}$, $\{x: \phi_2(x) \leq \alpha_2\}$ be bounded for any $\alpha_1, \alpha_2 < \infty$.

It is required to find a control $u^o(t)$ for which the trajectory $x(t) = x(x(t_o),u(\cdot),t)$ of Eq.(118) satisfies the condition

$$\phi_2(x(x(t_o),u^o(\cdot),t_1)) = \min_{u(\cdot)\varepsilon U(\cdot)} \phi_2(x(x(t_o),u(\cdot),t_1));$$

$$\phi_1(x(t_o)) \leq \delta_1 \quad . \tag{119}$$

By virtue of Cauchy's formula

$$x(t_1) = F(t_1)F^{-1}(t_o)x(t_o) + \int_{t_o}^{t_1} F(t_1)F^{-1}(\tau)b(u,\tau)d\tau. \tag{120}$$

We set

$$\int_{t_o}^{t_1} F(t_1)F^{-1}(\tau)b(u,\tau)d\tau = Su(\cdot) \quad ,$$

$$F(t_1)F^{-1}(t_o) = Q_1 \quad , \quad -E = Q_2 \quad .$$

From (120) we have

$$Su(\cdot) + Q_1x(t_o) + Q_2x(t_1) = 0 \quad .$$

Let $\delta_2 \geq 0$. We consider the problem

$$Su(\cdot) + Q_1x(t_o) + Q_2x(t_1) = 0 \ , \quad u(\cdot)\varepsilon U(\cdot) \ ,$$

$$\phi_1(x(t_o)) \leq \delta_1 \ , \quad \phi_2(x(t_1)) \leq \delta_2 \quad . \tag{121}$$

On the basis of Theorem 22 we conclude that problem (121) has a solution if and only if

$$\Lambda(\delta_2) = \max_{||g||=1} \Lambda(\delta_2,g) = \Lambda(\delta_2,g(\delta_2)) \leq 0 \quad , \tag{122}$$

where

$$\Lambda(\delta_2, g) = \min_{u(\cdot)\varepsilon U(\cdot)} g'Su(\cdot) + \min_{\phi_1(x)\leq\delta_1} g'Q_1x + \min_{\phi_2(x)\leq\delta_2} g'Q_2x \ .$$

Lemma 8. The function $\Lambda(\delta_2)$ is continuous on the right and non-increasing. If $\phi_2(x)$ is a continuous function, then

$$\Lambda(\delta_2) < \Lambda(\overline{\delta}_2) \quad , \quad \text{for } \overline{\delta}_2 > \delta_2 \quad .$$

Proof. The right-continuity of the function $\Lambda(\delta_2)$ is a corollary of the lower semicontinuity of $\Lambda_2(x)$. We show the monotonicity of $\Lambda(\delta_2)$. It is clear that if $\overline{\delta}_2 < \delta_2$, then

$$\min_{\phi_2(x)\leq\overline{\delta}_2} g'Q_2x \geq \min_{\phi_2(x)\leq\delta_2} g'Q_2x$$

and

$$\Lambda(\overline{\delta}_2, g(\overline{\delta}_2)) \geq \Lambda(\overline{\delta}_2, g(\delta_2)) \geq \Lambda(\delta_2, g(\delta_2)) \quad .$$

The lemma is proved.

Theorem 25. If there exists at least one control satisfying condition (121), then there exists an optimal control making the functional $\phi_2(x(x(t_o), u(\cdot), t_1))$ assume the value δ_2^o, which is the smallest number satisfying the inequality

$$\Lambda(\delta_2) = \Lambda(\delta_2, g(\delta_2)) \leq 0 \quad .$$

For the optimal control $u^o(\cdot)$, the endpoint of the optimal trajectory $x^o(t)$ satisfies the relations

$$-g'(\delta_2^o)F(t_1)F^{-1}(t)b(u^o(t),t)$$

$$= \max_{u\varepsilon U} [-g'(\delta_2^o)F(t_1)F^{-1}(t)b(u,t)] \quad , \tag{123}$$

$$\left.\begin{aligned} &g'(\delta_2^o)F(t_1)F^{-1}(t_o)x^o(t_o) \\ &\quad = \min_{\phi_1(x)\leq\delta_1} g'(\delta_2^o)F(t_1)F^{-1}(t_o)x \quad , \\ &g'(\delta_2^o)x^o(t_1) = - \min_{\phi_2(x)\leq\delta_2^o} g'(\delta_2^o)x \quad . \end{aligned}\right\} \tag{124}$$

Proof. The first part of the theorem follows from Lemma 9. Relations (123) (the maximum principle) and (124) (the transversality condition) are obtained from condition (122).

Remark. The function $\psi(t) = -[F^{-1}(t)]'F'(t_1)g(\delta_2^o)$ satisfies the equation

$$\frac{d\psi}{dt} = -A'(t)\psi \quad , \tag{125}$$

where

$$\psi(t_1) = -g(\delta_2^o) \quad .$$

Let $\phi_1(x) = ||x - c^1||$, $\phi_2(x) = ||x - c^2||$, where c^1, c^2 are fixed points in E_n. We study the problem of minimizing the functional

$$J(u) = ||x(t_1) - c^2|| \quad , \tag{126}$$

along trajectories of the equation

$$\frac{dx}{dt} = A(t)x + B(t)u + f(t) \quad , \quad u(\cdot)\varepsilon U_p(\cdot) \quad ,$$

where the initial state $x(t_0)$ satisfies the inequality $||x(t_0) - c^1|| \leq \delta_1$. We set

$$\left.\begin{aligned} &\int_{t_0}^{t_1} F(t_1)F^{-1}(\tau)B(\tau)u(\tau)d\tau = Su(\cdot) \quad , \quad F(t_1)F^{-1}(t_0) = Q_1 \\ &\int_{t_0}^{t_1} F(t_1)F^{-1}(\tau)f(\tau)d\tau = c \quad , \quad -E = Q_2 \quad . \end{aligned}\right\} \quad (127)$$

For $\delta_2 \geq 0$, we consider the equation

$$Su(\cdot) + Q_1x(t_0) + Q_2x(t_1) + c = 0 \quad , \quad u(\cdot)\varepsilon U_p(\cdot) \qquad (128)$$

for which

$$||x(t_0) - c^1|| \leq \delta_1 \quad , \quad ||x(t_1) - c^2|| \leq \delta_2 \quad .$$

By virtue of Theorem 23, problem (128) has a solution if and only if

$$\Lambda(\delta_2,t_1) = \min_{||g||=1} \Lambda(\delta_2,t_1,g) = \Lambda(\delta_2,t_1,g(\delta_2,t_1)) \leq 0 \qquad (129)$$

where

$$\Lambda(\delta_2,t_1,g) = g'(c + Q_1c^1 + Q_2c^2) - ||g'S|| - \delta_1||g'Q_1|| - \delta_2||g'Q_2|| \quad .$$

Lemma 9. The function $\Lambda(\delta_2,t_1)$ is continuous in δ_2 and t_1 and strictly decreasing in δ_2.

Proof. The continuity of $\Lambda(\delta_2,t_1)$ in δ_2 and t_1 follows from the continuity of $\Lambda(\delta_2,t_1,g)$ in δ_2 and t_1.

Let $\delta_2 < \overline{\delta}_2$. Then

$$\Lambda(\delta_2,t_1,g(\delta_2,t_1)) \geq \Lambda(\delta_2,t_1,g(\overline{\delta}_2,t_1))$$

$$> \Lambda(\overline{\delta}_2,t_1,g(\overline{\delta}_2,t_1)) \quad .$$

The lemma is proved.

Theorem 26. An optimal control exists for the problem (126) and gives the functional $\phi_2 = ||x(t_1) - c^2||$ the value

$$\delta_2^o = \min_{||g||\leq 1} \{g'(c + F(t_1)F^{-1}(t_o)c^1 - c^2) - ||g'S|| - \delta_1||g'F(t_1)F^{-1}(t_o)||\} = g_o'[c + F(t_1)F^{-1}(t_o)c^1 - c^2] - ||g_o'S|| - \delta_1||g_o'F(t_1)F^{-1}(t_o)|| \quad . \tag{130}$$

If $\delta_2^o > 0$, then the optimal control $u^o(\cdot)$ and the end of the optimal trajectory $x^o(\cdot)$ satisfy the relations

$$-g_o'Su^o(\cdot) = \min_{u(\cdot)\varepsilon U_p(\cdot)} [-g_o'Su(\cdot)] = -||g_o'S|| \quad , \tag{131}$$

$$g_o'F(t_1)F^{-1}(t_o)(x^o(t_o) - c^1) = -\delta_1||g_o'F(t_1)F^{-1}(t_o)||, \tag{132}$$

$$g_o'(x^o(t_1) - c^2) = \delta_2^o||g_o|| \quad . \tag{133}$$

Proof. The formula (130) follows from (129) by virtue of the monotonicity of $\Lambda(\delta_2, t_1, g)$ in δ_2. The relations (131)-(133) are a corollary of Theorem 24.

4. Minimal-Time Problems for Systems with a Nonlinear Input

We will call the function $\{u(t): t \varepsilon T\} \varepsilon U(\cdot)$ a T-admissible control for some problem if it generates a trajectory satisfying the imposed constraints.

Consider the equation

$$\frac{dx}{dt} = A(t)x + b(u,t) + f(t) \quad , \quad t \geq t_1 \quad . \tag{134}$$

We are given the disjoint sets Γ^1, Γ^2:

$$\Gamma' = \{x: \ ||x - c^1|| \leq \delta_1\} \quad , \quad \delta^2 = \{x: \ ||x - c^2|| \leq \delta_2\},$$

where the elements c^1, c^2 and the numbers δ_1, δ_2 are fixed with $\delta_1 \geq 0$, $\delta_2 \geq 0$. It is required to find a control $u(\cdot) \varepsilon U(\cdot)$ under which the trajectory of Eq. (134), beginning with the initial condition $x(t_o) \varepsilon \Gamma^1$, reaches Γ^2 in the minimal time.

The condition for the existence of a T-admissible control gives inequality (129). Since δ_2 is fixed, we write condition (129) in the form

$$\Lambda(t_1) = \max_{||g||=1} \Lambda(\delta_2, t_1, g) \equiv \Lambda(t_1, g(t_1)) \leq 0 \quad .$$

Theorem 27. *If there exists at least one T-admissible control then there exists an optimal control with minimal time $\tau^o = t_1^o - t_o$, where t_1^o is the smallest root of the equation*

$$\Lambda(t_1, g(t_1)) = 0 \quad . \tag{135}$$

The optimal control $u^o(\cdot)$ and the endpoints of the optimal trajectory $x^o(\cdot)$ satisfy the conditions

$$-g'(t_1^o)Su^o(\cdot) = \max_{u(\cdot)\varepsilon U(\cdot)} [-g'(t_1^o)Su(\cdot)] \quad ,$$

$$g'(t_1^o)F(t_1^o)F^{-1}(t_o)(x^o(t_o) - c^1) = -\delta_1 ||g'(t_1^o)F(t_1^o)F^{-1}(t_o)|| \quad ,$$

$$g'(t_1^o)(x^o(t_1^o) - c^2) = \delta_2 ||g(t_1^o)|| \quad .$$

Proof. By virtue of the continuity of $\Lambda(t_1)$ (Lemma 9), the existence of a T-admissible control implies the existence of an optimal control. On the basis of condition (129), the minimal time τ^o is such that the quantity $t_1^o = t_o + \tau^o$ is the smallest solution of Eq. (135). The second part of the assertion follows from Theorem 24.

5. Minimal Time Problems for Systems with State Variable Constraints

We consider the system

$$\frac{dx}{dt} = Ax + bu \tag{136}$$

with the initial condition $x(t_o) = x_o$. Let

$$U(\cdot) = \{u(t): \ |u(t)| \le 1, t\varepsilon T\} \quad .$$

We assign a region of variation for x by setting

$$G = \{x: \ e'x \le 1\} \quad , \tag{137}$$

where e is a fixed element in E_n, $||e|| \neq 0$. Let it be required to find a control $u^0(t)$, $t \varepsilon T$, transferring $x(t) \varepsilon G$, $t \varepsilon T$, to the origin in minimal time. We let $\tilde{U}(\cdot)$ denote the set of functions

$$\{u(t): \; |u(t)| \leq 1, e'x(t) \leq 1, t \varepsilon T\} \quad .$$

If $u(\cdot)$ is a T-admissible control, then

$$-x_o = \int_{t_o}^{t_1} \gamma(\tau)u(\tau)d\tau \quad , \quad \gamma(\tau) = F^{-1}(\tau)b \ ,$$

and

$$e'x(t) = e'F(t) \int_{t_1}^{t} F^{-1}(\tau)bu(\tau)d\tau \leq 1 \quad .$$

By virtue of the results of section 3 of this paragraph, a T-admissible control exists for this problem if and only if

$$\max_{||g||=1} g'x_o + \max_{u(\cdot)\varepsilon\tilde{U}(\cdot)} \int_{t_o}^{t_1} g'\gamma(\tau)u(\tau)d\tau \leq 0 \quad . \tag{138}$$

We find

$$\alpha = \max_{u(\cdot)\varepsilon U(\cdot)} \int_{t_o}^{t_1} g'\gamma(\tau)u(\tau)d\tau \quad . \tag{139}$$

If $u(t)$ is a solution of (139) and $x(t)$ is the corresponding trajectory (136), then by the Kuhn-Tucker theorem [134] there exists a nonnegative measure $d\mu(t)$ concentrated on $\{t: \; e'x(t) = 1\}$ such that

$$\int^{t_1} g'\gamma(\tau)u(\tau)d\tau + \int_{t_o}^{t_1} (e'x(\tau) - 1)d\bar{\mu}(\tau) \ ,$$

$$\leq \int_{t_o}^{t_1} g'\gamma(\tau)u(\tau)d\tau + \int_{t_o}^{t_1} (e'x(\tau) - 1)d\mu(\tau) \quad ,$$

$$\leq \int_{t_0}^{t_1} g'\gamma(\tau)\bar{u}(\tau)d\tau + \int_{t_0}^{t_1} (e'\bar{x}(\tau) - 1)d\mu(\tau) ,$$

for all $d\mu \geq 0$, $\bar{u}(t)$, $|\bar{u}(t)| \leq 1$. Consequently,

$$\alpha = \min_{u(\cdot)\varepsilon U(\cdot)} \max_{d\mu(\tau)\geq 0} \int_{t_0}^{t_1} g'\gamma(\tau)u(\tau)d\tau$$

$$+ \int_{t_0}^{t_1} (e'x(\tau) - 1)d\mu(\tau) \quad . \tag{140}$$

Since

$$\int_{t_0}^{t_1} e'F(\tau) \int_{t_1}^{\tau} \gamma(t)u(t)dtd\mu(\tau) ,$$

$$= - \int_{t_0}^{t_1} \gamma'(t) \int_{t_0}^{t} F'(\tau)e'd\mu(\tau)u(t)dt \quad ,$$

from (140) we have

$$\alpha = \max_{d\mu\geq 0} \; -\int_{t_0}^{t_1} \left|\gamma'(t)\; g - \int_{t_0}^{t} F(\tau)ed\mu(\tau)\right| dt - \int_{t_0}^{t_1} d\mu(\tau) \quad .$$

Thus, a T-admissible control exists if and only if

$$\left.\begin{aligned} \Phi(t_1) &= \max_{||g||=1, d\mu\geq 0} \Phi(t_1,g,d\mu) \\ &= \Phi(t_1,g(t_1),d\mu_{t_1}) \leq 0 \quad , \\ \Phi(t_1,g,d\mu) & \\ &= g'x_0 - \int_{t_0}^{t_1} \left|\gamma'(t)\; g - \int_{t_0}^{t} F'(\tau)ed\mu(\tau)\right| dt - \int_{t_0}^{t_1} d\mu(t). \end{aligned}\right\} \tag{141}$$

Lemma 10. The function $\Phi(t)$ is continuous from the left at the point $t = t_1$ if $\Phi(t_1) \leq 0$.

Proof. If $\Phi(t_1) \leq 0$, then there exists a T-admissible control $\tilde{u}(\cdot)$ of the form

$$\tilde{u}(t) = -\operatorname{sign} \gamma'(t) \left[g - \int_{t_0}^{t} F'(\tau) e d\mu(\tau) \right] , \tag{142}$$

transferring the trajectory to the origin in time $\tau = t_1 - t_0$. By the definition of the measure $d\mu$, there exists an $\varepsilon > 0$ such that $d\mu(\tau) = 0$ for $\tau \varepsilon (t_1 - \varepsilon, t_1]$. Let $t' \varepsilon (t_1 - \varepsilon, t_1]$. We let g, $d\mu$ and $\bar{g}, d\bar{\mu}$ denote solutions to the problem (141) corresponding to the times t_1, and t'. We set

$$d\mu'(t) = \begin{cases} d\bar{\mu}(t) & , \quad t \leq t' , \\ 0 & , \quad t > t' . \end{cases}$$

From (141) we have

$$\Phi(t_1) \geq \bar{g}'x_0 - \int_{t_0}^{t_1} \left| \gamma'(t)\, \bar{g} - \int_{t_0}^{t} F'(\tau) e du'(\tau) \right| dt - \int_{t_0}^{t_1} d\mu'(t) ,$$

$$\Phi(t') \geq g'x_0 - \int_{t_0}^{t'} \left| \gamma'(t)\, g - \int_{t_0}^{t} F'(\tau) e d\mu(\tau) \right| dt - \int_{t_0}^{t'} d\mu(t) .$$

Hence

$$\Phi(t') - \Phi(t_1) \leq \int_{t_0}^{t_1} \left| \gamma'(t) \left[\bar{g} - \int_{t_0}^{t} F'(\tau) e d\mu'(\tau) \right] \right| dt$$

$$- \int_{t_0}^{t'} \left| \gamma'(t) \left[\bar{g} - \int_{t_0}^{t} F'(\tau) e d\bar{\mu}(\tau) \right] \right| dt ,$$

$$\Phi(t') - \Phi(t_1) \geq \int_{t_0}^{t_1} \left| \gamma'(t) \left[g - \int_{t_0}^{t} F'(\tau) e d\mu(\tau) \right] \right| dt$$

$$- \int_{t_0}^{t'} \left| \gamma'(t) \left[g - \int_{t_0}^{t} F'(\tau) e d\mu(\tau) \right] \right| dt .$$

The right side of these inequalities converge to zero as $t' \to t_1$. The assertion follows.

We write condition (141) in another form. The maximum in (141) is assumed by an element g such that $g'x_0 \geq 0$. Therefore,

$$\left\{ \int_{t_0}^{t_1} \left| \gamma'(t) \left[g - \int_{t_0}^{t} F'(\tau) e d\mu(\tau) \right] \right| dt + \int_{t_0}^{t_1} d\mu(t) \right\} [g'x_0]^{-1} \geq 1$$

for all $g, d\mu \geq 0$, $g'x_0 \geq 0$, i.e. a T-admissible control exists if and only if

$$\Psi(t_1) = \min_{g, d\mu \geq 0} \Psi(t_1, g, d\ )$$

$$= \Psi(t_1, g(t_1), d\mu_{t_1}) \geq 1 \quad ,$$

$$\Psi(t_1, g, d\mu) \tag{143}$$

$$= \left\{ \int_{t_0}^{t_1} \left| \gamma'(t) \left[g - \int_{t_0}^{t} F(\tau) e d\mu(\tau) \right] \right| dt \right.$$

$$\left. + \int_{t_0}^{t_1} d\mu(t) \right\} [g'x_0]^{-1} \quad .$$

<u>Lemma 11. Let $\Psi(t_1) \geq 1$. Then $\Psi(t)$ is continuous at the point $t = t_1$.</u>

<u>Proof</u>. The existence of the T-admissible control (142) follows from condition (143). Let $t' \varepsilon (t_1 - \varepsilon, t_1]$, where ε is chosen as in the proof of Lemma 10. If g, $d\mu$, $\bar{g}$, $d\bar{\mu}$ are solutions of problem (143) for the moments t_1 and t', then

$$|\Psi(t') - \Psi(t_1)| = \left| \left\{ \int_{t_0}^{t'} \left| \gamma'(t)\ \bar{g} - \int_{t_0}^{t} F'(\tau) e d\bar{\mu}(\tau) \right] \right| dt \right.$$

$$+ \int_{t_0}^{t'} d\bar{\partial}(t) \Big\} [\bar{g}' x_0]^{-1} - \left\{ \int_{t_0}^{t_1} \left| \gamma'(t) \left[g - \int_{t_0}^{t} F'(\tau) e d\mu(\tau) \right| dt \right. \right.$$

$$+ \int_{t_0}^{t_1} d\mu(t) \quad [g' x_0]^{-1}$$

$$\leq \int_{t_0}^{t_1} \left| \gamma'(t)\ g - \int_{t_0}^{t} F'(\tau) e d\mu(\tau) \right| dt [g' x_0]^{-1} \quad .$$

Consequently, $\Psi(t') \to \Psi(t_1)$ as $t' \to t_1$.

Let $t'' > t_1$. We set

$$d\mu''(t) = \begin{cases} d\mu(t) & , \quad t_0 \leq t \leq t_1 \quad , \\ 0 & , \quad t_1 < t \leq t'' \quad . \end{cases}$$

Then

$$|\Psi(t'') - \Psi(t_1)| \leq \left| \left\{ \int_{t_0}^{t''} \left| \gamma'(t) \left[g - \int_{t_0}^{t} F'(\tau) e d\mu''(\tau) \right] \right| dt \right. \right.$$

$$+ \int_{t_0}^{t''} d\mu''(t) - \int_{t_0}^{t_1} \left| \gamma'(t) \left[g - \int_{t_0}^{t} F'(\tau) e d\mu(\tau) \right] \right| dt$$

$$- \int_{t_0}^{t_1} d\mu(t) \Big\} [g' x_0]^{-1}$$

$$\leq \int_{t_1}^{t''} \left| \gamma'(t) \left[g - \int_{t_0}^{t} F'(\tau) e d\mu(\tau) \right] \right| dt [g' x_0]^{-1} \quad .$$

Therefore, $\Psi(t'') \to \Psi(t_1)$ as $t'' \to t_1 + 0$ and the lemma is proved.

Remark. If $\Psi(t_1) \geq 1$, then $\Psi(t_1) \leq 0$. Conversly, the approach taken above is not completely rigorous because of the possibility that $g' x_0 = 0$. This situation arises when

$$\left|\gamma'(t)\left[g - \int_{t_o}^{t} F'(\tau)e\,d\mu(\tau)\right]\right| - \frac{d\mu(t)}{dt} \equiv 0 \quad . \qquad (144)$$

In the problem under consideration, the relation (144) will not be satisfied if $x_o'e < 1$ and the defining equation is nonsingular at $t = t_1$.

Theorem 28. <u>If there exists at least one T-admissible control, then there exists an optimal control with the minimal time being $\tau^o = t_1^o - t_o$, where t_1^o is the smallest root of the equation</u>

$$\Phi(t_1, g(t_1), d\mu_{t_1}) = 0 \quad (\Psi(t_1, g(t_1), d\mu_{t_1}) = 1) \ .$$

The optimal control is found from the relation

$$u^o(t) = - \operatorname{sign} c'(t)(F^{-1}(t))'[g(t_1^o) - \int_{t_o}^{t} F'(\tau)e\,d\mu_{t_1^o}(\tau)] \ . \qquad (145)$$

Proof. The first part of the assertion follows from Lemmas 10,11, and the inequalities (141),(143). We obtain (145) from (138) and (140).

6. Minimization of Mean-Square Error [106]

Let the system dynamics be given by the equation

$$\frac{dx}{dt} = A(t)x + b(t)u \quad , \qquad (146)$$

where x is an n-vector, u is a scalar control, and the elements of the matrix A(t) and the vector b(t) are continuous functions of t. We are given the following numbers: t_1, a moment of time, $\delta > 0$, and points $x_o = x(t_o)$, x_1. We will consider the functional

$$J(u) = \int_{t_o}^{t_1} [x'Mx + u^2]dt \quad ,$$

along trajectories of Eq.(146), starting at $t = t_0$ in the state x_0. Here M is a given diagonal matrix with nonnegative elements which are continuous functions of t. Let the function $u(t)$, $t\varepsilon T$, $T = [t_0, t_1]$ belong to the class $U(\cdot)$ of square-integrable functions on T.

Problem: To find a control $u^o(t)$ such that

$$J(u^o) = \min_{u(\cdot)\varepsilon U(\cdot)} J(u)$$

and

$$||x(t_1) - x_1|| \leq \delta \quad .$$

As above, we first write the solution of Eq.(146) by Cauchy's formula

$$x(t) = F(t)F^{-1}(t_0)x_0 + \int_{t_0}^{t} F(t)F^{-1}(\tau)b(\tau)u(\tau)d \quad ,$$

where $F(t)$ is the fundamental matrix for the solution of Eq.(146) for $u(t) \equiv 0$. We set $t = t_1$ and introduce the notation

$$s(t,x) = F(t)F^{-1}(t_0)x \quad , \quad S(t,\tau) = F(t)F^{-1}(\tau)b(\tau) \quad ,$$

$$Su(\cdot) = \int_{t_0}^{t_1} S(t_1,\tau)u(\tau)d\,\tau$$

The trajectory $x(t)$ must satisfy the condition

$$x(t_1) = x(t_1,x_0) + Su(\cdot) \quad , \quad ||x(t_1) - x_1|| \leq \delta . \qquad (147)$$

at $t = t_1$. Let ε be some positive number. We consider the relation (147) along those functions $u(t)$, $t\varepsilon T$, which satisfy the inequality

$$J(u) \leq \varepsilon \quad . \qquad (148)$$

We let $U_1(\cdot) \subset U(\cdot)$ denote the class of functions $u(t)$, $t \varepsilon T$, for which condition (148) is satisfied. Since the set

$$\Omega = \{x: \quad x = Su(\cdot), u(\cdot) \varepsilon U_1(\cdot)\}$$

is closed and convex, repeating the arguments of the proof of Theorem 22 (Section 2), we are led to the conclusion: relation (147), under the constraint (148), occurs if and only if

$$\Lambda = \max_{||g||=1} [g's(t_1, x_0) - g'x_1 - \delta||g|| + \min_{u(\cdot)\varepsilon U_1(\cdot)} g'Su(\cdot)] \leq 0 \quad . \tag{149}$$

We find min $g'Su(\cdot)$ under the condition (148). Since this minimum is achieved on those elements which unconditionally minimize the functional $g'Su(\cdot) + \lambda J(u)$, where the number λ is determined from condition (148), we deal with the problem in the following way.

First we transform $J(u)$. We substitute

$$x(\tau) = F(\tau)F^{-1}(t_0)x_0 + \int_{t_0}^{\tau} S(\tau,t)u(t)dt \quad ,$$

into the expression for $J(u)$ and interchange the order of integration. Then the expression for $J(u)$ assumes the form

$$J(u) = \int_{t_0}^{t_1} \int_{t_0}^{t_1} K(\tau,t)u(\tau)d\tau + u(t)]u(t)dt$$
$$+ 2\int_{t_0}^{t_1} u(t)n'(t)x_0 dt + \int_{t_0}^{t_1} s'(t,x_0)Mx(t,x_0)dt, \tag{150}$$

where

$$n(t) = \int_{t}^{t_1} (F^{-1}(t_o))'F'(\tau)MS(\tau,t)d\tau \quad ,$$

$$K(\tau,t) = \begin{cases} \int_{t}^{t_1} S'(\nu,\tau)MS(\nu,t)d\nu \quad , & t \geq \tau \quad , \\ \int_{\tau}^{t_1} S'(\nu,\tau)MS(\nu,t)d\nu \quad , & t \leq \tau \quad . \end{cases}$$

We let $L(u)$ denote the expression $L(u) = g'Su(\cdot) + \lambda J(u)$. We have

$$L(u) = \int_{t_o}^{t_1} g'S(t_1,t)u(t)dt$$

$$+ \lambda\left\{\int_{t_o}^{t_1}\left[\int_{t_o}^{t_1} K(\tau,t)u(\tau)d\tau + u(\tau)\right]u(t)dt\right.$$

$$\left. + 2\int_{t_o}^{t_1} u(t)n'(t)x_o dt + \int_{t_o}^{t_1} s'(t,x_o)Ms(t,x_o)dt\right\} \quad .$$

The first variation δL of the functional $L(u)$ has the form

$$\delta L = \int_{t_o}^{t_1}\left\{g'S(t_1,t)\right.$$

$$\left. + 2\lambda\left[n'(t)x_o + \int_{t_o}^{t_1} K(\tau,t)v(\tau)d\tau + v(t)\right]\right\}\delta u(t)dt \quad .$$

Let $v(\cdot)\varepsilon U(\cdot)$ satisfy the condition

$$L(v) = \max_{u(\cdot)\varepsilon U(\cdot)} L(u) \quad .$$

From the conditions for extremality of the functional $L(v)$, we must have $\delta L = 0$. Hence, by virtue of the fundamental lemma of the calculus of variations, we obtain

$$v(t) + \int_{t_0}^{t_1} K(\tau,t)v(\tau)d\tau = -\frac{1}{2\lambda} g'S(t_1,t) - n'(t)x_o , \quad (151)$$

where the multiplier λ is determined from condition (148). Thus, for determination of the extremal control $v(t)$ we obtain the integral equation (151), where λ must be calculated on the basis of (148). The kernel $K(\tau,t)$ of Eq. (151) is such that $K(\tau,t) = K(t,\tau)$ and

$$\int_{t_0}^{t_1}\int_{t_0}^{t_1} |K(\tau,t)|^2 dt\, d\tau < \infty .$$

Properties of the solution of the integral equation may be obtained from the following assertions.

1. The solution of the integral equation

$$u(t) + \int_{t_0}^{t_1} K(\tau,t)u(\tau)d\tau = f(t) ,$$

exists and is unique for any (square-integrable) function $f(t) \not\equiv 0$.

Proof. We assume that this is not the case. Then, according to the Fredholm Alternative [96], the homogeneous equation ($f(t) \equiv 0$) has a nontrivial solution. Let $\tilde{u}(t) \not\equiv 0$ be one such solution. We multiply both sides of this equation (where $f(t) \equiv 0$) by $u(t)$ and integrate from $t = t_0$ to $t = t_1$:

$$\int_{t_0}^{t_1}\left[\int_{t_0}^{t_1} K(\tau,t)u(\tau)d\tau + \tilde{u}(t)\;\; \tilde{u}(t)dt = 0\right] .$$

The left side of this relation equals $J(\tilde{u})$ for $x_o = 0$. This easily follows from (150) for $u = \tilde{u}$ and $x_o = 0$. But, we always have $J(\tilde{u}) > 0$, $\tilde{u}(\cdot) = 0$. This contradiction proves the lemma.

2. The solution of Eq.(151) can be given in the form

$$v(t) = -\frac{1}{2\lambda} v_1(t) = v^2(t), \quad v_1(t) = g'z(t), \quad v_2(t) = y'(t)x_o,$$

when the vectors z(t) and y(t) satisfy the following integral equations

$$z(t) + \int_{t_o}^{t_1} K(\tau,t)z(\tau)d\tau = S(t_1,t) \quad ,$$

$$y(t) + \int_{t_o}^{t_1} K(\tau,t)y(\tau)d\tau = n(t) \quad .$$

The proof of this assertion follows from the superposition principle since Eq.(150) is linear in v. We assume that the solutions z(t), y(t) are known. We substitute the expression for v(t) into condition (148) in order to obtain an explicit form for the constraint on λ. This constraint has the form

$$J(u) = \left(\frac{a}{2\lambda}\right)^2 + c + d \leq \varepsilon^2 \quad , \tag{152}$$

where

$$a^2 = \int_{t_o}^{t_1} \int_{t_o}^{t_1} K(\tau,t)v_1(\tau)d\tau + v_1(t) \; v_1(t)dt$$
$$= \int_{t_o}^{t_1} g'z(t)g'S(t_1,t)dt,$$

$$c = -\int_{t_o}^{t_1} v_2(t)n'(t)x_o dt = -\int_{t_o}^{t_1} y'(t)x_o n'(t)x_o dt \quad ,$$

$$d = \int_{t_o}^{t_1} s'(t,x_o)Mx(t,x_o)dt \quad .$$

It is possible to prove that

$$a^2 = J(v_1)\Big|_{x_o=0} = J(-v_1)\Big|_{x_{o=0}}, \quad \frac{a^2}{4\lambda^2} = J\,-\!\left(\frac{1}{2\lambda}\,v_1\right)\Big|_{x_{o=0}} ,$$

It is evident that a solution v of the problem

$$\max_{u(\cdot)\varepsilon U_1(\cdot)} g'Su(\cdot) = g'Sv(\cdot) \tag{153}$$

exists if and only if $c + d < \varepsilon^2$. Satisfaction of condition (149) necessarily implies the existence of a solution to problem (153); therefore, if condition (149) is satisfied then we have $c + d < \varepsilon^2$. Then we obtain

$$\lambda = \lambda^* = -\frac{|a|}{2\sqrt{\varepsilon^2 - c - d}} .$$

Consequently,

$$\max_{u(\cdot)\varepsilon U_1(\cdot)} g'Su(\cdot) = g'Sv(\cdot) = -\frac{1}{2\lambda}\, g'Sv_1(\cdot) - g'Sv_2(\cdot) ,$$

$$g'Sv_1(\cdot) = \int_{t_o}^{t_1} g'S(t_1,t)v_1(t)dt = \int_{t_o}^{t_1} g'S(t_1,t)g'z(t)dt = a^2,$$

$$g'Sv_2(\cdot) = \int_{t_o}^{t_1} g'S(t_1,t)y'(t)x_o dt .$$

Therefore,

$$g'Sv(\cdot) = -\frac{a^2}{2\lambda} - \int_{t_o}^{t_1} g'S(t_1,t)y'(t)x_o dt ,$$

where λ is determined from (152). Thus, the problem of maximizing the functional $g'Su(\cdot)$ over all $u(\cdot)\varepsilon U_1(\cdot)$, has been reduced (after finding the form of v) to the problem of finding $\max_{\lambda} g'Sv(\cdot)$ or $\max_{\lambda} (-a^2/2\lambda)$ under the condition (152) (the term $g'Sv_2(\cdot)$ is independent of λ).

Clearly, $\max_{\lambda} (-a^2/2\lambda)$ is assumed at a point $\lambda = \lambda^*$ since a^2 is positive. Since the second variation $\delta^2 L$ of the functional $L(u)$ has the form

$$\delta^2 L = 2\lambda \int_{t_o}^{t_1} \left[\int_{t_o}^{t_1} K(\tau,t)\delta u(\tau)d\tau + \delta u(t) \right] \delta u(t)dt \quad ,$$

or, with regard for (150), $\delta^2 L = 2\lambda J(\delta u)\Big|_{x_o=0}$; $\lambda < 0$, $J(\delta u)\Big|_{x_o=0} > 0$ if $\delta u(t) \not\equiv 0$. Thus, $\delta^2 L < 0$. Consequently, $g'Sv(\cdot) = \max_{u \varepsilon U_1} g'Su(\cdot)$.

On the basis of the above results, we write the final expression for $g'Sv(\cdot)$:

$$\max_{u(\cdot)\varepsilon U_1(\cdot)} g'Su(\cdot) = g'Sv(\cdot)$$

$$= (\varepsilon^2 - c - d)^{1/2} \left[\int_{t_o}^{t_1} g'S(t_1,t)g'z(t)dt \right]^{1/2}$$

$$- \int_{t_o}^{t_1} g'S(t_1,t)y'(t)x_o dt \quad .$$

Remark. The condition $c + d < \varepsilon^2$ distinguishes in E_n the set Δ of those initial conditions x_o for which problem (153) has a solution, i.e. the class $U_1(\cdot)$ of controls $u(t)$, satisfying condition (148) for any $x_o \varepsilon \Delta$ is not empty. Under this condition, we have $J(v)\Big|_{x_o} = \varepsilon^2$, which follows from (152) upon substitution of the value λ^*. The quantity $c + d$ is a quadratic form in x_{i_o}, $i = 1,\ldots,n$, and, since $c + d = J(-v_2)$ and $J(-v_2) > 0$, $||x_o|| \not\equiv 0$, the condition $c + d < \varepsilon^2$ defines an open ellipsoid in E_n.

We substitute the expression obtained for $g'Sv(\cdot)$ into (149). Then condition (149) assumes the form

$$\Lambda = \max_{||g||=1} \Big\{ g's(t_1,x_0) - g'x_1 - \delta||g|| - (\varepsilon^2 - c - d)^{1/2} \left[\int_{t_0}^{t_1} g'S(t_1,t)g'z(t)dt\right]^{1/2} + \int_{t_0}^{t_1} g'S(t_1,t)y'(t)x_0 dt \Big\} \leq 0 .$$

We introduce the notation

$$\mu = (\varepsilon^2 - c - d)^{1/2} , \qquad (154)$$

$$\bar{x} = s(t_1,x_0) + \int_{t_0}^{t_1} S(t_1,t)y'(t)x_0 dt , \quad q = -x_1 + \bar{x} .$$

Then

$$\Lambda = \max_{||g||=1} \Big\{ g'q - \delta||g|| - \mu \left[\int_{t_0}^{t_1} g'S(t_1,t)g'z(t)dt\right]^{1/2} \Big\} \leq 0 . \qquad (155)$$

It is not difficult to show that (155) is the condition for solvability of (147),(148) for $x_o = 0$, $x_1 = q$, in the class $U_\mu(\cdot)$ of controls $u(\cdot)$ such that $J(u) \leq \mu^2$. We call this problem problem A. From the foregoing, it follows that if condition (149) is satisfied so is condition (155), where μ, q are defined by the formulas (154). Clearly, the converse is also true: if condition (155) is satisfied and if x_0, x_1, and ε are such that relation (154) holds, then condition (149) is also satisfied. Thus, we have been led to the following assertion: The problem (147),(148) is solvable if and only if problem

A is solvable and if $x_o, x_1, \varepsilon, \mu, q$ are connected by the formulas (154).

It is not difficult to see that the problem of minimizing mean-square error under passage of the system (146) from the point x_o to a δ-neighborhood of the point x_1, leads to the problem of minimizing the mean-square error under passage of (146) from the point $x = 0$ to a δ-neighborhood of the point q.

We pass now to the study of an optimal control problem. We let ε^o denote the smallest value of ε satisfying (149), and $U^o(\cdot)$ denotes the class of admissible controls such that $J(u) \leq \varepsilon^o$.

Theorem 29. *The minimal value of ε, $\varepsilon = \varepsilon^o$, exists and is unique. If g_o is the solution of problem (149) for $\varepsilon = \varepsilon^o$, then the optimal control $u^o(t)$ is determined from the condition*

$$g_o' S u^o(\cdot) = \max_{u(\cdot) \varepsilon U^o(\cdot)} g_o' S u(\cdot)$$

and equals

$$u^o(t) = -\frac{1}{2\lambda_o} v_1^o(t) - v_2(t) \quad ,$$

where

$$v_1^o(t) = g_o' z(t) \quad , \qquad \lambda_o = \frac{\left[\int_{t_o}^{t_1} g_o' z(t) g_o' S(t_1, t) dt\right]^{1/2}}{2[(\varepsilon^o)^2 - c - d]^{1/2}} ,$$

$x^o(t_1)$ satisfies the condition $g_o'(x^o(t_1) - x') = -\delta||g_o||$.

Proof. The problem of minimizing ε is equivalent to the problem of minimizing μ(cf. (154),(155)). Thus, we consider condition (155), where μ and δ enter. This means that Λ from (155), for fixed t_1, δ, q, is monotonically decreasing and continuous in μ, where we always have $\Lambda\big|_{x_o=0} > 0$. Thus, the minimal value of $\mu = \mu^o$ exists and is defined by the condition $\Lambda\big|_{\mu=\mu^o} = 0$. It is not difficult to show that

$$\mu^o = [(\varepsilon^o)^2 - c - d]^{1/2} = \max_{||g||=1} \frac{g'q - \delta||g||}{\left[\int_{t_o}^{t_1} g'S(t_1,t)g'z(t)dt\right]^{1/2}} .$$

This completes the proof.

7. The Problem of Optimal Control in Systems with a Delay

Let the motion of the object be described by the differential equation with a deviating argument

$$\dot{x}(t) = A(t)x(t) + A_1(t)x(t-h) + B(t)u(t) \quad . \tag{156}$$

Here x is an n-vector in the space E_n; u(t) is an n-vector of piecewise-continuous control functions lying in the set of admissible controls $U(\cdot)$; A(t), $A_1(t)$, and B(t) are continuous matrix functions with B(t) nonsingular for all $t \varepsilon T = [t_o, t_1]$; $h > 0$ is a constant delay. Initial conditions are given: the vector function $\Phi(t)$ with values in E_n defines x(t) for $t_o - h \leq t < t_o$ and $x_o = x(t_o)$.

In analogy with section 4, we define a T-admissible control as those functions u(t), $t \varepsilon T$, for which the trajectories of (156) satisfy the condition

$$x(t) \equiv 0 \quad , \quad t_1 - h \leq t \leq t_1 \quad . \tag{157}$$

In E_n, we consider the set G of vectors

$$s = s(t_1 - h, x_o(\cdot))$$

$$= \int_{t_o - h}^{t_1} F(t_1 - h, \tau + h) A_1(\tau + h) \Phi(\tau) d\tau + F(t_1 - h, t_o) x_o \ .$$

Here $F(t,\tau)$ is the fundamental matrix for the homogeneous equation corresponding to (156), where $F(\tau,\tau) = E$, $F(t,\tau) = 0$, $\tau > t$. The set $G_1 \subset G$ of vectors s, formed with the help of the function $\Phi(t)$ and the vector x_o, for which there exists a T-admissible control is the domain of attainability. We let $U_1(\cdot) \subset U(\cdot)$ denote the set of T-admissible controls. The set G_1 is convex. Actually, by definition, if $s \varepsilon G_1$ then there exists a $u(\cdot) \varepsilon U_1(\cdot)$, such that $x(t) \equiv 0$ for $t_1 - h \leq t \leq t_1$, where

$$x(t) = s(t, x_o(\cdot)) + \int_{t_o}^{t} F(t,\tau) B(\tau) u(\tau) d\tau \quad .$$

We consider two elements of the set G_1: s_1 and s_2 and the controls $u_1(\cdot)$ and $u_2(\cdot)$ corresponding to them. We show that the element $s = \alpha s_1 + (1 - \alpha) s_2$, $0 \leq \alpha \leq 1$, also lies in the set G_1, where the control corresponding to it, ensuring satisfaction of condition (157), has the form

$$u(t) = \alpha u_1(t) + (1 - \alpha) u_2(t) \quad .$$

Indeed,

$$x(t) = s(t, x_o(\cdot)) + \int_{t_o}^{t} F(t,\tau) B(\tau) u(\tau) d\tau$$

$$= \alpha [s(t, x_{10}(\cdot)) + \int_{t_o}^{t} F(t,\tau) B(\tau) u_1(\tau) d\tau]$$

$$+ (1-\alpha)[s(t,x_{20}(\cdot)) + \int_{t_o}^{t} F(t,\tau)B(\tau)u_2(\tau)d\tau]$$

$$= \alpha x_1(t) + (1-\alpha)x_2(t)$$

But, since $x_1(t) \equiv 0$ and $x_2(t) \equiv 0$ for $t_1 - h \le t \le t_1$, then $x(t) \equiv 0$ for $t_1 - h \le t \le t_1$. The convexity of G_1 is proven. From Eq.(156) and condition (157), it follows that each T-admissible control u(t) on the interval $t_1 - h \le t \le t_1$ has the form

$$u(t) = -B^{-1}(t)A_1(t)x(t-h) \quad .$$

<u>Theorem 30. A T-admissible control exists* if and only if</u>

$$\max_{||g||=1} g's(t_1 - h, x_o(\cdot))$$

$$+ \inf_{u(\cdot)\varepsilon U_1(\cdot)} \int_{t_o}^{t_1-h} g'F(t_1 - h,\tau)B(\tau)u(\tau)d\tau \le 0. \quad (158)$$

<u>Proof</u>. We proceed by the scheme of section 2. (Necessity). Let the control $u = u(t)$, $u(\cdot)\varepsilon U(\cdot)$, for given x_o, $\Phi(t)$, t_1, and h, ensure satisfaction of condition (157). We prove that inequality (158) is satisfied. By condition (157), trajectories generated by the control u(t), reach the origin at time $t = t_1 - h$, i.e. $x(t_1 - h) = 0$ or

$$s(t_1 - h, x_o(\cdot)) + \int_{t_o}^{t_1-h} F(t_1 - h,\tau)B(\tau)u(\tau)d\tau = 0 \quad .$$

---

* Admitting a generalized optimal control.

We take the scalar product of all terms with the arbitrary vector g, $||g|| = 1$. We have

$$g's(t_1 - h, x_0(\cdot)) + \int_{t_0}^{t_1-h} g'F(t_1 - h, \tau)B(\tau)u(\tau)d\tau = 0 \quad .$$

We consider the functional

$$P(u) = g's(t_1 - h, x_0(\cdot)) + \int_{t_0}^{t_1-h} g'F(t_1 - h, \tau)B(\tau)u(\tau)d\tau$$

on the set of functions $u(\cdot)\varepsilon U_1(\cdot)$. Since the given control $u(\cdot)\varepsilon U_1(\cdot)$ and $P(u) = 0$, then $\inf_{u\varepsilon U_1} P(u) \leq 0$. But, the last inequality if satisfied for any vector g, $||g|| = 1$; consequently,

$$\max_{||g||=1} \{g's(t_1 - h, x_0(\cdot)) + \inf_{u(\cdot)\varepsilon U_1(\cdot)} \int_{t_0}^{t_1-h} g'F(t_1 - h, \tau)B(\tau)u(\tau)d\tau\} \leq 0 \quad ,$$

which was what we needed to prove.

<u>(Sufficiency)</u>. For given x_0, $\Phi(t)$, t_1, let condition (158) be satisfied. We prove that there exists at least one control $u(t)$, $u(\cdot)\varepsilon U_1(\cdot)$ ensuring satisfaction of condition (157).

We assume the contrary: there does not exist a control $u(t)$, $u(\cdot)\varepsilon U_1(\cdot)$ such that the trajectories generated by this control satisfy condition (157). This means that the vector $s(t_1 - h, x_0(\cdot))$, for given x_0 and $\Phi(t)$, does not lie in the closure $\overline{G}_1$ of the domain of attainability. Then by virtue of the convexity of $\overline{G}_1$, there exists a hyperplane and a vector g, $||g|| = 1$, normal to this hyperplane, such that the inequality

$$g's(t_1 - h, x_0(\cdot)) > g's$$

is satisfied for all $s \varepsilon \bar{G}_1$. But, since s is a point of the domain of attainability, there exists a control u(t), $u(\cdot) \varepsilon U_1(\cdot)$, such that

$$s + \int_{t_o}^{t_1-h} F(t_1 - h, \tau) B(\tau) u(\tau) d\tau = 0 \quad .$$

Hence,

$$s = - \int_{t_o}^{t_1-h} F(t_1 - h, \tau) B(\tau) u(\tau) d\tau \quad .$$

Then

$$g' s(t_1 - h, x_o(\cdot)) + \int_{t_o}^{t_1-h} g' F(t_1 - h, \tau) B(\tau) u(\tau) d\tau > 0 \quad .$$

Since s is an arbitrary point of the set G_1, the last inequality is satisfied for all $u(\cdot) \varepsilon U_1(\cdot)$; therefore,

$$g' s(t_1 - h, x_o(\cdot)) + \inf_{u(\cdot) \varepsilon U_1(\cdot)} \int_{t_o}^{t_1-h} g' F(t_1 - h, \tau) B(\tau) u(\tau) d\tau > 0.$$

By the condition of the theorem, inequality (158) is satisfied for given x_o and $\Phi(t)$ which means that for any vector $\bar{g}$, $||\bar{g}|| = 1$, we have the inequality

$$\bar{g}' s(t_1 - h, x_o(\cdot)) + \inf_{u(\cdot) \varepsilon U_1(\cdot)} \int_{t_o}^{t_1-h} \bar{g}' F(t_1 - h, \tau) B(\tau) u(\tau) d\tau \leq 0.$$

However, by our construction with the vector g, $||g|| = 1$, the inequality is satisfied in the opposite sense. This contradiction proves the theorem.

In what follows, we will assume that the set $U(\cdot)$, of admissible controls, has the form

$$U(\cdot) = u(\cdot): \int_{t_o}^{t_1} u'(t) u(t) dt \leq 1 \quad . \qquad (159)$$

We write condition (158) in another form:

$$\max_{||g||=1} \{g's(t_1 - h, x_0(\cdot)) + \mu(g)\} \leq 0 \quad ,$$

$$\mu(g) = \inf_{u(\cdot)\varepsilon U(\cdot)} \mu(g,u)$$

$$= \inf_{u(\cdot)\varepsilon U(\cdot)} \int_{t_0}^{t_1} g'F(t_1 - h,\tau)B(\tau)u(\tau)d\tau$$

under the condition

$$\int_{t_0}^{t_1-h} u'(t)u(t)dt$$

$$+ \int_{t_1-h}^{t_1} [B^{-1}(t)A_1x(t-h)]'B^{-1}(t)A_1(t)x(t-h)dt \leq 1 \ .$$

We transform the last inequality, substituting $x(t-h)$ by Cauchy's formula (§1.15):

$$\int_{t_0}^{t_1-h} u(t)u(t)dt + \int_{t_1-h}^{t_1} \Bigg[B^{-1}(t)A_1(t)\Big\{s(t - h,x_0(\cdot))$$

$$+ \int_{t_0}^{t-h} F(t-h,\tau)B(\tau)u(\tau)d\tau\Big\}\Bigg]'B^{-1}(t)A_1(t)\Big\{s(t-h,x_0(\cdot))$$

$$+ \int_{t_0}^{t-h} F(t - h,\tau)B(\tau)u(\tau)d\tau\Big\}dt \leq 1 \quad .$$

Removing the parentheses in the second integral and interchanging the order of integration, we obtain

$$\int_{t_0}^{t_1-h} u'(\tau)u(\tau)d\tau + N + 2\int_{t_0}^{t_1-h} f'(\tau)u(\tau)d\tau$$

$$+ \int_{t_0}^{t_1-h}\int_{t_0}^{t_1-h} u'(\tau)\Phi(\tau,\theta)u(\theta)d\theta\ d\tau \leq 1 \quad .$$

Here,

$$N = \int_{t_1-2h}^{t_1-h} [B^{-1}(t + h)A_1(t + h)s(t,x_o(\cdot))]' \cdot B^{-1}(t + h)A_1(t + h)s(t,x_o(\cdot))dt ,$$

$$f(\tau) = \int_{t_1-2h}^{t_1-h} [B^{-1}(t + h)A_1(t + h)F(t,\tau)B(\tau)]' \cdot B^{-1}(t + h)A_1(t + h)s(t,x_o(\cdot))dt ,$$

$$\Phi(\tau,\theta) = \int_{t_1-2h}^{t_1-h} [B^{-1}(t + h)A_1(t + h)F(t,\tau)B(\tau)]' \cdot B^{-1}(t + h)A_1(t + h)F(t,\theta)B(\theta)dt .$$

Introducing a Lagrange multiplier, we pass to a problem with an unconditional extremum. We seek a minimum of the functional $\Lambda(u)$:

$$\Lambda(u) = \left[\int_{t_o}^{t_1-h} g'F(t_1 - h,\tau)B(\tau)u(\tau)d\tau + \lambda \int_{t_o}^{t_1-h} u'(\tau)u(\tau)d\tau + N + 2\int_{t_o}^{t_1-h} f'(\tau)u(\tau)d\tau + \int_{t_o}^{t_1-h}\int_{t_o}^{t_1-h} u'(\tau)\Phi(\tau,\theta)u(\theta)d\theta\right]d\tau .$$

We calculate the variation of the functional

$$\delta\Delta(u) = \left.\frac{\partial\Lambda(u + \alpha v)}{\partial\alpha}\right|_{\alpha=0} = \int_{t_o}^{t_1-h} v'(\tau)L(\tau)g + \lambda v'(\tau)\ 2u(\tau) + 2f(\tau) + \int_{t_o}^{t_1-h} K(\tau,\theta)u(\theta)d\theta\ d\tau ,$$

where

$$L(\tau) = [F(t_1 - h,\tau)B(\tau)]' \quad , \quad K(\tau,\theta) = \Phi(\tau,\theta) + \Phi'(\theta,\tau) \ .$$

Since the variation of the functional $\delta\Delta(u)$ must equal zero for any value of the variation function, we have

$$L(\tau)g + \lambda\ 2u(\tau) + 2f(\tau) + \int_{t_0}^{t_1-h} K(\tau,\theta)u(\theta)d\theta = 0 \ .$$

The solution of the obtained integral equation may be represented in the form of a sum of two solutions:

$$u(t) = u_1(t) + u_2(t) = u_1(t) + \frac{1}{\lambda} R(t)g \ ,$$

where the vector $u_1(t)$ and matrix $R(t)$ satisfy the following integral equations

$$2u_1(t) + 2f(t) + \int_{t_0}^{t_1-h} K(t,\theta)u_1(\theta)d\theta = 0 \qquad (160)$$

and

$$2R(t) + L(t) + \int_{t_0}^{t_1-h} K(t,\theta)R(\theta)d\theta = 0 \ . \qquad (161)$$

The multiplier λ is found from the condition that the sought control lies in the set $U_1(\cdot)$. We have

$$\int_{t_0}^{t_1-h} u_1(\tau) + \frac{1}{\lambda} R(\tau)g\Big]'\Big[u_1(\tau) + \frac{1}{\lambda} R(\tau)g\Big]d\tau + N$$

$$+ 2\int_{t_0}^{t_1-h}\Big[f'(\tau)\ u_1(\tau) + \frac{1}{\lambda} R(\tau)g \quad d\tau$$

$$+ \int_{t_0}^{t_1-h}\int_{t_0}^{t_1-h} \Big[u_1(\tau) + \frac{1}{\lambda} R(\tau)g\Big]'$$

$$\cdot\ \Phi(\tau,\theta)\ u_1(\theta) + \frac{1}{\lambda} R(\theta) g\ d\theta\ d\tau = 1 \quad .$$

Hence, for λ we obtain the algebraic equation

$$a \frac{1}{\lambda^2} + 2b \frac{1}{\lambda} + c = 0 \quad . \tag{162}$$

Here

$$a = \int_{t_0}^{t_1-h} g' R'(\tau) R(\tau) g\ d\tau$$

$$+ \int_{t_0}^{t_1-h} \int_{t_0}^{t_1-h} g' R'(\tau) \Phi(\tau,\theta) R(\theta) g\ d\theta\ d\tau$$

$$= -\frac{1}{2} \int_{t_0}^{t_1-h} L'(\tau) R(\tau) g\ d\tau \quad ,$$

$$b = \int_{t_0}^{t_1-h} u_1'(\tau) R(\tau) g\ d\tau + \int_{t_0}^{t_1-h} f'(\tau) R(\tau) g\ d\tau$$

$$+ \frac{1}{2} \int_{t_0}^{t_1-h} \int_{t_0}^{t_1-h} g' R'(\tau) K(\tau,\theta) u_1(\theta) d\theta\ d\tau = 0 \quad ,$$

$$c = \int_{t_0}^{t_1-h} u_1'(\tau) u_1(\tau) d\tau + N + 2 \int_{t_0}^{t_1-h} f'(\tau) u_1(\tau) d\tau$$

$$+ \int_{t_0}^{t_1-h} \int_{t_0}^{t_1-h} u_1'(\tau) \Phi(\tau,\theta) u_1(\theta) d\theta\ d\tau - 1$$

$$= N - 1 + \int_{t_0}^{t_1-h} f'(\tau) u_1(\tau) d\tau \quad .$$

From the two roots of Eq.(162), a finite minimum of the functional $\Lambda(u)$ is attained by the positive root. Hence, condition (158) for the set of admissible controls (159) has the form

$$\max_{||g||=1} \; g's(t_1 - h, x_o(\cdot))$$

$$+ \int_{t_o}^{t_1 - h} g'F(t_1 - h, \tau)B(\tau)u(\tau, g)\,d\tau \leq 0 \qquad (163)$$

or, more concisely,

$$\max_{||g||=1} \Psi(g, t_1) \leq 0 \quad .$$

Here,

$$u(\tau, g) = u_1(\tau) + \frac{1}{\lambda} R(\tau)g \quad ,$$

where $u_1(\tau)$ and $R(\tau)$ are solutions of the integral equations (160) and (161), while λ is the positive root of Eq.(162).

Problem. In the class of admissible controls, find a control $u^o(t)$ which, for given x_o, $\Phi(t)$, solves problem (157) in minimal time.

We will call the control $u^o(t)$ *optimal*, while the time of passage t_1^o of the process corresponding to this control will be called the *minimal time*.

From Theorem 30 follows

Theorem 31. The minimal time t_1^o is the smallest number for which inequality (163) is satisfied. The optimal control $u^o(t)$ is defined by the relation

$$u^o(t) = \begin{cases} u(t, g_o) \quad , & t_o \leq t \leq t_1^o - h \quad , \\ -B^{-1}(t)A_1(t)x^o(t-h) \quad , & t_1^o - h \leq t \leq t_1^o \quad , \end{cases}$$

where g_o is the element for which the function $\Psi(g, t_1^o)$ is maximized.

Example 7.

$$\dot{x}(t) = -x(t - 1) + u(t) ,$$

$$\Phi(t) \equiv 0 , \quad -1 \le t < 0 , \quad x(0) = 10 , \quad t_1 = 3 , \qquad (164)$$

$$\int_0^3 u^2(t)dt \le M .$$

To simplify the computation, instead of the problem of minimal time, we consider another problem connected with the first in a natural way. We seek a control from the class of admissible controls in order that the trajectory corresponding to a given initial condition and the chosen control satisfies the condition $x(t) \equiv 0$, $2 \le t \le 3$, while the functional $\int_0^3 u^2(t)dt$ is minimized.

We write the solution of (164) using Cauchy's formula. We find $u(t,g)$ through the integral equations (160), (161), and we compute the multiplier λ. As a result, we see that the optimal control has the form

$$u^o(\tau) = \begin{cases} e^{\frac{\sqrt{3}}{2}\tau}\left(2.6521 \sin \frac{\tau}{2} - 2.5115 \cos \frac{\tau}{2}\right) \\ \quad -e^{-\frac{\sqrt{3}}{2}\tau}\left(2.3118 \sin \frac{\tau}{2} + 4.6325 \cos \frac{\tau}{2}\right), \quad 0 \le \tau \le 1; \\ e^{\frac{\sqrt{3}}{2}\tau}\left(1.1407 \sin \frac{\tau}{2} - 1.0301 \cos \frac{\tau}{2}\right) \\ \quad +e^{-\frac{\sqrt{3}}{2}\tau}\left(12.2625 \sin \frac{\tau}{2} + 1.0359 \cos \frac{\tau}{2}\right), \quad 1 \le \tau \le 2; \\ e^{\frac{\sqrt{3}}{2}\tau}\left(0.4901 \sin \frac{\tau}{2} - 0.4219 \cos \frac{\tau}{2}\right) \\ \quad -e^{-\frac{\sqrt{3}}{2}\tau}\left(17.2647 \sin \frac{\tau}{2} - 23.6179 \cos \frac{\tau}{2}\right) - 8.3334 + 19.9980, \\ \qquad 2 \le \tau \le 3 . \end{cases}$$

8. Statistical Problems of Optimal Control

Different applications of the theorem on the separation of convex sets are discovered when we study problems of optimization for systems whose behavior is complicated by the action of stochastic forces. Questions similar to these are considered in §15.

9. Pursuit Problems

Up to now we have considered situations in which antagonistic goals have not been assumed in defining the behavior of the system. In the theory of differential games, optimal control problems are studied for dynamical systems in those cases when aspects of antagonistic goals are present. Development of methods of functional analysis for this new circle of questions is discussed in §16.

§11. The L-Problem of Moments in the Theory of Optimal Processes

In this paragraph, we apply one of the fundamental theorems of functional analysis (the Hahn-Banach theorem or the extension of linear operators [46]) for the solution of several problems in the theory of optimal processes. Its concrete application in the L-problem of moments [4] played an obvious role in the first years of the development of the theory of optimal processes [74n]. For formulation and understanding of the essence of the L-problem, certain knowledge of the theory of normed linear spaces is required. We do not mean, however, to exaggerate the difficulties connected with the new circle of questions. If desired, all problems considered below may be formulated and solved in terms of finite-dimensional spaces. First we give a traditional statement of the

L-problem of moments, then we formulate it in terms of finite-dimensional spaces, and finally we solve this problem using the facts already proved.

1. The L-Problem of Moments and its Solution

We are given the linearly independent elements $x_1(\cdot),\ldots,x_n(\cdot)$ in the normed linear space $E(\cdot)$. We wish to find necessary and sufficient conditions for the numbers $c_1,\ldots,c_n, L, (\sum_{i=1}^{n} c_i^2 > 0,\ L > 0)$ in order that there exist a linear functional $f(\cdot)$ satisfying

$$f(x_i(\cdot)) = c_i \quad , \quad ||f|| \leq L \quad , \quad i = 1,\ldots,n \quad . \tag{165}$$

We give this problem another formulation. The numbers $c_1,\ldots,c_n$ may be treated as the components of a vector c from the n-dimensional space E_n. Then the values $x_i = f(x_i(\cdot))$, $i = 1,\ldots,n$, which the functional assumes on the given elements $x_i(\cdot)$, $i = 1,\ldots,n$, of the normed space $E(\cdot)$, are the components of an n-vector $x \varepsilon E_n$. Different vectors $x \varepsilon E_n$ correspond to different linear functionals f. We let R(L) denote the set

$$R(L) = \{x: \quad x = f(x(\cdot)), \ ||f|| \leq L\}$$

of n-vectors x, generated by the linear functionals $f(\cdot)$, $f(\cdot) \varepsilon E_n'$, defined on the given elements $x(\cdot) = \{x_1(\cdot),\ldots, x_n(\cdot)\}$ lying in the sphere $||f(\cdot)|| \leq L$. The relation $x = f(x(\cdot))$ may now be treated as an operator equation in which the functional $f(\cdot) \varepsilon E'(\cdot)$ is transformed into the vector $x \varepsilon E_n$. Clearly, the L-problem of moments (165) has a solution if and only if the set R(L) in the finite-dimensional space E_n contains the vector c.

the equations $f(x(\cdot)) = x$, $g(x(\cdot)) = y$ are satisfied. Then any point $x^\lambda = \lambda x + (1 - \lambda)y$, $0 \leq \lambda \leq 1$, of the in- interval containing the points x and y also lies in R(L). In fact, the functional $f_\lambda(\cdot) = \lambda f(\cdot) + (1 - \lambda)g(\cdot)$ satisfies the inequality $||f_\lambda(\cdot)|| = ||\lambda f(\cdot) + (1 - \lambda)g(\cdot)|| \leq \lambda||f(\cdot)|| + (1 - \lambda)||g(\cdot)|| \leq L$, while its value on $x(\cdot)$ (by virtue of the linearity of the operators) is $f_\lambda(x(\cdot)) = [\lambda f(\cdot) + (1 - \lambda)g(\cdot)](x(\cdot)) = \lambda x + (1 - \lambda)y$, which is equivalent to our assertion.

As the set Q figuring in §9,10, we take the set consisting of the single point c. Now it is clear that we may apply the minimax theorem or the theorem on separation of convex sets to the new problem (165). The set R(L) will contain the point c if and only if its diameter at the point c

$$\rho = \inf_{x \varepsilon R(L)} ||c - x||$$

equals zero. Recalling that $||x|| = \max_{||g|| \leq 1} g'x$ and using the minimax theorem, we have

$$\rho = \inf_{x \varepsilon R(L)} \max_{||g|| \leq 1} g'(c - x) = \max_{||g|| \leq 1} \inf_{x \varepsilon R(L)} g'(c - x)$$

$$= \max_{||g|| \leq 1} \inf_{||f|| \leq L} g'(c - f(x(\cdot)))$$

$$= \max_{||g|| \leq 1} [g'c - \sup_{||f|| \leq L} g'f(x(\cdot))]$$

$$= \max_{||g|| \leq 1} [g'c - \sup_{||f|| \leq L} f(g'x(\cdot))] \quad . \tag{166}$$

By definition, the quantity $\sup_{||f|| \leq 1} f(x(\cdot))$ is the norm of $x(\cdot) \varepsilon E(\cdot)$:

$$||x(\cdot)|| = \sup_{||f|| \leq 1} f(x(\cdot)) \quad . \tag{167}$$

The expression $g'x(\cdot) = \sum_{i=1}^{n} g_i x_i(\cdot)$ represents an element of $E(\cdot)$. Therefore (and by virtue of the linearity of the functional),

$$\sup_{||f(\cdot)||\leq L} f(g'x(\cdot)) = L||g'x(\cdot)|| \quad . \tag{168}$$

Substituting this expression into (166), we obtain a necessary and sufficient condition for the solvability of the L-problem of moments:

$$\rho(L) = \max_{||g||\leq 1} \{g'c - L||g'x(\cdot)||\} = 0 \quad .$$

If the elements $x_1(\cdot),\ldots,x_n(\cdot)$ are linearly independent (which is assumed in the formulation of the L-problem), then $||g'x(\cdot)|| \neq 0$ for all $||g|| \neq 0$. From this condition and from the fact that $g'c - L||g'x(\cdot)||$ is monotomically decreasing in L, it is not difficult to show that the minimal L for which $\rho(L) = 0$ is

$$\lambda = \max_{||g||\leq 1} \frac{g'c}{||g'x(\cdot)||} \quad .$$

From the homogeneity in g of the numerator and the denominator, it follows that

$$\lambda = \max_{g} \frac{g'c}{||g'x(\cdot)||} = \max_{||g'x(\cdot)||=1} g'c \quad ,$$

or

$$\frac{1}{\lambda} = \max_{g'c=1} ||g'x(\cdot)|| \quad . \tag{169}$$

Basic Proposition of the Theory of the L-Problem of Moments

The problem (165) has a solution if and only if $L \geq \lambda$, where the quantity λ is determined from the solution of the finite-dimensional problem (169).

From the method of proof, we find those functionals (extremals) $f(\cdot)$, for which problem (165) has a solution with minimal $L = \lambda$. The extremal functionals satisfy Eq.(168)*, where g is the solution of problem (166). To summarize: in order that some optimal control problem be solvable using the L-problem, it is necessary 1) to formulate it in the form of problem (165), 2) to solve problem (169), and 3) to find the desired optimal quantities from (168).

This general scheme will be made more concrete below by application to different optimal control problems. In relation (165), the functional $f(\cdot)$ does not always represent a control; sometimes it is useful to interpret it in a wider sense (Section 4). The constraint $||f(\cdot)|| \leq L$ arises from the conditions of the problem. In many problems the form of the norm $||f(\cdot)||$ is very complex. This property is the main obstacle for the solution of problem (169) because on the right side of (169) we have the norm of $x(\cdot)$ which must be in agreement by (167) with the norm $||f(\cdot)||$. In standard cases, of course, the computation $||x(\cdot)||$ by means of $||f(\cdot)||$ poses no difficulty. For example if

1) $$||f(\cdot)|| = \sup_{t\varepsilon T} \operatorname{vrai} f(t), \qquad ||x(\cdot)|| = \int_{t_o}^{t_1} |x(t)|dt;$$

*The extremal property of the functional is defined in Theorem 24.

2) $||f(\cdot)|| = \left(\int_{t_0}^{t_1} |f(t)|^q dt\right)^{1/q}$, $||x(\cdot)||$

$$= \left(\int_{t_0}^{t_1} |x(t)|^p dt\right)^{1/p} , \frac{1}{p} + \frac{1}{q} = 1 , 1 < q < \infty .$$

In more complicated cases, determination of $||x(\cdot)||$ represents a separate problem.

Remarks: 1) Above we have shown only the convexity of the set R(L). Under the conditions of problem (165) it must also be closed. This is a corollary of the definition of the space $E'(\cdot)$ of functionals on $E(\cdot)$. Briefly, the space $E'(\cdot)$ is defined so that the problem (167) of computation of the norm $||x(\cdot)||$ always has a solution in this space. In concrete problems, this property often leads to the necessity to extend the given class of functions.

2) Condition (169) for the solvability of the L-problem of moments is obtained under the additional assumption on the linear independence of the elements $x_1(\cdot),\ldots,x_n(\cdot)$. The case when these elements are linearly dependent is discussed in §9.6. It turns out that if the elements $x_1(\cdot),\ldots,x_n(\cdot)$ are connected by the relations $\sum_{k=1}^{n} \xi_{ik} x_k(\cdot) = 0$, $i = 1,\ldots,N$, then the numbers c_i for which the L-problem of moments has a solution, must also be connected by relations $\sum_{i=1}^{n} \xi_{ik} c_k = 0$, $i = 1,\ldots,N$. In the language of the minimal time problem, this means that if the elements $x_1(\cdot),\ldots,x_n(\cdot)$ are linearly dependent, then the initial states x_0 which may be transferred to the origin lie in some subspace of the state space.

2. Two-Point Boundary Conditions for Linear Minimal Time Problems

We are given the system

$$\frac{dx}{dt} = A(t)x + b(t)u + f(t) \quad , \quad t\varepsilon T = [t_0, t_1] \quad , \tag{170}$$

where $x\varepsilon E_n$ and the elements of the matrix $A(t)$ and the vectors $b(t)$, $f(t)$ are continuous functions of t.

Problem: Among all measurable controls $u(t)$, $t\varepsilon T$, $|u(t)| \leq L$, find a control $u^o(t)$ such that

$$x(t_0) = x_0 \ , \ x(t_1^o) = 0, \ t_1^o = \min_{|u(t)| \leq L} \{t_1 : x(t_1) = 0\}. \tag{171}$$

Here x_0 is a given initial condition. The symbol P_i denotes the ith row of the matrix P.

Writing the solution of the inhomogeneous equation (170) by Cauchy's formula, it is not difficult to see that the control $u(t)$ transfer's x_0 to the origin if and only if

$$c_i = \int_{t_0}^{t_1} F_i^{-1}(\tau)b(\tau)u(\tau)d\tau = \int_{t_0}^{t_1} \gamma_i(\tau)u(\tau)d\tau = f(x_i(\cdot)). \tag{172}$$

Here

$$c_i = \{-x_0 - \int_{t_0}^{t_1} F^{-1}(\tau)f(\tau)d\tau\}_i \quad ,$$

$$x_i(\cdot) = \{\gamma_i(\tau), \tau\varepsilon T\} \quad , \quad f(\cdot) = \{u(t), t\varepsilon T\} \quad ,$$

$F(t)$ is the fundamental solution matrix for (170) (with $u(t) \equiv f(t) \equiv 0$). If the norm of the functional f is defined by the formula

$$||f|| = \underset{t\varepsilon T}{\text{vrai max}} \ |u(t)| \quad ,$$

then the control constraint in the original problem may be written in the form

$$||f(\cdot)|| \leq L \quad . \tag{172}$$

Combining (172) and (172'), we see that the linear minimal time problem (171) has been reduced to seeking the smallest t_1 for which the L-problem of moments (165) has a solution. Now the pathway to the solution is clear. We solve the L-problem for fixed t_1, then from these numbers we choose the smallest.

We assume that the functions

$$\gamma_i(t) = F_i^{-1}(t)b(t) \quad , \quad t\varepsilon T \quad , \tag{173}$$

are linearly independent, i.e. there does not exist a vector g, $||g|| = 1$, such that $g'\gamma(t) = 0$ almost everywhere on T.

Remark. The functions (173) are linearly independent if the defining equations of the system (170) are nondegenerate for some $\bar{t}\varepsilon T$.

By virtue of (169), we conclude: a T-admissible control exists if and only if $\lambda(c,t_1) \geq L^{-1}$, where

$$\lambda(c,t_1) = \min_{g'c=1} \lambda(c,t_1,g) = \lambda(c,t_1,g(t_1)) \quad ,$$
$$\lambda(c,t_1,g) = \int_{t_0}^{t_1} |g'\gamma(t)|dt \quad . \tag{174}$$

Lemma 12. The function $\lambda(c,t_1)$ is continuous in c,t_1, and $\lambda(c,t_0) = 0$.

The continuity of the function $\lambda(c,t_1)$ in its arguments is a natural corollary of the continuity of the function $\lambda(c,t_1,g)$ in those same arguments which clearly is true.

Theorem 32. If there exists at least one T-admissible control, then there exists an optimal minimal time control with the time t^o being the smallest root of the equation

$$\lambda(c, t_o + \tau) = L^{-1} . \tag{175}$$

The optimal control $u^o(t)$ is defined by the relation

$$u^o(t) = L \text{ sign } g'(t_o + \tau^o) F^{-1}(t) b(t)$$

almost everywhere on $\sigma_{t_1^o} = \{t: g'(t_o + t^o)F^{-1}(t)b(t) \neq 0\}$.

Proof. From Lemma 12 and condition (169) we obtain relation (175). Let $g(t_1)$ be the solution of problem (174). Then the function $\zeta(t) = g'(t_1)\gamma(t)$, $t \in T$, is an extremal element (cf. (168)) for any solution of system (165) with norm equalling $\lambda^{-1}(c, t_1)$. Therefore,

$$\left| \int_{t_o}^{t_1} g'(t_1)\gamma(t)u(t)dt \right| = \lambda^{-1}(c, t_1) \int_{t_o}^{t_1} |g'(t_1)\gamma(t)| dt . \tag{176}$$

It is not difficult to see that relation (176) occurs if and only if

$$u(t) = \lambda^{-1}(c, t_1) \text{ sign } g'(t_1)\gamma(t) \text{ almost everwhere on } \sigma_{t_1^o} .$$

This completes the proof of the theorem.

Remark. Relation (176) defines the control $u^o(t)$ only at points where $g'(t_1)\gamma(t) \neq 0$. It is possible to show (§7.0) that at the remaining points the function $u^o(t)$ may be defined so that it remains piecewise-constant on T. Thus, having sought the optimal control in the class of measurable functions, we have found it in the narrower class of piecewise-constant controls. "Knowing" such a result, the linear minimal time problem may be formulated within a narrower class of controls. In this case, passage from the minimal time problem to the L-problem begins by having to broaden the class of admissible controls. Finally, the result turns out to be the same. The imbedding of the original problem into a more general problem, which has a solution that is often easier to obtain, and the extraction of the solution of the original problem from the general problem is sufficient for our current needs. By the above procedure, we also solve the question of the existence of a solution to the original problem within the given class of admissible controls.

We consider system (170) for $f(t) \equiv 0$. Now the relation (172) has the form

$$-x_{i_o} = \int_{t_o}^{t_1} \gamma_i(t)u(t)dt \quad ,$$

the functions $\lambda(c,t_1) = \lambda(x_o,t_1)$:

$$\lambda(x_o,t_1) = \min_{g'x_o=-1} \int_{t_o}^{t_1} |g'\gamma(t)|dt \quad .$$

Lemma 13. The function $\lambda(x_o,t_1)$ is continuous in x_o and t_1 and is non-decreasing in t_1.

Proof. The continuity of $\lambda(x_o,t_1)$ is obvious (cf. Lemma 12). We show that $\lambda(x_o,t_1)$ is non-decreasing in t_1. Let $t_1 < t^1$. Then we have

$$\lambda(x_o,t_1,g(t_1)) \leq \lambda(x_o,t_1,g(t^1)) \leq \lambda(x_o,t^1,g(t^1)) \quad ,$$

which proves the assertion.

Remark. For the stationary system (136), Lemma 13 may be strengthened. In this case, the function

$$\lambda(x_o,t_1) = {}_{g'x_o=-1} \int_{t_o}^{t_1} |g'F^{-1}(t)b|\,dt$$

is identically zero or is strictly increasing in t_1. In fact, we consider the function $\xi(t,g) = g'F^{-1}(t)b$. From §1.3, it follows that $\xi(t,g)$ is the solution of the equation

$$\xi^{(n)}(t,g) = \sum_{s=1}^{n} \alpha_s(-1)^{n-s+1}\xi^{(-s+1)}(t,g)\ , \quad ||\alpha|| \neq 0,\ t\varepsilon T\ .$$

Therefore, if $\xi(t^1,g) = 0$ for some $t = t^1$, then $\xi(t,g) \equiv 0$. Now, from the definition of the function $\lambda(x_o,t_1)$ it follows that $\lambda(x_o,t_1) \equiv 0$ or is strictly increasing in t_1.

We say that the functions (173) are completely linearly independent if for each g, $||g|| = 1$, the function $\zeta(t) = g'\gamma(t)$, $t_o \leq t \leq \infty$, does not take on the value zero on a set of positive measure.

Lemma. 14. If the functions (173) are completely linearly independent, then $\lambda(x_o,t_1)$ is strictly increasing in t_1.

The proof is analagous to the proof of Lemma 13.

It is clear that for the system (170) with $f(t) \equiv 0$, the existence of a T-admissible control implies the existence of an optimal control. However, Theorem 32, for (170) with $f(t) \equiv 0$, may be strengthened.

Theorem 33. The optimal time τ^o is defined by the relation

$$\tau^o = \max_g \tau(g) = \tau(\tilde{g}) \quad ,$$

where $\tau(g)$ is the smallest root of the equation

$$\int_{t_o}^{t_o+\tau} |g'F^{-1}(t)b(t)|dt = \frac{1}{L} \quad . \tag{177}$$

The optimal control is as defined in Theorem 32.

Remark. If the functions (173) are completely linearly independent, then $\tau^o = \max_g \tau(g)$, where $\tau(g)$ is the root of Eq. (177).

The results just obtained may be transferred to the system

$$\frac{dx(t)}{dt} = A(t)x(t) + A_1(t)x(t-h) + B(t)u(t) + f(t) \quad .$$

Let it be required to find a control $u(t)$, $t \varepsilon T$, $u(\cdot) \varepsilon U(\cdot)$, such that

$$x(t_1^o) = 0 \quad , \quad t_1^o = \min_{|u(t)| \leq 1} \{t_1 : \ x(t_1) = 0\} \quad .$$

The Cauchy formula (§1.15) is the starting point for the application of the L-problem.

For convenience, as in sections 4,5,7,§10, we introduce the definition of T-relatively admissible controls. We shall call the function $u(t)$, $t \varepsilon T$, a T-relatively admissible control if it generates a trajectory $x(t)$ with the property $x(t_1) = 0$. By virtue of Theorem 31, we conclude that a T-relatively admissible control exists if and only if

$$\Lambda(\tilde{c},t_1) = \max_{||g||=1} \{g'\tilde{c} - L||g'\tilde{\gamma}(\cdot)||\} = \tilde{\Lambda}(t_1,g(t_1)) \leq 0 .$$

Here

$$\left.\begin{aligned} &\tilde{\gamma}(\cdot) = \{\gamma_i(\cdot), i = 1,\ldots,n\} \quad , \quad \gamma_i(\cdot) = \{\gamma_i(t), t\varepsilon T\}, \\ &\gamma_i(t) = F_i(t_1,t)b(t) \quad , \\ &c = -F(t_1,t_0)x_0 - \int_{t_0-h}^{t_1} F(t_1,\tau+h)A_1(\tau+h) \\ &\quad \cdot \Phi(\tau)d\tau - \int_{t_0}^{t_1} F(t_1,\tau)B(\tau)f(\tau)d\tau \quad . \end{aligned}\right\} \quad (178)$$

Lemma 12 is true for the function $\Lambda(\tilde{c},t_1)$. Therefore, it is possible to prove an assertion analagous to Theorem 32.

<u>Theorem 34. If there exists at least one T-relatively admissible control, then there exists an optimal control with minimal time $\tau^o = t_1^o - t_o$, where t_1^o is the smallest root of the equation $\tilde{\Lambda}(t_1,g(t_1)) = 0$.</u>

<u>The optimal control $u^o(t)$ is defined by the relation</u>

$$u^o(t) = L \text{ sign } g'(t_1^o)\gamma(t)$$

<u>almost everywhere on $\sigma_{t_i^o} = \{t: \; g'(t_1^o)\gamma(t) \neq 0\}$.</u>

<u>Remark</u>. It is possible to show that, in the stationary case, the function $\lambda(t) = g'F(t_1,t)b$, $||g|| \neq 0$, is either almost always different from zero or is identically zero.

3. <u>Optimal Processes with Cyclic Constraints</u>

Let the motions of the control system be described by the equation

$$\dot{x}(t) = A(t)x(t) + b(t)u \quad , \tag{180}$$

where $x = \{x_1,...,x_n\}$ is a vector, $A = \{a_{ij}\}$ is an nxn matrix, $b = \{b_1,...,b_n\}$ is a vector, and $u(t)$ is the control function. The problem of optimal control which we will go into below is the following: we are given an initial state $x(o)$. It is required to find a function $u(t)$ such that the point $x(t)$, moving along the trajectory of (179), is transferred to the origin in minimal time t_1^o. The control function must satisfy the following contraints: we give a number $\omega > 0$ and we define a number N from the condition $N\omega < t_1^o \leq (N + 1)\omega$. We set $u(t) = 0$ for $t_1^o <_ t \leq (N + 1)\omega$; let

$$\max_{0\leq k\leq N} \left(\int_{t=k\omega}^{(k+1)\omega} |u(t)|^p dt \right)^{1/p} \leq L \ , \quad p \geq 1 \quad . \tag{180}$$

Problem (179),(180) differs from the usual minimal time problems in that the constraint on the control has a cyclic character (ω is the duration of the cycle). It is possible to give a physical interpretation of the given problem when $p = 1$. Some process is regulated by the quantity of fuel that is supplied. The fuel is given directly into the process from a supply point which is periodically replenished (each ω unit of time) from another reservoir which, for practical purposes, is infinite. Under such conditions, it is necessary to define optimal from the point of view of the minimal time system's fuel expenditure. If we assume that the process under study may be described by equations of the type (179), then we clearly have problem (179),(180).

The number of problems of the type (179),(180) may be expanded greatly, but we confine out attention to methods for its solution which, without essential changes, may be applied to other problems. The fact that each minimal time problem with fixed endpoints is equivalent to a problem of moments was shown in Section 2. Since the expression

$$\max_{0 \leq k \leq N} \left(\int_{t=k\omega}^{(k+1)\omega} |u(t)|^p dt \right)^{1/p}$$

satisfies all the conditions of a norm $||u(\cdot)||$: 1) $||u(\cdot)|| > 0$, if $u(\cdot) \neq 0$; 2) $||\lambda u(\cdot)|| = |\lambda| \, ||u(\cdot)||$; 3) $||u_1(\cdot) + u_2(\cdot)|| \leq ||u_1(\cdot)|| + ||u_2(\cdot)||$, the constraint (180) may be written in the form $||f(\cdot)|| \leq F$, with $f(\cdot) \equiv u(\cdot)$.

Thus, the optimal control problem with cyclic constraints is reduced to the L-problem of moments. For its effective solution, it is necessary to be able to compute the norm of the element $||x(\cdot)||$ (cf. (167)). In our case, this problem is easily solved:

$$||x(\cdot)|| = \sum_{k=0}^{N} \left(\int_{t=k\omega}^{(k+1)\omega} |x(t)|^q dt \right)^{1/q} , \quad \frac{1}{p} + \frac{1}{q} = 1 \quad .$$

It remains to show that the minimal time problem posed above is equivalent to the problem of moments (165). But, in problems of minimal time this has already been done in Section 2 using Cauchy's formula.

4. Linear Minimal Time Problems with Moving Boundary Conditions

For simplicity, we consider only the case of a moving right endpoint. Thus, it is required to find the minimal time necessary to transfer the trajectory of the system (179) from the point x_0 to the cube

$$|x_i - d_i| \leq \Delta \quad , \quad i = 1,\ldots,n \quad , \qquad (181)$$

using piecewise-continuous controls $u(t)$, $t \in T$, obeying the constraint

$$|u(t)| \leq L \quad . \tag{182}$$

From Cauchy's formula, we obtain

$$x(t_1) = F(t_1)F^{-1}(t_0)x_0 + \int_{t_0}^{t_1} F(t_1)F^{-1}(\tau)b(\tau)u(\tau)d\tau. \tag{183}$$

We set

$$c = F(t_1)F^{-1}(t_0)x_0 - d \;, \; \gamma(t) = F(t_1)F^{-1}(t)b(t) \quad ,$$

$$x_1(\cdot) = \{1,0,\ldots,0;\gamma_1(t),t\varepsilon T\} \quad ,$$

$$x_2(\cdot) = \{0,1,0,\ldots,0;\gamma_2(t),t\varepsilon T\},\ldots,$$

$$f(\cdot) = \{f_n;f(t),t\varepsilon T\} \quad ,$$

$$f_i \equiv x_i(t_1) - d_i \quad , \quad f(t) \equiv u(t) \quad .$$

Then Eq. (183), described in coordinate form, may be interpreted as the value c_i of the functional $f(\cdot)$ on the element $x_i(\cdot)$:

$$f(x_i(\cdot)) = c_i \quad , \quad i = 1,\ldots,n \quad . \tag{184}$$

We introduce the norm of the functional $f(\cdot)$ in the following form:

$$||f(\cdot)|| = \max\left\{\max_i \frac{L|f_i|}{\Delta} \quad , \sup_t |f(t)|\right\} \quad .$$

Then the constraints (181),(182) on the control and the right endoint of the trajectory assume the form $||f(\cdot)|| \leq L$. This condition, together with (184), shows that the original problem is reduced to the L-problem of moments. The norm of the element $x(\cdot) = \{x_i, \ldots, x_n; x(t), t \in T\}$ easily computed as

$$||x(\cdot)|| \overset{\Delta}{=} \left\{ \sum_{i=1}^{n} |x_i| + \int_{t_o}^{t_1} |x(t)|dt \right\} .$$

5. Minimal Time in Connected Systems

We are given the equations

$$\frac{dx}{dt} = A(t)x + B(t)u(t) \quad , \quad x(t_o) = x_o \quad , \tag{185}$$

$$\frac{dy}{dt} = D(t)y + C(t)u(t) \quad , \quad y(t_o) = y_o \quad , \tag{186}$$

where the continuous matrices $A(t)$, $B(t)$, $C(t)$, $D(t)$ have dimensions nxn, nxr, rxr, rxr, respectively, and the matrix $C(t)$ is nonsingular for $t \geq t_o$. We call the system (185),(186) connected.

We assume that the piecewise-continuous control $u(t)$ in (185),(186), may be a delta-function at different moments of time. Let it be required to find a control for which the point $x(t)$ of the trajectory of (185) is driven to the origin in minimal time, where

$$|y_i(t)| \leq 1 \quad , \quad i = 1,\ldots,r \quad . \tag{187}$$

From (185),(186) we have

$$x(t) = F(t)F^{-1}(t_o)x_o + B(t)C^{-1}(t)y(t)$$
$$- F(t)F(t_o)B(t_o)C^{-1}(t_o)y(t_o)$$
$$+ \int_{t_o}^{t} F(t)F^{-1}(\tau)[A(\tau)B(\tau)C^{-1}(\tau) - B(\tau)C^{-1}(\tau)D(\tau)$$
$$- \frac{d}{d\tau}(B(\tau)C^{-1}(\tau))]y(\tau)d\tau \quad . \tag{188}$$

Let $u(\cdot)$ be a T-admissible control. Denoting

$$F(t_1)F^{-1}(t_o)x_o = -c \ , \quad -F(t_1)F(t_oB(t_o)C^{-1}(t_o) = \Pi_i(t_o) \ ,$$

$$B(t_1)C^{-1}(t_1) = \Pi_2(t_1) \ ,$$

$$F(t_1)F^{-1}(\tau)[A(\tau)B(\tau)C^{-1}(\tau) - B(\tau)C^{-1}(\tau)D(\tau)$$

$$- \frac{d}{d\tau}(B(\tau)C^{-1}(\tau))] = \Pi_3(t_1,\tau) \ ,$$

from (188) we obtain

$$c = \Pi_1(t_1)y(t_o) + \Pi_2(t_1)y(t_1) + \int_{t_o}^{t_1} \Pi_3(t_1,\tau)y(\tau)d\tau \ , \qquad (189)$$

where $y(t)$ is the solution of (186) for $u = u(t)$. We set

$$x_1(\cdot) = \{\Pi_{1,11}(t_1),\ldots,\Pi_{1,1r}(t_1);\Pi_{2,11}(t_1),\ldots,\Pi_{2,r}(t_1);$$

$$\Pi_{3,11}(t_1,t),\ldots,\Pi_{3,1r}(t_1,t),t\varepsilon T\} \ ,$$

....

$$x_n(\cdot) = \{\Pi_{1,n1}(t_1),\ldots,\Pi_{1,nr}(t_1);\Pi_{2,n1}(t_1),\ldots,\Pi_{2,nr}(t_1);$$

$$\Pi_{3,n1}(t_1,t),\ldots,\Pi_{3,nr}(t_1,t),t\varepsilon T \ ,$$

$$f(\cdot) = \{f_{1i},\ldots,f_{1r};f_{21},\ldots,f_{2r};f_{31}(t),\ldots,f_{3r}(t),t\varepsilon T\} \ ,$$

$$f_{1i} = y_i(t_o) \ , \quad f_{2i} = y_i(t_1) \ , \quad f_{3i}(t) = y_i(t) \ ,$$

$$t_o < t < t_1 \ .$$

Then Eq. (189), in coordinate form, is nothing more than the value c_i, $i = 1,\ldots,n$ of the functional $f(\cdot)$ which is assumed on the element $x_i(\cdot)$, $i = 1,\ldots,n$. We introduce the norm

$$||f(\cdot)|| = \max\{\max_i |f_{1i}|, \quad \max |f_{2i}|, \quad \underset{t\varepsilon T}{\text{vrai sup}_i} |f_{3i}(t)|\} \tag{190}$$

which allows us to describe constraint (187) in the following form: $||f(\cdot)|| \leq 1$. Thus, the original problem is reduced to the L-problem of moments. For its solution it remains to describe the norm of the element $x(t) = \{x_{11},\ldots,x_{1r};\ x_{21},\ldots,x_{2r};\ x_{31}(t),\ldots,x_{3r}(t)\}$. It is easily obtained from (167),(190):

$$||x(\cdot)|| = \left[\sum_{i=1}^{r} |x_{1i}| + |x_{2i}| + \int_{t_0}^{t_1} |x_{3i}(t)|dt\right] .$$

Solving the L-problem of moments, we find a function $y^o(t)$, $t\varepsilon T$. Returning to Eqs. (185),(186), it is not difficult to compute $x^o(t)$ and the optimal control $u^o(t)$. As points of smoothness of $y^o(t)$, the control $u^o(t)$ is continuous, while at points of continuity of $y^o(t)$, the control $u^o(t)$ is a δ-function.

6. Optimal Control Problems with State Constraints

Again, we consider the connected system (185),(186). Let $x(t_0) = x_0, y(t_0) = 0$; the control $u(t)$ and the trajectory $y(t)$ satisfy the conditions

$$|u_j(t)| \leq 1; \quad j = 1,\ldots,r; \quad t\varepsilon T , \tag{191}$$

$$|y_j(\tau_k)| \leq 1, \quad \tau_k < \tau_{k+1}, \quad k = 1,\ldots,N ; \tag{192}$$

$$j = 1,\ldots,r .$$

Here τ_k is a given moment of time. It is required to find a control $u(t)$ under which the trajectory $x(t)$

reaches the point $x = 0$ in minimal time. If $u(t)$ is a T-admissible control, then

$$-x_o = \int_{t_o}^{t_1} F^{-1}(\tau)B(\tau)u(\tau)d\tau \quad , \tag{193}$$

and conversly. We let $\Phi(t)$ denote the fundamental solution matrix for (186), $u(\cdot) = 0$. We have

$$y(\tau_k) = \int_{t_o}^{\tau_k} \Phi(\tau_k)\Phi^{-1}(\tau)C(\tau)u(\tau)d\tau = \int_{t_o}^{\tau_k} \Phi(\tau_k,\tau)C(\tau)u(\tau)d\tau. \tag{194}$$

We set

$$c_1 = -x_{10},\ldots,c_n = -x_{n0}, c_{n+1} = 0,\ldots,c_{Nr+n} = 0 \quad ,$$

$$x_1(\cdot) = \{0,\ldots,0;\ldots;0,\ldots,0;[F^{-1}(t)B(t)]_{ij},\ j = 1,\ldots,r;\ t\varepsilon T\} \quad ,$$

$$\cdots\cdots\cdots\cdots\cdots\cdots\cdots\cdots\cdots\cdots\cdots\cdots$$

$$x_n(\cdot) = \{0,\ldots,0;\ldots;0,\ldots,0;[F^{-1}(t)B(t)]_{nj},\ j = 1,\ldots,r;\ t\varepsilon T\} \quad ,$$

$$x_{n+1}(\cdot)$$

$$= \{1,0,\ldots,0;0,\ldots,0;\ldots;0,\ldots,0;\alpha_1[\Phi(\tau_1)\Phi^{-1}(t)C(t)]_{ij} \quad , \ j = 1,\ldots,r;\ t\varepsilon T\} \quad ,$$

$$\cdots\cdots\cdots\cdots\cdots\cdots\cdots\cdots\cdots\cdots\cdots\cdots$$

$$x_{n+r}(\cdot)$$

$$= \{0,\ldots,0,1;0,\ldots,0;\ldots;0,\ldots,0;\alpha_1[\Phi(\tau_1)\Phi^{-1}(t)C(t)]_{rj} \quad , \ j = 1,\ldots,r;\ t\varepsilon T \quad ,$$

$$x_{n+r+1}(\cdot)$$

$$= \{0,\ldots,0;1,0,\ldots,0;0,\ldots,0;\ldots;0,\ldots,0;\alpha_2[\Phi(\tau_2)\Phi^{-1}(\tau)C(\tau)]_{ij} \ j = 1,\ldots,r;\ t\varepsilon T \quad ,$$

$$\cdots\cdots\cdots\cdots\cdots\cdots\cdots\cdots\cdots\cdots\cdots\cdots$$

$x_{2r+n}(\cdot)$

$=\{0,\ldots,0;0,\ldots,0,1;0,\ldots,0;\ldots;0,\ldots,0;\Phi_2[\alpha(\tau_2)\Phi^{-1}(t)C(t)]_{rj},$
$j = 1,\ldots,r;$
$t\varepsilon T\}$.

....

$x_{(N-1)r+n}(\cdot)$

$=\{0,\ldots,0;\ldots;0,\ldots,0;1,0,\ldots,0;\alpha_N[\Phi(\tau_N)\Phi^{-1}(t)C(t)]_{ij}$,
$j = 1,\ldots,r;$
$t\varepsilon T\}$,

....

$x_{Nr+n}(\cdot)$

$=\{0,\ldots,0;\ldots;0,\ldots,0;0,\ldots,0,1;\alpha_N[\Phi(\tau_N)\Phi^{-1}(t)C(t)]_{rj}$,
$j = 1,\ldots,r;$
$t\varepsilon T\}$,

$\alpha_1 = \alpha_1(t) = 1,\ t_o \leq t \leq \tau_1;\ \alpha_1(t) = 0,\ \tau_1 \leq t \leq t_1,\ldots;$

$\alpha_N = \alpha_N(t) = 1$,

$t_o \leq t \leq \tau_N;\ \alpha_N(t) \equiv 0,\ \tau_N \leq t \leq t_1$,

$f(\cdot) = \{f_{11},\ldots,f_{1r};f_{21},\ldots,f_{2r};\ldots;f_{N1},\ldots,f_{Nr};$

$f_1(t),\ldots,f_r(t),t\varepsilon T\}$,

$f_{1i} \equiv y_i(\tau_i),\ldots,f_{Ni} \equiv y_i(\tau_N),f_i(t) \equiv u_i(t)$,

$i = 1,\ldots,r;\ t\varepsilon T$,

$$||f(\cdot)|| = \max\{\max_i |f_{1i}|,\ldots,\max_i |f_{Ni}|;\ \operatorname{vrai}\sup_{t\varepsilon T}{}_i|f_i(t)|\}.$$

In this notation, relations (193),(194) with the constraints (191),(192) may be treated as the L-problem of moments

$$f(x_v(\cdot)) = c_v \quad , \quad v = 1,\dots,Nr + n \quad . \tag{195}$$

The norm of the element $x(\cdot) = \{x_{11},\dots,x_{1r};\dots;x_{N1},\dots,f_{Nr};x_1(t),\dots,x_r(t),t\varepsilon T\}$ has the form

$$||x(\ )|| = \sum_{i=1}^{N}\left[\sum_{j=1}^{r}|x_{ij}| + \int_{t_o}^{t_1}|x_j(t)|dt\right] .$$

Solving the $(Nr+n)$-dimensional L-problem of moments under these conditions, we obtain the optimal control $u^o(t)$ and the value $y^o(t)$ at the points $\tau_1,\dots,\tau_N$.

Remarks.

1) We have reduced the original problem to the $(Nr + n)$-dimensional L-problem of moments (195). The dimensionality of the L-problem of moments may be made equal to n if we find the norm of the element $x(\cdot) = \{x_1(t),\dots,x_r(t);t\varepsilon T\}$ corresponding to the norm of the functional $f(\cdot) = \{f_1(t),\dots,f_r(t);t\varepsilon T\}$:

$$||f(\cdot)|| = \max\{\underset{t\varepsilon T}{\text{vrai sup}}_i|f_i(t)|, \ \max_{i,k}|y_i(\tau_k)|\}. \tag{196}$$

Here $y(\tau_k)$ is the value of the solution $y(t)$ of Eq. (186), $u(t) = f(t)$ at the moment $t = \tau_k$. The fact that the expression (196) actually gives a norm follows immediately. Such an approach to the problem of this section is taken in [21d].

2) It is possible to generalize the problem formulated above by requiring the condition

$$|y_i(t)| \le 1 \quad , \quad i = 1,\dots,r; \quad t\varepsilon T \quad .$$

instead of (192). Then the expression

$$||f(\cdot)|| = \max\{\text{vrai}\sup_{t\varepsilon T} {}_i |f_i(t)|, \max_{t\varepsilon T} {}_i |y_i(t)|\}$$

again gives a norm. However, in this case the computation of the norm $||x(\cdot)||$ is nontrivial. The form of the norm $||x(\cdot)||$ may be found using the results of §6.10. The corresponding formula is derived in [21d]. This formula contains the operation of maximization over measures which represents a difficulty in the general problem of higher order.

7. Minimal Time in Systems with Parameters

We are given the equation

$$\frac{dx}{dt} = A(t)x + B(t)u + C(t)w , \quad x(t_o) = x_o , \qquad (197)$$

where $u = u(t)$ is the control constrained by condition (191), and the values $w = w(t)$ may be measured at the given moments τ_k, $\tau_k < \tau_{k+1}$, $k = 1,\dots,N$, where $|w_j(\tau_k)| \leq 1$, $j = 1,\dots,q$. Using $u(t)$ and $w(t)$, it is required to transfer the trajectory of the system (197) to the origin in minimal time. In this problem, the T-admissible control $u(\cdot)$ satisfies the equation

$$-x_o = \int_{t_o}^{t_1} \Pi(\tau)u(\tau)d\tau + \sum_{k=1}^{N+1} \pi(k)w(k), \; \Pi(\tau) = F^{-1}(\tau)B(\tau) ,$$

$$\pi(k) = \int_{\tau_{k-1}}^{\tau_k} F^{-1}(\tau)C(\tau)d\tau, \quad k = 1,\dots,N + 1; \; \tau_o = t_o .$$

Therefore, a definite function $u(t)$ leads to the L-problem of moments (by the scheme of Section 4).

§12. Application of the Theorem on the Existence of a Supporting Hyperplane to Optimal Control Problems

1. Existence of a Supporting Hyperplane to a Convex Surface

From the theorem on the separation of convex sets, then follows the assertion:

Ley $y = \phi(x)$, $x \varepsilon E_n$, $y \varepsilon E_1$ be a convex surface, $\{x,y\}$ being the points on this surface. Then there exists a supporting hyperplane passing between $\{x,y\}$, i.e. there exists a vector g, $||g|| = 1$, and a number $f \neq 0$ such that

$$\inf_{\bar{x} \varepsilon E_n, \bar{y}=\phi(\bar{x})} [g'\bar{x} + f\bar{y}] = g'x + fy \quad .$$

2. Optimization of Convex Functionals Along Trajectories of Linear Systems

We consider the equations

$$\frac{dy}{dt} = A_1(t)y + B_1(t)z + C_1(t)u + f_1(t) \quad , \tag{198}$$

$$\frac{dz}{dt} = A_2(t)y + B_2(t)z + C_2(t)u + f_2(t) \quad , \tag{199}$$

where y is a p-vector, z is a q-vector, $y \varepsilon E_p$, $z \varepsilon E_q$; the continuous matrices $A_1(t)$, $A_2(t)$, $B_1(t)$, $B_2(t)$, $C_1(t)$, $C_2(t)$ have dimensions pxp, qxp, pxq, qxq, pxr, qxr, respectively; $u = \{u_1,\ldots,u_r\}$ is the control; $f_1(t) = \{f_{11}(t),\ldots,f_{1p}(t)\}$, $f_2(t) = \{f_{21}(t),\ldots,f_{2q}(t)\}$ are known functions. We are given numbers t_o, t_1, s_k, L_j ($t_1 > t_o$, $s_k > s_{k-1}$, $k = 3,4,\ldots,m-1$; $s_3 = t_o$, $s_{m-1} \leq t_1$; $j = 1,\ldots,m$; $L_j \geq 0$, $j = 1$), and the functional $\phi(u(\cdot), t_1)$ defining on the functions $u(\cdot) \varepsilon U_p(\cdot)$. ϕ is continuous in $u(\cdot)$, t_1 and convex in $u(\cdot)$. In the space E_q, we are given a curve $\gamma(t)$, $t \in T$, and we are also given fixed points y_0 and y_1 in E_p. We require that

$$\begin{aligned} &\phi(u,t_1) \le L_1 , \qquad ||y(t_0) - y_0|| \le L_2 , \\ &||y(t_1) - y_1|| \le L_3 , \quad ||z(s_k) - \gamma(s_k)|| \le L_k . \end{aligned} \tag{200}$$

For each fixed t, let the operators $F_{ij}(t,\tau)$ consist of the fundamental matrix $F(t,\tau)$ of the solution of the homogeneous systems (198),(199):

$$F(t,\tau) = \begin{Bmatrix} F_{11}(t,\tau) & F_{12}(t,\tau) \\ F_{21}(t,\tau) & F_{22}(t,\tau) \end{Bmatrix} ,$$

$$F(t,t) = E \quad ,$$

and

$$\left.\begin{aligned} y(t) &= F_{11}(t,t_0)y(t_0) + F_{12}(t,t_0)z(t_0) \\ &\quad + \int_{t_0}^{t} \{F_{11}(t,\tau)[C_1(\tau)u(\tau) + F_1(\tau)] \\ &\quad + F_{12}(t,\tau)[C_2(\tau)u(\tau) + f_2(\tau)]\}d\tau \quad , \\ z(t) &= F_{21}(t,\tau)y(t_0) + F_{22}(t,\tau)z(t_0) \\ &\quad + \int_{t_0}^{t} \{F_{21}(t,\tau)[C_1(\tau)u(\tau) + f_1(\tau)] \\ &\quad + F_{22}(t,\tau)[C_2(\tau)u(\tau) + f_2(\tau)\}d\tau \quad . \end{aligned}\right\} \tag{201}$$

We set

$$u = u^1 \;,\; y(t_o) - y_o = u^2 \;,\; y(t_1) - y_1 = u^3 \;,$$

$$z(t_k) - \gamma(t_k) = u^{4+k} \;,\; t_k = s_{k+3} \;,\; k = 0,1,\dots,m-4 \;,$$

$$S_{11}u^1 = \int_{t_o}^{t_1} [F_{11}(t_1,\tau)C_1(\tau) + F_{12}(t_1,\tau)C_2(\tau)]u^1(\tau)d\tau,$$

$$S_{12} = F_{11}(t_1,t_o) \;,\; S_{13} = -E \;,\; S_{14} = F_{12}(t_1,t_o) \;,$$

$$S_{1,4+i} = 0 \;,\; i \geq 1 \;,$$

$$h^1 = -y_1 + F_{11}(t_1,t_o)y_o + F_{12}(t_1,t_o)\gamma(t_o) + \int_{t_o}^{t_1} [F_{11}(t_1,\tau)f_1(\tau) + F_{12}(t_1,\tau)f_2(\tau)]d\tau \;,$$

$$S_{2+k,1}u^1 = \int_{t_o}^{t_{k+1}} [F_{21}(t_{k+1},\tau)C_1(\tau) + F_{22}(t_{k+1},\tau)C_2(\tau)]u^1(\tau)d\tau \;,$$

$$S_{2+k,2} = F_{21}(t_{k+1},t_o) \;,$$

$$S_{2+k,3} = 0 \;,\; S_{2+k,4} = F_{22}(t_{k+1},t_o) \;,\; S_{2+k,4+i} = -E(k = i-1) \;,\; S_{2+k,4+i} = 0 \;(k \neq i-1) \;,$$

$$h^{2+k} = -\gamma(t_{k+1}) + F_{21}(t_{k+1},t_o)y_o + F_{22}(t_{k+1},t_o)\gamma(t_o) + \int_{t_o}^{t_{k+1}} [F_{21}(t_{k+1},\tau)f_1(\tau) + F_{22}(t_{k+1},\tau)f_2(\tau)]d\tau.$$

The quantities u^j, $j = 1,\dots,m$ are called <u>controls</u>. We will call the controls u^j, $j = 1,\dots,m$, <u>optimal controls</u> if one of the numbers L_j, t_1 is minimal (under the condition that the other one remains fixed). By virtue of (200)-(202), the solution u^j of problem (200) satisfies the relation

$$\left.\begin{aligned} &\sum_{j=1}^{m} S_{ij} u^j + h^i = 0 \quad , \quad i = 1,\ldots,m-3 \quad , \\ &u^1 = u^1(\cdot)\varepsilon U_p(\cdot) \quad , \quad ||u^j|| \leq L_j \quad , \quad j > 1 \quad . \end{aligned}\right\} \tag{203}$$

Here u_j, $j \neq 1$, are elements of the finite-dimensional spaces E_p, E_q, the linear operators S_{ij}, $j > 1$ act from E_p, E_q into E_p, E_q. We set

$$\sum_{i=1}^{m-3} g_i' S_{i1} u^1(\cdot) + f\phi(u^1(\cdot),t_1) = H(g,f,u^1(\cdot)), \ f \geq 0 \ ,$$

$$\Lambda(L_1,\ldots,L_m,t_1,g,f) = \sum_{i=1}^{m-3} g_i' h^k - \sum_{j=2}^{m} L_j || \sum_{i=1}^{m-3} g_i' S_{ij} ||$$

$$- L_1 f + \min_{u^1(\cdot)\varepsilon U_p(\cdot)} H(g,f,u^1(\cdot)) \quad .$$

Theorem 35. Problem (203) has a solution if and only if

$$\max_{||\{g,f\}||=1, f\geq 0} \Lambda(L_1,\ldots,L_m,t_1,g,f) \leq 0 \quad . \tag{204}$$

Proof. Let the quantities u^j, $u^1(\cdot)\varepsilon U_p(\cdot)$, $||u^j|| \leq L_j$ be a solution of problem (203). Then

$$\sum_{i=1}^{m-3} g_i' [\sum_{j=1}^{m} S_{ij} u^j + h^i] = 0 = \sum_{i=1}^{m-3} g_i' S_{i1} u^1(\cdot)$$

$$+ f\phi(u^1(\cdot),t_1) + \sum_{j=2}^{m} \sum_{i=1}^{m-3} g_i' S_{ij} u^j - f(u^1(\cdot),t_1)$$

$$+ \sum_{i=1}^{m-3} g_i'h^i \geq \min_{u^1(\cdot)\varepsilon U_p(\cdot)} H(g,f,u^1(\cdot))$$

$$- \sum_{j=2}^{m} L_j || \sum_{i=1}^{m-3} g_i' S_{ij} || - L_1 f + \sum_{i=1}^{m-3} g_i' h^i$$

for any $g_i \, \varepsilon \, E_p$, $i = 1,\ldots,m-3$; $f \geq 0$. This means that relation (204) is valid.

Let condition (204) be satisfied. We show that there exists a solution of problem (203). By virtue of (204)

$$\sum_{i=1}^{m-3} g_i'h^i - \sum_{j=2}^{m} L_j || \sum_{i=1}^{m-3} g_i' S_{ij} || - L_1 f$$

$$+ \min_{u^1(\cdot)\varepsilon U_p(\cdot)} H(g,f,u^1(\cdot)) \leq 0 \qquad (205)$$

for any $g_i \varepsilon \; E_p$, E_q, $i = 1,\ldots,m-3$; $f \geq 0$. We consider the set

$$\chi(L_2,\ldots,L_m) = \{v\colon v = \{v^1,\ldots,v^{m-3}\}, \quad v^i = \sum_{j=1}^{m} S_{ij} u^j \quad ,$$

$$u^1(\cdot)\varepsilon U_p(\cdot), ||u^j|| \leq L_j, j = 2,\ldots,m\} \quad .$$

This set is convex and compact. If we follow the earlier scheme of §10, then it is necessary to introduce the set

$$Q(v^1,\ldots,v^{m-3},v^{m-2})$$

$$= \{v^i = \sum_{j=1}^{m} S_{ij} u^j, \; i = 1,\ldots,m-3; \; v^{m-2} = \phi(u^1(\cdot),t_1)\} \; .$$

But, since this set is not convex, in general it is necessary to use the theorem about the separation of convex sets. On $\chi(L_2,\ldots,L_m)$, we define the hyperplane $\delta = \delta(v^1,\ldots,v^{m-3})$ setting

$$\delta(v^1,\ldots,v^{m-3}) = \min_{u^1(\cdot)\varepsilon U_p(\cdot)} \phi(u^1(\cdot),t_1) \tag{206}$$

under condition (203). The hyperplane $\delta = \delta(v^1,\ldots,v^{m-3})$ is convex in $\{v^1,\ldots,v^{m-3}\}$. If problem (203) does not have a solution, then the intersection of the sets

$$R_1 = \{\zeta:\ \zeta = \{-h^1,\ldots,-h^{m-3},\xi\}\ ,\ \xi \leq L_1\}\ ,$$

$$R_2 = \{\zeta:\ \zeta = \{v^1,\ldots,v^{m-3},\ \phi(u^1(\cdot),t_1)\}\ ,$$

$$u^1(\cdot)\varepsilon U_p(\cdot),||u^j|| \leq L_j,\ j \neq 1\}$$

will be empty. By virtue of the theorem on the existence of a separating hyperplane (206), for some $g_i \varepsilon E_p$, $f \geq 0$, and ε, there occurs

$$\min_{u^j}\{\sum_{i=1}^{m-3} g_i' \sum_{j=1}^{m} S_{ij}u^j + f\phi(u^1(\cdot),t_1)\} \geq \varepsilon > -\sum_{i=1}^{m-3} g_i'h^i + L_1 f .$$

This means that

$$\min_{u^1(\cdot)\varepsilon U_p(\cdot)} H(g,f,u^1(\cdot)) - \sum_{j=2}^{m} L_j||\sum_{i=1}^{m-3} g_i'S_{ij}||$$

$$- L_1 f > -\sum_{i=1}^{m-3} g_i'h^i\ ,$$

which contradicts inequality (205). The theorem is proved.

Theorem 35 admits a generalization to the case of several constraints of the type $\phi_i(u(\cdot),t_1) \leq L_i$, $i = 1,\ldots,\alpha$, where $\phi_i(u(\cdot),t_1)$ is continuous and convex in $u(\cdot)$.

Lemma 15. The function $\phi(L_1,\ldots,L_m,t_1)$ is continuous in L_j, t_1 and is non-increasing in L_j.

The proof proceeds completely analagously to the proof of Lemma 13.

We set $t_1 = L_o$. We call the control $\{u^j\}$, $\tilde{L}_k$-admissible if it satisfies condition (200) for $L_k = \tilde{L}_k$, $L_j < \infty$, $j \neq k$.

Theorem 36. If there exists at least one $\tilde{L}_k$-admissible control, then there exists an optimal control $\{u^{jo}\}$ with the values L_k^o minimizing the quantities L_k being the smallest roots of the equation

$$\max_{\|\{g,f\}\|\leq 1, f\geq 0} \Lambda(L_1,\ldots,L_m,L_o,g,f) = \Lambda(L_1,\ldots,L_m,L_o,g_o,f_o) = 0 \quad .$$

The optimal control is defined by the relation

$$H(g_o,f_o,u^{10}(\cdot)) = \min_{u^1(\cdot)\varepsilon U_p(\cdot)} H(g_o,f_o,u^1(\cdot)) \quad ,$$

$$\sum_{i=1}^{m-3} g_{io}' S_{ij} u^{jo} = - L_j \left\| \sum_{i=1}^{m-3} g_{io}' S_{ij} \right\| \quad , \quad j \neq 1 \quad ,$$

where $L_k = L_k^o$.

This assertion follows from Lemma 15 and Theorem 35.

§13. Conditions for the Imbeddability of Convex Sets. Applications to Optimal Control Problems

We say that a convex set X_1 is imbeddable in another convex set X_2 if $X_1 \subset \text{int } X_2$.

1. Conditions for the Solvability of One Functional Problem

Let the linear operator S transform the r-dimensional measurable function $u(t)$, $t \varepsilon T$, into a vector $x \varepsilon E_n$. We are given Q_1, Q_2, two nonsingular linear transformations acting from E_n into E_n, elements c, c_1, $c_2 \varepsilon E_n$, and numbers Δ_1, Δ_2, L, $\Delta_1 \geq 0$, $\Delta_2 \geq 0$.

Problem. To find conditions under which

$$Su(\cdot) + Q_1 v + Q_2 w + c = 0 \quad , \quad u(\cdot) \varepsilon U_p^L(\cdot) \quad ,$$

$$||v - c_1|| \leq \Delta_1 \quad , \quad ||w - c_2|| \geq \Delta_2 \quad . \tag{207}$$

Theorem 37. Problem (207) has a solution if and only if

$$\min_{||g||=1} \{g'[-c - Q_1 c_1 - Q_2 c_2] - L||g'S|| - \Delta_1 ||g'Q_1|| + \Delta_2 ||g'Q_2||\} \leq 0 \quad . \tag{208}$$

Proof. Let $u(\cdot) \varepsilon U_p^L(\cdot)$, v,w be some solution of problem (207). Then there exists a $g \varepsilon E_n$ such that

$$\Delta_2 ||g'Q_2|| \leq ||w - c_2|| \; ||g'Q_2|| = g'Q_2(c_2 - w)$$

$$= g'[Su(\cdot) + Q_1 v + Q_2 c_2 + c]$$

$$\leq L||g'S|| + \Delta_1 ||g'Q_1|| + g'[c + Q_1 c_1 + Q_2 c_2] \quad .$$

This means that inequality (208) is satisfied. Let (208) occur. Then there exists at least one element $g \varepsilon E_n$ such that

$$-g'[c + Q_1 c_1 + Q_2 c_2] - L||g'S|| - \Delta_1 ||g'Q_1|| + \Delta_2 ||g'Q_2|| \leq 0 \quad . \tag{209}$$

We assume that problem (207) has no solution. Consider the sets

$$\beta(\Delta_2) = \{x: \ x = -Q_2 w, \ ||w - c_2|| \leq \Delta_2\} \quad ,$$

$$\beta(L, \Delta_1) = \{x: \ x = Su(\cdot) + Q_1 v + c, u(\cdot) \varepsilon U_p^L(\cdot), ||v - c_1|| \leq \Delta_1\}.$$

These sets are compact and convex. Since problem (207) has no solution, then $\beta(L, \Delta_1)$ is imbeddable into $\beta(\Delta_2)$. Therefore, for each $f \varepsilon E_n$ and number Δ_2, there is an element $y \varepsilon E_n$ such that $||y|| = \Delta_2$, $g'y = \Delta_2 ||f||$, and for each y, $||y|| \leq \Delta_2$, the inequality $f'y \leq \Delta_2 ||f||$ will be satisfied.

Let $f = Q_2' g$, $y = c_2 - 2$. Then $g'Q_2(c_2 - w) = ||g'Q_2||\Delta_2$ for $||w - c_2|| = \Delta_2$ and $g'x = ||g'Q_2||\Delta_2 - g'Q_2 c_2$, for $x = -Q_2 w$, $||w - c_2|| = \Delta_2$. If $x \varepsilon \beta(L, \Delta_1)$, then $g'x = g'[Su(\cdot) + Q_1 v + c] < ||g'Q_2||\Delta_2 - g'Q_2 c_2$ and

$$\max_{u(\cdot), v} g'[Su(\cdot) + Q_1 v + c] = g'(c + Q_1 c_1) + L||g'S|| + \Delta_1 ||g'Q_1|| < ||g'Q_2||\Delta_2 - g'Q_2 c_2 \quad .$$

The last inequality is valid for all $g \varepsilon E_n$ contradicting (209). The theorem is proved.

Let $||g'S|| > 0$ for all g, $||g|| \neq 0$. Condition (208) is equivalent to the inequality $L \geq \lambda_1$, where

$$\lambda_1 = \min_{||g'S||=1} \{-g'[c + Q_1c_1 + Q_2c_2] - \Delta_1||g'Q_1|| + \Delta_2||g'Q_2||\}. \tag{210}$$

<u>Theorem 38. In order that the element $g = g_o$ be a solution of problem (210), it is necessary and sufficient that for an arbitrary solution $u(\cdot)$, $||u(\cdot)|| = \lambda_1$, v, w, of problem (207), the equations</u>

$$g_o'Su(\cdot) = ||g_o'S|| \; ||u(\cdot)|| = \lambda_1 \quad , \tag{211}$$

$$g_o'Q_1(v - c_1) = \Delta_1||g_o'Q_1|| \quad , \tag{212}$$

$$g_o'Q_2(w - c_2) = -\Delta_2||g_o'Q_2|| \quad . \tag{213}$$

<u>be satisfied</u>.

<u>Proof</u>. <u>(Necessity)</u> Let g_o be a solution of problem (210); $u(\cdot)$, $||u(\cdot)|| = \lambda_1$, v, w a solution of (207). We have

$$\begin{aligned} \lambda_1 &= -g_o'[c + Q_1c_1 + Q_2c_2] - \Delta_1||g_o'Q_1|| + \Delta_2||g_o'Q_2|| \\ &= g_o'[Su(\cdot) + Q_1(v - c_1) + Q_2(w - c_2)] - \Delta_1||g_o'Q_1|| \\ &+ \Delta_2||g_o'Q_2|| \leq ||g_o'S|| \; ||u(\cdot)|| + g_o'Q_1(v - c_1) \\ &+ g_o'Q_2(w - c_2) - \Delta_1||g_o'Q_1|| + \Delta_2||g_o'Q_2|| \quad . \end{aligned} \tag{214}$$

Hence,

$$g_0'[Q_1(v-c_1)+Q_2(w-c_2)] \geq \Delta_1||g_0'Q_1|| - \Delta_2||g_0'Q_2|| \quad . \tag{215}$$

It is clear that $g_0'Q_1(v - c_1) \leq \Delta_1||g_0'Q_1||$. If the inequality $g_0'Q_2(w - c_2) < -\Delta_2||g_0'Q_2||$, then from (215) we immediately have $g_0'Q_1(v - c_1) > \Delta_1||g_0'Q_1||$. Therefore, we set $g_0'Q_2(w - c_2) = -\Delta_2||g_0'Q_2|| + \varepsilon$, $\varepsilon > 0$. It is not difficult to conclude that there exists a solution $u(\cdot)$, v,w, such that $||u(\cdot)|| < \lambda_1$. This is impossible. This means that condition (213) is satisfied. From (215) we obtain Eq. (212). Using (214), we have (211).

(Sufficiency) Let $u(\cdot)$, $||u(\cdot)|| = \lambda_1, v, w$ be a solution of problem (207) and for some $g \varepsilon E_n$ let conditions (211)-(213) be satisfied. Then

$$-g'[c + Q_1c_1 + Q_2c_2] - \Delta_1||g'Q_1|| + \Delta_2||g'Q_2||$$

$$= g'[Su(\cdot) + Q_1(v - c_1) + Q_2(w - c_2)] - \Delta_1||g'Q_1||$$

$$+ \Delta_2||g'Q_2|| = \lambda_1||g'S|| \quad .$$

This means that g is a solution of problem (210). The theorem is proved.

2. The Problem of Maximizing the Norm of the Terminal State

In the state space of the equation

$$\frac{dx}{dt} = A(t)x + B(t)u \tag{216}$$

let points c_1, c_2 be given. For given numbers t_1, L, Δ_1, $t_1 > t_0$, $\Delta_1 \geq 0$, $L > 0$, it is required to find a control

$u^o(t)$, $t \varepsilon T$, $||u^o(\cdot)|| \leq L$ which ensures the condition

$$||x^o(t_1) - c_2|| = \max_{||u(\cdot)|| \leq L} ||x(t_1) - c_2|| \ , \ ||x(t_0) - c_1|| \leq \Delta_1 . \tag{217}$$

This problem, in the earlier notation (see (127)) is equivalent to the requirement

$$Su(\cdot) + Q_1 x(t_0) + Q_2 x(t_1) + c = 0 \ ,$$

$$||x(t_1) - c_2|| = \max \ , \quad ||x(t_0) - c_1|| \leq \Delta_1 \ .$$

We consider the auxiliary problem

$$Su(\cdot) + Q_1 v + Q_2 w + c = 0 \ , \quad ||v - c_1|| \leq \Delta_1 \ , \qquad (218)$$
$$||w - c_2|| \leq \Delta_2 \ ,$$

where $Q_2 = -E$, $u(\cdot) \varepsilon U_p^L(\cdot)$, and Δ_2 is a nonnegative number. By virtue of Theorem 37, problem (218) has a solution if and only if

$$\Lambda_1(\Delta_2,t_1) = \min_{||g||=1} \Lambda_1(\Delta_2,t_1,g)$$
$$= \Lambda_1(\Delta_2,t_1,g(\Delta_2,t_1)) \leq 0 \ ,$$
$$(219)$$
$$\Lambda_1(\Delta_2,t_1,g) = -g'[c + Q_1 c_1 + Q_2 c_2]$$
$$- L||g'S|| - \Delta_1||g'Q_1|| + \Delta_2||g'Q_2|| \ .$$

<u>Lemma 16. The function $\Delta_1(\Delta_2,t_1)$ is continuous in Δ_2,t_1 and strictly increasing in Δ_2.</u>

<u>Proof</u>. The continuity of the function $\Lambda_1(\Delta_2,t_1)$ is obvious. We show that $\Lambda_1(\Delta_2,t_1)$ is strictly increasing

in Δ_2. Let $\Delta_2 < \Delta_2^*$. We have

$$\Lambda_1(\Delta_2^*, t_1, g(\Delta_2^*, t_1)) > \Lambda_1(\Delta_2, t_1, g(\Delta_2^*, t_1))$$

$$> \Lambda_1(\Delta_2, t_1, g(\Delta_2, t_1)) \quad .$$

The lemma is proved.

Theorem 39. An optimal control exists for problem (217) and gives the functional $\phi_2 = ||x(t_1) - c_2||$ the value

$$\Delta_2^o = \max_{||g||\leq 1} \{L||g'S|| + \Delta_1||(F^{-1}(t_o))'F'(t_1)g|| + g'[c + F(t_1)F^{-1}t_o)c_1 - c_2]\}$$

$$= L||g_o'S|| + \Delta_1||g_o'F(t_1)F^{-1}(t_o)|| + g_o'[c + F(t_1)F^{-1}(t_o)c_1 - c_2]. \tag{220}$$

If $\Delta_2^o > 0$, then the relations

$$g_o'Su^o(\cdot) = \max_{u(\cdot)\varepsilon U_p^L(\cdot)} [g_o'Su(\cdot)] = L||g_o'S|| \quad , \tag{221}$$

$$\left.\begin{aligned} g_o'F(t_1)F^{-1}(t_o)[x^o(t_o) - c_1] &= \Delta_1||g_o'F(t_1)F^{-1}(t_o)||, \\ g_o'[x^o(t_1) - c_2] &= \Delta_2^o||g_p|| \quad , \end{aligned}\right\} \tag{222}$$

are satisfied by the optimal control $u^o(\cdot)$ and the ends of the optimal trajectory $x^o(\cdot)$.

Proof. Relation (220) is obtained from (219) if we consider that $\Lambda_1(\Delta_2, t_1)$ is monotonic in Δ_2. Eqs. (221) and (222) follow from Theorem 38.

3. The Minimal Time Problem

We consider the sets

$$\Gamma^1 = \{x: ||x - c_1|| \leq \Delta_1\} , \quad \Gamma^2 = \{x: ||x - c_2|| \geq \Delta_2\} ,$$

where c_1, c_2 are fixed points, and Δ_1, Δ_2 are given nonnegative numbers such that the set $\Gamma^1 \cap \Gamma^2$ is empty. It is required to determine a control $u^o(\cdot)\varepsilon U_p^L(\cdot)$ under which the trajectory of Eq. (216), with the initial condition $x^o(t_o)\varepsilon\Gamma^1$, reaches Γ^2 in minimal time.

By virtue of the above investigations, there exists a T-admissible control in this problem only if inequality (219) is satisfied. We set

$$\Lambda_1(t_1,g(t_1)) = \min_{||g||\leq 1} \Lambda_1(\Delta_2,t_1,g) .$$

Theorem 40. If there exists at least one T-admissible control, then there exists an optimal control with minimal time $\tau^o = t_1^o - t_o$, where t_1^o is the smallest root of the equation

$$\Lambda_1(t_1,g(t_1)) = 0 .$$

The optimal control $u^o(\cdot)$ and the endpoints of the optimal trajectory $x^o(\cdot)$ satisfy the conditions

$$g'(t_1^o)Su^o(\cdot) = L||g'(t_1^o)S|| ,$$

$$g'(t_1^o)F(t_1^o)F^{-1}(t_o)[x^o(t_o) - c_1] = \Delta_1||g'(t_1^o)F(t_1^o)F^{-1}(t_o)|| ,$$

$$g'(t_1^o)[x^o(t_1^o) - c_2] = \Delta_2 .$$

The proof is analogous to that of Theorem 27.

§14. The Generalized Neyman-Pearson Lemma in the Theory of Optimal Processes

As indicated in §9,10, the variational problem of optimal control may be considered as a finite-dimensional problem if the computation

$$\mu(g) = \inf_{u(\cdot)\varepsilon U(\cdot)} \int_{t_o}^{t_1} g'S(t_1,\tau,u(\tau))d\tau$$

is sufficiently simple to carry out for a given family $U(\cdot)$ of admissible controls. Up to now, this problem has been successfully dealt with either by a direct computation (§9), or using a Lagrange multiplier (§10) or by use of the Kuhn-Tucker theorem (§10). In this paragraph we investigate still one more approach to the calculation (223) for the case when the family $U(\cdot)$ has a special form.

Let it be required to find

$$\mu(g) = \inf_{u(\cdot)\varepsilon U(\cdot)} \int_{t_o}^{t_1} g'S(t_1,\tau)u(\tau)d\tau$$

$$= \inf_{u(\cdot)\varepsilon U(\cdot)} \int_{t_o}^{t_1} a'(t)u(t)dt \quad , \tag{223}$$

$$U(\cdot) = \{u(t): \quad |u_j(t)| \leq 1 \quad , \quad t\varepsilon T \quad ,$$

$$\int_{t_o}^{t_1} \left[\sum_{j=1}^{r} |b_j(t)u_j(t) - \phi_j(t)|^p dt \right] \leq k\} \quad . \tag{224}$$

Here $b_j(t)$, $\phi_j(t)$, $j = 1,\ldots,r$; $t\varepsilon T$, are given piecewise-continuous functions, p, k, $p \geq 1$, $k > 0$ are given numbers. A similar problem was considered in [12] where the integral constraint has the form $\int_{t_o}^{t_1} b'(t)u(t)dt \leq k$. The

Neyman-Pearson lemma gives the solution to the latter problem. An analagous argument is used below.

Initially, we assume $p > 1$. Clearly, the problem has a solution if the first inequality (224) is satisfied for

$$u_i(t) = \text{Sat}\left[\frac{b_i(t)}{\phi_i(t)}\right] , \quad i = 1,\dots,r; \; t\varepsilon T \quad , \tag{225}$$

where

$$\text{Sat}\ \gamma = \begin{cases} 1 , & \gamma \geq 1 , \\ \gamma , & -1 \leq \gamma \leq 1 , \\ -1 , & \gamma \leq -1 . \end{cases}$$

Otherwise the inequalities (224) are not satisfied simultaneously because for the u(t) defined by (225), the integral in (224) has minimal value. We introduce the functions $\psi_i(\tau,\lambda)$:

$$\psi_i(\tau,\lambda) = -\left|\frac{a_i(t)}{\lambda_p b_i(t)}\right|^{\frac{1}{p-1}} \text{sign}\ a_i(t) + \frac{\psi_i(t)}{b_i(t)} \quad , \tag{226}$$

$$\lambda b_i(t) \neq 0 \quad , \quad i = 1,\dots,r ,$$

where λ is a real parameter. We set

$$u_i(t,\lambda) = \begin{cases} \text{Sat}\ \psi_i(t,\lambda) , & \lambda b_i(t) \neq 0, \\ -\ \text{sign}\ a_i(t) , & \lambda b_i(t) = 0,\ a_i(t) \neq 0, \\ \text{arbitrary functions} & \\ u_i(t) ,\ |u_i(t)| \leq 1 , & \lambda b_i(t) = a_i(t) = 0. \end{cases}$$

(227)

Let λ_o be a lower bound of nonnegative λ for which the corresponding functions $u(t,\lambda)$ satisfy the second inequality (224).

Generalized Neyman-Pearson Lemma. The functions $u^*(\tau)$, $\tau \varepsilon T$, $i = 1,\ldots,r$ minimizing the functional (223), are determined by the formula

$$u^*(\tau) = u(\tau,\lambda_o) \quad . \tag{228}$$

Proof. The lemma is clearly true for $\lambda_o = 0$. For λ decreasing from $+\infty$ to 0, the quantity $\int_{t_o}^{t_1} \sum_{i=1}^{r} |b_i(t)u_i(t,\lambda) - \phi_i(t)|^p dt$ is monotonically increasing from its minimal value $\int_{t_o}^{t_1} \sum_{i=1}^{r} |b_i(t) \text{ Sat } \frac{\phi_i(t)}{b_i(t)} - \phi_i(t)|^p dt$. Therefore

$$\int_{t_o}^{t_1} \sum_{i=1}^{r} |b_i(t)u_i(t,\lambda_o) - \phi_i(t)|^p dt = k \text{ for } \lambda_o > 0 \quad .$$

For any control $u(\cdot)\varepsilon U(\cdot)$ we have

$$\lambda_o \left[\int_{t_o}^{t_1} \sum_{i=1}^{r} |b_i(t)u_i^*(t) - \phi_i(t)|^p dt - \int_{t_o}^{t_1} \sum_{i=1}^{r} |b_i(t)u_i(t) - \phi_i(t)|^p dt \right] \geq 0 \; , \tag{229}$$

$$\Delta = \int_{t_o}^{t_1} \sum_{i=1}^{r} a_i(t)[u_i(t) - u_i^*(t)]dt$$

$$\geq \int_{t_o}^{t_1} \sum_{i=1}^{r} \{a_i(t)[u_i(t) - u_i^*(t)]$$

$$+ \lambda_o[b_i(t)u_i(t) - \phi_i(t)|^p - |b_i(t)u_i^*(t) - \phi_i(t)|^p]\}dt.$$

If the functions $u_i^*(t)$, $t \varepsilon T$, $i = 1,\ldots,r$, are defined by (228), (227), (226), then the expression in curly brackets in (229) is nonnegative for all $u(\cdot) \varepsilon U(\cdot)$. Therefore, $\Delta \geq 0$ where equality holds only for $u(\cdot) = u^*(\cdot)$. This completes the proof.

A limiting case of the lemma is obtained for $p = 1$. The $u^*(t) = u(t,\lambda_o)$ and $u_i(t,\lambda) =$

$$\begin{cases} \begin{cases} -\operatorname{sign} a_i(t), & |a_i(t)| > |\lambda b_i(t)|; \\[2ex] \operatorname{Sat} \dfrac{\phi_i(t)}{b_i(t)} & \begin{cases} |a_i(t)| < |\lambda b_i(t)|, \text{ or} \\ |a_i(t)| = |\lambda b_i(t)|, \ \left|\dfrac{\phi_i(t)}{b_i(t)}\right| \geq 1, \\ \operatorname{sign} a_i(t) = -\operatorname{sign}\left|\dfrac{\phi_i(t)}{b_i(t)}\right| \end{cases} \end{cases} \\[4ex] \text{arbitrary functions } z_i(t), \text{ if} \\[1ex] \dfrac{\phi_i(t)}{b_i(t)} \leq z_i(t) \leq -\operatorname{sign} a_i(t) \quad |a_i(t)| = |\lambda b_i(t)|, \ \dfrac{\phi_i(t)}{b_i(t)} < 1; \\[2ex] \text{arbitrary functions } y_i(t), \text{ if} \quad |a_i(t)| = |\lambda b_i(t)|, \ \dfrac{\phi_i(t)}{b_i(t)} \geq 1 \\[2ex] \qquad |y_i(t)| \leq 1, \qquad \operatorname{sign} a_i(t) = \operatorname{sign} \dfrac{\phi_i(t)}{b_i(t)} \quad . \end{cases}$$

The number λ_o is obtained as the lower bound of all nonnegative λ for which, in the set of corresponding functions $u_i(t,\lambda)$, there occur functions satisfying the condition

$$\int_{t_o}^{t_1} \sum_{i=1}^{r} |b_i(t) u_i(t,\lambda) - \phi_i(t)| \, dt \leq k \quad .$$

We examine one of the possible interpretations of the problem with the constraints (224). Let $b_i(t) \equiv 1$,

$i = 1,\ldots,r$. We assume that the functions $\phi_i(t)$, $i = 1,\ldots,r$, $t \varepsilon T$, denote controls computed for a system of several arguments (for example, from the optimality of a system in some sense). Then the problem with constraints (244) may be considered as the determination of a control $u^o(t)$, $|u_1^o(t)| \leq 1$ which has a small deviation from the programmed control $\phi(t)$ and which minimizes the norm of the terminal state.

§15. Statistical Problems of Optimal Control (Functional Analysis Approach)

In recent years, in the theory of optimal process a large role has been played by control problems for stochastic systems. The most general results in this direction have been obtained by L.S. Pontryagin, E.F. Mischenko [97a], R. Bellman [11c], A.A. Fel'dbaum [124c], N.N. Krasovskii [74h], and R.L. Stratonovich [120]. Below, we use the techniques of functional analysis to construct optimal controls in systems influenced by stochastic effects. Statistical problems are reduced to equivalant deterministic problems; properties of optimal solutions are studied.

1. The Problem of Minimizing the Mean Value of a Function of the Terminal State (Discrete Probability Distributions)

(1) Problem Statement. Let the n-dimensional stochastic process $x(t)$ be given at $t = t_1$ by the relation

$$x(t_1) = s(t_1,x_o,f(t_1)) + \int_{t_o}^{t_1} S(t_1,t)u(t)dt, \quad x(t_o) = x_o,$$

where the elements of the vector $s(z_1,z_2,z_3)$ and the matrix $S(z_1,z_4)$ are continuous in all arguments while $f(t)$

is a function describing the external stochastic error, and $u(t) = \{u_1(t), \ldots, u_r(t)\}$ is the controlling function subjected to the condition

$$\int_{t_0}^{t_1} \sum_{i=1}^{r} |u_j(t)|^p dt \leq L^p \quad , \quad p > 1 \quad .$$

We write the last constraint in the form $u(\cdot)\varepsilon U_p^L(\cdot)$.
We set

$$s(t_1, x_0, f(t_1)) = \tilde{c} \quad , \quad \int_{t_0}^{t_1} S(t_1,t)u(t)dt = Su(\cdot) \quad ,$$

$$x(t_1) = \tilde{x} \quad .$$

Then we have

$$\tilde{x} = Su(\cdot) + \tilde{c} \quad . \tag{230}$$

Here $\tilde{x}$ is an element of E_n. We will study this equation in the sequel. We assume that the matrices $S(t_1,t)$ are such that the functions $[S(t_1,t)]_{ij}$, $i = 1,\ldots,n$, are linearly independent for at least one $j = 1,\ldots,r$. Then it is not difficult to show that $||g'S|| > 0$ for all g, $||g|| \neq 0$.

We let Mf denote the expected value of the stochastic quantity f. Let a positive function $\phi(x)$ be given. It is required to find a control $u^o(t)$, $t\varepsilon T$, for which

$$M\phi(x(\tilde{c},u^o(\cdot),t_1)) = \min_{u(\cdot)\varepsilon U_p^Z(\cdot)} \{M\phi(x(\tilde{c},u(\cdot),t_1))\} = \delta \quad .$$

We will let e denote the mean value of the stochastic vector $\tilde{c}$, while d denotes the measure of variance (variances), where

$$d = M\phi(\tilde{c} - e) = \min_{x\varepsilon E_n} \{M\phi(\tilde{c} - x)\} \quad .$$

We assume that

$$\phi(x) = ||x||_{\bar{p}} = \left(\sum_{i=1}^{n} |x_i|^{\bar{p}}\right)^{1/\bar{p}}, \quad \bar{p} > 1 . \qquad (232)$$

At first, we consider the case when $\tilde{c}$ assumes only a finite number of values c_j, $j = 1,\ldots,m$, $||c_j|| < \infty$. Let $p_j (p_j > 0, \sum_{j=1}^{m} p_j = 1)$ be the probability that $\tilde{c} = c_j$.

(2) Optimal Controls. We let E_{nm} denote the direct sum of m copies of E_n. Let

$$||z|| = \sum_{j=1}^{m} p_j ||x_j|| \quad , \quad z \varepsilon E_{nm}, \ x_j \varepsilon E_n \quad .$$

If x_j is the value of $\tilde{x}$ corresponding to a realization c_j of the vector $\tilde{c}$, then from (230) we have

$$z = Pu(\cdot) + \eta \ , \quad P = \{\underbrace{S,\ldots,X}_{m}\} \quad , \quad \eta = \{c_1,\ldots,c_m\} \ .$$

This means that problem (231), (232) is reduced to determination of the quantity $\tilde{\delta} = \min_{u(\cdot)\varepsilon U_p^L(\cdot)} ||z||$. It is clear that

$$f'P = \sum_{j=1}^{m} f_j'S \ , \quad f \varepsilon E_{nm} \ , \quad ||f|| = \max_j \{||f_j||/p_j\} \quad .$$

Therefore,

$$\tilde{\delta} = \max_{||f||=1} \{f'\eta - L||f'P||\} = \max_{\max_j ||f_j||=1} \sum_{j=1}^{m} p_j f_j' c_j - L|| \sum_{j=1}^{m} p_j f_j' S||\}$$

$$= \sum_{j=1}^{m} p_j f_{jo}' c_j - L|| \sum_{j=1}^{m} p_j f_{jo}' S|| , \quad (233)$$

$$\sum_{j=1}^{m} p_j f'_{jo} S u^o(\cdot) = \min_{u(\cdot)\varepsilon U_p^L} \left[\sum_{j=1}^{m} p_j f_{jo} S u(\cdot) \right] . \tag{234}$$

The extremal elements f_{jo} of problem (233) may be uniquely associated to one of the two cases:

α) $f_{jo} = \hat{g}, \ ||\hat{g}|| = 1, \ j = 1,\ldots,m;$

β) $\left|\left| \sum_{j=1}^{m} p_j f_{jo} \right|\right| < 1 \ , \ ||f_{jo}|| = 1 \ .$

Lemma 17. In order that $f_{jo} = \hat{g}$, $j = 1,\ldots,m$, it is necessary and sufficient that

$$\max_{||g||=1} \{g'c_j - L||g'S||\} = \hat{g}'c_j - L||\hat{g}'S||, \quad j = 1,\ldots,m \ . \tag{235}$$

Proof. (Necessity) Let $f_{jo} = g$, $j = 1,\ldots,m$. Then $\hat{\delta} = \sum_{j=1}^{m} p_j \hat{g}\, c_j - L||\hat{g}'S||$. If

$$\max_{||g||=1} \{g'c_k - L||g'S||\} = g_k'c_k - L||g_k'S||, \quad g_k \neq g \ ,$$

then

$$g_k'c_k - L||\hat{g}_k'S|| > \hat{g}'c_k - L||\hat{g}'S|| \ ,$$

and therefore

$$\sum_{j=1, j\neq k}^{m} p_j \hat{g}'c_j + p_k g_k'c_k - L\left|\left| \sum_{j=1, j\neq k}^{m} p_j \hat{g}'S + p_k \hat{g}_k'S \right|\right|$$

$$\sum_{j=1, j\neq k}^{m} p_j [\hat{g}'c_j - L||\hat{g}'S||] + p_k g_k'c_k - p_k ||g_k'S||$$

$$> \sum_{j=1}^{m} p_j \hat{g}'c_j - L||\hat{g}'S|| \ .$$

which is not possible. This means condition (235) is satisfied.

(Sufficiency) If the extremal elements in problem (235) are equal to each other for all j then, by virtue of Theorem 26, the system (230) may be reduced to the point x = 0 from all states x_{jo} for each c_j using the same control. But

$$\min_{u(\cdot)\varepsilon U_p^L(\cdot)} \left[\sum_{j=1}^{m} p_j ||Su(\cdot) + c_j||\right] \geq \sum_{j=1}^{m} p_j \min_{u(\cdot)\varepsilon U_p^L(\cdot)} ||x_j|| .$$

Consequently, there does not exist an element f_{jo} for which

$$\hat{\delta} < \sum_{j=1}^{m} p_j \hat{g}' c_j - L||g'S|| .$$

This means $f_{jo} = \hat{g}$, j = 1,...,m, and the lemma is proved.

<u>Lemma 18. Let condition (β) be satisfied. Then there exists a vector λ, $||\lambda|| \neq 0$, $\lambda\varepsilon E_n$, such that</u>

$$\left.\begin{aligned} &\max_{||f_j||\leq 1} \sum_{j=1} p_j f_j' c_j - L|| \sum_{j=1} p_j f_j' S|| \\ &\qquad = \sum_{j=1}^{m} p_j ||c_j - \lambda|| + g'\lambda - L||g'S|| , \\ &\sum_{j=1}^{m} p_j f_{jo} - g = 0 , \quad f_j'(c_j - \lambda) = ||c_j - \lambda|| . \end{aligned}\right\} \quad (236)$$

<u>Proof</u>. In the (n + 1)-dimensional space $\{v, v_{n+1}\}$, we consider the set $G = \{v: v = \sum_{j=1}^{m} p_j f_j,\ v_{n+1} = \sum_{j=1}^{m} p_j f_j' c_j - L|| \sum_{j=1}^{m} p_j f_j' S||,\ ||f_j|| \leq 1\}$. The set G is

compact. The problem formulated in the conditions of the lemma consists in the determination of points of G at which $v = g$, $v_{n+1} = \max$. We consider the concave function

$$z(f_1,\ldots,f_m) = \sum_{j=1}^{m} p_j f_j' c_j - L\left|\left|\sum_{j=1}^{m} p_j f_j' S\right|\right| ,$$

defined on the convex set $\sigma = \{\{f_1,\ldots,f_m\}: \ ||f_j|| \leq 1\}$. Since g is an interior point of σ, there exists a vector $\{\ell,\ell_{n+1}\}$, $\ell_{n+1} > 0$, such that

$$-\ell' g + \ell_{n+1} z(f_{10},\ldots,f_{mo})$$

$$= \max_{||f_j||\leq 1} \{-\ell' \sum_{j=1}^{m} p_j f_j + \ell_{n+1}[\sum_{j=1}^{m} p_j f_j' c_j - L||g'S||]\} ,$$

i.e.

$$z(f_{1o},\ldots,f_{mo}) = g\ell/\ell_{n+1} - L||g'S||$$

$$+ \max_{||f_j||\leq 1} \sum_{j=1}^{m} p_j f_j' (c_j - \ell/\ell_{n+1}) = g'\lambda - L|| + \sum_{j=1}^{m} p_j ||c_j - \lambda|| .$$

Here

$$\lambda = \ell/\ell_{n+1} , \quad g = \sum_{j=1}^{m} p_j f_{jo} , \quad \sum_{j=1}^{m} p_j f_{jo}' (c_j - \lambda)$$

$$= \sum_{j=1}^{m} p_j ||c_j - \lambda|| .$$

The assertion is proved.

Lemma 19. The vector λ satisfies relation (236) if and only if

$$\min_x \{g'x + \sum_{j=1}^{m} p_j ||c_j - j||\} = g'\lambda + \sum_{j=1}^{m} p_j ||c_j - \lambda|| . \quad (237)$$

Proof. Since

$$\sum_{j=1}^{m} p_j f'_{jo}(c_j - x) \leq \sum_{j=1}^{m} p_j ||c_j - x||$$

for any $x \varepsilon E_n$, then

$$\sum_{j=1}^{m} p_j f_{jo} c_j \leq g'x + \sum_{j=1}^{m} p_j ||c_j - x|| \quad .$$

On the other hand, by virtue of (236) we have the equation

$$\sum_{j=1}^{m} p_j f'_{jo} c_j = g'\lambda + \sum_{j=1}^{m} p_j ||c_j - x|| \quad .$$

This means that condition (237) is satisfied. We assume that Eq. (237) is true. Then

$$\min_{x} \{g'x + \max_{||f_j|| \leq 1} \sum_{j=1}^{m} p_j f_j (c_j - x)\}$$

$$= g'\lambda + \sum_{j=1}^{m} p_j f_{jo}(c_j - \lambda) = g'\lambda + \sum_{j=1}^{m} p_j ||c_j - \lambda|| \ ,$$

i.e. the second relation from (236) is satisfied. Obviously, the equation

$$g'\lambda + \sum_{j=1}^{m} p_j f_{jo}(c_j - \lambda) = \min \{g'x + \sum_{j=1}^{m} p_j f_{jo}(c_j - x)\}$$

$$= \min_{x} \{x'[g - \sum_{j=1}^{m} p_j f_{jo}] + \sum_{j=1}^{m} p_j f'_{jo} c_j\} \ ,$$

is valid. Therefore, it is necessary that $g = \sum_{j=1}^{m} p_j f_{jo}$ and the theorem is proved.

Theorem 41. There exists an optimal control in problem (230)-(232).

If condition (235) is satisfied, then

$$\tilde{\delta} = \max_{||g||=1} \{g'M\tilde{c} - L||g'S||\} = \hat{g}'Mc - L||\hat{g}'S|| \quad . \tag{238}$$

For $\tilde{\delta} > 0$, the optimal control $u^o(\cdot)$ is determined by the relation

$$\hat{g}'Su^o(\cdot) = \min_{u(\cdot)\varepsilon U_p^L(\cdot)} \left[\hat{g}'Su(\cdot) = -L||\hat{g}'S||\right] \quad . \tag{239}$$

If condition (235) is not satisfied, then

$$\tilde{\delta} = \max_{g} \min_{x} \{ \sum_{j=1}^{m} p_j||c_j - x|| + g'x - L||g'S||\}$$

$$= \sum_{j=1}^{m} p_j||c_j - x^o|| + g_o'x^o - L||g_o'S|| . \tag{240}$$

For $\tilde{\delta} > 0$, the optimal control is determined by (239), where $\hat{g} = g_o$.

Proof. The formulas (238), (239) follow from (233), (234) if condition (235) is satisfied. The second part of the assertion is also obtained from relations (233), (234) since, in case (β) Lemmas 18, 19 are true.

Thus, if condition (α) is satisfied, the optimal control in the statistical problem (230)-(232) coincides with the solution of the problem of minimizing the norm of the terminal state of the system

$$x = Su(\cdot) + M\tilde{c} \quad .$$

In case (β), the function $u^o(\cdot)$ is the same as that obtained in the problem of minimizing the norm of the terminal state for the system

$$x = Su(\cdot) + x^o \quad .$$

Here x^o is a saddle-point of the game (240).

3. Several Properties of the Solution. 1) From the definition of the saddle-point of the game (240), we have

$$\sum_{j=1}^{m} p_j||c_j - x^o|| + g'x^o - L||g'S||$$
$$\leq \sum_{j=1}^{m} p_j||c_j - x^o|| + g_o' - L||g_o'S||$$
$$\leq \sum_{j=1}^{m} p_j||c_j - x|| + g_o'x - L||g_o'S|| \tag{241}$$

for all $x, g \varepsilon E_n$. Let $x = x_o$. Then the first inequality in (241) leads to the relation

$$g_o'x^o - L||g_o'S|| = \max_{||g|| \leq ||g_o||} \{g'x^o - L||g'x||\} \, . \tag{242}$$

Setting $g = g_o$, from the second inequality (241) we have

$$\sum_{j=1}^{m} p_j||c_j - x^o|| + g_o'x^o = \min_x \{\sum_{j=1} p_j||c_j - x|| + g'x\} \, . \tag{243}$$

Conversely, if the vectors x^o, g_o satisfy conditions (242), (243), then they are saddle-points of the game (240).

2) The point $\{g_o, x^o\}$ satisfies the condition

$$g_o'x^o - L||g_o'S|| = \beta(g_o, x^o) = 0 \quad . \tag{244}$$

In fact, from the first inequality of (241) it is clear that the inequality $\beta(g_o, x^o) > 0$ is impossible. Setting $g = g_o/||g_o||$, we see that $\beta(g_o, x^o)$ cannot be positive.

3) The element g_o of the saddle-point $\{g_o, x^o\}$ satisfies the condition

$$\min_{x\varepsilon G_1} g_o'x = \max_{x\varepsilon G_2} g_o'x \quad , \tag{245}$$

where

$$\begin{aligned} G_1 &= \{x: \sum_{j=1}^{m} p_j||c_j - x||\} \leq \tilde{\delta} \quad , \\ G_2 &= \{x: \quad x = -Su(\cdot), u(\cdot)\varepsilon U_p^L(\cdot)\} \quad . \end{aligned} \tag{246}$$

Actually, from (240) we have

$$\tilde{\delta} = \min_{x} \{ \sum_{j=1}^{m} p_j||c_j - x|| + \max_{||g||\leq||g_o||} [g'x - L||g'S||]\} \quad .$$

Taking into account (244), we obtain

$$\tilde{\delta} = \min_{x} \sum_{j=1}^{m} p_j||c_j - x|| \tag{247}$$

with the conditions

$$\Delta(x) = \max_{||g||\leq||g_o||} \{g'x - L||g'S||\} \leq 0 \quad . \tag{248}$$

But, the set $Q = \{x: \Delta(x) \leq 0\}$ coincides with G_2, which proves the assertion.

4) Let e be the mean, d the variance of the stochastic vector $\tilde{c}$. There exists a number q, $0 \leq q \leq 1$, such that

$$\tilde{\delta} = d + q\{ \max_{||g||\leq||g_o||} [g'e - L||g'S||]\} \quad . \tag{249}$$

In fact,

$$\tilde{\delta} \geq \min_{x \varepsilon E_n} \sum_{j=1}^{m} p_j ||c_j - x|| = \sum_{j=1}^{m} p_j ||c_j - e|| = d . \quad (250)$$

By virtue of (241),(243), we have

$$\tilde{\delta} \leq \sum_{j=1}^{m} p_j ||c_j - e|| + \max_{||g|| \leq ||g_o||} [g'e - L||g'S||] ,$$

which, together with (250), proves the validity of (249).

5) If $\Delta(e) < 0$, then $\tilde{\delta} = d$, $x^o = e$, $g_o = 0$. This assertion is a consequence of condition (247). The optimal control in case 5) may be determined by one of the following approaches: 1) decreasing L to $L = L_1$, until the equation $\Delta(e)$ is satisfied; 2) determine (for fixed L) the time t_1, when $\Delta(e) = 0$. The minimal value δ is achieved using the control solving one of the following deterministic problems:

1) $x = Su(\cdot) + e , \quad \min_{u(\cdot) \varepsilon U_p^L(\cdot)} ||x|| = 0 ,$

2) $x = Su(\cdot) + e , \ x|_{t_1^o} = 0 , \ t_1^o = \min_{u(\cdot) \varepsilon U_p^L(\cdot)} \{t : x|_t = 0\}.$

Below it is assumed that $\Delta(e) > 0$.

6) The saddle-point $\{g_o, x^o\}$ is such that the inequalities

$$g_o' x^o > 0 \quad , \quad g_o' e > 0 \quad . \quad (251)$$

are satisfied.

The first inequality follows from (244). We show the second inequality. Assume that $g_o'e < 0$. Then

$$g_o'x^o + \sum_{j=1}^{m} p_j||c_j - x^o|| > g_o'e + \sum_{j=1}^{m} p_j||c_j - e|| \quad ,$$

which contradicts (243). The assertion is proved.

7) If $n = 1$, the optimal control is completely determined by the mean e. Actually, the assertion is clear if condition (235) is satisfied. If not, the assertion follows from inequality (251).

8) Let $n > 1$. If

$$\min_{x} \sum_{j=1}^{m} p_j||c_j - x|| + g'x = \sum_{j=1}^{m} p_j||c_j - e|| + g'e$$

under the conditions $\{g: \ g'e > 0, ||g|| = \ell\}$ and $\max \ell = ||g_o||$, then

$$\tilde{\delta} = d + ||g_o|| \max_{||g|| \leq ||g_o||} \{g'e - L||g'S||\} \quad .$$

In this case, the optimal control is completely determined by the mean e.

(4) The Minimization Problem. The functional $M\gamma(x) = (\sum_{j=1}^{m} p_j(x_j'x_j)^{p/2})^{1/p}$, $p > 1$. Let $p = 2$.

Theorem 42. An optimal control $u^o(\cdot)$ exists and makes the functional $M\gamma(x)$ assume the value ε,

$$\varepsilon^2 = d^2 + \Delta_1^2(M\tilde{c}) \quad , \tag{252}$$

where

$$\Delta_1(M\tilde{c}) = \max_{||g||\leq 1} \{g'Mc - L||g'S||\} = g'Mc - L||g'S||. \quad (253)$$

<u>The optimal control $u^o(\cdot)$ is determined by the relation (239) for $g = \hat{g}$, i.e. it is completely characterized by the quantity $M\tilde{c}$.</u>

<u>Proof</u>. We find the mean and variance. We have

$$\min_x \left(\sum_{j=1}^{m} p_j (c_j - x)'(c_j - x)\right)^{1/2}$$

$$= \left[\sum_{j=1}^{m} p_j (c_j - M\tilde{c})'(c_j - M\tilde{c})\right]^{1/2} . \quad (254)$$

This means $e = M\tilde{c}$. From (254) we have

$$d^2 = M\tilde{c}'\tilde{c} - [M\tilde{c}]'M\tilde{c} \quad ,$$

i.e. the variance equals the variance of the stochastic vector $\tilde{c}$. If condition (235) is satisfied, then, by virtue of formula (233), it is possible to make the conclusion: the extremal element g of problem (233) is determined by the quantity $M\tilde{c}$, i.e. $g = \hat{g}$. It is not difficult to see that $\varepsilon = d$.

Let $g_k \nrightarrow g$. We show that relation (252) holds. We set

$$\min_{u(\cdot)\varepsilon U_p^L(\cdot)} \{ \sum_{j=1}^{m} p_j [Su(\cdot) + c_j]'[Su(\cdot) + c_j]\}^{1/2} = \varepsilon \quad .$$

By virtue of (247),(248) we have

$$\varepsilon^2 = \max_{||g||\leq 1} \min_x \sum_{j=1}^{m} p_j (c_j - x)'(c_j - x)$$

under the conditions $g'x - L||g'S|| = 0$. We find the minimum over all x using a Lagrange multiplier. Let

$$H(x) = \sum_{j=1}^{m} p_j (c_j - x)'(c_j - x) + \lambda(g'x - L||g'S||) .$$

Then

$$\text{grad } H(x) = 2 \sum_{j=1}^{m} p_j (x - c) + \lambda g , \quad x = Mc - \frac{\lambda}{2} g .$$

Consequently,

$$\varepsilon^2 = \max_{||g|| \leq 1} \{ \sum_{j=1}^{m} p_j [c_j - M\tilde{c} - (g'M\tilde{c} - L||g'S||)$$

$$\cdot (g'g)^{-1} g]'[-(g'M\tilde{c} - L||g'S||)(g'g)^{-1} g + c_j - M\tilde{c}]\}$$

$$= d^2 + [\max_{||g|| \leq 1} \{g'M\tilde{c} - L||g'S||\}]^2 .$$

The theorem is proved.

Let $p = 4$. From (245), we obtain

$$\min_{\sum_{j=1}^{m} p_j [(c_j - x)'(c_j - x)]^2 \leq \delta^4} g'x = \max_{\Delta(x) \leq 0} g'x .$$

We set

$$h(x) = g'x + \lambda \sum_{j=1}^{m} p_j [(c_j - x)'(c_j - x)]^2 .$$

Then

$$\text{grad } h(x) = g - 4\lambda \sum_{j=1}^{m} p_j [c_j' c_j - 2x'c_j + x'x](c_j - x) . \quad (255)$$

We expand the vectors c_j in elements ℓ_j of some basis $\{\ell_1,\ldots,\ell_n\}$ in E_n. From (255), we have

$$\text{grad } h(x) = g + 4\lambda x\{\sum_{s,k=1}^{n} Ma_s a_k(\ell_s'\ell_k) - 2\sum_{i=1}^{n} Ma_i\ell_i'x + x'x\}$$

$$- 4\lambda\{\sum_{s,k,i=1}^{n} Ma_s a_k a_i \ell_s'\ell_k\ell_i - 2\sum_{i,x=1}^{n} Ma_i a_s \ell_i' x\ell_s + \sum_{i=1}^{n} Ma_i x'x\},$$

$$c_j = \sum_{i=1}^{n} a_{ji}\ell_i \quad .$$

It is clear that the x arising from the condition grad $h(x) = 0$ is found as a function of λ, g, Ma_x, $Ma_x a_k$, $Ma_s a_k a_i$. This means the quantity g_0 will, in general, depend on Ma_s, $Ma_s a_k$, $Ma_s a_k a_i$; $s,k,i = 1,\ldots,n$. Thus, the optimal control for $p = 4$ is completely characterized by the quantities $M\tilde{c}$, $M\tilde{c}'\tilde{c}$, $M\tilde{c}\tilde{c}'\tilde{c}$, $M\tilde{c}\tilde{c}_k$, where c_{jk} is the <u>kth</u> component of the <u>jth</u> realization.

<u>Remark</u>. For a given arbitrary norm in E_n, it is necessary, generally speaking, to find the characteristics of a stochastic vector, the number of which is less than the number of realizations: it is possible to construct examples when the optimal control depends on the probability of each realization.

(5) <u>Systems with Stochastic Parameters</u>. Let there be given the equation

$$\frac{dx}{dt} = A(t)x + \alpha(t)b(t)u(t) + r(t) \quad , \qquad (256)$$

where $x \varepsilon E_n$; $b(t)$, $r(t)$ are known continuous vector functions; $A(t)$ is a given continuous matrix. It is assumed that the scalar function $\alpha(t)$ describes a stochastic process of the following type: a change of value of $\alpha(t)$ is possible at the moments $\tau_k (k = 1,\ldots,s)$, $\tau_s = t_1$, where on each interval $[\tau_{k-1},\tau_k)$ the function $\alpha(t)$ assumes one of the given values α_i, $i = 1,\ldots,m$.

We let $p(i_1,\ldots,i_s)$ denote the probability of the event consisting of the realization for the first interval of the value α_{i_1}, for the second of α_{i_2}, and so on. It is required to determine a control $u^o(\cdot)\varepsilon U_p^L(\cdot)$ minimizing the functional (231),(232).

The fundamental solution matrix $F(t)$ of the system corresponding to (256) defines operators S_k:

$$S_k u_k(\cdot) = \int_{\tau_{k-1}}^{\tau_k} F(t_1)F^{-1}(\tau)b(\tau)u(\tau)d\tau \ , \ k = 1,\ldots,s \ .$$

We set

$$F(t_1)F^{-1}(t_o)x_o + \int_{t_o}^{t_1} F(t_1)F^{-1}(\tau)r(\tau)d\tau = c \ .$$

Let

$$\min_{\{v_k\}_1^s} \{ \sum_{\{i_k=1\}_1^s}^{m} p(i_1,\ldots,i_s)||c - \sum_{k=1}^{s} \alpha_{i_k} v_k||\}$$

$$= \sum_{\{i_k=1\}_1^s}^{m} p(i_1,\ldots,i_s)||c - \sum_{k=1}^{s} \alpha_{i_k} e_k|| \ ,$$

$$\eta_k = \{v_k: \max_{||g_k||=1} [\sum_{k=1}^{s} g_k' v_k - L||g_k' S_k||] \leq 0\} \ .$$

<u>Theorem 43. An optimal control exists for the problem (230)-(232).</u>

<u>If $e_k \notin \text{int } \eta_k$ then</u>

$$= \max_{x} \min_{\{z_k\}_1^s} \{ \sum_{\{i_k=1\}_1^s}^{m} p(i_1,\ldots,i_s) ||c - \sum_{k=1}^{s} \alpha_{i_k} z_k|| + \sum_{k=1}^{s} [g_k' z_k - L||g_k' S_k||] \}$$

$$= \sum_{\{i_k=1\}_1^s}^{m} p(i_1,\ldots,i_s) ||c - \sum_{k=1}^{s} \alpha_{i_k} z_k^o|| + \sum_{k=1}^{s} [g_{ko}' z_k^o - L||g_{ko}' S_k||].$$

For $\tilde{\delta} > 0$, the optimal control $u_k^o(\cdot)$ satisfies the equation

$$\sum_{k=1}^{s} g_{ko}' S_k u_k^o(\cdot) = \min_{u(\cdot) \varepsilon U_p^L(\cdot)} \sum_{k=1}^{s} g_{ko}' S_k u_k(\cdot) \quad .$$

Proof. The proof is analogous to that of Theorem 41.

Remarks. 1) Results related to system (256), when x_o is a stochastic quantity and, $r(t)$, $\alpha(t)$ are stochastic functions, are omitted on account of the unwieldy nature of the corresponding formulas. Methods for studying such systems are constructed by the scheme described in sub-sections (1)-(4).

2) The problem of determining the vectors g_{ko} is simplified when $M\phi(\tilde{x}) = (\sum_{j=1}^{m} p_j (x_j' x_j))^{1/2}$. In this case, if $s = 2$, then

$$\tilde{\delta}^2 = M[\omega^{-1} \alpha(t_o) \xi + \alpha(t_1) \eta - \omega]^2 c'c - \max_{g_1, g_2} [\omega^{-1} c' (\xi g_1 + \eta g_2) - L||g_1' S_1|| - L||g_2' S_2||] M^{-1} \gamma,$$

where

$$\omega = M\alpha^2(t_0)M\alpha^2(t_1) - M\alpha(t_0)\alpha(t_1) \quad ,$$

$$\xi = M\alpha(t_0)M\alpha^2(t_1) - M\alpha(t_1)M\alpha(t_0)\alpha(t_1) \quad ,$$

$$\eta = M\alpha(t_1)M\alpha^2(t_0) - M\alpha(t_0)M\alpha(t_0)\alpha(t_1) \quad ,$$

$$\gamma = [\alpha(t_1)g_1 - \alpha(t_0)g_2]'[\alpha(t_1)g_1 - \alpha(t_0)g_2]\omega^{-1} \quad .$$

Computation of the elements g_{ko} may be further simplified if constraints are imposed upon the stochastic process $\alpha(t)$.

We assume that the values which the stochastic function $\alpha(t)$ takes on in different intervals are independent and, further, that $M\alpha(t) = 0$, $t\varepsilon[\tau_{k-1},\tau_k)$, $k \geq 2$. Then, for example, when $s = r$ we have

$$\tilde{\delta}^2 = D^2\alpha(t_0)M^{-1}\alpha^2(t_0)c'c + \max_{g_1} \{M\alpha(t_0)M^{-1}\alpha^2(t_0)g_1'c$$

$$- L||g_1'S_1|| - LD\alpha(t_1)[1 - g_1'g_1M^{-1}\alpha^2(t_0)]^{1/2}||S_2||\} \quad ,$$

$$||S_2|| = \max_{||g||\leq 1} ||g'S_2|| = ||g_{20}'S_2||$$

$$(D\tilde{x} = M\tilde{x}'\tilde{x} - M'\tilde{x}M\tilde{x}) \quad .$$

3) The statistical problem of Section 1 may be interpreted as the following deterministic problem. Let a set $\sigma \subset E_n$ be given, where σ consists of the m points $c_1,\ldots,c_m$. It is required to transfer the trajectory of Eq. (216) in time $\tau = t_1 - t_0$ from the point $x(t_0) = x_0$

to the point x = a, such that

$$\sum_{j=1}^{m} p_j ||c_j - a|| = \min_{u(\cdot)\varepsilon U_p^L} \sum_{j=1}^{m} p_j ||c_j - x(x_o, u(\cdot), t_1)||.$$

2. Reduction of the Statistical Optimal Control Problem to a Game

(1) On One Problem of Optimal Control. Let x_μ be an n-vector in E_n depending upon the parameter $\mu\varepsilon\Omega$. We set $x(\cdot) = \{x : \mu\varepsilon\Omega\}$, and we denote the set of elements $x(\cdot)$ by $E_n(\cdot)$. We are given the relation

$$x(\cdot) = Su(\cdot) + c(\cdot) \ , \quad u(\cdot)\varepsilon U_p^L(\cdot) \quad . \tag{258}$$

Here S is a linear operator, $u(\cdot)$ is the control, $c(\cdot)$ is a given element of $E_n(\cdot)$. We call the functional $f(x(\cdot))$, defined on the elements $x(\cdot)$, quaisiconcave if the set

$$\{z: \ f(c(\cdot) - z) \geq \varepsilon\}$$

is convex for each ε. Let $f(x(\cdot))$ be a quasiconcave upper semicontinuous functional.

Problem. To find a function $u^o(\cdot)\varepsilon U_p^L(\cdot)$ such that

$$f(x^o(\cdot)) = \max f(x(\cdot)) \ , \quad x^o(\cdot) = Su^o(\cdot) + c(\cdot) \ . \tag{259}$$

We set $y = c(\cdot) - x(\cdot)$. Then from (258),(259), we have

$$f(x^o(\cdot)) = \max_{y=-Su(\cdot),u(\cdot)\varepsilon U_p^L(\cdot)} f(c(\cdot) - y)$$

$$= f(c(\cdot) - y^o) = \varepsilon^o \ . \tag{260}$$

We let e,d denote a vector and a scalar satisfying the condition

$$\max_{y \varepsilon E_n} f(c(\cdot) - y) = f(c(\cdot) - e) = d \quad ,$$

and we assume that

$$\Delta(e) = \max_{||g||=1} \{g'e - L||g'S||\} \geq 0^* \quad . \tag{261}$$

Since the set

$$G(y) = \{y: \; f(c(\cdot) - y) \geq \varepsilon^o\}$$

is convex and closed, y^o is an element of the boundary of this set (cf. (246)). From (248), it follows that the boundary points satisfy the condition

$$\Delta_1(y) = \max_{||g|| \leq 1} \{g'y - L||g'S||\} = - \max_{||g|| \leq 1} \{L||g'S|| - g'y\}.$$

Therefore, from (260) we have

$$f(x^o(\cdot)) = \max_{\Delta_1(y)=0} f(c(\cdot) - y) \quad . \tag{262}$$

<u>Theorem 44. The maximum of (262) is assumed on the saddle-point $y = y^o$ of the game</u>

$$\max_y \min_g \{f(c(\cdot) - y) - g'y + L||g'S||\}$$
$$= \min_g \max_y \{f(c(\cdot) - y) - g'y + L||g'S||\} = \varepsilon^o. \tag{263}$$

---

* The case $\Delta e < 0$ is studied as in 5) section 1 (3).

<u>There exists a number $k > 0$ such that</u>

$$\varepsilon^o = d - k\Delta_1(e) \quad . \tag{264}$$

<u>Proof</u>. If g_o, y^o is a saddle-point of the game (263), then

$$f(c(\cdot) - y) - g_o'y + L||g_o'S|| \leq f(c(\cdot) - y^o) - g_o'y^o$$
$$+ L||g_o'S|| \leq f(c(\cdot) - y^o) - g'y^o + L||g'S|| \tag{265}$$

for any g, y. From (265) follows the inequality

$$-g_o'y^o + L||g_o'S|| \leq -g'y^o + L||g'S|| \quad ,$$

which is valid for any g. But the latter is possible only under the condition

$$\min_g \{-g'y^o + L||g'S||\} = 0 \quad . \tag{266}$$

Therefore, from the left-side of inequality (265) we have

$$f(c(\cdot) - y^o) \geq f(c(\cdot) - y) - g_o'y + L||g_o'S||$$
$$\geq f(c(\cdot) - y) + \min_g \{L||g'S|| - g'y\} \quad ,$$

which is equivalent to relation (262). We show that condition (264) is satisfied. From (265),(266), and the definition of the quantities e, d, we have

$$f(c(\cdot) - y^o) - g_o'y^o + L||g_o'S|| \geq f(c(\cdot) - e)$$
$$- g_o'e + L||g_o'S|| \geq d - \max_{||g|| \leq 1} \{g'e - L||g'S||\} \quad .$$

But, clearly $f(c(\cdot) - y^o) < d$. This completes the proof.

Remark. We have the inequalities

$$g_o' y^o > 0 \quad , \quad g_o' e > 0 \quad .$$

The first property of the saddle-point $\{g_o, y^o\}$ follows from (266) since $||g_o' S|| \neq 0$, by convention. The second is proved by contradiction, as in the second inequality of (251).

(2) The Problem of Minimizing the Mean-Value of a Function of the Terminal State. Continuity of Types of Probability Distributions. Let $\Phi_x(s)$, $s \varepsilon E_n$, be a function of the distribution of the stochastic vector x from (230). It is required to find a control $u^o(\cdot)$, generating a solution of (230), which minimizes

$$M||\tilde{x}|| = \int_{E_n} ||s|| d\Phi_{\tilde{x}}(s) \quad . \tag{267}$$

We find the mean and variance:

$$\min_{x \varepsilon E_n} M||\tilde{c} - x|| = M||\tilde{c} - e|| = d \quad .$$

Let the mean e be such that condition (264) is satisfied. Since the function (267) is convex, by virtue of the results of paragraph 1 we have

$$\min M||x(\tilde{c}, u(\cdot), t_1)||$$

$$= \tilde{\delta} = \max_g \min_z \{M||\tilde{c} - z|| + g'z - L||g'S||\} \quad . \tag{268}$$

The optimal control $u^o(\cdot)$ is determined from condition (239), where $\hat{g}$ is an element of the saddle-point $\{\hat{g}, \hat{x}\}$ of the game (268). From (268), we obtain the following estimate for $\tilde{\delta}$:

$$d \le \tilde{\delta} \le d + \Delta_1(e) \quad .$$

Remark. We have the inequalities

$$\hat{g}'e > 0 \quad , \quad \hat{g}'\hat{x} > 0 \quad .$$

Formula (268) assumes a more compact form if we minimize the quantity

$$\int_{E_n} s'sd_{\tilde{x}}\Phi(s) \quad . \tag{269}$$

It is possible to show that Theorem 42 is valid in the case (269).

3. The Problem of Maximizing the Probability of Hitting a Given Neighborhood

In Eq. (256), let the function r(t) describe a stochastic process, where $x(t_0) = x_0$ is a stochastic vector and $\alpha(t) \equiv 1$ (cf. (230),(257)). We give a point $x_1 \varepsilon E_n$ and its neighborhood $||x - x_1|| \le \varepsilon$. For each control $u(\cdot)\varepsilon U_p^L(\cdot)$, Eq. (256) determines a stochastic process $\tilde{x}(t)$, $t \ge t_0$. A cross-section of the stochastic process $\tilde{x}(t)$ at a fixed moment gives a stochastic quantity $\tilde{x}$ for which we will assume the probability is known:

$$[||\tilde{x} - x_1|| \le \varepsilon] = P(||\tilde{x} - x_1|| \le \varepsilon) \quad .$$

It is required to find a control $u(\cdot)\varepsilon U_p^L(\cdot)$ maximizing the functional

$$J(u) = P(||\tilde{x}(\tilde{x}_0, u(\cdot), t_1) - x_1|| \le \varepsilon) \quad . \tag{270}$$

We will say that condition A is satisfied if the probabilities of the quantities x_0, $r(t)$, $t \varepsilon T$, are such that for any given $\varepsilon > 0$ and for any β, $0 \leq \beta \leq 1$, the set

$$Q(\varepsilon,\beta) = \{z: P(||\tilde{a} - z|| \leq \varepsilon) \geq \beta,\ \tilde{a} = \tilde{c} - x_1\}$$

is closed and convex. Let

$$\max_{x \varepsilon E_n} P(||\tilde{a} - x|| \leq \varepsilon) = P(||\tilde{a} - e|| \leq \varepsilon) = d < \infty$$

and assume inequality (261) is valid.

Theorem 45. If condition A is satisfied, then the optimal control gives the value

$$\begin{aligned}\beta^o &= \min_g \max_z \{P(||\tilde{a} - z|| \leq \varepsilon) - g'z + L||g'S||\} \\ &= P(||\tilde{a} - z^o|| \leq \varepsilon) - g_o'z^o + L||g_o'S|| \quad ,\end{aligned}$$

to the functional (270).

The optimal control $u^o(\cdot)$ is determined by relation (239) with $\hat{g} = g_o$.

The assertion follows from Theorem 44.

Remarks. 1) By virtue of Theorem 44, we have

$$d - \Delta_1(e) \leq \beta^o \leq d \quad .$$

From (266), it follows that $g_o'e > 0$.

2) In order that the set $Q(\varepsilon,\beta)$ be convex for any $\varepsilon \geq 0$, $\beta \varepsilon [0,1]$, it is necessary that the probability density function of the stochastic vector $\tilde{a}$ be unimodal

(has a single maximum). For a one-dimensional vector $\tilde{a}$, the property of unimodality is a necessary and sufficient condition for convexity (simple-connectivity) of the set $Q(\varepsilon,\beta)$ for $\varepsilon \geq 0$, $\beta\varepsilon[0,1]$.

If the probability density of the vector $\tilde{a}$ is not unimodal, but the set is vertex bounded, then for each β, $0 \leq \beta \leq 1$, there exists an ε_o such that the set $Q(\varepsilon,\beta)$ is convex for $\varepsilon \geq \varepsilon_o$.

The set $Q(\varepsilon,\beta)$ is closed if the probability density is continuous.

If condition A is not satisfied, then two paths of solution are possible: 1) to minimize $M||x||$ and to apply Chebyshev's inequality; 2) to replace the probability density function for $\tilde{a}$ by another for which condition A is satisfied.

Among other statistical problems studied from the position of functional analysis, we note especially the problem of minimizing the mean-value of the time for the trajectory to hit the origin [74g].

§16. On Several Differential Games

Below we study pursuit problems in linear systems [1,108b] using tools of functional analysis.

1. Reduction of the Problem of Programmed Pursuit to a Functional Problem

(1) Problem Statement. Let the equations

$$\frac{dx}{dt} = A_1(t)x + B_1(t)u + f_1(t) \quad ,$$

$$\frac{dy}{dt} = A_2(t)y + B_2(t)v + f_2(t) \quad , \qquad (271)$$

be given in E_n. Here $u(\cdot)$, $v(\cdot)$ are controls satisfying the conditions

$$u(\cdot)\varepsilon U_p^{\ell_1}(\cdot) \quad , \quad v(\cdot)\varepsilon U_p^{\ell_2}(\cdot) \quad . \tag{272}$$

We impose conditions on the matrices $A_i(t)$, $B_i(t)$ and the vector functions $f_i(t)$, which ensure representation of the solutions by Cauchy's formula. We call x the pursuer, y the evader. If the controls assume values $u(t,\theta)$, $v(t,\theta)$ at time t, $t \leq \theta \leq s$, we denote the positions of x,y at time s by $x(t,s,u(\cdot))$, $y(t,s,v(\cdot))$. Let

$$x(t,t,u(\cdot)) = x(t) \quad , \quad y(t,t,v(\cdot)) = y(t) \quad .$$

Problem A. To find the smallest time $s = t_1$ and the corresponding controls $u^1(\cdot)$, $v^1(\cdot)$ for which

$$||x(t,t_1,u^1(\cdot)) - y(t,t_1 v^1(\cdot))||$$

$$= \min_{s\geq t} \max_{v(\cdot)\varepsilon U_p^{\ell_2}(\cdot)} \min_{u(\cdot)\varepsilon U_p^{\ell_1}(\cdot)} ||x(t,s,u(\cdot)) - y(t,s,v(\cdot))||. \tag{273}$$

Problem B. To determine the smallest time $s = t_2$ and the corresponding controls $u^2(\cdot)$, $v^2(\cdot)$ when

$$||x(t,t_2,u^2(\cdot)) - y(t,t_2,v^2(\cdot))||$$

$$= \min_{s\leq 1} \min_{u(\cdot)\varepsilon U_p^{\ell_1}(\cdot)} \max_{v(\cdot)\varepsilon U_p^{\ell_2}(\cdot)} ||x(t,s,u(\cdot)) - y(t,s,v(\cdot))||. \tag{274}$$

The solutions of these problems, the functions

$u^i(\cdot)$, $v^i(\cdot)$, we call <u>optimal controls</u>. We introduce the sets

$$\Omega_x(t,s) = \{x: \quad x = x(t,s,u(\cdot))/x(t,t,u(\cdot)) = x(t) \ , \quad u(\cdot)\varepsilon U_p^{\ell_1}(\cdot)\} \ ,$$

$$\Omega_y(t,s) = \{y: \quad y = y(t,s,v(\cdot))/y(t,t,v(\cdot)) = y(t) \ , \quad v(\cdot)\varepsilon U_p^{\ell_2}(\cdot)\} \ .$$

We set

$$||x(t,s,u(\cdot)) - y(t,s,v(\cdot))|| = \ell_s(x,y) \ ,$$

$$\beta(s) = \beta_s(\Omega_y,\Omega_x) = \sup_{y\varepsilon\Omega_y(t,s)} \ \inf_{x\varepsilon\Omega_x(t,s)} \ \rho_s(x,y) \ ,$$

$$\delta_x(s) = \sup_{y\varepsilon\Omega_y(t,s)} \rho_s(x,y) \ .$$

The quantity $\beta(s)$ is the distance of the set $\Omega_y(t,s)$ from the set $\Omega_x(t,s)$. The quantity $\delta_x(s)$ is the distance of the set $\Omega_y(t,s)$ from the point x. Consequently, problem A is a problem defining the least time when the sets $\Omega_y(t,s)$ and $\Omega_x(t,s)$ assume their minimal distance. Problem B is the problem describing the least time and the function $u(\cdot)$ when the point x is at a minimal distance from the set $\Omega_y(t,s)$.

Below it is assumed that in system (271) we have $f_1(t) = f_2(t) \equiv 0$, so that the defining equation (110) is nonsingular for almost all t on the considered time interval. This assumption (see §7.8) guarantees the uniqueness of the optimal controls $u^i(\cdot)$, $v^i(\cdot)$.

(2) Optimal Controls. We introduce the fundamental matrices $F(x,\tau)$, $\Phi(x,\tau)$ of the solutions of the homogeneous equations corresponding to (271), and we set

$$\int_t^s F(s,\tau)B_1(\tau)u(\tau)d\tau = Su(\cdot) \quad ,$$

$$-\int_t^s \Phi(s,\tau)B_2(\tau)v(\tau)d\tau = Pv(\cdot) \quad ,$$

$$c = F(s,t)x(t) - \Phi(s,t)y(t)$$

$$-\int_t^s [F(s,\tau)f_1(\tau) - \Phi(s,\tau)f_2(\tau)]d\tau \quad .$$

It is not difficult to see that the vector $z = x-y$ at time s satisfies the relation

$$z = Su(\cdot) + Pv(\cdot) + c \quad . \tag{275}$$

Theorem 46. The minimal deviation of the sets $\Omega_y(t,s)$ and $\Omega_x(t,s)$ equals

$$\beta^o(t_1) = \min_{s\leq 1 \quad ||g||\leq 1} \{g'c + \ell_2||g'P|| - \ell_1||g'S||\}$$
$$= g'(t_1)c + \ell_2||g'(t_1)P|| - \ell_1||g'(t_1)S|| . \tag{276}$$

If $\beta^o(t_1) > 0$, then the functions $u^1(\cdot)$, $v^1(\cdot)$, for which the distance $\beta(s)$ is minimal, are defined by the relations

$$g'(t_1)Su^1(\cdot) = \min_{u(\,)\varepsilon U_p^{\ell_1}(\cdot)} g'(t_1)Su(\cdot) ,$$
$$g'(t_1)Fv^1(\cdot) = \max_{v(\cdot)\varepsilon U_p^{\ell_2}(\cdot)} g'(t_1)Pv(\cdot) . \tag{277}$$

Proof. We fix $v(\cdot)$ and find

$$\beta(x,v(\cdot)) = \min_{x\varepsilon\Omega_x(t,s)} \rho_s(x,y) .$$

By virtue of Theorem 26, Eq. (275), and constraint (272), we have

$$\beta(s,v(\cdot)) = \max_{||g||\leq 1} \{g'c + g'Pv(\cdot) - \ell_1||g'S||\} .$$

Now, from (273) it follows that at each fixed moment s, the distance $\beta(s)$ equals

$$\beta(s) = \max_{||g||\leq 1} \{g'c + \ell_2||g'P|| - \ell_1||g'S||\} . \tag{278}$$

This means that formula (276) is valid. Relations (277) follow from Theorem 26 and Eq. (278).

We introduce the set

$$Q(s,\ell_1) = \{\xi:\ \ \xi = -Su(\cdot) - c,\ s \geq t,\ u(\cdot)\varepsilon U_p^{\ell_1}(\cdot)\} .$$

Let

$$\max_{v(\cdot)\varepsilon U_p^{\ell_2}(\cdot)} ||Pv(\cdot) - \xi|| = \Psi(s,\xi) \tag{279}$$

and

$$\min_{\xi} \Psi(s,\xi) = \Psi(s,e) = d \quad . \tag{280}$$

Theorem 47. If $e \notin \operatorname{int} Q(s,\ell_1)$ for all $s \geq t$, then the minimal distance from the set $\Omega_y(t,s)$ to the point x equals

$$\delta_x^o(t_2) = \min_{s \geq t} \min_{\xi} \max_{g} \{\Psi(s,\xi) + g'[c + \xi] - \ell_1 ||g'S||\}$$

$$= ||Pv^1(\cdot) - \xi(t_2)|| + g'(t_2)[c + \xi(t_2)] - \ell_1 ||g'(t_2)S|| . \tag{281}$$

We have the estimate

$$d \leq \min_{x \varepsilon \Omega_x(t,s)} \delta_x(s) \leq d + \max_{||g|| \leq 1} \{g'[c+e] - \ell_1 ||g'S||\}. \tag{282}$$

The function $u^2(\cdot)$ is defined by the relation

$$g'(t_2)Su^2(\cdot) = \min_{u(\cdot) \varepsilon U_p^{\ell_1}(\cdot)} g'(t_2)Su(\cdot) \quad .$$

The function $v^2(\cdot)$ satisfies the condition

$$||Pv^2(\cdot) - \xi(t_2)|| = \Psi(t_2, \xi(t_2)) \quad .$$

Proof. We consider the distance $\delta_x(s)$ at the fixed moment $s = \tau$. We have

$$\delta_x(\tau) = \max_{v(\cdot) \varepsilon U_p^{\ell_2}(\cdot)} ||Su(\cdot) - Pv(\cdot) + c|| \quad .$$

We set $Su(\cdot) + c = -z$. According to (279), we have

$$\delta_x(\tau) = \Psi(\tau,z) \quad .$$

Clearly, $\Psi(\tau,z)$ is convex in z. Therefore, applying Theorem 44 we have

$$\min_{x\varepsilon\Omega_x(t,\tau)} \delta_x(\tau) = \max_g \min_z \{\Psi(\tau,z) + g'[c + z] - \ell_1||g'S||\}.$$

By virtue of (274), we obtain (281) from the last relation.

We show (282). We have

$$\min_{x\varepsilon\Omega_x(t,\tau)} \delta_x(\tau) = \min_z \max_{||g||\leq 1} \{\Psi(\tau,z) + g'[c+z] - \ell_1||g'S||\}$$

$$\leq \max_{||g||\leq 1} \{\Psi(\tau,e) + g'[c + e] - \ell_1||g'S||\}$$

$$= \Psi(\tau,e) + \max_{||g||\leq 1} \{g'[c + e] - \ell_1||g'S||\} \quad .$$

Consequently, the right-side of the inequality (282) is valid. The left inequality of (282) follows from (280). The relations for $u^2(\cdot)$, $v^2(\cdot)$ are obtained from (279), (281). This proves the assertion.

Remark. The case $e \varepsilon$ int Q is studied in 5) of paragraph 1(3), §15.

We formulate a problem which is closely connected with the preceding one. Let numbers ε_1, ε_2 be given.

Problem A_1. Determine the shortest time $t_1(\varepsilon_1)$ and functions $u^1(\cdot)$, $v^1(\cdot)$, such that the distance $\beta(s) = \varepsilon_1$.

Problem B_1. Determine the shortest time $t_2(\varepsilon_2)$ and functions $u^2(\cdot)$, $v^2(\cdot)$, such that the distance $\delta_x(s) = \varepsilon_2$.

It is clear that solutions to problems A_1, B_1 exist only when $\varepsilon_1 \geq \beta^o(t_1)$, $\varepsilon_2 \geq \delta_x^o(t_1)$. From Theorem 46 it follows that the solution of problem A_1 gives the smallest number $s = \tau^o$, satisfying the inequality

$$\max_{||g||=1} \{g'c + \ell_2||g'P|| - \ell_1||g'S|| - \varepsilon_1||g||\} \leq 0 \ . \qquad (283)$$

If $\tilde{g}$ is a solution of problem (283) for $s = \tau^o$, then the optimal controls in this problem are determined from relations (277) with $g = \tilde{g}$. The solution approach to problem B_1 is given by Theorem 47.

Problem A_1, for $\varepsilon_1 = 0$, is considered in [42]. In [74i], this last variant of problem A_1 is called the problem of absorption. Below we will confine ourselves to this last variant. In paragraph (3) we establish the connection between the problem of absorption and the problem of [62]. We formulate this latter problem.

Problem C. Let $T(u,v)$ be the time at which the point $x(t,s,u(\cdot))$ reaches $y(t,s,v(\cdot))$ using the controls $u(\cdot)$, $v(\cdot)$. It is required to determine $u^o(\cdot)$, $v^o(\cdot)$ for which

$$T(u^o,v^o) = T^o = \max_{v(\cdot)\varepsilon U_p^{\ell_2}(\cdot)} \ \min_{u(\cdot)\varepsilon U_p^{\ell_1}(\cdot)} T(u,v) \ .$$

(3) Conditions for the Equivalence of Problems A, B and A_1, C. It isn't difficult to see that the inequalities

$$\alpha)\ \beta^o(t_1) \leq \delta_x^o(t_1) \quad , \quad \beta)\ T^o \leq t_1(0) \ . \qquad (284)$$

hold.

Let values $\beta(t_1)$, $\delta_x(t_1)$ of the distances $\beta(s)$, $\delta_x(s)$ be assumed at $s = t_1$, using functions $u^1_{t_1}(\cdot)$, $v^1_{t_1}(\cdot)$ and $u^2_{t_1}(\cdot)$, $v^2_{t_1}(\cdot)$, respectively.

Theorem 48. If

$$\max_{y \in \Omega_y} \rho_{t_1}(x(t,t_1,u^1_{t_1}(\cdot)),\ v(t,t_1,v(\cdot))) = \beta(t_1) ,$$

where $\beta(t_1) = \delta_x(t_1)$ and

$$\begin{aligned} &\rho_{t_1}(x(t,t_1,u^1_{t_1}(\cdot)),\ y(t,t_1,v(\cdot))) \\ &\quad \leq \rho_{t_1}(x(t,t_1,u^1_{t_1}(\cdot)),\ y(t,t_1,v^1_{t_1}(\cdot))) \\ &\quad\quad \leq \rho_{t_1}(x(t,t_1,u(\cdot)),\ y(t,t_1,v^1_{t_1}(\cdot))) \end{aligned} \tag{285}$$

for all admissible $u(\cdot)$, $v(\cdot)$. If

$$\min_{x \in \Omega_x} \rho_{t_1}(x(t,t_1,u(\cdot)),\ y(t,t_1,v^2_{t_1}(\cdot))) = \delta_x(t_1),$$

where $\delta_x(t_1) = \beta(t_1)$ then

$$\begin{aligned} &\rho_{t_1}(x(t,t_1,u^2_{t_1}(\cdot)),\ y(t,t_1,v(\cdot))) \\ &\quad \leq \rho_{t_1}(x(t,t_1,u^2_{t_1}(\cdot)),\ y(t,t_1,v^2_{t_1}(\cdot))) \\ &\quad\quad \leq \rho_{t_1}(x(t,t_1,u(\cdot)),\ y(t,t_1,v^2_{t_1}(\cdot))) \end{aligned} \tag{286}$$

for all admissible $u(\cdot)$, $v(\cdot)$.

Proof. It is not difficult to see that

$$\max_{y\in\Omega_y} \min_{x\in\Omega_x} \rho_{t_1}(x,y) \leq \min_{x\in\Omega_x} \max_{y\in\Omega_y} \rho_{t_1}(x,y)$$

$$\leq \max_{y\in\Omega_y} \rho_{t_1}(x(t,t_1,u^1_{t_1}(\cdot)), y(t,t_1,v(\cdot))) .$$

Hence we have

$$\beta(t_1) \leq \min_{x\in\Omega_x(t,t_1)} \max_{y\in\Omega_y(t,t_1)} \rho_{t_1}(x,y) \leq \beta(t_1) ,$$

which means

$$\max_{y\in\Omega_y(t,t_1)} \min_{x\in\Omega_x(t,t_1)} \rho_{t_1}(x,y) = \min_{x\in\Omega_x(t,t_1)} \max_{y\in\Omega_y(t,t_1)} \rho_{t_1}(x,y),$$

i.e. $\beta(t_1) = \delta_x(t_1)$ and inequality (285) holds. We prove the second part of the assertion. Since

$$\min_{x\in\Omega_x} \max_{y\in\Omega_y} \rho_{t_1}(x,y) \geq \max_{y\in\Omega_y} \min_{x\in\Omega_x} \rho_{t_1}(x,y)$$

$$\geq \min_{x\in\Omega_x} \rho_{t_1}(x(t,t_1,u(\cdot)), y(t,t_1,v^2_{t_1}(\cdot))) ,$$

then we have

$$\delta_x(t_1) \geq \max_{y\in\Omega_y} \min_{x\in\Omega_x} \rho_{t_1}(x,y) \geq \delta_x(t_1) .$$

This means $\delta_x(t_1) = \beta(t_1)$, and inequality (286) holds. The theorem is proved.

We let $\tilde{u}(\cdot)$, $\tilde{v}(\cdot)$ denote the solution of the absorption problem.

Theorem 49. Let $t_1(0) < +\infty$ and

$$\Lambda(x,\tilde{v}(\cdot)) = \max_{||g||=1} \{g'[c + Pv(\cdot)] - \ell_1||g'S||\} > 0 ,$$

$$s < t_1 .$$

Then $t_1(0) = t^o$ and $\tilde{u}(t,s) = u^o(t,s)$, $\tilde{v}(t,s) = v^o(t,s)$, $s \geq t$.

Proof. By assumption, $t_1(0)$ is the first zero of the function $\Lambda(s,\tilde{v}(\cdot))$, and $\Lambda(s,\tilde{v}(\cdot)) > 0$ for $s\varepsilon[t_1,t_1(0))$. Consequently, $T^o \geq t_1(0)$. But, by virtue of (284), β) is possible only if we have the opposite inequality. This means $t_1(0) = t^o$.

Since the functions $u^o(\cdot)$, $v^o(\cdot)$, $\tilde{u}(\cdot)$, $\tilde{v}(\cdot)$ have a uniquely determined form by the conditions of the problem (cf. part (1)), $\tilde{u}(t,s) \equiv u^o(t,s)$, $\tilde{v}(t,s) \equiv v^o(t,s)$. The assertion is proved.

Theorem 50. If $T^o < \infty$, then

$$\Lambda(T^o) = \max_{||g||=1} \{g'c + \ell_2||g'P|| - \ell_1||g'S||\} \leq 0 .$$

If θ is the first moment for which $\Lambda(\theta,\tilde{v}(\cdot)) \leq 0$, $\theta < \infty$, and $\Lambda(T) \leq 0$ for some $t\varepsilon[t,\theta]$, then $\theta = T^o$ and $\tilde{u}(t,s) \equiv u^o(t,s)$, $\tilde{v}(t,s) = v^o(t,s)$ for $s\varepsilon[t,\theta]$.

Proof. By virtue of the conditions of the theorem $t_1(0) \leq T^o$. But from (284), β), it follows that only the equality $t_1(0) = T^o$ is possible. Then $u^o(t,s) \equiv \tilde{u}(t,s)$, $v^o(t,s) \equiv \tilde{v}(t,s)$ for $s\varepsilon[t,T^o]$. The second part of the assertion is obvious.

2. The Optimal Pursuit of a Group of Targets

In the space E_n, let the motion of a group of m points

$y_1, \ldots, y_m$ be given at time $s = t_1$ by the equation

$$y_i(S_i v_i(\cdot) + c_i) \quad , \quad v_i(\cdot) \varepsilon U_p^{\ell_i}(\cdot) \quad . \tag{287}$$

Here, the transformations S^i and the vectors c^i have the same meanings as S and c have in relation (275). We call the points y_i <u>evaders</u>. We call the point x a <u>pursuer</u>. Its motion is given by Eq. (230), where c is a fixed vector in E_n, $u(\cdot) \varepsilon U_p^{\ell}(\cdot)$. Let numbers $\alpha_i \geq 0$ be given. The quantity $\varepsilon_i = \alpha_i ||x - y_i||$ will be called the <u>shield</u> of the point y_i. Naturally, the quantity

$$\sum_{i=1}^{m} \alpha_i ||x - y_i|| \tag{288}$$

characterizes the shield of the group. Let it be required to find controls $u^o(\cdot)$, $v_i^o(\cdot)$, and a number δ^o such that

$$\sum_{i=1}^{m} \alpha_i ||x(t, t_1, u^o(\cdot)) - y_i(t, t_1, v_i^o(\cdot))||$$

$$= \max_{v_i(\cdot) \varepsilon U_p^{\ell_i}(\cdot)} \min_{u(\cdot) \varepsilon U_p^{\ell}(\cdot)} \sum_{i=1}^{m} \alpha_i ||x - y_i|| = \delta^o \ . \tag{289}$$

<u>Remark</u>. Here we do not cite other problems in the pursuit of a group of targets (see [65]).

Let

$$\min_{z} \sum_{i=1}^{m} \alpha_i ||c - c_i - S_i v_i(\cdot) - z||$$

$$= \sum_{i=1}^{m} \alpha_i ||c - c_i - S_i v_i(\cdot) - \tilde{e}|| = \tilde{d}$$

and

$$\max_{||g||=1} \{g'\tilde{e} - \ell||g'S||\} = \Delta(\tilde{e}) \geq 0 \quad .$$

Theorem 51. The optimal controls $u^o(\cdot)$, $v_i^o(\cdot)$ give the functional (288) the value

$$\delta^o = \max_g \max_{||v_i||=1} \min_z \{\sum_{i=1}^m \alpha_i(\gamma_i'[c - c_i - z] + \ell_i||\gamma_i'S_i||)$$

$$+ g'z - \ell||g'S||\} = \sum_{i=1}^m \alpha_i(\gamma_{io}'[c - c_i - z^o] + \ell_i||\gamma_{io}'S_i||)$$

$$+ g_o'z^o - \ell||g_o'S|| \quad . \qquad (290)$$

The functions $u^o(\cdot)$, $v_i^o(\cdot)$ are determined by the equations

$$g_o'Su^o(\cdot) = \min_{u(\cdot)\varepsilon U_p^\ell(\cdot)} g_o'Su(\cdot) \quad ,$$

$$\gamma_{io}'S_iv_i^o(\cdot) = \max_{v_i(\cdot)\varepsilon U_p^{\ell_i}(\cdot)} \gamma_{io}'S_iv_i(\cdot) \quad . \qquad (291)$$

Proof. We fix functions $v_i(\cdot)$ and find

$$\delta(v_1,\ldots,v_m) = \min_{u(\cdot)\varepsilon U_p^\ell(\cdot)} \sum_{i=1}^m \alpha_i||x - y_i|| \quad .$$

From (287), (230), and Theorem 44, it follows that

$$\delta(v_1,\ldots,v_m)$$

$$= \max_z \min_g \{\sum_{i=1}^m \alpha_i||c - c_i - S_iv_i(\cdot) - z|| + g'z - \ell||g'S||\} \quad . \qquad (292)$$

But

$$||c - c_i - S_i v_i(\cdot) - z|| = \max_{||\gamma_i||\leq 1} \gamma_i [c - c_i - S_i v_i(\cdot) - z].$$

Therefore,

$$\delta(v_1, \ldots, v_m)$$

$$= \max_{g, ||\gamma_i||=1} \min_z \{ \sum_{i=1}^m \alpha_i \gamma_i' [c - c_i - S_i v_i(\cdot) - z] + g'z - \ell ||g'S|| \}.$$

From (289), it follows that

$$\delta^o = \max_{g, ||\gamma_i||=1} \min_z \{ \sum_{i=1}^m \alpha_i (\gamma_i' [c - c_i - z])$$

$$+ \max_{v_i(\cdot) \varepsilon U_p^{\ell_i}(\cdot)} \gamma_i' S_i v_i(\cdot)) + g'z - \ell ||g'S|| \} ,$$

i.e. Eq. (290) holds. We obtain relation (291) from (290), (292). The theorem is proved.

§17. Development of the Method of Increments for Optimization Problems in Gaming Situations

The introduction into control processes of two (or more) sides with conflicting (non-coincident) interests appreciably generalizes the theory of optimal processes, dividing it from those complicated ordinary problems with a single player. Although the statement of optimization problems in gaming situations requires that the control be of feedback type, there is more emphasis than before on solutions in programmed form. Conceptually, programmed control may be used by one side to construct its strategies (feedback controls).

Below we shall indicate how to apply the method of increments to obtain necessary conditions of optimality in problems with conflicting objectives.

Let the motion of an object be described by the equation

$$\frac{dx}{dt} = f(x,v,w), \quad x(0) = x_o, \quad t \varepsilon T = [0,t_1] . \qquad (293)$$

The two players control the perturbations v(t), w(t). Their controls are chosen from the class of piecewise-continuous functions (for v, p-measurable, for w, q-measurable) with values in U, a bounded set in (p+q)-dimensional space:

$$u(t) = \{v(t),w(t)\} \varepsilon U \quad , \quad t \varepsilon T \quad . \qquad (294)$$

We assume that the players act according to the functional

$$J(u) = J(v,w) = \phi(x(t_1)) \quad , \qquad (295)$$

where $\phi(x)$ is a given continuously-differentiable function. The controls $v^o(t)$, $w^o(t)$ are called optimal if they satisfy condition (294) and

$$J(v^o,w) \leq J(v^o,w^o) \leq J(v,w^o)$$

for all pairs $\{v^o(t),w(t)\} \varepsilon U$, $\{v(t), w^o(t)\} \varepsilon U$. Applying the method of increments (§1) to each of the inequalities

$$J(v^o,w^o) \leq J(v,w^o) \quad , \quad J(v^o,w^o) \geq J(v^o,w)$$

we obtain the following result.

Theorem 52. For the optimal controls $v^o(t)$, $w^o(t)$, and the corresponding trajectory $x^o(t)$, $t\varepsilon T$, in problem (293)-(295), at each moment we have

$$H(x^o(t),\psi(t),v^o(t),w) \geq H(x^o(t),\psi(t),v^o(t),w^o(t))$$

$$\geq H(x^o(t),\psi(t),v,w^o(t)), \quad t\varepsilon T \quad , \tag{296}$$

for all v,w such that $\{v^o(t),w\}\,\varepsilon\,U$, $\{v,w^o(t)\}\,\varepsilon\,U$. Here $H(x,\psi,v,w) = \psi' f(x,v,w)$, while $\psi(t)$ is the solution of the equation

$$\frac{d\psi}{dt} = - \frac{\partial H(x^o(t),\psi(t),v^o(t),w^o(t))}{\partial x} \quad ,$$

$$\psi(t_1) = - \frac{\partial\phi^{(x^o(t_1))}}{\partial x} \quad .$$

Corollary. If v(t), w(t) are constrained by

$$v(t)\varepsilon V \quad , \quad w(t)\varepsilon W \quad ,$$

where V,W are bounded sets in p,q-dimensional space, then the necessary conditions for optimality may be written in the form:

$$H(x^o(t),\psi(t),v^o(t),w^o(t)) = \max_{v\varepsilon V} H(x^o(t),\psi(t),v,w^o(t)) \quad ,$$

$$H(x^o(t),\psi(t),v^o(t),w^o(t)) = \min_{w\varepsilon W} H(x^o(t),\psi(t),v^o(t),w) \quad .$$

Commentary on Chapter VI

1. Two properties lie at the basis of the proof of Theorem 1: the formula for the increment of the functional (3) and the use of the special control increment (6). The idea of such an approach in the theory of optimal processes was first applied in [114a,b], although in the classical calculus of variations [16], Weierstrass conditions were obtained by the method of increments, while the "needle" variations (6) were used in [190a]. In [114b] results analogous to Theorem 1 were obtained under more stringent conditions.

2. In [32d], an analysis of the proof of Theorem 1 is given from which it follows that the scheme used in §1 allows us to prove the maximum principle for the case when the functions $f(x,y,t)$, $\partial f(x,y,t)/\partial x$ are continuous in x,u and measurable in t, while the class of admissible controls is taken as the measurable functions. The conditions imposed on the function $\phi(x)$ can be weakened. Differentiability may be replaced by a condition insuring the existence of a directional derivative.

We say that the function $\phi(x)$ has a derivative $\partial\phi/\partial g$ in the direction g at the point x if

$$\lim_{\alpha\to 0,\alpha>0} \frac{\phi(x + \alpha g) - \phi(x)}{\alpha}$$

exists and equals $\partial\phi/\partial g$. A convex function is differentiable in any direction [111]. In [48b], it is proved that the function $\phi(x) = \max_{y\in\Omega} f(x,y)$ is differentiable in any direction if $f(x,y)$ is a function which is continuous together with $\partial f(x,y)/\partial x$, and Ω is a closed set. The existence of a directional derivative allows us to write

$$\phi(x + \alpha g) - \phi(x) = \alpha \frac{\partial\phi(x)}{\partial g} + o(\alpha) \quad .$$

This equation may be used as a basis for those transformations which were produced in §1 with the expansion

$$\phi(x + \Delta x) - \phi(x) = \Delta x' \frac{\partial\phi(x)}{\partial x} + o(||\Delta x||) \quad .$$

As a result, we obtain a maximum principle in which the vector $\psi(t_1)$ will not be determined uniquely but will be analogous in form to Theorem 2. Incidentally, the latter theorem may be proved by this method. Details of the proof and a formulation of the corresponding results are omitted in this monograph because they are of interest only for a narrow class of specialists to whom they are already sufficiently well known.

3. In [32d], necessary conditions for optimality obtained from (4),(5) are presented. However, these results are poorly formulated and we may obtain stronger results for singular controls (cf. §5). Nevertheless, formulas (4) and (5), by virtue of their compactness, may turn out to be useful for other reasons, in particular for the establishment of sufficent conditions for optimality.

4. Theorem 1 is formulated for the problem of minimizing the functional (2). In problem (1),(2) is easily reduced to the problem of minimizing the functional

$$J(u) = \phi(x(t_1)) + \int_{t_0}^{t_1} f_{n+1}(x(t),u(t),t)dt \quad ,$$

over (1). For this it suffices to augment (1) by adding the equation

$$\frac{dx_{n+1}}{dt} = f_{n+1}(x,u,t), \quad x_{n+1}(t_0) = 0 \quad , \tag{297}$$

and considering the problem of minimizing the quantity

$$\phi(x(t_1)) + x_{n+1}(t_1) \tag{298}$$

for the system (1), (297). It is not difficult to assert that from the special form of (1), (297), it follows that $\dot{\psi}_{n+1}(t) \equiv 0$, while from (298) we have $\psi_{n+1}(t_1) = -1$ by virtue of Theorem 1, i.e. $\psi_{n+1}(t) = -1$, $t\varepsilon T$.

5. Lemma 4 may be proved without conditions (2),(3).

6. Theorem 5 may be generalized in several directions. First of all, the smoothness conditions on $\phi(x)$, $g(x)$ may be replaced by conditions of quasiconvextity and

lower-semicontinuity. Secondly, several constraints of the type (18) may be considered. Finally, constraints of the type (18) may be imposed at interior points of the interval T.

7. The terminology "variational derivative of second order" is not completely correct; however, we retain it in the sequel for the sake of ease of writing.

8. In [32e], a new form of the necessary conditions for optimality is developed for linear systems depending on the controls and their derivatives. This result is of interest for the study of the nonlinear case.

9. Formulation of optimal control problems in the language of integral equations possesses advantages over the analagous formulation in terms of differential equations in the sense that Eqs. (29)-(30) completely describe the motions $x(t)$, $\psi(t)$ and eliminate the necessity to compute boundary conditions. The particular form of the criteria (38) is clearly not a restriction.

10. Analyzing the theorems of §3, it is possible to conclude that the form (28)-(31) of the necessary conditions of optimality is invariant with respect to a wide class of systems described by functions of a single independent variable. Theorem 9 shows that an analagous form of necessary conditions occurs for systems described by the equations of mathematical physics. It is possible to show that the form (43)-(45) is retained for very general systems of integral equations for functions of two independent variables.

11. Passage from the operator of the left part of expression (43) to the operator of the left part of expression (44) corresponds to the rule given in (29),(30) for the case of functions of a single variable and here extends the known results [119] to functions of two variables.

12. The method of proof based on the study of the increment of functionals allows us to very simply obtain other conditions of optimality differing from the maximum principle. For simplicity, our discussion is confined to problem (1),(2) and we assume in addition to the conditions of Theorem 1, that the function $f(x,y,t)$ is differentiable in u, while the set U is convex. Then, from (9) it directly follows that the optimal control $u^o(t)$, $t \varepsilon T$, satisfies the condition

$$\frac{\partial H'(x^o(t),\psi(t),u^o(t),t)}{\partial u} u^o(t) = \max_{u\varepsilon U} \frac{\partial H'(x^o(t),\psi(t),u^o(t),t)}{\partial u} u \, . \tag{299}$$

Hence, necessary conditions of optimality are obtained in the form

$$\frac{\partial H(x^o(t),\psi(t),u^o(t),t)}{\partial u} = 0 \quad , \quad t\varepsilon T \quad , \tag{300}$$

if $u^o(t)\varepsilon$ int U, $t\varepsilon T$. Formula (3) allows us to find conditions for the maximum in problem (1),(2) in the case when the class of admissible controls is a convex set of piecewise-continuous, r-dimensional vector functions u(t), defined on T. Retaining the assumptions on the continuity of $\partial f/\partial u$, from (3), in place of (299) we obtain

$$\int_{t_o}^{t_1} \frac{\partial H'(x^o(t),\psi(t),u^o(t),t)}{\partial u} u^o(t)dt$$
$$\geq \int_{t_o}^{t_1} \frac{\partial H'(x^o(t),\psi(t),u^o(t),t)}{\partial u} u(t)dt$$

for all $u(\cdot)\varepsilon U(\cdot)$. In the case $u^o(\cdot)\ \varepsilon$ int $U(\cdot)$, conditions (300) are obtained. Similar conditions may be obtained for all problems considered in this chapter.

13. In [36c], Theorem 10 is proved for systems with a delay. It should be noted that in [105] the conditions obtained from Theorem 10, called necessary conditions of optimality, are clarly obvious and are satisfied for any trajectories of the system (1),(8).

14. The methods developed in §5 are easily extended to the case when

$$\frac{\partial^\ell \phi(x^o(t_1))}{\partial_{x_{j_1}} \cdots \partial_{x_j}} = 0 \quad , \quad \frac{\partial \phi^{k+1}\phi(x^o(t_1))}{\partial x_{j_1} \cdots \partial x_{j_{k+1}}} \neq 0 \quad .$$

For this, the dimensionality is increased and the number of arguments of the auxiliary functions $m_{j_1\cdots j_\ell}(t_{j_1},\ldots,t_{j_\ell})$ and the order of the equations for these functions are also increased.

15. If we assume convexity of the set U and differentiability of f(x,y,t) in u, then from condition (52) it follows that

$$m_{ij}(t,t)\ \frac{\partial f_i(x^o(t),u^o(t),t)}{\partial u_\mu} \cdot \frac{\partial f_j(x^o(t),u^o(t),t)}{\partial u_\mu}$$

$$\cdot\ (u^*_\nu - u^o_\nu(t))'(u^*_\mu - u^o_\mu(t)) \leq 0 \quad ,$$

$$t\varepsilon T \quad , \quad u^*\varepsilon U \quad , \quad \nu = 1,\ldots,r;\ \mu = 1,\ldots,r \quad .$$

16. If $u^o(t)\ \varepsilon$ int U(t), $t\varepsilon T$, then along the optimal control $u^o(t)$, the quadratic form with matrix

$$m_{ij}(t,t)\ \frac{\partial f_i(x^o(t),u^o(t),t)}{\partial u_\nu} \quad \frac{\partial f_j(x^o(t),u^o(t),t)}{\partial u}$$

is nonnegative.

17. The cases when the set U is convex or open in Theorem 12 are studied in the same way as was done in the remarks following Theorem 11.

18. The concept of a singular optimal control first appeared in [114b]. There, by a simple example it was shown how to find a singular control beginning from its definition. Methods for the determination of singular controls based on direct use of the definition of such controls were then considered in a series of works. The study [63a] apparently was the first in which necessary conditions for the optimality of singular controls were given. This work is of interest in connection with the introduction of new control variations into the arsenal of techniques in the theory of optimal processes. The role of the "needle" variations (6) in the theory of necessary conditions of optimality is well known. If it is possible to consider "needle" variations as an analogue of the Dirac δ-function, then the variations used in [63a] are analagous to the derivative of the δ-function. The methods of [63a] were developed in [26,71,113].

Giving great value to the necessary conditions of optimality of singular controls obtained in the cited works, we note that the theory of necessary conditions in the singular case turns out to be analagous to those which arise in the classical calculus of variations in the study of minimal time problems in automatic control systems [109,124a]. The variations used in both cases assume the openness of the range of values of these functions. To overcome this constraint, which is essential for consideration of optimal regulator systems, the theory of the maximum principle was developed based on a direct study of optimization problems with arbitrary sets for the values of the variation functions. Afterwards, for several problems of the new circle, techniques of the classical calculus of variations were developed [92,176].

In this chapter, the method of study of singular controls does not require the openness of the set U. The scheme of the proof is based on the formula for the increment of scalar functions obtained in §5 and uses the variation (6), which is typical in the theory of nonsingular optimal control problems.

19. Interest in singular controls may be explained by several reasons. First, in the theory of optimization of aerospace devices, singular regimes frequently appear. The first nontrivial example of a singular control was found in [89], augmented by the examples presented in [26]. Second, optimal singular regimes, in several senses, are direct satellites of optimal chattering regimes. In fact, according to [38e] any optimization problem for which it is not known whether a solution exists, is replaced by an auxiliary problem. For example, consider problem (5.1)-(5.3). From this problem we move to problem (5.11)-(5.13) which trivially (Theorem 5.1) has a solution. However, to find the optimal control $v^{o}_{\nu\nu}(t)$, w^{o}_{ν}, $t \in T$, the maximum principle turns out to be ineffective since the control $w^{o}_{\nu}(t)$, $t \in T$, is always singular by virtue of property (5.16). Therefore, in the consideration of any complex optimization problem it may be necessary to consider singular controls.

20. The conditions of optimality for nonsingular controls are usually formulated with the help of the auxiliary vector function $\psi(t)$ (impulses, adjoint variables, Lagrange multipliers). In Theorem 13, for the formation of first order necessary conditions of optimality for singular controls, we introduce the additional matrix function $\Psi(t)$. If the dual variable $\psi(t)$ has

the property

$$\psi'(t)z(t) \equiv \text{const} \quad , \quad t\varepsilon T \quad ,$$

where z(t) is the solution of the variational equation

$$\dot{z} = \frac{\partial f(x(t),u(t),t)}{\partial x} z \quad ,$$

then $\Psi(t)$ satisfies the identity

$$z'(t)\Psi(t)z(t) = -\frac{1}{2}\psi'(t)\frac{\partial^2 f(x(t),u(t),t)}{\partial x^2} z'(t)z(t) + \text{const}, \qquad t\varepsilon T \quad ,$$

which, for systems linear in x and for singular controls reduces to the simple property

$$z'(t)\Psi(t)z(t) \equiv \text{const} \quad , \quad t\varepsilon T \quad .$$

21. The formula (3) for the increment allows us to obtain first-order necessary conditions for the optimality of singular controls expressed only through the variable $\psi(t)$. The idea of this approach is to use a control variation $\Delta u(t)$ for which $||\Delta x(t)|| \sim \varepsilon^2$ for $t \geq \theta + \varepsilon$ and $\Delta x(t) \equiv 0$ for $t_0 \leq t \leq \theta$. If the set U is open, then for this objective a suitable increment is used in [63a]. For an arbitrary set U, the required variation may not exist.

22. The convexity and openness of the set U in Theorem 13 may be dropped as was done in commentary 15, 16 to Theorem 11.

23. Theorem 14 imposes sufficiently strong restrictions on the class of admissible controls. However, this deficiency may be dropped in Theorem 15 which uses singular controls "rougher" than in Theorem 14.

24. Optimization problems with parameters are considered in the book [18d].

25. Basic investigations of linear control systems using the tools of functional analysis are carried out in the monographs [21d, 74n, 80]. In all these works, the starting point is the L-problem of moments. We have presented other paths to the solution of optimal control problems.

The minimax theorem, the theorem on the separation of convex sets, and the theorem on the existence of separating hyperplanes for convex surfaces studied in this monograph, allow us to bypass elementary considerations from linear algebra. It should be noted that in a series of cases application of the latter saves the duality approach, which is based on the problem of moments, from cumbersome transformations.

26. The minimax theorem (in its proof we follow [60]) was first used in the study of optimal processes in [33]. Also mentioned was the connection between game theory and the theory of optimal processes. Another approach to this question is given in [74].

27. The theorem on the separation of convex sets (our proof follows [60]) was the first result from functional analysis which was effectively applied in the theory of optimization of linear systems [12,138]. Later, this approach was developed in the works [83,88b,101,135, 194a]. A more general scheme was presented in [34f]. We note that the report [34d] and the works [34e,f] are based upon a result from the theory of linear inequalities.

28. The problem of moments in the theory of optimal processes was first studied in [74a,b]. This direction was further developed in [8b,c,21c,64a,140,189]. For a long time, only two-point minimal time problems with standard control constraints were considered. Beginning with [32b], in a series of works [32c,34a-c,81a] the L-problem was developed for a wider circle of problems. Some of these problems were considered in sections 3-7 of §11. It is of interest to note that in the first works, conditions of linear independence of the elements $x_1(\cdot),\ldots,x_n(\cdot)$ were constantly involved. The cases when these conditions are not satisfied is studied in [32a].

29. The theorem on the existence of a separating hyperplane for convex functions for the solution of problems of optimizing a convex functional was used in [34g].

30. Conditions for the imbeddability of convex sets were proved in [34f]. Extensions of these results to optimal control problems are also presented there.

31. The Neyman-Pearson lemma [12] in general form for optimization with two constraints on the control function is given in [33]. Similar problems are solved by other methods in [74k].

32. The possibility of transferring the method of increments to gaming-type problems is investigated in the report [65].

CHAPTER VII

Sufficient Conditions for Optimality. Uniqueness of Optimal Controls. Well-Posedness of the Statement of Several Problems in the Theory of Optimal Processes.

§1. Optimization Problems with Functionals which are Convex Relative to the State Variables.

1. The Inequality for the Increment of a Functional.

Let the trajectory of the system be described by the equation

$$\left.\begin{aligned} \frac{dx(t)}{dt} &= f(x(t), x(t-h), u, t), \quad t \varepsilon T, \quad x \varepsilon E_n, \quad u \varepsilon E_r, \\ x(t) &= \Phi(t), \quad t \varepsilon S_o, \quad h = h(x,u,t) \geq 0, \quad u(\cdot) \varepsilon D_1. \end{aligned}\right\} \tag{1}$$

We minimize the functional

$$J(u) = \phi(x(t_1)) + \int_{t_o}^{t_1} f_{n+1}(x(t), x(t-h)(x,u,t)), u(t), t)\,dt \tag{2}$$

along the trajectory of (1) under the additional constraint on the variable $x(t)$:

$$g(x(t),t) \leq 0 \quad , \quad t \varepsilon \sigma \subset T \quad . \tag{3}$$

We assume that 1) the functions $f(x,y,u,t)$, $\phi(x)$, $f_{n+1}(x,y,u,t)$, $h(x,u,t)$, $g(x,t)$, $\Phi(t)$ are defined and continuous together with the functions $\partial f/\partial x$, $\partial f/\partial y$, $\partial\phi/\partial x$, $\partial f_{n+1}/\partial x$, $\partial f_{n+1}/\partial y$, $\partial h/\partial x$, $\partial h/\partial u$, $\partial h/\partial t$, $\partial g/\partial x$, $\partial\Phi/\partial t$; 2) $h(x,u,t) \geq 0$; 3) along the trajectory $x(t)$ and the control $u(t)$, the function $\tau(t) = t - h(x(t),u(t),t)$ is strictly increasing, where $d\tau/dt(x,u,t) \geq \alpha > 0$ at points where the derivative exists.

We let $\Delta J(u)$ denote the increment $\Delta J(u) = J(u+\Delta u) - J(u)$ of the functional (2) along the control $u(\cdot)\varepsilon D_1$, resulting from the increment $\Delta u(t)$, $t\varepsilon T$, from the same class D_1. The trajectories of the equation corresponding to the controls $u(t)$, $\tilde{u}(t) = u(t) + \Delta u(t)$, are denoted by $x(t)$, $\tilde{x}(t) = x(t) + \Delta x(t)$. Let $d\mu(t)$ be some nonnegative measure concentrated on the set $\sigma_1 = \{t\colon\ g(x(t),t) \neq 0,\ t\varepsilon\sigma\}$. We introduce the function $\psi(t)$, $t\varepsilon T$, by the equation

$$\frac{d\psi(t)}{dt} = -\frac{\partial H(x,y,\psi,u,t)}{\partial x} - \left.\frac{\partial H(x(s),y(s),\psi(s),u(s),s)}{\partial y}\right|_{s=r(t)} \cdot \frac{dr(t)}{dt} + \left.\frac{\partial H'(x,y,\psi,u,t)}{\partial y} \cdot \frac{dx(x)}{ds}\right|_{s=\tau(t)} \cdot \frac{\partial h(x,u,t)}{\partial x} + \frac{\partial g(x,t)}{\partial x} \cdot \frac{d\mu(t)}{dt}\,,\quad t\varepsilon[t_0,t']\,,$$

$$\frac{d\psi(t)}{dt} = \frac{\partial H(x,y,\psi,u,t)}{\partial x} + \left.\frac{\partial H'(x,y,\psi,u,t)}{\partial y} \cdot \frac{dx(s)}{ds}\right|_{s=\tau(t)} \cdot \frac{\partial h(x,u,t)}{\partial x} + \frac{\partial g(x,t)}{\partial x} \cdot \frac{d\mu(t)}{dt}\,,\quad t\varepsilon[t',t_1],\quad \psi(t_1) = -\frac{\partial\phi(x(t_1))}{\partial x}$$

Here $H(x,y,\psi,u,t) = \psi' f(x,y,u,t) - f_{n+1}(x,y,u,t)$, $r(t)$ is the solution of the equation $t = \tau(r)$, $t' = t_1 - h(x(t_1), u(t_1),t_1)$, $y = x(t-h)$. If both trajectories $x(t)$, $\tilde{x}(t)$, generated by the controls $u(t)$, $\tilde{u}(t)$, satisfy condition (3), then the increment of the functional (2) satisfies the inequality

$$
\begin{aligned}
\Delta J(u) \geq & -\int_{t_o}^{t_1} \Delta_{\tilde{u}} H(x,y,\psi,u,t)\,dt \\
& - \int_{t_o}^{t_1} \Delta x'(t) \frac{\partial}{\partial x} \Delta_{\tilde{u}} H(x,y,\psi,u,t)\,dt \\
& - \int_{t_o}^{t_1} \Delta x'[\tau(\tilde{x},\tilde{u},t)] \frac{\partial}{\partial y} H(x,x[\tau(x,\tilde{u},t)],\psi,\tilde{u},t)\,dt \\
& + \int_{t_o}^{t_1} \Delta x'[\tau(x,u,t)] \frac{\partial}{\partial y} H(x,y,\psi,u,t)\,dt \\
& + \int_{t_o}^{t_1} \frac{\partial}{\partial y} H(x,x[\tau(x,\tilde{u},t)],\psi,u,t) \int_0^1 \left. \frac{dx(s)}{ds} \right|_{\substack{s=\tau(x,\tilde{u},t) \\ -[h(\tilde{x},\tilde{u},t) \\ -h(x,\tilde{u},t)]\theta}} \\
& \cdot d\theta \frac{\partial h'(x,u,t)}{\partial x} \Delta x(t)\,dt \\
& - \int_{t_o}^{t_1} \frac{\partial}{\partial y} H(x,u,\psi,u,t) \left. \frac{dx(s)}{ds} \right|_{s=\tau(t)} \frac{\partial h'(x,u,t)}{\partial x} \Delta x(t)\,dt \\
& + o_1(||\Delta x(t_1)||) + \int_{t_o}^{t_1} o_2(||\Delta x(t)|| + ||\Delta x[\tau(\tilde{x},\tilde{u},t)]||)\,dt \\
& + \int_{t_o}^{t_1} \frac{\partial}{\partial y} H(x,x[\tau(x,\tilde{u},t)],\psi,\tilde{u},t) \int_0^1 \left. \frac{dx(s)}{ds} \right|_{\substack{s=\tau(x,\tilde{u},t) \\ -[h(\tilde{x},\tilde{u},t) \\ -h(x,\tilde{u},t]\theta}} \\
& \cdot d\theta o_3(||\Delta x(t)||)\,dt - \int_{t_o}^{t_1} o_4(||\Delta x(t)|| + ||\Delta x[\theta(\tilde{x},\tilde{u},t)]||)\,dt \\
& + \int_{t_o}^{t_1} o_5(||\Delta x(t)||)\,d\mu(t) \quad . \qquad (4)
\end{aligned}
$$

Here the quantities o_1-o_5 satisfy the relations

$$\phi(x + \Delta x) - \phi(x) = \frac{\partial \phi'(x)}{\partial x} \Delta x + o_1(||\Delta x||) \quad ,$$

$$f_{n+1}(\tilde{x},\tilde{x}[\tau(\tilde{x},\tilde{u},t)],\tilde{u},t) - f_{n+1}(x,x[\tau(x,\tilde{u},t)],\tilde{u},t)$$

$$= \frac{\partial f'_{n+1}(x,x[\tau(x,\tilde{u},t)],\tilde{u},t)}{\partial x} \Delta x$$

$$+ \frac{\partial f'_{n+1}(x,x[\tau(x,\tilde{u},t)],\tilde{u},t)}{\partial y} (x[\tau(\tilde{x},\tilde{u},t)] - x[\tau(x,\tilde{u},t)])$$

$$+ o_2(||\Delta x(t)|| + ||\Delta x[\tau(\tilde{x},\tilde{u},t)]||) \quad ,$$

$$h(\tilde{x},\tilde{u},t) - h(x,\tilde{u},t) = \frac{\partial h'(x,\tilde{u},t)}{\partial x} \Delta x + o_3(||\Delta x||) \quad ,$$

$$H(\tilde{x},\tilde{x}[\tau(\tilde{x},\tilde{u},t)],\psi,\tilde{u},t) - H(x,x[\tau(x,\tilde{u},t)],\psi,\tilde{u},t)$$

$$= \frac{\partial H'(x,x[\tau(x,\tilde{u},t)],\psi,\tilde{u},t)}{\partial x} \Delta x(t)$$

$$+ \frac{\partial H'(x,x[\tau(x,\tilde{u},t)],\psi,\tilde{u},t)}{\partial y} (\tilde{x}[\tau(\tilde{x},\tilde{u},t)] - x[\tau(x,u,t)])$$

$$+ o_4(||\Delta x(t)|| + ||\Delta x[\tau(\tilde{x},\tilde{u},t)]||) \quad ,$$

$$g(x + \Delta x,t) - g(x,t) = \frac{\partial g(x,t)}{\partial x} \Delta x + o_5(||\Delta x||) \quad ,$$

$$H(x,y,\psi,u,t) = \psi' f(x,y,u,t) \quad .$$

The proof of inequality (4) is analagous to the scheme of obtaining the formula for the increment of a scalar function studied in §6.1, and presented in [36e]. A particular case of the inequality (4) is when $\tau(x,u,t) = t - h(t)$:

$$\Delta J(u) \geq -\int_{t_0}^{t_1} \Delta_{\tilde{u}} H(x,u,\psi,u,t)\,dt$$

$$- \int_{t_0}^{t_1} \frac{\partial}{\partial x} \Delta_{\tilde{u}} H'(x,y,\psi,u,t)\,\Delta x(t)\,dt$$

$$-\int_{t_0}^{t_1} \frac{\partial}{\partial y} \Delta_{\tilde{u}} H'(x,y,\psi,u,t)\; x(t - h(t))\,dt + o_1(||\Delta x(t_1)||)$$

$$+\int_{t_0}^{t_1} o_2(||\Delta x(t)|| + ||\Delta x(t - h(t))||)\,dt$$

$$- \int_{t_0}^{t_1} o_4(||\Delta x(t)|| + ||\Delta x(t - h(t))||)\,dt$$

$$+ \int_{t_0}^{t_1} o_5(||\Delta x(t)||)\,d\mu(t) \quad .$$

2. Sufficient Conditions of Optimality

Let the system (1) have the form

$$\left.\begin{aligned} \frac{dx(t)}{dt} &= f_1(x(t),x(t - h(t)),t) + f_2(u(t),t), \\ x(t) &= \Phi(t),\; t\varepsilon S_0 = \{t:\; t - h(t) \leq t_0,\; t\varepsilon T\}. \end{aligned}\right\} \quad (5)$$

Along trajectories of system (5), we consider the functional

$$J(u) = \phi(x(t_1)) + \int_{t_o}^{t_1} [f_{1,n+1}(x(t), x(t - h(t)), t) + f_{2,n+1}(u(t), t)]dt. \quad (6)$$

We call the function $\alpha(x)$ strongly convex with constant N if it is differentiable and the quantity $o(||\Delta x||)$ in the expansion

$$\alpha(x + \Delta x) - \alpha(x) = \frac{\partial \alpha'(x)}{\partial x} \Delta x + o(||\Delta x||)$$

satisfies the inequality

$$o(||\Delta x||) \geq N||\Delta x||^2 \quad , \quad N \geq 0 \quad .$$

Theorem 1. Let the following conditions be satisfied: 1) the function $f_{1,n+1}(x,y,t)$ is strongly convex in $\{x,y\}$ with constant N; 2) the functions $\phi(x)$, $g(x,t)$ are convex in x; 3)

$$\left\|\frac{\partial f_1(x,y,t)}{\partial x}\right\| \leq L_1, \quad \left\|\frac{\partial f_1(x,y,t)}{\partial y}\right\| \leq L_1, \quad \left\|\frac{\partial f_{1,n+1}(x,u,t)}{\partial x}\right\| \leq L_1,$$

$$\left\|\frac{\partial f_{1,n+1}(x,y,t)}{\partial y}\right\| \leq L_1, \quad \left\|\frac{\partial \phi(x)}{\partial x}\right\| \leq L_1, \quad \left\|\frac{\partial g(x,t)}{\partial x}\right\| \leq L_2,$$

$$\left\|\frac{\partial^2 f_1(x,y,t)}{\partial x^2}\right\| \leq L_3,$$

$$\left\|\frac{\partial^2 f_1(x,y,t)}{\partial x\, \partial y}\right\| \leq L_3, \quad \left\|\frac{\partial^2 f_1(x,y,t)}{\partial y^2}\right\| \leq L_3, \quad \left|\frac{dh(t)}{dt}\right| \leq 1 - \alpha, \alpha > 0.$$

Then, in problem (5),(6),(3), for $N \geq M_1$, where

$$M_1 = \frac{1}{2} L_3\left[L_1(t_1 - t_0)\left(1 + \frac{1}{\alpha}\right) + L_2L_4 + L_1\right]$$

$$\cdot \exp\left\{L_1\left(1 + \frac{1}{\alpha}\right)(t_1 - t_0)\right\} \quad ,$$

each admissible control $u(t)$, $t\varepsilon T$, from the class D is optimal if $g(x,t) \leq 0$, $t\varepsilon\sigma$, and $u(t)$ satisfies the condition

$$H(x(t),x(t - h(t)),\psi(t),u(t),t)$$

$$= \max_{u\varepsilon U} H(x(t),x(t - h(t)),\psi(t),u,t) \quad ,$$

where the function $\psi(t)$, $t\varepsilon T$, is a solution of the equation

$$\left.\begin{aligned}
\frac{d\psi(t)}{dt} &= - \frac{\partial(x,y,\psi,u,t)}{\partial x} \\
&\quad - \left.\frac{\partial H(x(s),y(s),\psi(s),u(s),s)}{\partial y}\right|_{s=r(t)} \cdot \frac{dr(t)}{dt} \\
&\quad + \frac{\partial g(x(t),t)}{\partial x} \cdot \frac{d\mu(t)}{dt} \quad , \quad t\varepsilon[t_0,t'] \quad , \\
\frac{d\psi(t)}{dt} &= - \frac{\partial H(x,y,\psi,u,t)}{\partial x} \\
&\quad + \frac{\partial g(x(t),t)}{\partial x} \cdot \frac{d\mu(t)}{dt} \quad , \quad t\varepsilon[t',t_1] \quad , \\
\psi(t_1) &= - \frac{\psi\phi(x(t_1))}{\partial x} \quad ,
\end{aligned}\right\} \quad (7)$$

and the function of bounded variation $\mu(t)$ is such that $\int_\sigma d\mu(t) \leq L_4$.

Proof. For the problem (5),(6),(3), inequality (4) assumes the form

$$\Delta J(u) \geq - \int_{t_o}^{t_1} \Delta_{\tilde{u}} H(x,y,\psi,u,t)\,dt + o_1(||\Delta x(t_1)||)$$
$$+ \int_{t_o}^{t_1} o_2(||\Delta x(t)|| + ||\Delta x(t - h(t))||)\,dt$$
$$- \int_{t_o}^{t_1} o_4(||\Delta x(t)|| + ||\Delta x(t - h(t))||)\,dt$$
$$+ \int_{t_o}^{t_1} o_5(||\Delta x(t)||)\,d\mu(t).$$

According to the conditions of the theorem

$$\Delta_{\tilde{u}} H(x,y,\psi,u,t) \leq 0,\ o_1(||\Delta x(t_1)||) \geq 0,\ o_5(||\Delta x(t)||) \geq 0,$$

$$o_2(||\Delta x(t)|| + ||\Delta x(t-h(t))||) \geq N[||\Delta x(t)|| + ||\Delta x(t-h(t))||]^2.$$

The quantity o_4 satisfies the inequality

$$o_4 \leq \frac{1}{2}\,||\psi(t)||L_3[||\Delta x(t)|| + ||\Delta x(t - h(t))||]^2 \quad .$$

From the equation for $\psi(t)$, we obtain as an estimate for $||\psi(t)||$:

$$||\psi(t)|| \leq \left[L_1(t_1 - t_o)\left(1 + \frac{1}{\alpha}\right) + L_2 L_4 + L_1\right]$$
$$\cdot \exp\left\{L_1\left(1 + \frac{1}{\alpha}\right)(t_1 - t_o)\right\} \quad .$$

Substituting these estimates into the expression for $\Delta J(u)$ and taking into account the inequality $N \geq M_1$, we obtain $\Delta J(u) \geq 0$, i.e. $u(t)$, $t \in T$, is an optimal control.

3. Systems Linear in the State

We consider a particular case of system (5):

$$\left.\begin{aligned} \frac{dx(t)}{dt} &= A(t)x(t) + A_1(t)x(t-h(t)) + b(u(t),t), \\ x(t) &= \Phi(t) \quad , \quad t \in S_o \quad . \end{aligned}\right\} \quad (8)$$

In analogy with Theorem 1, we obtain the following result.

Theorem 2. Let the functions $\psi(x)$, $f_{1,n+1}(x,y,t)$, $g(x,t)$, $\Phi(t)$, $f_{2,n+1}(u(t),t)$, $h(t)$, $A(t)$, $A_1(t)$, $b(u,t)$ be defined and continuous together with the functions $\partial\phi/\partial x$, $\partial f_{1,n+1}/\partial x$, $\partial f_{1,n+1}/\partial y$, $\partial g/\partial x$ and, moreover, let the functions $\phi(x)$, $f_{1,n+1}(x,y,t)$, $g(x,t)$ be convex in $\{x,y\}$ and let $|dh(t)/dt| \leq 1 - \alpha$, $\alpha > 0$. Then every piecewise-continuous control $u(t)$, $u(t) \varepsilon U$, in problem (8),(6),(3) is optimal if $g(x,t) \leq 0$ and $u(t)$, $t \varepsilon T$, satisfies the maximum condition

$$\psi'(t)b(u(t),t) - f_{2,n+1}(u(t),t)$$

$$= \max_{u \varepsilon U} [\psi'(t)b(u,t) - f_{2,n+1}(u,t)]$$

with the function $\psi(t)$ being a solution of the system

$$\frac{d\psi(t)}{dt} = - A'(t)\psi(t) - A_1'(r(t))\psi(r(t)) \frac{dr(t)}{dt}$$

$$+ \frac{\partial f_{1,n+1}(x(t),x(t-h(t)),t)}{\partial x} + \left.\frac{\partial f_{1,n+1}(x(\bar{t}),x(\bar{t}-h(\bar{t})),\bar{t})}{\partial y}\right|_{\bar{t}=r(t)}$$

$$\cdot \frac{dr(t)}{dt} + \frac{d\mu(t)}{dt} \cdot \frac{\partial g(x(t),t)}{\partial x} \quad , \quad t \varepsilon [t_o,t'] \quad ,$$

$$\frac{d\psi(t)}{dt} = - A(t)\psi(t) + \frac{\partial f_{1,n+1}(x(t),x(t - h(t)),t)}{\partial x}$$

$$+ \frac{d\mu(t)}{dt} \cdot \frac{\partial g(x(t),t)}{\partial x}, \quad t\varepsilon[t',t_1], \quad \psi(t_1) = - \frac{\partial \phi(x(t_1))}{\partial x}.$$

4. Uniqueness of Optimal Trajectories

In Theorem 1, let the strict inequality $N > M_1$ be satisfied. Then, repeating the scheme of proof for this theorem, we can see that the increment of the functional $\Delta J(u)$ is positive if $\Delta x(t) \not\equiv 0$. Thus, if $N > M_1$ in the conditions of Theorem 1, then the optimal trajectory is unique. The optimal control will be unique if, along the optimal trajectory, the system of equations (5) has a unique solution at each point $t\varepsilon T$, with respect to the parameter u_ν.

Under the conditions of Theorem 2, for uniqueness of the optimal trajectory it is sufficient to require strict convexity in $\{x,y\}$ of the function $f_{1,n+1}(x,y,t)$.

§2. Sufficient Conditions for Optimality in Problems with Functionals Convex in the Controls

In this paragraph, the admissible controls are defined as the set of piecewise-continuous functions $u(t)$, $t\varepsilon T$, with values in some bounded, convex set U.

1. Nonlinear Systems

We consider the system

$$\left.\begin{aligned} \frac{dx(t)}{dt} &= f(x(t),x(t - h(t)),u(t),t) \\ x(t) &= \Phi(t), \quad t\varepsilon S_o, \end{aligned}\right\} \tag{9}$$

and the functional

$$J(u) = \phi(x(t_1)) + \int_{t_0}^{t_1} f_{n+1}(x(t), x(t - h(t)), u(t), t)\,dt. \quad (10)$$

Theorem 3. In addition to conditions 2) and 3) of Theorem 1, let the following conditions be satisfied: 1) the function $f_{n+1}(x,y,u,t)$ is strongly convex in u with constant N_1;

2) $$\left\|\frac{\partial^2 f_{n+1}(x,u,u,t)}{\partial x^2}\right\| \le L_3, \quad \left\|\frac{\partial^2 f_{n+1}(x,u,u,t)}{\partial x \partial x}\right\| \le L_3,$$

$$\left\|\frac{\partial^2 f_{n+1}(x,u,u,t)}{\partial x \partial u}\right\| \le L_6, \quad \left\|\frac{\partial^2 f(x,u,u,t)}{\partial y \partial u}\right\| \le L_6$$

$$\left\|\frac{\partial^2 f_{n+1}(x,y,u,t)}{\partial x \partial u}\right\| \le L_6, \quad \left\|\frac{\partial^2 f_{n+1}(x,u,u,t)}{\partial \dot{y} \partial u}\right\| \le L_6,$$

$$\left\|\frac{\partial^2 f(x,y,u,t)}{\partial u^2}\right\| \le L_7 .$$

Then in problem (9),(10),(3) for

$$N_1 \ge 2M_2 + 2M_3 + M_4 , \quad (11)$$

where

$$M_2 = L_5 L_6 \left\{1 + \left[L_1 + L_1(t_1 - t_0)\left(1 + \frac{1}{\alpha}\right) + L_2 L_4\right]\right\}$$
$$\cdot (t_1 - t_0) \exp\left(L_1\ 3 + \frac{1}{\alpha}\right)(t_1 - t_0) ,$$

$$M_3 = 2L_3L_5^2\left[L_1 + L_1(t_1 - t_0)\left(1 + \frac{1}{\alpha}\right) + L_2L_4\right]$$
$$\cdot \exp L_1\left(5 + \frac{1}{\alpha}\right)(t_1 - t_0) \quad ,$$

$$M_4 = \frac{1}{2} L_7\left[L_1 + L_1(t_1 - t_0)\left(1 + \frac{1}{\alpha}\right) + L_2L_4\right]$$
$$\cdot \exp L_1\left(1 + \frac{1}{\alpha}\right)(t_1 - t_0) \quad ,$$

each admissible control u(t) is optimal if

$$\frac{\partial H'(x(t),y(t),\psi(t),u(t),t)}{\partial u} u(t)$$

$$= \max_{u \in U} \frac{\partial H'(x(t),y(t),\psi(t),u(t),t)}{\partial u} u \ , \tag{12}$$

$$H'(x,y,\psi,u,t) = \psi' f(x,y,u,t) - f_{n+1}(x,y,u,t) \ ,$$

$$y(t) = x(t - h(t)), \tag{13}$$

the function $\psi(t)$ satisfying system (7) with H from (13).

Proof. The proof proceeds according to the scheme of proof of Theorem 1 and consists of the reduction of inequality (4) to the form

$$\Delta J(u) \geq -\int_{t_0}^{t_1} \frac{\partial H'(x,y,\psi,u,t)}{\partial u} u(t)dt + N_2 \int_{t_0}^{t_1} ||\Delta u(t)||^2 dt.$$

The first integral is nonpositive by virtue of condition (12). The coefficient N_2 of the second integral equals $N_1 - 2M_2 - 2M_3 - M_4$ which, by (11), is nonnegative. Thus, $\Delta J(u) \geq 0$, which proves the optimality of the control u(t).

2. Linear Systems

If Eq. (9) has the form

$$\frac{dx(t)}{dt} = A(t)x(t) + A_1(t)x(t - h) + b(t)u \quad ,$$

where $A(t)$, $A_1(t)$, $b(t)$ are piecewise-continuous functions, and the functions $\phi(x)$, $f_{n+1}(x,y,u,t)$ are continuous, together with $\partial\phi/\partial x$, $\partial f_{n+1}/\partial x$, $\partial f_{n+1}/\partial y$, $\partial f_{n+1}/\partial u$ and, moreover, the functions $\phi(x)$, $f_{n+1}(x,y,u,t)$ are convex in $\{x,y,u\}$, then each control $u(t)$, $t\varepsilon T$, satisfying condition (12), is optimal.

3. Uniqueness of the Optimal Control

If, under the conditions of Theorem 3, we have strict inequality in (11), then the optimal control $u^o(t)$ in problem (9),(10),(3) is unique.

§3. Optimality of Controls Satisfying the Maximum Principle

In general, the maximum principle is insufficient for the optimality of controls as is seen by the following example.

Example 1. $\dot{x} = u$, $x(0) = 0$, $U = \{u: |u| \le 1\}$, $T = [0,1]$, $J(u) = - \int_0^t x^2(t)dt$.

The maximum condition in this case has the form

$$\psi(t)u(t) = \max_{|u|\le 1} \psi(t)u \quad , \tag{14}$$

where $\psi(t)$ is the solution of the equation $\dot{\psi} = -2x$, $\psi(1) = 0$. The control $u(t) = 0$, $t\varepsilon T$, satisfies condition (14) although it is clearly not optimal. In addition to the trivial control u(t) = 0, condition (14) is satisfied by a countable number of nontrivial controls (cf. §6.6). This example also shows that even in the simplest cases the optimal trajectories and controls may turn out to be non-unique.

We illustrate the basic approach to studying the sufficiency of the maximum principle by the simple problem (6.1),(6.2). This question was investigated for problems (6.12),(6.2) in [36e] by the same scheme.

Thus, for minimization of the functional

$$J(u) = \phi(x(t_1)) \tag{15}$$

along trajectories of the system

$$\frac{dx}{dt} = f(x,u,t) \quad , \quad x(t_o) = x_o \quad , \tag{16}$$

let there be found a piecewise-continuous function $u(t) \varepsilon U$, $t \varepsilon T$, for which

$$\psi'(t) f(x(t),u(t),t) = \max_{u \varepsilon U} \psi'(t) f(x(t),u,t)$$

with the function $\psi(t)$ being a solution of the equation

$$\frac{d\psi}{dt} = - \frac{\partial f'(x(t),u(t),t)}{\partial x} \psi \ , \ \psi(t_1) = - \frac{\partial \phi(x(t_1))}{\partial x} \ .$$

Let the functions $\Delta_{\tilde{u}} H(x,\psi,u,t)$ be different from zero on the set $t \varepsilon \sigma = [\theta_1, \theta_2]$, $\tilde{u} \varepsilon U$, U compact, and let there be found a constant $\beta_1 > 0$ for which

$$\Delta_{\tilde{u}} H(x(t),\psi(t),u(t),t) \le -\beta_1 ||\Delta_{\tilde{u}} f(x(t),u(t),t)||, \ t \varepsilon \sigma. \tag{17}$$

Analyzing the procedure for obtaining the estimate for $||\Delta_{\varepsilon\theta}x(t)||$ (see Ch. VI), we note that

$$||\Delta x(t)|| \leq \beta_2 \int_{\theta_1}^{\theta_2} ||\Delta_{\tilde{u}} f(x(\tau),u(\tau),\tau)||d\tau, \quad t\varepsilon T, \ \beta_2 = \text{const}, \tag{18}$$

if

$$\Delta u(t) = \tilde{u}(t) - u(t) = 0 \quad , \quad t\bar{\varepsilon}\sigma \quad . \tag{19}$$

The increment of the functional (15) along the control increment (19) satisfies, by virtue of (17), the inequality

$$\Delta J(u) \geq \beta_1 \int_{\theta_1}^{\theta_2} ||\Delta_{\tilde{u}} f(x(t),u(t),t)||dt$$

$$- \int_{\theta_1}^{\theta_2} \frac{\partial \Delta_{\tilde{u}} H'(x(t),\psi(t),u(t),t)}{\partial x} \Delta x(t)dt$$

$$- \int_{\theta_1}^{t_1} o_1(||\Delta x(t)||)dt + o_4(||\Delta x(t_1)||) \quad . \tag{20}$$

Let $\left|\left|\frac{\partial f}{\partial x}\right|\right| \leq L_1$, $||\psi(t)|| \leq K_1$, $o_1(\alpha) \leq K_2\alpha^2$, $o_4(\alpha) \leq K_3\alpha^2$.

Then

$$\left|\int_{\theta_1}^{\theta_2} \frac{\partial \Delta_{\tilde{u}} H'(x(t),\psi(t),u(t),t)}{\partial x} \Delta x(t)dt\right|$$

$$\leq 2(\theta_2 - \theta_1)K_1L_1\beta_2 \int_{\theta_1}^{\theta_2} ||\Delta_{\tilde{u}} f(x(\tau),u(\tau),\tau)||d\tau \quad ,$$

$$\left|\int_{\theta_1}^{t_1} o_1(||\Delta x(t)||)dt\right|$$

$$\leq (t_1 - t_0)K_2\beta_2^2\left[\int_{\theta_1}^{\theta_2} ||\Delta_{\tilde{u}}f(x(\tau),u(\tau),\tau)||d\tau\right]^2 ,$$

$$|o_4(||\Delta x(t_1)||)| \leq K_3\beta_2^2\left[\int_{\theta_1}^{\theta_2} ||\Delta_{\tilde{u}}f(x(\tau),u(\tau),\tau)||d\tau\right]^2 ,$$

and inequality (20) assumes the form

$$\Delta J(u) \geq [\beta_1 - 2(\theta_2 - \theta_1)K_1L_1\beta_2] \int_{\theta_1}^{\theta_2} ||\Delta_{\tilde{u}}f(x(\tau),u(\tau),\tau)||d\tau$$

$$- \beta_2^2(t_1 - t_0)K_2 + K_3\left[\int_{\theta_1}^{\theta_2} ||\Delta_{\tilde{u}}f(x(\tau),u(\tau),\tau)||d\tau\right]^2 . \quad (21)$$

We will say that on the set $\sigma = [\theta_1,\theta_2]$, the control $u(t)$, $t\varepsilon T$, satisfies the strong maximum condition with constant β_1 if inequality (17) is satisfied for all $u\varepsilon U$, $t\varepsilon\sigma$. We call the control $u(t)\varepsilon U$, $t\varepsilon T$, optimal relative to perturbations which are small in the mean on σ if there exists a number $\varepsilon > 0$, such that $J(u + \Delta u) - J(u) \geq 0$ for all $\Delta u(t)$ such that $u(t) + \Delta u(t)\varepsilon U$, $\Delta u(t) = 0$ for $t\notin\sigma$, and $u(t) + \Delta u(t)\varepsilon U$, $\Delta u(t) = 0$ for $t\varepsilon\sigma$ and

$$\int_{\theta_1}^{\theta_2} ||f(x(t), u(t) + \Delta u(t),t) - f(x(t),u(t),t)||dt \leq \varepsilon.$$

The control $u(t)\varepsilon U$, $t\varepsilon T$, is called optimal relative to perturbations concentrated on σ if $J(u + \Delta u) - J(u) \geq 0$ for all $\Delta u(t)$ such that $u(t) + \Delta u(t)\varepsilon U$ and $\Delta u(t) = 0$ for $t\notin\sigma$.

Theorem 4. Let the functions $f(x,u,t)$, $\phi(x)$ be defined and continuous together with the functions

$\partial f(x,u,t)/\partial x$, $\partial\phi/\partial x$, $\partial^2 f(x,u,t)/\partial x^2$, $\partial^2\phi/\partial x^2$. Then in problem (16),(15), for $\beta_1 > 2(\theta_2 - \theta_1)K_1L_1\beta_2$, each admissible control u(t) of class D satisfying a strong maximum condition with constant β_1 on $\sigma = [\theta_1,\theta_2]$, is optimal with respect to perturbations small in the mean on σ. The indicated control will be optimal relative to perturbations concentrated on σ if the difference $\theta_2 - \theta_1$ is sufficiently small and the set U is bounded.

§4. On the Optimality of Singular Controls

Although the maximum principle is not, in general, a sufficient condition for optimality, in systems which are linear in the state it contains all conditions which ensure optimality of the control (Theorem 2). From Example 1, it follows that not every singular control is optimal. To find sufficient conditions for the optimality of singular controls, we use the formula for the kth order increment of functionals in the same way as it is used to investigate nonsingular controls (§1-3). We do not consider the general case of this question, but indicate the general arguments on a problem in which the optimality of singular controls follows from their existence.

Let there be given the equation

$$\frac{dx}{dt} = A(u,t)x + b(u,t) \quad , \tag{22}$$

with continuous functions A(u,t), b(u,t). We assume that the functional

$$J(u) = \phi(x(t_1)) \quad ,$$

is minimized among controls of class D, where $\phi(x)$ is a convex, differentiable function. From the formula for the increment of the functional (6.3), it follows that

$$\Delta J(u) \geq - \int_{t_0}^{t_1} \psi'(t)\Delta_{\tilde{u}}[A(u(t),t)x(t) + b(u(t),t)]dt$$

$$- \int_{t_0}^{t_1} \psi'(t)\Delta_{\tilde{u}}A(u(t),t)\Delta x(t)dt \quad .$$

Therefore, if the control u(t), $t\varepsilon T$, is such that

$$\left.\begin{aligned} &\psi'(t)\Delta_{u*}[A(u(t),t)x(t) + b(u(t),t)] \equiv 0, \ t\varepsilon T, \ u^*\varepsilon U \ , \\ &\psi'(t)\Delta_{u*}A(u(t),t) \equiv 0 \quad , \end{aligned}\right\} \quad (23)$$

then it is optimal. We show that conditions (23) are satisfied for first-order singular controls if the system (22) is equivalent to the equation

$$\left.\begin{aligned} x^{(n)} + a_n(u,t)x^{(n-1)} + \cdots + a_1(u,t)x = b(u,t), \\ x^{(i-1)}(t_0) = x_{io} \quad . \end{aligned}\right\} \quad (24)$$

In this case, the matrix A has the form

$$\begin{Bmatrix} 0 & 1 & 0 & & \cdots & 0 \\ 0 & 0 & 1 & 0 & \cdots & 0 \\ \cdots & \cdots & \cdots & \cdots & \cdots & \cdots \\ 0 & \cdots & \cdots & \cdots & 0 & 1 \\ -a_1 & \cdots & \cdots & \cdots & \cdots & -a_n \end{Bmatrix},$$

and the maximum principle reduces to the condition

$$\psi_n(t)[-\sum_{i=1}^{n} a_i(u^o(t),t)x_i^o(t) + b(u^o(t),t)]$$

$$= \max_{u\varepsilon U} \psi_n(t)[-\sum_{i=1}^{n} a_i(u,t)x_i^o(t) + b(u,t)] \quad ,$$

where $x_i(t) = x^{(i-1)}(t)$, $\psi_n(t)$ is the last component of the vector function $\psi(t) = \{\psi_1(t),\ldots,\psi_n(t)\}$, satisfying the equation

$$\dot{\psi}(t) = -A'(u^o(t),t)\psi(t) \quad , \quad \psi(t_1) = -\frac{\partial\psi(x^o(t_1))}{\partial x} \quad .$$

A typical practical case of the system (24) is the singular control which arises when

$$\psi_n(t) \equiv 0 \quad , \quad t \in T \quad . \tag{25}$$

It is not difficult to calculate that the identity in (23) follows from (25).

Theorem 5. In the problem (24),(15), each singular control for which (25) is satisfied, is optimal.

We may also obtain the following result.

Theorem 6. Along trajectories of Eq.(24), let the functional

$$J(u) = \phi(x(t_1)) + \int_{t_o}^{t_1} f_{n+1}(x(t),u(t),t)dt \quad ,$$

be minimized, where $\phi(x)$, $f_{n+1}(x,u,t)$ are continuous functions together with $\partial\phi/\partial x$, $\partial f_{n+1}(x,u,t)/\partial x$, and $\phi(x)$, $f_{n+1}(x,u,t)$ are convex in x. The piecewise-continuous control $u(t)$, $t \in T$, is optimal if the condition

$$\dot{\psi}(t) = -A'(u(t),t)\psi(t) + \frac{\partial f_{n+1}(x(t),u(t),t)}{\partial x} \quad ,$$

$$\psi(t_1) = -\frac{\partial\phi(x(t_1)}{\partial x} \quad ,$$

$$\psi_n(t) \equiv 0, \quad f_{n+1}(x(t),u(t),t)$$

$$\leq f_{n+1}(x(t),u,t) \quad , \; t\varepsilon T \; , \; u\varepsilon U \; .$$

is satisfied along u.

§5. Sufficient Conditions for Optimality

1. Controls Generating Optimal Trajectories.

We desire to minimize the functional

$$J(u,v) = \int_{t_0}^{t_1} f_{n+1}(x(t),u(t))dt \quad ,$$

along admissible controls $u(t)\varepsilon U$ and parameters $v\varepsilon V$, connected with x by the system

$$\frac{dx}{dt} = f(x,u)$$

having initial conditions $x(t_o) = g(v)$. We let M denote the set of admissible controls $u^*(t)$ and parameters $v^*\varepsilon V$ which generate the optimal trajectory $x^o(t)$. Let

$$H(x,\psi,u) = \psi' f(x,u) - f_{n+1}(x,u), \quad h(\psi,v) = \psi' g(v) \quad ,$$

and let the function $\psi(t)$ be defined by the equation

$$\frac{d\psi(t)}{dt} = - \frac{\partial H(x(t),\psi(t),u(t))}{\partial x} \quad , \quad \psi(t_1) = -c \quad .$$

Theorem 7. In order that

$$J(\bar{u},\bar{v}) \leq J(u,v) \quad , \quad \{u,v\}\varepsilon M \quad ,$$

it suffices that $\bar{u}(t)$, $t\varepsilon T$, $\bar{v}$ satisfy the conditions

$$H(x^o(t), \psi(t), \bar{u}(t)) = \max_{u \in U} H(x^o(t), \psi(t), u) \quad , \tag{26}$$

$$h(\psi(t_o), \bar{v}) = \max_{v \in V} h(\psi(t_o), v) \quad . \tag{27}$$

Proof. Generalizing Lemma 6.5 to the problem under consideration, we obtain

$$J(u,v) = -\psi'(t_o)g(v) - \int_{t_o}^{t_1} H(x(t), \psi(t), u(t))dt$$

$$+ \int_{t_o}^{t_1} \frac{\partial H'(x(t), \psi(t), u(t))}{\partial x} x(t)dt \quad . \tag{28}$$

But, from the condition $\{u^o, v^o\} \varepsilon M$, $\{\bar{u}, \bar{v}\} \varepsilon M$, it follows that $f(x^o(t), u^o(t)) = f(x^o(t), \bar{u}(t))$ and, moreover, from the maximum principle and condition (26), for $u^o(t)$ we have $f_{n+1}(x^o(t), u^o(t)) = f_{n+1}(x^o(t), \bar{u}(t))$. Substituting these equalities into the equation for $\psi(t)$, we obtain

$$\psi(t) \Big|_{u=u^o(\cdot), v^o} = \psi(t) \Big|_{u=\bar{u}(\cdot), \bar{v}} \quad ,$$

which, together with (27) and (28), gives $J(\bar{u}, \bar{v}) \leq J(u^o, v^o)$ contradicting (29). The theorem is proved.

2. Homogeneous Functionals

Theorem 8. Along trajectories of the system

$$\frac{dx}{dt} = f(u,w) \quad , \quad x(t_o) = g(v) \quad , \quad u(t) \varepsilon U \quad , \quad v \varepsilon V \quad , \quad w \varepsilon W \quad , \quad t \varepsilon T \quad ,$$

let the functional

$$J(u,v,w) = \int_{t_o}^{t_1} f_{n+1}(x,u,w)dt \quad ,$$

be minimized, where

$$f_{n+1}(\lambda x,u,w) = \lambda^m f_{n+1}(x,u,w) \quad , \quad m > -1 \quad .$$

Then for the optimality of $u^o(t)$, $t\varepsilon T$, and the parameters v^o, w^o, it is necessary and sufficient that

$$H(x^o(t),\psi(t),u^o(t),w^o) = \max_{u\varepsilon U} H(x^o(t),\psi(t),u,w^o) \quad ,$$

$$h_{v^o}(t_o) + (t_1-t_o)H_{u^o,w^o}(t_1) \tag{30}$$

$$= \max_{\{u,v,w\}\varepsilon\Omega} [h_v(t_o) + (t_1 - t_o)H_{u,w}(t_1)] \quad .$$

Here

$$H_{u,w}(t) = \psi'(t)f(u(t),w) - f_{n+1}(x(t),u(t),w) \quad ,$$

$$h_v(t) = \psi'(t)g(v) \quad ,$$

Ω is the set of controls $u(t)$, $t\varepsilon T$, and parameters v,w satisfying condition (30).

Proof. Under the conditions of the theorem, we have

$$\frac{\partial H'(x,\psi,u,w)}{\partial x} x = - \frac{\partial f'_{n+1}(x,u,w)}{\partial x} x = -mf_{n+1}(x,u,w) \quad .$$

Therefore, from (28) and (30) it follows that

$$J(u,v,w) = - \psi'(t_o)g(v) - (t_1 - t_o)H_{u,w}(t_1)$$

$$- mJ(u,v,m) \quad ,$$

which proves the theorem.

By virtue of this theorem, the control $u^o(t) = \pm 1$ is optimal in Example 1 of §6.6.

§6. Sufficiency of the Maximum Principle for Linear Systems

We will say that the function $\bar{u}(t)$, $t \in T$, is an extremal control for the equation

$$\frac{dx}{dt} = A(t)x + B(t)u(t) \quad ,$$

where u(t) is restricted by the condition $|u_\nu(t)| \leq L$, if

$$\psi'(t)B(t)\bar{u}(t) = \max_{|u_\nu| \leq L} \psi'(t)B(t)u \ , \quad \nu = 1,\ldots,r \ . \qquad (32)$$

Here $\psi(t)$ is some nontrivial solution of the equation

$$\frac{d\psi}{dt} = -A'(t)\psi \quad .$$

We consider the solution δ_2^o, $u^o(t)$, $t \in T$, of the problem of minimizing the norm $||x(t_1) - c_2||$ of the terminal state of Eq.(31), under the condition that $||x(t_o) - c_1|| \leq \delta_1$, where c_1, c_2 are given points of the state space (31), while δ_1 is a fixed, nonnegative number. The following assertion can be proved.

<u>Theorem 9. Let $\bar{u}(t)$ be an extremal control.</u>
<u>If the trajectory $x(x_o,\bar{u}(\cdot),t)$ satisfies the conditions</u>

$$\psi'(t_o)[x(t_o) - c_1] = \delta_1||\psi(t_o)|| \quad ,$$

$$\psi'(t_1)[c_2 - x(x(t_o),\bar{u}(\cdot),t_1)] = \delta_2||\psi(t_1)|| \quad ,$$

then $\bar{u}(t)$ is an optimal control.

This assertion follows from Theorem 6.24.

Now we assume that system (31) is stationary. For it, we consider the problem of minimal time (6.171) with the fixed terminal condition $x(t_1) = 0$, t_1 being the final moment of time. We have

Theorem 10. If the T-admissible control $\bar{u}(\cdot)$ is extremal, then $\bar{u}(\cdot)$ is an optimal control.

Proof. By virtue of the remarks following Lemma 6.13, the extremal control is unique. But, $u^o(\cdot)$ is an extremal control. This means $u(\cdot) = u^o(\cdot)$ for almost all $t \geq t_o$. This proves the theorem.

The uniqueness of the optimal control in problem (6.171),(31) is connected with the properties of the defining equation (1.110) when $h = 0$, $A_1 = 0$.

Theorem 11. Let the T-admissible control $\bar{u}(\cdot)$ in problem (6.171),(31), be extremal. If the defining equations of the system (1.110) is nonsingular at $t = t_1$ for $A_1 = 0$, then $\bar{u}(\cdot)$ is optimal.

Proof. It suffices to show that t_1 is the smallest root of Eq.(6.175).

Since the defining equation of the system is non-singular at $t = t_1$, for some ε we have

$$(x_o,t) < \lambda(x_o,t_1) \quad , \quad t\varepsilon(t_1 - \varepsilon,t_1) \quad .$$

By virtue of Lemma 6.13 and the above inequality, we see that t_1 is the minimal root for (6.175). The assertion is proved.

Remark. Assertions analagous to Theorems 9.11 are valid for problems of minimizing the norm of the terminal state and minimal time problems for the system with a delay (6.156), if the function $\psi(t)$ in the definition of an extremal control is a nontrivial solution of the equation

$$\frac{d\psi(t)}{dt} = -A^*(t)\psi(t) - \eta(t)A_1^*(t+h)\psi(t+h) \quad ,$$

$$\eta(t) = 1 \ , \quad t_0 \le t \le t_1 - h \ , \quad \eta(t) = 0 \ , \quad t_1 - h \le t \le t_1 \ .$$

Let there be given system (6.136)

$$\frac{dx}{dt} = A(t)x + b(t)u \quad .$$

The admissible region of $x(t)$ in (6.137) is

$$G(x) = \{x: \ e'x \le 1\} \quad , \quad e \neq 0 \quad ,$$

where the control $u(t)$ is constrained by the condition $|u(t)| \le 1$.

We consider the minimal time problem (6.171). In the definition of an extremal control (cf. (32)) the function $\psi(t)$ satisfies the equation

$$\frac{d\psi}{dt} = -A'(t)\psi + e\,\frac{d\mu(t)}{dt} \quad .$$

The following assertion is valid.

Theorem 12. Let the T-admissible control $\bar{u}(\cdot)$ in problem (6.171),(6.136),(6.137) be extremal and let $\psi(t_1) \neq 0$. Then $\bar{u}(\cdot)$ is an optimal control.

The assertion is proved as in Theorem 11 since the function (6.141) possesses the following properties.

Lemma 1. Let $(t_1) = 0$. Then $\Phi(t') \leq \Phi(t_1)$ for all $t' > t_1$; there exists an $\varepsilon > 0$, such that $\Phi(t") > \Phi(t_1)$ for all $t" \varepsilon (t_1 - \varepsilon, t_1)$.

Proof. The first part of the assertion follows from the fact that the trajectory of the system may be held at the origin. We prove the second part.

Let ε be defined as in the proof of Lemma 6.10. Then

$$\Phi(t_1) = g'x_0 - \int_{t_0}^{t_1} \left| \gamma'(t) \left[g - \int_{t_0}^{t'} F'(\tau) e d\mu(\tau) \right] \right| dt - \int_{t_0}^{t_1} d\mu(t)$$

$$\leq \max_{||g||=1, d\mu \geq 0} \left\{ g'x_0 - \int_{t_0}^{t"} \left| \gamma'(t) \left[g - \int_{t_0}^{t} F'(\tau) e d\mu(\tau) \right] \right| dt \right.$$

$$\left. - \int_{t_0}^{t"} d\mu(t) \right\} - \int_{t"}^{t_1} \left| \gamma'(t) \left[g - \int_{t_0}^{t} F(\tau) e d\mu(\tau) \right] \right| dt$$

$$= \Phi(t") - \int_{t"}^{t_1} \left| \gamma'(t) \left[g - \int_{t_0}^{t} F(\tau) e d\mu(\tau) \right] \right| dt \quad .$$

The last integral equals zero only in the event $g = \int_{t_0}^{t} F'(\tau) e d\mu(\tau)$, $t \leq t_1$. But this is impossible since

$$\psi(t_1) = -(F^{-1}(t_1))' \left[g - \int_{t_0}^{t_1} F'(t) e d\mu(t) \right] \neq 0 \ .$$

The lemma is proved.

Theorem 13. Let the T-admissible control $\bar{u}(\cdot)$ in problem (6.171),(31),(6.136) be extremal and let $\psi(t_1) \neq 0$. If the defining equation (1.110) with $A_1 = 0$ is nonsingular, then $\bar{u}(\cdot)$ is an optimal control.

The assertion is proved analogously to Theorem 11.

Remark. The assumption $\psi(t_1) \neq 0$ in Theorems 12,13 becomes redundant if there exists a T-admissible control for which

$$e'x(t) < 1 \quad , \quad t\varepsilon T \quad .$$

In this case, it is possible to prove the results studying the function $\Phi(t_1)$ or using the arguments from [50a].

§7. Toward Sufficient Conditions for Optimality in Gaming Problems

At first we consider a terminal control problem. Let the motion under consideration be

$$\frac{dx}{dt} = Ax + b(v,w) \quad , \quad x(0) = x_o \quad , \quad t\varepsilon T = [0,t_1] \quad , \qquad (33)$$

where the control functions are $v(t)\varepsilon V$, $w(t)\varepsilon W$, $t\varepsilon T$. We assume that there is found controls $\bar{v}(t)$, $\bar{w}(t)$ such that

$$\psi'(t)b(\bar{v}(t),\bar{w}(t)) = \max_{v\varepsilon V} \psi'(t)b(v,\bar{w}(t)) \quad ,$$

$$\psi'(t)b(\bar{v}(t),\bar{w}(t)) = \min_{w\varepsilon W} \psi'(t)b(\bar{v}(t),w) \quad ,$$

$$x(t) = x(t,\bar{v}(\cdot),\bar{w}(\cdot)) \quad ,$$

where $x(t)$ is a solution of Eq.(33) for $v = \bar{v}$, $w = \bar{w}$, and $\psi(t)$ is a solution of the equation

$$\frac{d\psi}{dt} = -A'\psi \quad , \quad \psi(t_1) = -c \quad .$$

Then $\{\overline{x}(\cdot),\overline{w}(\cdot)\}$ is a saddlepoint of the functional

$$J(v,w) = c'x(t_1) \quad ,$$

i.e. for all $v(t)\varepsilon V$, $w(t)\varepsilon W$,

$$J(\overline{v},w) \leq J(\overline{v},\overline{w}) \leq J(v,\overline{w}) \quad .$$

The proof of this proposition is completely analagous to the proof from §1.

Let there be given the equations of the pursuer

$$\frac{dy}{dt} = Ay + b_1 v \quad , \quad y(0) = y_o \quad , \tag{34}$$

and the evader

$$\frac{dz}{dt} = Az + b_2 w \quad , \quad z(0) = z_o \quad . \tag{35}$$

We will assume that (34) is controllable and that the piecewise-continuous controls $v(t)$, $w(t)$ are constrained by the conditions

$$|v| \leq 1 \quad , \quad |w| \leq 1 \quad .$$

We assume that we have also found controls $\overline{v}(t)$, $\overline{w}(t)$, $t\varepsilon T$, such that the corresponding trajectories $\overline{y}(t)$, $\overline{z}(t)$, of the systems (34),(35) satisfy the conditions $\overline{y}(t_1) = \overline{z}(t_1)$ and

$$\psi'(t)b_1 v(t) = \max_{|v|\leq 1} \psi'(t)b_1 v \quad , \tag{36}$$

$$\psi'(t)b_2 w(t) = \min_{|w|\leq 1} \psi'(t)b_2 w \quad . \tag{37}$$

Here $\psi(t)$ is a nontrivial solution of the equation

$$\frac{d\psi}{dt} = -A'\psi \quad . \tag{38}$$

Then the functions $\overline{v}(t)$, $\overline{w}(t)$, $t \varepsilon T$, are elements of the saddlepoint of the functional at the moment of contact

$$t_1(\overline{v}(\cdot), w(\cdot)) \leq t_1(\overline{v}(\cdot), \overline{w}(\cdot)) \leq t_1(v(\cdot), \overline{w}(\cdot)). \tag{39}$$

For proof of this assertion, we use the identity from §6.4. The function $x(t) = y(t) - z(t)$ satisfies the differential equation

$$\frac{dx}{dt} = Ax + b_1 v - b_2 w \quad , \quad x(0) = y_0 - z_0 \quad .$$

Therefore, along the trajectories $x(t)$ and $\psi(t)$ from (38) we have the identity

$$\psi'(t_1)x(t_1) - \psi'(0)x(0) = \int_0^{t_1} \psi'(t)b_1 v(t)dt - \int_0^{t_1} \psi'(t)b_2 w(t)dt.$$

At first, we show the validity of the second inequality in (39). We assume it is not satisfied, i.e. there exists a control $v(t)$, $0 \leq t \leq \theta$, $|v(t)| \leq 1$, under which the pursuing object achieves its goal in time $\theta < t_1$. According to the condition

$$\psi'(t)b_1\overline{v}(t) \geq \psi'(t)b_1 v(t) \quad ,$$

we have

$$\psi'(\theta)\overline{x}(\theta)$$

$$= [\psi'(\theta)x(\theta) - \psi'(0)x(0)] - [\psi'(\theta)x(\theta) - \psi'(0)x(0)]$$

$$= \int_0^\theta \psi'(t) b_1 \bar{v}(t) dt - \int_0^\theta \psi'(t) b_1 v(t) dt \geq 0 \quad . \tag{40}$$

On the other hand,

$$\psi'(t) b_1 \bar{v}(t) = \max_{|u| \leq 1} \psi'(t) b_1 u = |\psi'(t) b_1| \not\equiv 0 \quad , \tag{41}$$

$$\psi'(t) b_2 \bar{w}(t) = \min_{|u| \leq 1} \psi'(t) b_2 u = - |\psi'(t) b_2| \leq 0 \ . \tag{42}$$

Therefore,

$$\psi'(\theta)\bar{x}(\theta) = \psi'(\theta)\bar{x}(\theta) - \psi'(t_1)\bar{x}(t_1)$$

$$= -\int_\theta^{t_1} \psi'(t) b_1 \bar{v}(t) dt + \int_\theta^{t_1} \psi'(t) b_2 \bar{w}(t) dt < 0 \quad ,$$

which contradicts (40). The first inequality in (39) is proved analagously. Under the control $u(t)$, $|w(t)| \leq 1$, $0 \leq t \leq \sigma$, let the evading object avoid contact up until the moment $\sigma > t_1$. We extend the controls $\bar{v}(t)$, $\bar{w}(t)$ to the interval $[t_1, \sigma]$ using conditions (41),(42). Then

$$\psi'(t) b_2 \bar{w}(t) \leq \psi'(t) b_2 w(t) \quad , \quad 0 \leq t \leq \sigma \quad ,$$

and, therefore, $(x(t) = \bar{y}(t) - z(t))$

$$\psi'(\sigma)\bar{x}(\sigma) = [\psi'(\sigma)\bar{x}(\sigma) - \psi'(0)\bar{x}(0)]$$

$$- [\psi'(\sigma)x(\sigma) - \psi'(0)x(0)]$$

$$= \int_0^\sigma \psi'(t) b_2 \bar{w}(t) dt - \int_0^\sigma \psi'(t) b_2 w(t) dt \leq 0 \ . \tag{43}$$

But, since properties (41),(42) hold for all $0 \leq t \leq \sigma$,

$$\psi'(\sigma)\bar{x}(\sigma) = \psi'(\sigma)\bar{x}(\sigma) - \psi'(t_1)\bar{x}(t_1).$$

$$= \int_{t_1}^{\sigma} \psi'(t)b_1\bar{v}(t)dt - \int_{t_1}^{\sigma} \psi'(t)b_2\bar{w}(t)dt > 0 \quad ,$$

which contradicts (43). The validity of (39) is established.

Remarks. 1) The assumption on the controllability of the pursuing object may be dropped.

2) If the objects are nonstationary, then we must further assume that at $t = t_1$, the conditions of Theorem 1.19, with $A_1 = 0$, are satisfied for the pursuing object.

3) The existence of controls $\bar{v}(t)$, $\bar{w}(t)$ satisfying the formulated conditions is proved in §6.16. An approach to their computation is indicated in the same section.

§8. Uniqueness of Optimal Controls for Linear Systems

Below we study the minimal time problem (6.171). As first, we consider the solution $u^o(t)$ of the minimal time problem for the stationary system

$$\frac{dx}{dt} = Ax + bu \quad .$$

Theorem 14. If an optimal control $u^o(\cdot)$ exists, then it is unique.

Proof. For an optimal control $u^o(t)$, we have

$$u^o(t) = L \operatorname{sign} \xi(t,g(t_1^o)) \quad . \tag{44}$$

From the existence of $u^o(\cdot)$, it follows that the function $\xi(t,g(t_1^o)) = g'(t_1^o)F^{-1}(t)b$ will not be invertible at zero on a set of positive measure (cf. the remark to Lemma 6.13). This means that relation (44) defines the function $u^o(\cdot)$ uniquely. The assertion is proved.

Theorem 15. Let system (6.170)

$$\frac{dx}{dt} = A(t)x + B(t)u(t)$$

be such that the defining equation (1.110) with $A_1 = 0$, is nonsingular for almost all $t \in T$. If an optimal minimal time control exists in problem (6.171),(6.170), then it is unique.

Proof. The assertion follows from relation (44) and the remarks to Lemma 6.13.

Let there be given equation (6.156) with a delay.

$$\frac{dx(t)}{dt} = Ax(t) + A_1 x(t - h) + bu(t) \quad .$$

Theorem 16. If an optimal control exists in problem (6.171),(6.156), then it is unique.

The function $g'(t_1^o)F(t_1,t)b$ is not invertible at zero on a set of positive measure (cf. the remark to Lemma 6.34). This means that the relation of Theorem 6.34 defines $u^o(t)$ uniquely, which is what we were required to prove.

Theorem 17. Let system (6.156) be such that the defining equation (1.110) is nonsingular for almost all $t \in T$. If an optimal minimal time control exists in problem (6.171),(6.156), then it is unique.

This assertion is a corollary of the remarks to Theorem 6.34.

Theorems 14-17 establish the uniqueness of optimal controls in their entire domain of definition. If the conditions of Theorem 15,17 are not satisfied, then the optimal control may still be uniquely defined by Theorems 6.32, 6.34 for some initial states. We clarify this question for a simple minimal time problem [40b].

We are given the system

$$\frac{dx}{dt} = A(t)x + b(t)u \quad , \quad x(t_o) = x_o \quad .$$

Using a piecewise-continuous control $u(t)$,

$$|u(t)| \leq L \quad ,$$

it is required to minimize the time $t_1^o - t_o$ of passage of the trajectory $x(t)$ from the point x_o to the origin. From the results of §6.10, it follows that this problem is solvable if, for some t_1, the function

$$\rho(t_1) = \lambda(t_1,g_1) = \max_{||g||=1} \lambda(t_1,g) \quad ,$$

$$\lambda(t_1,g) = g'S(t_1,x_o) - L\int_{t_o}^{t_1} |g'S(t_1,t)|dt \ ,$$

is nonpositive.

The smallest t_1^o for which $\rho(t_1^o) \leq 0$ corresponds to the minimal time $t_1^o - t_o$. The optimal control $u^o(t)$ satisfies the relation

$$u^o(t) = -L \operatorname{sign} g_1'S(t_1^o,t) \quad ,$$

which uniquely defines $u^o(t)$ on the set

$$\sigma_1 = \{t: \ g_1'S(t_1^o,t) \not\equiv 0, t\varepsilon T\} \quad .$$

We introduce the control

$$u_1(t) = \begin{cases} u^o(t) & , t\varepsilon\sigma_1 , \\ 0 & , t\varepsilon T - \sigma_1 . \end{cases}$$

This control transfers the point x_o to the point $x_1 = S(t_1^o, x_o) + \int_{\sigma_1}^{\sigma} S(t_1^o, t)u_1(t)dt$. By definition of the control $u_1(t)$, for the points x_o, x_1 we have $g_1'[x_1 - x_o] = 0$, i.e. the point x_o lies in the hyperplane ℓ_1: $g_1'x = g_1'x_1$. Therefore, the original problem is reduced to finding a control $u(t)$, $|u(t)| \leq L$, $t\varepsilon T - \sigma_1$, for which

$$0 = x_1 + \int_{T-\sigma_1} S(t_1^o, t)u(t)dt \quad . \tag{45}$$

But, since the set $\{x: \quad x = \int_{T-\sigma_1} S(t_1^o, t)u(t)dt, |u(t)| \leq L\}$ lies entirely in ℓ_1, the problem has dimensionality one less than the original. We find

$$\rho_1(t_1^o, \alpha) = \lambda_1(t_1^o, \alpha, g_2) = \max_{||g||=1, g'g_1=0} \lambda_1(t_1, \alpha, g),$$
$$\lambda_1(t_1^o, \alpha, g) = g'x_1 - \alpha \int_{T-\sigma_1} |g'S(t_1^o, t)|dt \quad . \tag{46}$$

Clearly, $\rho_1(t_1^o, L) \leq \lambda(t_1^o, g_1) = 0$. Therefore, there is an L_1, $0 \leq L_1 \leq L$, such that $\rho_1(t_1^o, L_1) = 0$. Then one of the controls solving problem (45) has the form

$$u_1^o(t) = - L_1 \text{ sign } g_2'S(t_1^o, t) \quad , t\varepsilon T - \sigma_1 \; , \tag{47}$$

where g_2 is a vector solving problems (46) for $\alpha = L_1$. Relation (47) uniquely defines $u_1^o(t)$ on the set

$\sigma_2 = \{t: g_2' S(t_1^o, t) \neq 0, t \in T - \sigma_1\}$. We set

$$u_2(t) = \begin{cases} u^o(t) & , \quad t \in \sigma_1 \quad , \\ u_1^o(t) & , \quad t \in \sigma_2 \quad , \\ 0 & , \quad t \in T - \sigma_1 - \sigma_2 \quad . \end{cases}$$

This control transfers the point x_o to the point $x_2 = S(t_1^o, x_o) + \int_{\sigma_1+\sigma_2} S(t_1^o, t) u_2(t) dt$. It is not difficult to see that the set $\{x: x = \int_{T-\sigma_1-\sigma_2} S(t_1^o, t) u(t) dt, |u(t)| \leq L\}$ lies in the (n-2)-dimensional hyperplane ℓ_2: $g_1'x = g_1'x_1$, $g_2'x = g_2'x_2$. The original problem has now been reduced to the solution of the problem

$$0 = x_2 + \int_{T-\sigma_1-\sigma_2} S(t_1^o, t) u(t) dt \quad . \tag{48}$$

We form the function

$$\left. \begin{aligned} \rho_2(t_1^o, \alpha) &= \lambda_2(t_1^o, \alpha, g_3) = \max_{\substack{||g||=1 \\ g'g_1=0, g'g_2=0}} \lambda_2(t_1^o, \alpha, g), \\ \lambda_2(t_1^o, \alpha, g) &= g'x_2 - \alpha \int_{T-\sigma_1-\sigma_2} |g'S(t_1^o, t)| dt \quad . \end{aligned} \right\} \tag{49}$$

It is clear that $\rho_2(t_1^o, L_1) \leq 0$. Therefore, there is a number L_2, $0 \leq L_2 \leq L_1$, such that $\rho_2(t_1^o, L_2) = 0$. One of the controls solving problem (48) has the form

$$u_2^o(t) = - L_2 \operatorname{sign} g_3' S(t_1^o, t) \quad , \quad t \in T - \sigma_1 - \sigma_2 \quad ,$$

where the vector g_3 is a solution of problem (49) for $\alpha = L_2$. Continuing this process, we arrive at one of the following cases.

1) mes $\sum_{i=1}^{k} \sigma_i$ = mes T for some $k \leq n$. Then the control

$$u_k(t) = \begin{cases} u^o(t) , & t\varepsilon\sigma_1 , \\ u_1^o(t) , & t\varepsilon\sigma_2 , \\ \dots\dots\dots\dots\dots\dots & \\ u_{k-1}^o(t), & t\varepsilon\sigma_k , \end{cases}$$

uniquely defined on T, will solve the original problem.

2) mes $\sum_{i=1}^{n} \sigma_i <$ mes T. The control

$$u_n(t) = \begin{cases} u^o(t) , & t\varepsilon\sigma_1 , \\ u_1^o(t) , & t\varepsilon\sigma_2 , \\ \dots\dots\dots\dots\dots\dots\dots\dots & \\ u_{n-1}^o(t) , & t\varepsilon\sigma_n , \\ u(t) , & t\varepsilon T - \sum_{i=1}^{n} \sigma_1 , \end{cases}$$

where $u(t)$, $|u(t)| \leq L$, is an arbitrary function on $T - \sum_{i=1}^{n} \sigma_i$, solves the original problem of transferring the point x_1 to the origin.

From the construction of the controls $u_i(t)$, we have the following assertion: The minimal time problem has a unique solution for the point x_o if and only if the following conditions are satisfied:

1) for some k, $1 \leq k \leq n$, mes $\sum_{i=1}^{k} \sigma_i$ = mes T,

2) $L_i = L$, $i = 1,\dots,k-1$.

Remark. The above scheme has an obvious generalization to the case of several inputs. The form of the scheme in no way uses the form of the original system. Therefore, these conclusions are valid for system (6.156). Transfer of the results to systems with nonlinear inputs is immediate if on each step, we isolate the set σ on which the controls $u_i^o(t)$ are uniquely defined by the vector g_{i+1}. This situation occurs in general.

§9. Optimization Problems with Hierarchical Criteria

For clarity, the essential questions considered in this section are restricted to the problems of the preceding paragraph.

If an optimization problem is solved by more than one control, then it is natural to choose from this set those for which the system is optimal with respect to supplementary criteria. If the new problem has several solutions, then a third criterion is introduced, and so on.

In general, we have a system of criteria [219]

$$J_1(u), J_2(u), \ldots,$$

and it is required to find an admissible control $u^o(t)$ under which the boundary conditions are satisfied and for which

$$J_1(u) = \min_{u(\cdot)\varepsilon U_1(\cdot)} J_1(u) , \quad J_2(u) = \min_{u(\cdot)\varepsilon U_2(\cdot)} J_2(u), \ldots .$$

Here $U_1(\cdot) = U(\cdot)$ is a given class of admissible controls, $U_2(\cdot) = \{u(\cdot): J_1(u) = J_1(u^o), u(\cdot)\varepsilon U_1(\cdot)\}$, and so on. This problem with hierarchical criteria may be generalized so that it makes sense in those cases when the optimization problem with a single criterion has a unique solution. Let there be given a control system, boundary conditions, class of admissible controls, an ordered system of functionals

$$J_1(u), J_2(u), \ldots, J_m(u), J_{m+1}(u)$$

and a set of numbers $\varepsilon_1, \ldots, \varepsilon_m$.

Problem. Among the admissible controls satisfying the boundary conditions, find a $u^o(t)$ such that

$$J_1(u^o) \leq \min_{u(\cdot)\varepsilon U_1(\cdot)} J_1(u) + \varepsilon_1,$$

$$J_2(u^o) \leq \min_{u(\cdot)\varepsilon U_2(\cdot)} J_2(u) + \varepsilon_2, \ldots,$$

$$J_m(u^o) \leq \min_{u(\cdot)\varepsilon U_m(\cdot)} J_m(u) + \varepsilon_m,$$

$$J_{m+1}(u^o) = \min_{u(\cdot)\varepsilon U_{m+1}(\cdot)} J_{m+1}(u) .$$

Here $U_1(\cdot) = U(\cdot)$ is a given class of admissible controls

$$U_2(\cdot) = \{u(\cdot): u(\cdot)\varepsilon U_1(\cdot),$$

$$J_1(u) \leq \min_{u(\cdot)\varepsilon U_1(\cdot)} J_1(u) + \varepsilon_1\}, \ldots,$$

$$U_{m+1}(\cdot) = \{u(\cdot): u(\cdot)\varepsilon U_m(\cdot),$$

$$J_m(u) \leq \min_{u(\cdot)\varepsilon U_m(\cdot)} J_m(u) + \varepsilon_m\} .$$

The numbers $\varepsilon_1, \ldots, \varepsilon_m$ may be given in advance or chosen during the course of the solution (depending upon the results of the previous steps). We show how to use the scheme of §9 for minimzing the system of functionals

$$J_1(u) = t_1 - t_0,\ J_2(u) = \int_{t_0}^{t_1} a_2(t)u(t)dt, \ldots,$$

$$J_{m+1}(u) = \int_{t_0}^{t_1} a_{m+1}(t)u(t)dt.$$

We find $u_1(t)$, σ_1, x_1, g, according to the scheme of §8. Further, instead of problem (45), we consider the problem

$$0 = x_1 + \int_{T-\sigma_1} S(t_1^o,t)u(t)dt,\ 0 = -\beta + \int_{T-\sigma_1} a_2(t)u(t)dt. \tag{50}$$

We form the function

$$\rho_1(\beta) = \beta_1(\beta,g_2,f_2) = \max_{|f|+||g||\leq 1,\, g'g_1=0} \lambda_1(\beta,g,f),$$

$$\lambda_1(\beta,g,f) = g'x_1 - f\beta - L \int_{T-\sigma_1} |g'S(t_1^o,t) + fa_2(t)|dt.$$

If $L < L_1$ in problem (45), then we find the minimal β_1 for which $\rho_1(\beta_1) = 0$. Clearly, the control

$$u_1^o(t) = -L \operatorname{sign} [g_2'S(t_1^o,t) + f_2a_2(t)]$$

minimizes the functional $J_2(u)$. Let

$$\sigma_2 = \{t:\ g_2'S(t_1^o,t) + f_2a_2(t) \not\equiv 0,\ t\varepsilon T - \sigma_1\} .$$

We set

$$u_2(t) = \begin{cases} u^o(t), & t\varepsilon\sigma_1, \\ u_1^o(t), & t\varepsilon\sigma_2, \\ 0, & t\varepsilon T - \sigma_1 - \sigma_2 . \end{cases}$$

Further, instead of problem (48) we solve the problem

$$\left.\begin{aligned} 0 &= x_2 + \int_{T-\sigma_1-\sigma_2} S(t_1^o, t) u(t) dt \quad , \\ 0 &= -\beta + \int_{T-\sigma_1-\sigma_2} a_3(t) u(t) dt \quad . \end{aligned}\right\} \tag{51}$$

If it turns out that $L_2 < L$ in problem (48), then we find min $J_3(u)$ according to the described scheme, and so on. If $L_1 = L$ on the first step, then we form relation (50) instead of (51) and find $J_2(u)$. The remainder of the sequence is obvious. As a result, we obtain a relay control optimizing the hierarchical system of functionals.

<u>Remark</u>. In §8, the optimal control transferring x_o to the origin was obtained as a piecewise-constant function having levels $L_1, L_2, \ldots$. From the results of this paragraph, it follows that in linear systems there always exists a relay control ($u(t) = \pm L$) transferring the initial point to the origin in minimal time. In order to find this control, it is necessary to solve the minimal time problem with some hierarchical system of criteria.

The solution of a general optimization problem with a hierarchical system of criteria of the type in §6.9-6.11 proceeds according to the scheme that was described in this paragraph. There arise no major difficulties in this respect.

§10. <u>The Well-Posed Statement of Optimal Control Problems</u>

In this paragraph we study the properties of the solutions of several optimal control problems in their dependence on initial data and parameters. We confine ourselves to cases when the control functions u(t) are measurable and bounded almost everywhere, satisfying the inequality

$$\operatorname*{vrai\,sup}_{t \varepsilon T} |u_\nu(t)| \leq L, \ \nu = 1,\ldots,r; \ L = \text{const.}$$

Let $J(z_o,\mu_o)$ be the value of the optimized functional for the initial state $z_o \varepsilon E_n$ and the value μ_o of the parameters $\mu \varepsilon \Delta$, Ω is the region of state and parameter space $\{z_o,\mu\}$ on which is defined the optimal control $u = u^o(z_o,\mu,t)$ and the function $J(z_o,\mu)$.

Definition 1. We call the optimal solution $\{J(z_o,\mu), u^o(z_o,\mu,t)\}$ continuous at $\{z_o,\mu\}$ if for any $\varepsilon > 0$, $\sigma > 0$ there exists a $\delta > 0$ such that

$$|J(z_o,\mu) - J(\bar{z}_o,\bar{\mu})| < \varepsilon \quad , \quad \text{mes } \varepsilon_k < \varepsilon \quad ,$$

$$\varepsilon_k = \{t: \ |u_k^o(z_o,\mu,t) - u_k^o(\bar{z}_o,\mu,t)| \geq \sigma\}$$

are satisfied when $||z_o - \bar{z}_o|| + |\mu - \bar{\mu}| < \delta$.

We will say that the optimal control problem is well-posed if the optimal control is unique for $\{z_o,\mu\}\varepsilon\Omega$ and the solution $J(z_o,\mu)$, $u^o(z_o,\mu,t)$ depends continuously on $\{z_o,\mu\}$.

1. Application of Methods of Functional Analysis

1. Linear Systems. The Problem of Minimzing the Norm of the Terminal State. Let the right side of Eq.(31) depend on the numerical parameter μ, i.e.

$$\frac{dx}{dt} = A(t,\mu)x + B(t,\mu)u(t) + f(t,\mu), \quad t \geq t_o \ , \tag{52}$$

where $\mu\varepsilon\Delta$, $\Delta = \{\mu: \mu_1 < \mu < \mu_2\}$, the elements of the matrices $A(t,\mu)$, $B(t,\mu)$ are continuous in t and μ and have (n-1)-continuous derivatives in t, while the function $f(t,\mu)$ is continuous in t and μ. We fix μ. For ease of formulation, we introduce a definition.

Definition 2. If the defining equation (1.110) with $A_1 = 0$,

$$q_k(t) = A(t,\mu)q_{k-1}(t) - \dot{q}_{k-1}(t), \; q_0(t) = b^i(t,\mu),$$

$$i = 1,\ldots,r \;, \; t \geq t_0,$$

where $b^i(t,\mu)$ is the ith column of the matrix $B(t,\mu)$ of the system

$$\frac{dx}{dt} = A(t,\mu)x + B(t,\mu)u(t) \quad , \qquad (53)$$

is nonsingular for each $i = 1,\ldots,r$ for almost all $t \geq t_0$, then we call the system (53) normal.

Let system (53) be normal. Let it be required to minimize the norm $||x(t_1) - c_2||$ under the conditions that

$$|u\ (t)| \leq L \;, \quad ||x(t_0) - c_1|| \leq \delta_1 \;, \qquad = 1,\ldots,r \;.$$

Here t_1 is a fixed moment of time, δ is a given number, and c_1, c_2 are given points. We let $\delta^o(c_1,c_2,\mu)$, $u^o(c_1,c_2,\mu,t)$ denote the solution to this problem.

Theorem 18. *The functions $u^o(c_1,c_2,\mu,t)$, $\delta^o(c_1,c_2,\mu)$ are continuous in c_1,c_2,μ at each point of $E_n \times E_n \times \Delta$, if $\delta^o > 0$.*

Proof. In the case under consideration, the quantity c, the transformation S, and the matrix F from (6.127) depend on μ. Therefore, from now on will write them with indices: c_μ, S_μ, F_μ. By virtue of Theorem 6.26, the quantity $\delta^o = \delta^o(c_1, c_2, \mu)$ is defined by the condition

$$\delta^o(c_1,c_2,\mu) = \max_{||g||\leq 1} \Lambda(c_1,c_2,\mu,g) = \Lambda(c_1,c_2,\mu,g_o), \quad (54)$$

$$\Lambda(c_1,c_2,\mu,g) = g[c_\mu + F_\mu(t_1)F_\mu^{-1}(t_o)c_1 - c_2]$$
$$- L||g'S_\mu|| - \delta_1||g'F_\mu(t_1)F_\mu^{-1}(t_o)|| . \quad (55)$$

Continuity of $\mu^o(c_1,c_2,\mu)$ in c_1, c_2, μ is a corollary of the continuity of the function (55) in these variables. By virtue of Theorem 15, the function $u^o(c_1,c_2,\mu,t)$ is uniquely determined by the relation

$$u^o(c_1,c_2,\mu,t) = -L \text{ sign } g_o'S_\mu \quad . \quad (56)$$

We show convergence in measure of $u^o(c_1,c_2,\mu,t)$. Let $\{c_1,c_2,\mu\}_k \to \{c_1,c_2,\mu\}$ as $k \to \infty$. It is clear that the set of all vectors g_k defining the optimal controls $u^o(c_{1k},c_{2k},\mu_k,t)$ are uniformly bounded in k. If $g_{k_m} \to g$, then

$$\tilde{g}'[c_\mu + F_\mu(t_1)F_\mu^{-1}(t_o)c_1 - c_2] - L||\tilde{g}'S||$$
$$- \delta_1||\tilde{g}'F_\mu(t_1)F_\mu^{-1}(t_o)|| \leq g_o'[c_\mu + F_\mu(t_1)F_\mu^{-1}(t_o)c_1 - c_2]$$
$$- L||g_o'S_\mu||-\delta_1||g_o'F_\mu(t_1)F^{-1}(t_o)|| \quad . \quad (57)$$

On the other hand,

$$g'k_m[c_{\mu_{k_m}} + F_{\mu_{k_m}}(t_1)F_{\mu_{k_m}}^{-1}(t_o)c_{1k_m} - c_{2k_m}]$$
$$- L||g_{k_m}'S_{\mu_{k_m}}||-\delta_1||g_{k_m}'F_{\mu_{k_m}}(t_1)F_{\mu_{k_m}}^{-1}(t_o)||$$

$$\geq g_o'[c_{\mu_{k_m}} + F_{\mu_{k_m}}(t_1 F_{\mu_{k_m}}^{-1}(t_o)c_{1k_m} - c_{2k_m}]$$

$$- L||g_o' S_{\mu_{k_m}}|| - \delta_1||g_o' F_{\mu_{k_m}}(t_1)F_{\mu_{k_m}}^{-1}(t_o)|| \quad . \tag{58}$$

Passage to the limit in (58) and equating it with (57) shows that

$$\tilde{g}'[c_\mu + F_\mu(t_1)F_\mu^{-1}(t_o)c_1 - c_2] - L||\tilde{g}'S_\mu||$$
$$- \delta_1||\tilde{g}'F_\mu(t_1)F_\mu^{-1}(t_o)|| = \delta^o.$$

By virtue of the normality of the system (53), we have sign $g_o'S_\mu$ = sign $\tilde{g}'S_\mu$, almost everywhere on T. Thus, each sequence of controls $u_\nu^o(c_{1k}, c_{2k}, \mu_k, t)$, defined by relation (56), contains a subsequence $u_\nu^o(c_{1k_m}, c_{2k_m}, \mu_{k_m}, t)$ which converges to $u_\nu^o(c_1, c_2, \mu, t)$ almost everywhere. This means [112] that the functions $u_\mu^o(c_{1k}, c_{2k}, \mu_k, t)$ converge to $u_\nu^o(c_1, c_2, \mu, t)$ in measure. The assertion is proved.

2. Linear Systems. The Minimal Time Problem (6.171). We let $T(x_o, \mu)$, $u^o(x_o, \mu, t)$ denote the solution of problems (6.171) (here and below, $T(x_o, \mu)$ is the minimal time). Let system (52) be given.

Theorem 19. In order that the function $T(x_o, \mu)$ be continuous in x_o, μ, it is necessary and sufficient that for each sufficiently small $\sigma > 0$, there exists a neighborhood Δ_σ of the point $\{0, \mu\}$ such that all points $\{x_o', \mu'\} \varepsilon \Delta_\sigma$ have $(t_o + \sigma)$-admissible controls.

Proof. (Necessity) Obvious.

(Sufficiency) Let

$$\beta_\delta = \{\{x_o', \mu'\}: ||x_o' - x_o|| + |\mu' - \mu| \leq \delta\} .$$

We consider the set of trajectories for (52) for $\{x_o', \mu'\} \varepsilon \beta_\delta$, generated by the control $u = u^o(x_o, \mu, t)$. Let $\xi(x_o', \mu')$ be the intersection of this set with the hyperplane $t = t_o + T(x_o, \mu)$. We fix σ and choose $\delta' \leq \delta$ such that $\xi(x_o', \mu') \varepsilon \Delta_\sigma$. This is possible on the basis of the continuous dependence of the solution $x(x_o, \mu, u^o(x_o, \mu, t), t)$ on x_o and μ. If $u(x_o, \mu, t)$ is a $(t_o + \sigma)$-admissible control for the point $\{x_o, \mu\} \varepsilon \xi(x_o', \mu')$, then setting

$$u(x_o', \mu', t) = \begin{cases} u^o(x_o, \mu, t) , & t_o \leq t \leq t_o + T(x_o, \mu) , \\ u_\xi(x_o, \mu, t) , & t_o + T(x_o, \mu) < t , \end{cases}$$

we obtain a T-admissible control $(T = t_o + T(x_o, \mu) + \sigma)$ for the points $\{x_o', \mu'\} \varepsilon \beta_{\delta'}$. By virtue of Theorem 6.32, for these points there exists an optimal control $u^o(x_o', \mu', t)$. By construction

$$T(x_o', \mu') \leq T(x_o, \mu) + \sigma . \quad (59)$$

We assume that

$$\inf_{\{x_o', \mu'\} \varepsilon \beta_{\delta'}} T(x_o', \mu') = T' < T(x_o, \mu) . \quad (60)$$

We extract a weakly convergent (in the $L_2(t_o, t_o + T')$ sense) subsequence $u^o(x_o^k, \mu_k, t)$ from the set $\{u^o(x_o', \mu', t), \{x_o', \mu'\} \varepsilon \beta_{\delta'}\}$ and set $x_o = x_o^k$, $\mu = \mu_k$, $u = u^o(x_o^k, \mu_k, t)$ in (52). It is not difficult to see (a passage to the limit as $k \to \infty$ in (52)) that the point $\{x_o, \mu\}$ may be transferred to $x = 0$ in time $\tau < T(x_o, \mu)$. This is impossible. Consequently, $T' = T(x^o, \mu)$.

By virtue of the arbitrariness of σ, we see that the function $T(x_o, \mu)$ is continuous at the points $\{x_o, \mu\}$. The theorem is proved.

Theorem 20. The minimal time problem (6.171) for the system

$$\frac{dx}{dt} = Ax + bu$$

is well-posed.

Proof. If the system is normal, then for each σ there exists a neighborhood Δ_σ of the point $x = 0$ satisfying the conditions of Theorem 19. Therefore, the function $T(x_o)$ is continuous in x_o. From Theorem 14, it follows that the control $u^o(x_o, t)$ is unique throughout the region Ω.

If the defining equation of this system (cf.(1.110), $A_1 = 0$) is singular, then the function $T(x_o)$ is defined in a region Ω_k of a k-dimensional subspace and, clearly, is continuous in Ω_k. By virtue of Theorem 14, the control $u^o(x_o, t)$ is uniquely determined in Ω_k. Continuity of the function $u^o(x_o, t)$ is proved as in Theorem 18.

From Theorem 19, the following assertion is valid. Problem (6.171) for the stationary system

$$\frac{dx}{dt} = Ax + Bu$$

with boundary conditions $x(t_o) = x_o$, $x(t_1) = a$, $||a|| \neq 0$, is well-posed if and only if the set

$$\{v: \quad v = Aa + Bu, \ |u_\nu| \leq L, \ \nu = 1,\ldots,r\}$$

contains the point $v = 0$.

Theorem 21. If the system (53) is normal, then the minimal time problem (6.171), (53) is well-posed.

Proof. Continuity of the function $T(x_o,\mu)$ follows from Theorem 19 (cf. §1.15).

By virtue of Theorem 15, the functions $u^o(x_o,\mu,t)$ are determined uniquely in Ω. We convince ourselves of the continuity of $u^o(x_o,\mu,t)$ as in the proof of Theorem 18.

3. Nonlinear Systems. We study problem (6.171). Let there be given the equation

$$\frac{dx}{dt} = f(x,u,\mu,t) \quad , \tag{61}$$

where $x \varepsilon E_n$, $u \varepsilon E_r$, $\mu \varepsilon \Delta$, $u(\cdot) \varepsilon U_\infty^L(\cdot)$. We assume that the functions $f(x,u,\mu,t)$, $\partial f/\partial x$, $\partial f/\partial u$ are continuous in $E_n \times E_r \times E_1 \times E_1$, and satisfy a uniform Lipschitz condition in u,μ in E_n, with $f(0,0,\mu,t) = 0$.

Let $x(x_o,\mu,u(\cdot),t)$ be a continuous solution of Eq.(61) generated by the control $u(\cdot) \varepsilon U_\infty^L(\cdot)$. We denote the matrices $\partial f/\partial x$, $\partial f/\partial u$ by $P(x_o,\mu,t)$ and $P_u(x_o,\mu,t)$, respectively, and write the variation of equations (61) along the trajectory $x(t) = x(x_o,\mu,u(\cdot),t)$. We have

$$\frac{d\delta x}{dt} = P(x_o,\mu,t)\delta x + P_u(x_o,\mu,t)\delta u \quad . \tag{62}$$

We let $F(x_o,\mu,t)$ denote the fundamental matrix for the solution of system (62) when $\delta u \equiv 0$. We say that condition A is satisfied if for all continuous curves being solutions of (62) for $u(\cdot)\varepsilon U_\infty^L(\cdot)$ or limits of uniformly convergent sequences of continuous solutions of (61) also generated by controls from $U_\infty^L(\cdot)$, the function

$$[F^{-1}(x_o,\mu,t)Pu(x_o,\mu,t)]_j$$

is totally linearly independent for all $j = 1,\ldots,r$.

Condition A is equivalent to the requirement: system (62) is normal for all continuous solutions of (61) and on limits of uniformly convergent sequences of trajectories of (61). Let condition A be satisfied in the entire space $E_n \times \Delta$.

Theorem 22. If there exists a solution $u^o(x_o,\mu,t)$ of problem (6.171), (61) at the point $\{x_o,\mu\}$, then it is possible to find a δ-neighborhood of the point $\{x_o,\mu\}$ for which the problem also has a solution. The function $T(x_o,\mu)$ is continuous at the point $\{x_o,\mu\}$.

Proof. Since condition A is satisfied, for a given $\tau^o = t_o + T(x_o,\mu)$, $\sigma > \tau^o$, as was shown in [64e], there exists a neighborhood Δ_σ of the point $\{x_o,\mu\}$ such that for $\{x_o',\mu'\}\varepsilon\Delta_\sigma$, it is possible to construct a σ-admissible control $u_\sigma(x_o',\mu',t)$.

Remarks. As in the proof of sufficiency in Theorem 19, lead us to conclude the existence of a neighborhood β_δ of the point $\{x_o,\mu\}$ such that inequality (59) is satisfied, where σ is a sufficiently small number. Since the family of trajectories

$$\{x(x_o,\mu,u(\cdot),t),\ \{x_o,\mu\}\varepsilon\beta_\delta,\ u(\cdot)\varepsilon U_\infty^L(\cdot)\}$$

is uniformly bounded in $x_o,\mu,u(\cdot)$, and uniformly continuous in t(§5.1), inequality (60) is impossible. The theorem is proved.

Corollary. The set Ω of points $\{x_o, \mu\}$ for which the minimal time problem (6.171) has a solution, is open.

Remark. If some conditions are satisfied ensuring the uniqueness of the optimal trajectory for (61), then it is possible to prove continuity of the function $u^o(x_o, \mu, t)$.

2. Differentiability of the Bellman Function

1. On the Existence of a Smooth Bellman Function.

Let the matrix $B(t,\mu)$ in system (53) having dimension $n\times n$, be nonsingular for $t \geq t_o$, $\mu\varepsilon\Delta$, with the control $u(\cdot)$ satisfying the condition

$$u(\cdot)\varepsilon U(\cdot) = \{u(\cdot): \sum_{i=1}^{n} u_i^2(t) \leq L^2, \ t\varepsilon T\} \quad . \tag{63}$$

We consider the following problem. Determine a control $u^o(\cdot)$ for which

$$||x(x_o, \mu, u^o(\cdot), t_1)|| = \min_{u(\cdot)\varepsilon\tilde{U}(\cdot)} ||x(x_o, \mu, u(\cdot), t_1)|| = \delta(x_o, \mu),$$

$$x(t_o) = x_o \neq 0 \quad , \quad ||x|| = (x'x)^{1/2} \quad .$$

We assume that the matrices $A(t,\mu)$, $B(t,\mu)$ are continuously differentiable in μ up to the kth order.

Theorem 23. The function $\delta(x_o, \mu)$ has a continuous derivative of any order in x_o, and a continuous derivative of kth order in μ.

Proof. Let δ be some positive number. We find conditions for which the trajectory of Eq.(53) satisfies the relation

$$x(t_0) = x_0, \quad ||x(x_0,\mu,u(\cdot),t_1)|| \leq \delta, \quad \delta > 0, \quad u(\cdot)\varepsilon\tilde{U}(\cdot). \tag{64}$$

Since, by Cauchy's formula

$$x(x_0,\mu,u(\cdot),t_1) = F_\mu(t_1)F_\mu^{-1}(t_0)x_0 + \int_{t_0}^{t_1} F_\mu(t_1)F_\mu^{-1}(\tau)B(\tau,\mu)u(\tau)d\tau \quad ,$$

conditions (64) (cf. §6.10) are satisfied if and only if

$$\max_{||g||=1} \quad g'c_\mu(x_0) - L\int_{t_0}^{t_1} ||g'S(t_1,t)||dt - \delta||g|| \leq 0. \tag{65}$$

Here

$$c_\mu(x_0) = F_\mu(t_1)F_\mu^{-1}(t_0)x_0 \quad , \quad S(t_1,\mu) = F_\mu(t_1)F_\mu^{-1}(\tau)B(\tau,\mu) \quad .$$

We set

$$\lambda(\delta,t_1,\mu,g) = \delta||g|| + L\int_{t_0}^{t_1} ||g'S(t_1,t)||dt \quad . \tag{66}$$

Condition (65) is equivalent to the inequality

$$\min_g \lambda(\delta,t_1,\mu,g) = \lambda(\delta,t_1,\mu,g_0) \geq 1 \quad , \tag{67}$$

for

$$g'c_\mu(x_0) = 1 \quad . \tag{68}$$

It is not difficult to verify that $\lambda(\delta,t_1,x_o,\mu,g_o)$ is continuous in the variables δ,t_1,x_o,μ. Hence, the function $\delta(x_o,\mu)$ is a solution of the problem

$$\min_g \lambda(\delta,t_1,\mu,g) = \lambda(\delta,t_1,\mu,g_o) = 1 \quad , \tag{69}$$

under condition (68). We assume that $[c_\mu(x_o)]_1 = 0$, and seek the variable $g_1 = [g]_1$ in (67),(68). Then

$$\lambda(\delta,t_1,\mu,g) = \lambda(\delta,t_1,\mu,\tilde{g}) \quad ,$$

$$\tilde{g} = \left\{\left(1 - \sum_{i=2}^{n} g_i[c_\mu(x_o)]_i\right)[c_\mu(x_o)]_1^{-1},\ g_2,\ldots,g_n\right\} \ .$$

The quantities g_{oi}, $i \geq 2$, are determined from the conditions

$$p_i(\delta,t_1,\mu,x_o,g_2,\ldots,g_n) = \frac{\partial\lambda(\delta,t_1,\mu,\tilde{g})}{\partial g_i} = 0 \ . \tag{70}$$

Using (66), it is not difficult to see that the functional determinant for system (70) is different from zero. But, since the functions $p_i(\cdot)$, $i = 2,\ldots,n$, have continuous derivatives of all orders in x_{i_o} and of order k in μ, so do the functions $g_o = g_o(x_o,\mu)$. From (69), we have $\partial\lambda/\partial\delta = ||g|| > 0$. The function $\lambda(\delta,t_1,x_o,\mu,g_o)$ is continuously differentiable in x_o,μ (just as the functions $P_i(\cdot)$).

As above, by virtue of known theorems on implicit functions, we find that the function $\delta(x_o,\mu)$ has a continuous derivative of all orders in x_o and of order k in μ. The assertion is proved.

2. A Theorem on the Existence of a Piecewise-Smooth Bellman Function. For the normal system

$$\frac{dx}{dt} = A(t)x(t) + b(t)u(t) \quad ,$$

where $u(\cdot)\varepsilon U_\infty^L(\cdot)$, we study the solution $u^0(x_0,t)$, $T(x_0)$ of the minimal time problem (6.171). Let $\pi(x,t)$ be the hyperplane in the space $E_n \times t$, $t \leq t_1 + T(x_0)$, where $u^0(x_0,t)$ changes sign.

Theorem 24. The function $T(x_0)$ is continuously differentiable (an arbitrary number of times in x_0, $||x_0|| \neq 0$) for any sought point $\{x_0,t_0\} \varepsilon \pi(x,t)$.

Proof. In (53), let the nonsingular matrix $B(t,\mu)$ have the form

$$B(t,\mu) = \begin{Bmatrix} b_{11}(t) & \mu b_{12}(t) & \cdots & \mu b_{1n}(t) \\ \cdots & \cdots & \cdots & \cdots \\ b_{n1}(t) & \mu b_{n2}(t) & & \mu b_{nn}(t) \end{Bmatrix} , \quad \mu_1 \leq \mu \leq \mu_2 ,$$

$\mu_1\mu_2 < 0$, $u(\cdot)\varepsilon\tilde{U}(\cdot)$. We denote the solution of problem (6.171),(63) by $\tilde{T}(x_0,\mu)$, $\tilde{u}(x_0,\mu,t)$. By virtue of relation (67) and the continuity of the function (65) in t_1, the function $\tilde{T}(x_0,\mu)$ satisfies the condition

$$\left.\begin{aligned} &\lambda(t_1,x_0,\mu) = L^{-1} , \quad t_1 = t_0 + \tilde{T}(x_o,\mu) , \\ &\lambda(t_1,x_0,\mu) = \min_{g'c_\mu(x_0)=-1} \int_{t_0}^{t_1} ||g'S(t_1,t)||dt . \end{aligned}\right\} \quad (71)$$

From the nonsingularity of the matrix $B(t,\mu)$, it follows that the system under consideration is normal and, therefore, applying Theorem 21 we have

$$\lim_{\mu\to 0} \tilde{T}(x_o,\mu) = T(x_o) \quad . \tag{72}$$

Since the function $T(x_o,\mu)$ has [74f] continuous derivatives of all orders in x_o, $||x_o|| \neq 0$,

$$\frac{\partial \tilde{T}(x_o,\mu)}{\partial x} = - \frac{\partial \lambda(t_1,x_o,\mu)}{\partial x} \Bigg/ \frac{\partial \lambda(t_1,x_o,\mu)}{\partial t_1} \quad . \tag{73}$$

Let Δ_{x_o} be a neighborhood of the point x_o such that the set $\Delta_{x_o} \cap \pi(x,t_o)$ is empty. From (71),(73), it is clear that the function (73) is uniformly bounded in μ, $\mu_1 \leq \mu \leq \mu_2$.

Since the functions $\partial^2\lambda(0,t_1,x_o,\mu,g)/\partial x^2$, $\partial^2\lambda(0,t_1,x_o,\mu,g)/\partial t_1 \ \partial x$ are uniformly bounded in x_o and μ, the function (73) is uniformly continuous in $x_o \varepsilon \Delta_{x_o}$. Let

$$\lim_{\mu\to 0} \frac{\partial \tilde{T}(x_o,\mu)}{\partial x_o} = \sigma(x_o) \quad .$$

From (72),(73), we have $\sigma(x_o) = \partial T(x_o)/\partial x_o$, $x_o \varepsilon \Delta_{x_o}$. It is clear that the function $\sigma(x_o)$ is continuous. We can easily see that $\tilde{T}(x_o,\mu)$ is uniformly bounded in μ and uniformly continuous in x_o in the region Δ_{x_o}, with derivations of all orders.

The proof is now complete.

3. On the Well-Posedness of Optimization Problems for Systems with a Delay.

Below, we study the solution of problem (7.171) for the system (6.156)

$$\frac{dx(t)}{dt} = A(t)x(t) + A_1(t)x(t - h) + b(t)u(t)$$

and the system

$$\frac{dx(t)}{dt} = f(x(t),x(t-h),u(t),t) \ , \quad h = \text{const.} \qquad (74)$$

We will say that the problem (6.171) is well-posed with h = const, if the optimal control $u^o(x_o(\cdot),t)$ is unique in the region of definition Ω and the functions $t(x_o(\cdot))$, $u^o(x_o(\cdot),t)$ depend continuously (in the sense of Definition 1) on $x_o(\cdot)$.

Theorem 25. If system (6.156) is normal, then the minimal time problem (6.171) is well-posed.

Proof. By virtue of the results of §1.15, system (6.156) is relatively controllable. Therefore, for any $\sigma > t_o$, there exists a neighborhood Δ_σ of the origin $x(\cdot) = 0$, for whose points we may construct σ-relatively admissible controls. As in the proof of Theorem 19, we see that $T(x_o(\cdot))$ is continuous in $x_o(\cdot)$.

By virtue of Theorem 17, the optimal control is uniquely determined by Theorem 6.34. Continuity of the function $u^o(x_o(\cdot),t)$ is proved as in Theorem 18.

Remark. The problem (6.171) is well-posed for the stationary system (6.156). This is proved analogously to Theorem 20.

We consider system (74).

Theorem 26. If the linear system (74) is normal, then the function $T(x_o(\cdot))$ in problem (6.171),(74), is continuous in $x_o(\cdot)$.

For system (74), a theorem analogous to Theorem 19 is valid. Therefore, the proof may be carried out as in the proof of sufficiency in Theorem 19.

Commentary on Chapter VII

1. As already noted above, the scheme of studying the increment of functionals is traditionally used in the calculus of variations to establish sufficient conditions for an extremum. In the theory of optimal processes, this method was first used in [114a] where a series of results on the optimality of controls satisfying the maximum principle was obtained. The sufficiency of the maximum condition for optimization problems which are linear in the state was proved in [114b,77c].

Other methods for obtaining sufficient conditions for optimality are based upon dynamic programming. These procedures reduce to the study of the Bellman equation [11b] or its generalization [77c]. A series of results in this direction was obtained in [18d], where conditions are also formulated under which the maximum principle is a sufficient condition for optimality.

2. Problems of sufficient conditions for optimality arise every time the optimal control is sought from those controls satisfying necessary conditions for optimality. One scheme of proof of the optimality of a control satisfying the necessary conditions for optimality consists in the verification of two properties of the considered problem: 1) the problem has a solution, 2) the solution of the problem is unique. In general, both of these properties are difficult to check. Thus, there arises the problem of seeking other conditions ensuring the optimality of concrete controls.

3. The functions involved in the formulation of Theorem 1 are assumed to be defined and continuous. The inequalities figuring in condition 3) of Theorem 1 are assumed to be satisfied in some neighborhood of the trajectory x(t) for $t \in T$, which contains all trajectories of Eq.(5) generated by all admissible controls.

4. Theorem 2, for a system of ordinary differential equations and for the functional (1) with $\phi(x) = c'x$, $f_{n+1}(x,y,u,t) = 0$, has been proved many times [114b,77c].

5. In [36e], more general sufficient conditions for optimality based upon formula (4) are proved.

6. Just as in Theorem 1, all functions entering into the formulation of Theorem 3 are assumed to be defined and continuous. Conditions 1) and 2) are assumed to be satisfied for $t \in T$, $u \in U$, and for all x,y lying in

the region of possible values of x(t), $t \varepsilon T$, generated by admissible controls.

7. The number ε_1, characterizing Theorem 4 the optimality of a control relative to small perturbations in the mean on σ, satisfies, by virtue of (21), the inequality

$$\varepsilon_1 \leq \frac{\beta_1 - 2(\theta_2 - \theta_1)K_1 L_1 \beta_2}{\beta_2^2[K_2(t_1 - t_o) + K_3]} .$$

To estimate the quantity $\theta_2 - \theta_1$ under which an admissible control from Theorem 4 is optimal relative to perturbations concentrated on s, we deal with the following approach.

Let $G = \sup_{\tilde{u}\varepsilon U, t\varepsilon T} ||\Delta_u f(x,u,t)||$. Then

$\int_{\theta_1}^{\theta_2} ||\Delta_{\tilde{u}} f(x(t),u(t),t)||dt \leq G(\theta_2 - \theta_1)$ and, from (21), it follows that

$$\theta_2 - \theta_2 \leq \frac{\beta}{2K_1 L_1 \beta_2 + \beta_2^2[K_2(t - t_o) + K_3]G} .$$

8. The constants entering into the formulation of Theorem 4, are easily expressed by the characteristics of system (9) and the functional (10):

$$K_1 \leq \left|\left|\frac{\partial \phi(x(t_1))}{\partial x}\right|\right| \exp L_1(t_1 - t_o),\ K_2 \leq K_1 \left|\left|\frac{\partial^2 f(x,u,t)}{\partial x^2}\right|\right| ,$$

$$K_3 \leq \left|\left|\frac{\partial^2 \phi(x)}{\partial x^2}\right|\right| ,\quad \beta_2 \leq \exp L_1(t_1 - t_o) .$$

9. Inequality (17), which lies at the basis of the proof of Theorem 4, follows directly by virtue of the definition of the function $\Delta_{\tilde{u}} H(x,\psi,u,t)$ and property (18) for the increment of the trajectory. In [114a], for the proof of sufficient conditions of optimality additional constraints are imposed on the control u(t), $t\varepsilon T$, which are required by the method of proof.

10. Sufficiency of the maximum principle in minimal time problems for linear, stationary systems was proved in [18d,38b,186], under the assumption that the system is controllable. The general case without this assumption (with regard for state constraints) was shown in [38b,163]. Different conditions for the uniqueness of optimal controls were obtained in the works [64a,74a,n].

CHAPTER VIII

Computational Problems of Optimal Control

§1. Two Methods for the Improvement of Admissible Controls

1. First Approach. Let there be given the system

$$\frac{dx(t)}{dt} = f(x(t), x(t-h(x,u,t)), u(t), t), \quad t \in T = [t_o, t_1], \qquad (1)$$

and let it be required to minimize the functional

$$J(u) = \phi(x(t_1)) \qquad (2)$$

along trajectories $x(t)$, $t \in T$, generated by the initial conditions $x(t) = \Phi(t)$, $t \in S_o$, and functions $u(t)$ of class D_1. We will assume that the moments t_o, t_1 are fixed. We assume that the constraint

$$g(x(t_1)) \leq 0 \quad , \qquad (3)$$

is imposed upon the trajectory $x(t)$. Relative to the functions $f(x,y,u,t)$, $h(x,u,t)$, $\phi(x)$, $g(x)$, $\Phi(t)$, we assume that they satisfy the conditions of Theorem 6.5. In §6.2, it is shown that if the minimum of the functional (2) is assumed on a piecewise-smooth admissible control $u(t)$, then the optimal control $u^o(t)$ satisfies the maximum condition

$$\Delta_{u*} H(x^o(t), y^o(t, u^o(t)), \psi(t), u(t), t) \leq 0 \ , \qquad (4)$$

at each $t \varepsilon T$, for all $u^* \varepsilon U$, where

$$H(x,y,\psi,u,t) = \psi' f(x,y,u,t) \ ,$$

$$y(t,u) = x(t-h(x,u,t)), \quad \frac{d\psi(t)}{dt} = -\frac{\delta\pi(x,\psi,u)}{\delta x(t)} \ ,$$

$$\psi(t_1) = -\mu\frac{\partial\phi(x(t_1))}{\partial x} - \lambda\frac{\partial g(x(t_1))}{\partial x}, \ \lambda,\mu \geq 0, \ \lambda + \mu = 1 \ , \tag{5}$$

$$\pi(x,\psi,u) = \int_{t_0}^{t_1} \psi' f(x,y,u,t) dt \ .$$

Let $u(t)$ be an admissible piecewise-smooth control generating a trajectory $x(t)$, which satisfies condition (3) and gives the value $J(u)$ to the functional (2). If $u(t)$ does not satisfy the maximum condition (4), then obviously it is not an optimal control.

We pose the problem of refining the control $u(t)$, i.e. to find an admissible control $u_1(t)$, $t \varepsilon T$, for which $J(u_1) < J(u)$ and $g(x_1(t_1)) \leq 0$. From the results of §7.1, it follows that the increment of the functional

$$\Delta J(u) = J(u_1) - J(u)$$

along the trajectories $x(t)$ and $x_1(t)$, satisfying the conditions

$$g(x(t_1)) \leq 0 \ , \quad g(x_1(t_1)) \leq 0 \ ,$$

has the form

$$\Delta J(u) = -\int_{t_o}^{t_1} \Delta_{u_1} H(x,y,\psi,u,t)dt$$

$$- \int_{t_o}^{t_1} \Delta x'(t) \frac{\partial}{\partial x} \Delta_{u_1} H(x,y,\psi,u,t)dt$$

$$- \int_{t_o}^{t_1} \Delta x'[\tau(x_1,u_1,t)] \frac{\partial}{\partial y} H(x,x[\tau(x,u_1,t)],\psi,u_1,t)dt$$

$$+ \int_{t_o}^{t_1} \Delta x'[\tau(x,u,t)] \frac{\partial}{\partial y} H(x,x[\tau(x,u,t)],\psi,u,t)dt$$

$$+ \int_{t_o}^{t_1} \frac{\partial}{\partial y} H'(x,x[\tau(x,u_1,t)],\psi,u_1,t) \int_0^1 \left.\frac{dx(s)}{ds}\right|_{\substack{s=\tau(x,u_1,t) \\ -[h(x_1,u_1,t) \\ -h(x,u_1,t)]\theta}}$$

$$\cdot\ d\theta \frac{\partial h'(x,u_1,t)}{\partial x} \Delta x(t)dt$$

$$- \int_{t_o}^{t_1} \frac{\partial}{\partial y} H'(x,x[\tau(x,u,t)],\psi,u,t) \left.\frac{dx(s)}{ds}\right|_{s=\tau(x,u,t)}$$

$$\cdot \frac{\partial h'(x,u,t)}{\partial x} \Delta x(t)dt + o_1(||\Delta x(t_1)||)$$

$$+ \int_{t_o}^{t_1} o(||\Delta x(t)|| + ||\Delta x[\tau(x_1,u_1,t)]||)dt , \qquad (6)$$

where the function $\psi(t)$ satisfies Eq.(5) with the following conditions at the right end:

$$\psi(t_1) = -\frac{\partial\phi(x(t_1))}{\partial x} - \lambda\frac{\partial g(x(t_1))}{\partial x} , \quad \lambda \geq 0 .$$

Under the control u(t), we compute the trajectory x(t) using (1). With the value $\tilde{\psi}(t_1) = \partial\phi(x(t_1))/\partial x$, we compute the solution $\tilde{\psi}(t)$ of Eq.(5) and find the function

$$\tilde{H}(t) = H(x(t), y(t,u(t)), \tilde{\psi}(t), u(t), t) \quad ,$$

$$\tilde{H}_1(t) = H(x(t), y(t,\tilde{u}(t)), \tilde{\psi}(t), \tilde{u}(t), t)$$

$$= \max_{u \in U} H(x(t), y(t,u), \tilde{\psi}(t), u, t) \quad .$$

We set $\sigma_1 = \left[\tau_1 - \frac{\delta_1}{2}, \tau_1 + \frac{\delta_1}{2}\right]$, where τ_1 is the point of maximum of the function $\Delta\tilde{H}(t) = \tilde{H}_1(t) - \tilde{H}(t)$, $t \in T$, while $\delta_1 = \delta_1(\beta)$ is such that

$$\Delta\tilde{H}(t) \geq \beta ||\Delta_{\tilde{u}} f(x(t), y(t,u(t)), u(t), t)|| \quad , \quad t \in \sigma_1 \quad .$$

We compute the number $\varepsilon_1 = \varepsilon_1(\beta)$:

$$\varepsilon_1 = 2\left\{\frac{\beta - [M_8\delta_1 + M_1(\delta_1) + M_2(\delta_1)]\, e^{-2L_1(2+ML_7)(t_1-\theta_1)}}{L_{12} + (\frac{1}{2} ML_{13}M_8 + M_9)(t_1 - \theta_1)e^{-L_1(2+ML_7)(t_1-\theta_1)} + L_{14}L_4}\right\} .$$

The constraints entering into the last formula are given in [36c,e]. We assume that $\beta - M_8\delta_1 - M_1(\delta_1) - M_2(\delta_1) > 0$. If not, we decrease δ_1 until this inequality is satisfied. Since $M_1(\delta_1)$ and $M_2(\delta_1)$ go to zero as $\delta_1 \to 0$, this is always possible. We choose β such that

$$\int_{\sigma_1} ||\Delta_{\tilde{u}} f|| dt < \varepsilon_1 \quad .$$

The control

$$u_1(t) = \begin{cases} u(t) , & t \overline{\varepsilon} \sigma_1 \\ \tilde{u}(t) , & t \varepsilon \sigma_1 \end{cases} ,$$

decreases the value of the functional $J(u)$. If

$$g(x_1(t_1)) \leq \varepsilon_2 ,$$

where ε_2 is the given accuracy, then this operation of refining the control $u(t)$ is completed. If the last inequality is not satisfied, then to compute $u_1(t)$ it is necessary to consider the possibility of violating the boundary condition $g(x(t_1)) \leq 0$. In this case, it is possible to deal with the following form. We calculate the functions $\tilde{\psi}$, $\tilde{H}$, $\tilde{H}_1$, $\Delta\tilde{H}$, and compute the functions $\overline{\psi}$, $\overline{H}$, $\overline{H}_1$, $\Delta\overline{H}$, where $\overline{\psi}(t)$ is the solution of Eq.(5) with the condition $\overline{\psi}(t_1) = - \partial g(x(t_1))/\partial x$;

$$\overline{H}(t) = H(x(t),y(t,u(t)),\overline{\psi}(t),u(t),t) ;$$

$$\overline{H}_1(t) = \max_{u \varepsilon U} H(x(t),y(t,u),\overline{\psi}(t),u,t)$$

$$= H(x(t),y(t,\overline{u}(t)),\overline{\psi}(t),\overline{u}(t),t) ;$$

$$\Delta\overline{H}(t) = \overline{H}_1(t) - \overline{H}(t) .$$

Not citing the cumbersome estimates for the general case, we note that by virtue of the results of [36e] there exists a number δ such that for each special increment of the control $\Delta_{\varepsilon\theta}u$ with $\varepsilon = \delta$, the increment of the functional $\Delta J(u)$ is defined by the value of $\Delta_{u*}H(x,y,\psi,u,t)$. Using the controls $\tilde{u}(t)$ and $\overline{u}(t)$, we calculate the functions

$$G_1(t) = H(x(t), y(t, \tilde{u}(t)), \tilde{\psi}(t), \tilde{u}(t), t) - \overline{H}(t) \quad ,$$

$$G_2(t) = H(x(t), y(t, \overline{u}(t)), \overline{\psi}(t), \overline{u}(t), t) - \tilde{H}(t) \quad .$$

We give the number $\delta_1 > 0$. We compute the functions

$$F_1(t, \delta_1) = \int_{t-\delta_1/2}^{t+\delta_1/2} [\Delta\tilde{H}(\theta) + \chi G_1(\theta)]d\theta \quad ,$$

$$F_2(t, \delta_1) = \int_{t-\delta_1/2}^{t+\delta_1/2} [\Delta\overline{H}(\theta)\ \chi + G_2(\theta)]d\theta \quad ,$$

where χ is some positive weighting coefficient, the value of which depends upon the degree of violation of the boundary conditions, the magnitude of the functional along the trajectory $x(t)$, and the values of the previous refinements $\Delta\phi(x(t_1))$, $\Delta g(x(t_1))$, if they were carried out. Generally speaking, χ is smaller if $g(x(t_1))$ is smaller, and it equals zero if $g(x(t_1))$ is negative. We find $\tau_1(\delta_1)$ and $\tau_2(\delta_1)$ from the conditions

$$F_1(\tau_1, \delta_1) = \max_{t \varepsilon T} F_1(t, \delta_1) \quad , \quad F_2(\tau_2, \delta_1) = \max_{t \varepsilon T} F_2(t, \delta_1) \quad .$$

Let δ_1 be sufficiently small. Then the control $u_1(t)$ is chosen according to the following rule:

$$u_1(t) = \begin{cases} u(t) \ , & t \overline{\varepsilon} \left[\tau_1 - \frac{\delta_1}{2} \ , \ \tau_1 + \frac{\delta_1}{2}\right] \ , \\ \tilde{u}(t) \ , & t \varepsilon \left[\tau_1 - \frac{\delta_1}{2} \ , \ \tau_1 + \frac{\delta_1}{2}\right] \ , \end{cases}$$

if $F_1(\tau_1,\delta_1) > F_2(\tau_2,\delta_1)$, while

$$u_1(t) = \begin{cases} u(t) , & t\bar{\varepsilon} \left[\tau_2 - \frac{\delta_1}{2} , \tau_2 + \frac{\delta_1}{2}\right] , \\ \bar{u}(t) , & t\varepsilon \left[\tau_2 - \frac{\delta_1}{2} , \tau_2 + \frac{\delta_1}{2}\right] , \end{cases}$$

if $F_1(\tau_1,\delta_1) \leq F_2(\tau_2,\delta_1)$.

2. Refinement of Admissible Controls for Linear Systems

Along the trajectories of the system

$$\left.\begin{aligned} \frac{dx(t)}{dt} &= A(t)x(t) + A_1(t)x(t - h(t)) + b(u,t) , \\ x(t) &= \Phi(t) , \quad t\varepsilon S_o , \end{aligned}\right\} \qquad (7)$$

is imposed the condition

$$g(x(t_1)) \leq 0 .$$

We desire to minimize the functional

$$J(u) = \phi(x(t_1)) ,$$

where $g(x)$ and $\phi(x)$ are differentiable, concave functions. In this case, the refinement algorithm described in the preceding section is essentially simplified.

Let $u(t)$, $t\varepsilon T$, be an admissible control, $x(t)$ the corresponding trajectory. We calculate the functions $\tilde{\psi}$, $\tilde{H}$, $\tilde{H}_1$, $\Delta\tilde{H}$, $\bar{\psi}$, $\bar{H}$, $\bar{H}_1$, $\Delta\bar{H}$, G_1, G_2. We find the numbers

$$F_1 = \int_{t_0}^{t_1} [\Delta\tilde{H}(t) + \chi G_1(t)]dt \ , \quad F_2 = \int_{t_0}^{t_1} [\chi\Delta\overline{H}(t) + G_2(t)]dt.$$

If $F_1 > F_2$, then we set

$$u_1(t) = \tilde{u}(t) \quad , \quad t\varepsilon T \quad ,$$

if $F_1 \leq F_2$, then

$$u_1(t) = \overline{u}(t) \quad , \quad t\varepsilon T \quad .$$

An analogous algorithm for the global refinement of an admissible control for system (7) is obtained for the functional

$$J(u) = \int_{t_0}^{t_1} \left[\int f_{1,n+1}(x(t),x(t - h(t)),t) + f_{2,n+1}(u(t),t) \right] dt \quad .$$

Such an algorithm for refining admissible controls may be used in those problems where system (7) is weakly nonlinear in x,y and the function $f_{1,n+1}(x,y,t)$ is strongly concave in $\{x,y\}$. If the function $f_{1,n+1}(x,y,t)$ is arbitrary, then the described global variation of the control does not lead to an improvement, even for linear systems.

Example 1. It is required to minimize the functional

$$J(u) = \int_0^1 x'x \, dt$$

along trajectories of the equation

$$\dot{x} = Ax + bu \ , \quad |u| \leq 1 \ , \quad x(0) = 0 \ .$$

If $u(t)$ is an admissible control then, setting $u_1(t) = \text{sign}\ \psi'(t)b$, where $\psi(t)$ is a solution of the equations $\dot{\psi} = - A'\psi + 2x$, $\psi(1) = 0$, we obtain

$$\Delta J(u) = - \int_0^1 \psi'(t)b[u_1(t) - u(t)]dt + \int_0^1 \Delta x'(t)\Delta x(t)dt \ ,$$

where $\Delta x(t)$ is the variation of the trajectory $x(t)$ due to the control $u_1(t)$. Since the second term is non-negative, we still have the possibility that $\Delta J(u) > 0$.

In fact, let $\dot{x} = u$, $x(0) = 0$, $|u| \leq 1$, $x \varepsilon E_1$,

$$J(u) = \int_0^1 x^2(t)dt \ .$$

We attempt to refine the admissible control $u(t) \equiv \alpha$, $t \varepsilon [0,1]$, $|\alpha| < 1$. We have

$$x(t) = \alpha t \ , \ J(u) = \frac{\alpha^2}{3} \ , \ \psi(t) = \alpha(t^2 - 1) \ , \ u_1(t) = - \text{sign}\ \alpha \ ,$$

$$x_1(t) = - t \text{ sign } \alpha \ , \quad J(u_1) = \frac{1}{3} > \frac{\alpha^2}{3} = J(u) \ ,$$

i.e. the control $u_1(t)$ is "worse" than $u(t)$.

3. The Second Approach.

Again we consider the problem of minimizing the functional (2) along trajectories x(t) of system (1), constrained by condition (3). We assume that in addition to the properties noted in paragraph 1, the function f(x,y,u,t) is differentiable in u, while the set U is convex. From the general formula (6.3) for the increment of a function, we have

$$\left.\begin{aligned} \Delta J(u) &= -\int_{t_0}^{t_1} \Delta u' \frac{\partial}{\partial u} H(x(t),y(t),\psi(t),u(t),t)dt \\ &\qquad + o(||\Delta u(\cdot)||) \ , \\ &o(||\Delta u(\cdot)||) \leq \alpha \int_{t_0}^{t_1} ||\Delta u(t)||^2 dt \ , \end{aligned}\right\} \quad (8)$$

where the function $\psi(t)$ is computed from Eq.(5) with

$$\psi(t_1) = -\frac{\partial\phi(x(t_1))}{\partial x} - \lambda\frac{\partial g(x(t_1))}{\partial x} \ , \quad \lambda \geq 0 \ .$$

Let u(t) be an admissible control not satisfying the necessary condition for optimality

$$\psi'(t) \frac{\partial f(x(t),y(t,u(t)),u(t),t)}{\partial u} u(t)$$

$$= \max_{u\varepsilon U} \psi'(t) \frac{\partial f(x(t),y(t,u(t)),u(t),t)}{\partial u} u \ .$$

We find a control $\tilde{u}(t)$ from the condition

$$\tilde{\psi}'(t)\ \frac{\partial f(x(t),y(t,u(t)),u(t),t)}{\partial u}\ \tilde{u}(t)$$

$$= \max_{u \in U} \tilde{\psi}(t)\ \frac{\partial f(x(t),y(t,u(t)),u(t),t)}{\partial u}\ u \quad ,$$

where $\tilde{\psi}(t)$ is a solution of Eq.(5) with

$$\tilde{\psi}(t_1) = -\frac{\partial \phi(x(t_1))}{\partial x} \quad .$$

From (8), it follows that, for sufficiently small ε, for the admissible control $u_1(t) = u(t) + \varepsilon[\tilde{u}(t) - u(t)]$, we have the inequality $J(u_1) < J(u)$. If $g(x_1(t_1)) \leq \varepsilon_2$, when ε_2 is a given tolerance on the boundary condition, then the refined control is admissible.

In general, the refinement process proceeds according to the following scheme. As above, we find $\tilde{u}(t)$, $\tilde{\psi}(t)$. We compute $\overline{\psi}(t)$, $\overline{u}(t)$:

$\overline{\psi}(t)$ is the solution of Eq.(5) with

$$\overline{\psi}(t_1) = -\frac{\partial g(x(t_1))}{\partial x}$$

The control $\overline{u}(t)$ is such that

$$\overline{\psi}'(t)\ \frac{\partial f(x(t),y(t,u(t)),u(t),t)}{\partial u}\ \overline{u}(t)$$

$$= \max_{u \in U} \overline{\psi}'(t)\ \frac{\partial f(x(t),y(t,u(t)),u(t),t)}{\partial u}\ u.$$

We calculate the numbers

$$E_1 = \int_{t_0}^{t_1} [\tilde{\psi}(t) + \chi\overline{\psi}(t)]' \frac{\partial f(x(t),y(t,u(t)),u(t),t)}{\partial u} \cdot [\tilde{u}(t) - u(t)]dt \quad ,$$

$$E_2 = \int_{t_0}^{t_1} [\tilde{\psi}(t) + \chi\overline{\psi}(t)]' \frac{\partial f(x(t),y(t,u(t)),u(t),t)}{\partial u} \cdot [\overline{u}(t) - u(t)]dt \quad .$$

(The weighting coefficient χ depends on the quantities $\phi(x(t_1))$, $g(x(t_1))$ and the previous refinements $\Delta\phi(x(t_1))$, $\Delta g(x(t_1))$, if they were obtained.) Let $E_1 > E_2$. Then we set $u_1(t) = u(t) + \varepsilon[\tilde{u}(t) - u(t)]$. In the opposite case, $u_1(t) = u(t) + \varepsilon[\overline{u}(t) - u(t)]$.

§2. Combined Refinement of Initial Controls

In the works [20,133], for optimization problems without constraints on the controls, an original method for combining refinements was used to refine the controls. An analogous idea may be applied in problems with constraints on the controls without preliminary passage to an open region of controls.

1. Use of Local Variations. Let us consider the problem of §1. Following the scheme of §1, we find the controls $\tilde{u}(t)$, $\overline{u}(t)$ and the functions $\tilde{H}(t)$, $\overline{H}(t)$, $\tilde{H}_1(t)$, $\overline{H}_1(t)$. We let θ_1 and θ_2 denote points at which the functions $\Delta\tilde{H}(t) = \tilde{H}_1(t) - \tilde{H}(t)$ and $\Delta\overline{H}(t) = \overline{H}_1(t) - \overline{H}(t)$ assume their maximal values.

We construct the control* $u_1(t)$

$$u_1(t) = \begin{cases} u(t) , & t \bar{\varepsilon} \left[\theta_1 - \frac{\delta_1}{2} , \theta_1 + \frac{\delta_1}{2}\right] \\ & \cdot \cup \left[\theta_2 - \frac{\delta_2}{2} , \theta_2 + \frac{\delta_2}{2}\right] , \\ \tilde{u}(t) , & t \varepsilon \left[\theta_1 - \frac{\delta_1}{2} , \theta_1 + \frac{\delta_1}{2}\right] , \\ \bar{u}(t) , & t \varepsilon \left[\theta_2 - \frac{\delta_2}{2} , \theta_2 + \frac{\delta_2}{2}\right] , \end{cases} \tag{9}$$

where δ_1, δ_2 are some positive numbers. The linear terms in δ_1 and δ_2 of the increment of $g(x(t_1))$, generated by the control (9), have the form

$$- a\delta_1 - b\delta_2 , \tag{10}$$

where

$$a = \bar{\psi}'(\theta_1)[f(x(\theta_1),y(\theta_1,\tilde{u}(\theta_1)),\tilde{u}(\theta_1),\theta_1)$$
$$- f(x(\theta_1),y(\theta_1,u(\theta_1)),u(\theta_1),\theta_1)] ,$$

$$b = \bar{\psi}'(\theta_2)[f(x(\theta_2),y(\theta_2,\bar{u}(\theta_2)),\bar{u}(\theta_2),\theta_2)$$
$$- f(x(\theta_2),y(\theta_2,u(\theta_2)),u(\theta_2),\theta_2)] .$$

Analogously to (6.3), we find that the linear terms in δ_1 and δ_2 of the increment of the functional $J(u)$ may be described in the form

---

* For simplicity, we consider only the case when the intervals in the definition of u_1 do not intersect.

$$- c\delta_1 - d\delta_2 \quad , \tag{11}$$

where

$$c = \tilde{\psi}'(\theta_1)[f(x(\theta_1),y(\theta_1,\tilde{u}(\theta_1)),\tilde{u}(\theta_1),\theta_1) - f(x(\theta_1),y(\theta_1,u(\theta_1)),u(\theta_1),\theta_1)] \; ,$$

$$d = \tilde{\psi}'(\theta_2)[f(x(\theta_2),y(\theta_2,\bar{u}(\theta_2)),\bar{u}(\theta_2),\theta_2) - f(x(\theta_2),y(\theta_2,u(\theta_2)),u(\theta_2),\theta_2)] \; .$$

From the definition of the controls $\tilde{u}(t)$, $\bar{u}(t)$, the functions $\tilde{\psi}(t)$, $\bar{\psi}(t)$, and the numbers θ_1 and θ_2, it follows that the coefficients a, b, c, d in expressions (10), (11), satisfy the conditions

$$b \geq 0 \; , \quad b \geq a \; , \quad c \geq 0 \; , \quad c \geq d \; . \tag{12}$$

We assume that along the control $u(t)$, $t \varepsilon T$, at least one of the functionals $g(x(t_1))$, $\phi(x(t_1))$ does not assume a stationary value. Then $b > 0$, $c > 0$ and, by virtue of condition (12), the inequalities $a\delta_1 + b\delta_2 \geq 0$, $c\delta_1 + d\delta_2 \geq 0$ have a positive solution with respect to δ_1, δ_2, with the exception of the case

$$d < 0 \; , \quad a < 0 \; , \quad \frac{c}{-d} < \frac{-a}{b} \quad . \tag{13}$$

<u>Remark</u>. If we admit a violation of the boundary condition by an amount ε, then for any $\varepsilon > 0$ we may find δ_1 and δ_2 leading to a reduction of the criterion outside of the dependence on (13).

2. Refinement of an Admissible Control Using Small Variations. Let the function $g(x)$ assume the value $g(x(t_1))$ along the trajectory $x(t)$. For a given ε, we construct a control $u_1(t)$ on which $\phi(x_1(t_1)) < \phi(x(t_1))$ and $g(x_1(t_1)) \leq g(x(t_1)) + \varepsilon$. We set

$$u_1(t) = u(t) + \varepsilon_1\varepsilon_3(\tilde{u}(t) - u(t)) + \varepsilon_2(1 - \varepsilon_3)(\bar{u}(t) - u(t)) , \tag{14}$$

where $0 < \varepsilon_3 < 1$, while the numbers $\varepsilon_1 \geq 0$, $\varepsilon_2 \geq 0$. For the principal part of the increment of the functional $g(x(t_1))$, we require that the following inequality be satisfied:

$$\int_{t_0}^{t_1} \psi'(t) \frac{\partial f(x(t),y(t,u(t)),u(t),t)}{\partial u} [\varepsilon_1\varepsilon_3(\tilde{u}(t) - u(t)) + \varepsilon_2(1 - \varepsilon_3)(\bar{u}(t) - u(t))]dt \leq \varepsilon . \tag{15}$$

If the control $u(t)$ does not yield a stationary value for the functional $g(x(t_1))$, then

$$\frac{1}{d_1} = \int_{t_0}^{t_1} \psi'(t) \frac{\partial f(x(t),y(t,u(t)),u(t),t)}{\partial u} \cdot [\bar{u}(t) - u(t)]dt > 0 .$$

Therefore, from inequality (15) it is possible to find ε_2:

$$\varepsilon_2 \leq c_1\varepsilon_1\left[\frac{\varepsilon_3}{1 - \varepsilon_3} + \frac{d_1\varepsilon}{2 - \varepsilon_3}\right] ,$$

where

$$c_1 = \int_{t_0}^{t_1} \psi'(t) \frac{\partial f(x(t),y(t,u(t)),u(t),t)}{\partial u} [\tilde{u}(t) - u(t)]dt .$$

We assume that $c_1\varepsilon_1\varepsilon_3 + d\varepsilon > 0$. For $\varepsilon > 0$, this inequality is satisfied if the quantity $\varepsilon_1\varepsilon_3$ is sufficiently small. The principal part of the increment of the functional $\phi(x(t_1))$ has the form

$$- \varepsilon_1\varepsilon_3 \int_{t_0}^{t_1} \tilde{\psi}'t \frac{\partial f(x(t),y(t,u(t)),u(t),t)}{\partial u} [\tilde{u}(t) - u(t)]dt$$

$$- \varepsilon_2(1 - \varepsilon_3) \int_{t_0}^{t_1} \tilde{\psi}'(t) \frac{\partial f(x(t),y(t,u(t)),u(t),t)}{\partial u}$$

$$\cdot [\bar{u}(t) - u(t)]dt ,$$

and is negative if ε_2 is sufficiently small. Hence, it follows that the control (14), for sufficiently small $\varepsilon_1\varepsilon_3$ and ε_2, reduces the value of the functional $\phi(x(t_1))$ without increasing the increment of $g(x(t_1))$ by more than ε. Thus, it is possible to refine the control $u(t)$ both in the direction of decreasing $\phi(x(t_1))$, and in the direction of decreasing $g(x(t_1))$. Variations in the arguments, which are necessary in the latter case, are obvious.

3. Free-Time Optimization Problems. All the approaches presented above for the refinement of controls are related to optimization problems over a fixed time horizon. Let us consider the problem of §1 with the moment t_1 being free. It is known (§6.2) that in this case, the necessary conditions for optimality must

be augmented by the new relation

$$H(x(t_1),x[t_1 - h(x(t_1),u(t_1),t_1)],\psi(t_1),u(t_1),t_1)$$

$$= \frac{\partial\phi(x(t_1),t_1)}{\partial t} \quad .$$

The change in the scheme of refining admissible controls due to this additional condition is the following: if $[t_0,t_1]$ is the interval on which the admissible control $u(t)$ is given, then we compute the quantity

$$\Gamma_1(t_1) = \frac{\partial\phi(x(t_1),t_1)}{\partial t}$$

$$- H(x(t_1),x[t_1 - h(x(t_1),u(t_1),t_1)],\psi(t_1),u(t_1),t_1) \quad .$$

If $\Gamma_1(t_1) \neq 0$, then we define the new interval $[t_0,t_1^1]$, where $t_1^1 = t_1 + \Delta t_1 \text{ sign } \Gamma_1(t_1)$, $\Delta t_1 > 0$. If $\Gamma_1(t_1) > 0$, then we set $u_1(t) = u(t_1)$, $t \geq t_1$. For sufficiently small Δt_1, we have

$$\phi(x_1(t_1^1),t_1^1) < \phi(x(t_1),t_1) \quad .$$

§3. Convergence Proofs for Two Methods of Approximating the Solution of Optimal Control Problems

1. On the Passage to the Limit in the Solution of a Problem with an Integral Constraint.

For the system

$$\frac{dx}{dt} = A(t)x + b(t)u \tag{16}$$

under the control constraint

$$\int_{t_o}^{t_1} |u(t)|^p \, dt \leq L^p \ , \quad L = \text{const} > 0 \ , \tag{17}$$

we study the solution $u^o(p,t)$, $T^o(p)$ of the minimal time problem (6.171). Let the system (16) be normal. From the results of §6.9, it follows that the function $u^o(p,t)$ is uniquely defined by the relation

$$u^o(p,t) = L^q |g'(q)\gamma(t)|^{\frac{q}{p}} \operatorname{sign} g'(q)\gamma(t) \ , \quad (p^{-1} + q^{-1} = 1) \ . \tag{18}$$

Here $\gamma(t)$ is the function (6.173) and $g(q)$ is the solution of the problem

$$\gamma(t_1^o, q) = L^{-1} \ ,$$

$$\gamma(t_1, q) = \min_{g'x_o = -1} \left(\int_{t_o}^{t_1^o} |g'\gamma(t)|^q \, dt \right)^{\frac{1}{q}}$$

$$= \left(\int_{t_o}^{t_1^o} |g'(q)\gamma(t)|^q \, dt \right)^{\frac{1}{q}} \quad \text{for } t_1^o = t_o + T^o(p). \tag{19}$$

Lemma 1. The function $\lambda(t_1, q)$ is continuous and strictly increasing in t_1.

This assertion is proved analogously to Lemma 6.14. Below, we let $\lambda(t_1)$ denote the function

$$\min_{g'x_o = -1} \int_{t_o}^{t_1} |g'\gamma(t)| dt \ ,$$

corresponding to system (16) with the initial condition $x(t_0) = x_0$.

Lemma 2. If $\lim t_1^{(q)} \to t_1$, then

$$\min_{q \to 1+0} \int_{t_0}^{t_1} \lambda(t_1^{(q)}, q) = \lambda(t_1) \quad .$$

Proof. Let $q_s > 1$, $q_s \to 1$. We set

$$\lambda(t_1) = \int_{t_0}^{t_1} |g_0' \gamma(t)| dt \quad ,$$

$$\lambda(t_1^{(q_s)}, q_s, g) = \left(\int_{t_0}^{t_1^{(q_s)}} |g' \gamma(t)|^{q_s} dt \right)^{\frac{1}{q_s}} \quad ,$$

$$\lambda(t_1^{(q_s)}, q_s, g(s)) = \min_{g' x_0 = -1} \lambda(t_1^{(q_s)}, q_s, g) \quad .$$

By virtue of the definition of $\lambda(t_1, q)$, we have

$$\lambda(t_1^{(q_s)}, q_s, g(s)) \leq \lambda(t_1^{(q_s)}, q_s, g_0) \quad . \tag{20}$$

We will assume, without loss of generality, that the sequence $g(s)$ converges: $g(s) \to \tilde{g}$ as $s \to \infty$. It is clear that

$$\lambda(t_1, 1, \tilde{g}) \geq \lambda(t_1, 1, g_0) \quad . \tag{21}$$

Passing to the limit as $q \to 1 + 0$ in (20) and recalling

inequality (21), we obtain

$$\lambda(t_1,1,\tilde{g}) = \lambda(t_1,1,g_o) = \lambda(t_1) \quad .$$

This argument is valid for any sequence q_s. Thus, the lemma is proved.

Let the control constraint in Eq.(16) have the form $|u(t)| \leq L$. We let $u^o(\cdot)$, T^o be the solution of the minimal time problem (6.171) for this case.

<u>Theorem 1. If there exists a solution $u^o(\cdot)$, T^o, then for any $\varepsilon > 0$, $\sigma > 0$, there is a $p_1 > 1$ such that for $p > p_1$, problem (6.171), (16), (17) has a solution and we have the inequality</u>

$$|T^o(p) - T^o| < \varepsilon \ , \ \text{mes}\ \{t: |u^o(p,t) - u^o(t)| \geq \sigma\} < \varepsilon \quad . \tag{22}$$

<u>Proof</u>. We begin by proving the first inequality. By virtue of Lemma 1, the function $\lambda(t_1,q)$ is strictly increasing in t_1. Therefore, for any $\varepsilon > 0$, there exists a $\beta > 0$ such that

$$\lambda(t_1 - \varepsilon,q) < L^{-1} - \beta \ , \quad \lambda(t_1 + \varepsilon,q) > L^{-1} + \beta \quad .$$

From Lemma 2, there exists a $\delta > 0$ such that

$$|\lambda(t_1 \pm \varepsilon,q) - \lambda(t_1 \pm \varepsilon)| < \beta$$

for $q - 1 < \delta$. Consequently, it is possible to find a moment of time $t_1 = t_o + \tau(p)$ when

$$\lambda(t_o + \tau(p),q) = L^{-1} \ , \quad q - 1 < \delta \quad .$$

Clearly, $|\tau(p) - T^o| < \varepsilon$. But, by virtue of Lemma 1 and relation (19), the quantity $\tau(p)$ is optimal for problem (6.171), (16), (17). This proves the first part of the assertion.

We consider the function (19). In the proof of Lemma 2, it was established that if $q_s \to 1 + 0$ and $g(q_s) \to \tilde{g}$, then the vector $\tilde{g}$ gives a solution of the problem with the computation of $\lambda(t_1)$. By virtue of the continuity of the function $T^o(p)$ in p and Lemma 2, for any $\varepsilon > 0$ it is possible to find a q_1 such that for $q < q_1$, the zeros of $u^o(p,t)$ lie in an ε-neighborhood of the zeros of the function $u^o(t)$. Since the quantity q_s/p_s may be made arbitrarily small, from (18) follows the existence of a $q_2 \leq q_1$ such that for $q < q_2$ the second inequality from (22) occurs. This proves the theorem.

Theorem 1 may be generalized to nonlinear systems.

2. One Approach to the Construction of Approximate Optimal Controls for Nonlinear Systems

Let there be given the system

$$\frac{dx}{dt} = f(x,t) + B(t,\mu)u(t) \quad , \tag{23}$$

where $x \varepsilon E_n$, while the continuous matrix $B(t,\mu)$ has order n and is nonsingular for $t \geq t_o$, $\mu_1 < \mu < \mu_2$. It is assumed that the function $f(x,t)$ is continuous in t, has a continuous, bounded derivative in x, and $f(0,t) = 0$.

For such a system, we consider the minimal time problem (6.171). Let the constraints on the control u(t) have the form

$$u'(t)u(t) \le L^2 \quad . \tag{24}$$

We let $\tilde{u}(\mu,t)$, $\tilde{T}(\mu)$ denote the solution to problem (6.171), (23), (24). Since the matrix $B(t,\mu)$ is nonsingular, by virtue of the results of §1.7, for any $\sigma > t_0$ it is possible to construct a neighborhood of the origin $x = 0$, such that there exist σ-admissible controls for the points of this neighborhood. We present (without proof) an existence theorem for optimal controls.

Theorem 2. In problem (6.171), (23), (24), let there exist a T-admissible control. Then there exists an optimal control.

If $x^0(\mu,t)$ is the corresponding optimal trajectory, then the optimal control has the form

$$\tilde{u}(\mu,t) = LB'(t,\mu)[F^{-1}_{x^0(\cdot)}(t)]'g_0[||B'(t,\mu)(F^{-1}_{x^0(\cdot)}(t))'g_0||]^{-\frac{1}{2}},$$

where $F_{x^0(\cdot)}(t)$ is the fundamental solution matrix for the system

$$\frac{dz}{dt} = P(x^0(\cdot),t)\ z = \frac{\partial f(x^0(\mu,t),t)}{\partial x}\ z \quad ,$$

and g_0 is the solution of problem (7.71) for $F = F_{x^0(\cdot)}$. We let $u^0(t)$, T^0 denote the solution of problem (6.171) for Eq.(23) with the scalar control $u(t)$:

$$\frac{dx}{dt} = f(x,t) + b(t)u(t) \ , \quad |u(t)| \le L \ . \tag{25}$$

Let $b_j(t,\mu) \to 0$, $j \neq 1$, $b_1(t,\mu) \to b(t)$ as $\mu \to \mu_o$, $\mu_o \varepsilon (\mu_1, \mu_2)$ and let condition A (cf. §7.10) be satisfied for the system

$$\frac{dz}{dt} = P(x(\cdot),t)z + b(t)u(t) \quad .$$

Theorem 3. If there exists a solution to problem (6.171), (23), (24), then problem (6.171), (25) also has a solution and $\lim_{\mu \to \mu_o} \tilde{T}(\cdot) = T^o$.

Proof. The function $u(\cdot) = \{u^o(\cdot), 0, \ldots, 0\}$ is a $(t_o + T^o)$-admissible control for the problem (6.171), (23), (24). By virtue of Theorem 2, for this problem there exists an optimal control $\tilde{u}(\mu,t)$ for all μ. Obviously,

$$\tilde{T}(\mu) \le T^o \quad . \tag{26}$$

Let $T^1 = \inf_{\mu} \{\tilde{T}(\mu)\}$. From the set $\{\tilde{u}(\mu,t), \mu_1 < \mu < \mu_2\}$, we extract a sequence $\tilde{u}(\mu_k,t)$ which possesses the following properties:

a) $\lim \tilde{u}(\mu_k,t) = u(t)$ weakly (in the $L_2(t_o, t_o + T^1)$ sense),

b) $\lim x(x_o, \mu_k, \tilde{u}(\mu_k,t),t) = x(t)$ uniformly in t, $t_o \le t \le t_o + T^1$,

c) $\lim T(\mu_k) = T^1$.

Such a choice of $\tilde{u}(\mu_k,t)$ is possible on the basis of the constraints imposed on the functions $\tilde{u}(\mu,t)$, $\partial f/\partial x$ and inequality (26).

If we have the inequality $T^1 < T^o$, then from a), b), c) it follows that the control $u(\cdot)$ generates a trajectory $x(\cdot)$ of system (25) for which $x(t_o + T^1) = 0$, which is impossible. Hence, $\lim \tilde{T}(\mu_k) = T^o$.

Remark. In (25), if the optimal control is unique, then $\lim_{\mu\to\mu_o} \tilde{u}(\mu,t) = u^o(t)$ in measure.

§4. Numerical Algorithm for the Solution of Several Optimal Control Problems

Below we describe several computational schemes based upon the results of §6.9 - 6.13, leading to detailed descriptions for most extensions of the problems; discussed are the problems of constructing a first approximation and the choice of the step size.

1. A General Scheme for the Computation of Optimal Controls

We consider problems from §6.12. As follows from Theorem 6.37, optimal controls are determined by the vector $\{g_o,f_o\}$ which is the solution of the finite-dimensional problem (6.204). We will assume that we have a method to determine the maximum of $H(g,f,u(\cdot))$ and we describe some of the possible computational schemes.

A_1. Minimization of the time T.

We choose $T(0) > 0$. If $\Lambda(L_1,\ldots,L_m,0) > 0$, then we find the zeroth approximation $\{g(0),f(0)\}$ from the condition

$$\Lambda(L_1,\ldots,L_m,T(0),g(0),f(0))$$

$$= \max_{||\{g,f\}||=1, f\geq 0} \Lambda(L_1,\ldots,L_m,T(0),g,f) = \Lambda(0) \quad .$$

If $T(0)$ is sufficiently small, then $\Lambda(0) \geq 0$ and $\Lambda(L_1,\ldots,L_m,T) > 0$ for $T < T(0)$. If $\Lambda(0) = 0$, then the problem is solved and $T(0) - t_o$ is the minimal time. Let $\Lambda(0) > 0$. For the first approximation, we choose the number $T(1) = T(0) + \Delta T(0)$, where $\Delta T(0) \geq 0$ is a small increment. The quantities $\{g(1),f(1)\}$ are determined from the condition

$$\Lambda(L_1,\ldots,L_m,T(1),g(1),g(1))$$

$$= \max_{||\{g,f\}||=1, f\geq 0} \Lambda(L_1,\ldots,L_m,T(1),g,f) = \Lambda(1) \quad .$$

If there does not exist a $\Delta T(0) > 0$ such that $\Lambda(1) > 0$, then $T^o = T(1)$. In the opposite case, for sufficiently small $\Delta T(0)$ we have $\Lambda(1) \geq 0$, $\Lambda(L_1,\ldots,L_m,T) > 0$, if $T < T(1)$. The number $\tau = T(1) - t_o$ is the minimal time when $\Lambda(1) = 0$. If $\Lambda(1) > 0$, the process is continued.

The function $\Lambda(L_1,\ldots,L_m,T,f,g)$ has no local maxima. The quantities $\Lambda(k)$ may be determined by the method of steepest ascent.

A_2. Let the functions $\Lambda(L_1,L_2,\ldots,L_m,T)$ be strictly increasing in T, when $(L_1,\ldots,L_m,0) > 0$.

We consider the minimal time problem. For the zeroth approximation, we choose the numbers $\{g(0),f(0)\}$,

$f(0) \geq 0$, $||\{g(0),f(0)\}|| = 1$. We compute $T(0)$ from the equation

$$\Lambda(L_1,\ldots,L_m,T(0),g(0),f(0)) = 0 \quad .$$

We find the first approximation from the condition

$$\Lambda(L_1,\ldots,L_m,T(0),g(1),f(1))$$

$$= \max_{||\{g,f\}||=1,f\geq 0} \Lambda(L_1,\ldots,L_m,T(0),g,f) = \Lambda(0) \quad .$$

If $\Lambda(0) = 0$, then $T(0)$ is the optimal time and $g_o = g(0)$, $f_o = f(0)$. If $\Lambda(0) > 0$, then the process is continued, i.e. we find $T(1)$ from the equation

$$\Lambda(L_1,\ldots,L_m,T(1),g(1),f(1)) = 0$$

and so on. Clearly, $T(k + 1) \geq T(k)$.

A_3. The time $T = T^o$ in case A_2 satisfies the condition.

$$T^o = \max_{g,f} \{T: \Lambda(L_1,\ldots,L_m,T,g,f) = 0\} \quad .$$

In this case, determination of g_o, f_o may be carried out as in [21d]. In this work, problem (6.171) is considered.

A_4. Methods A_1, A_2 valid for minimization of the quantities L_j.

If $\sum_{i=1}^{m-3} g_i' S_{ik} \neq 0$ for all $g_i \neq 0$, then

$$L_k^o = \max_{\{g_i, f\}} \left\{ L_k : \sum_{i=1}^{m-3} g_i h^i - L_1 f - \sum_{\substack{j=2 \\ j \neq k}}^{m} L_j || \sum_{i=1}^{m-3} g' S_{ij} || \right.$$

$$\left. + \min_{u(\cdot) \varepsilon U(\cdot)} H(g,f,u(\cdot)) \right\} / || \sum_{i=1}^{m-3} g' S_{ik} || = 1$$

$$f \geq 0 , \quad k \neq 1 \quad .$$

This means that scheme A_3 is acceptable in the last case.

2. The Problem of Minimizing the Norm of the Terminal State

Let it be required to find a control $u(\cdot) \varepsilon U_\infty^L(\cdot)$ which minimizes the functional

$$J(u) = ||x(x_o, u(\cdot), t_1)|| , \quad J(u^o) = \delta^o , \quad ||x||^2 = x'x \quad ,$$

along trajectories of the system (16) with $x(t_o) = x_o$ and where t_1 is a fixed moment of time, x_o a given point.

By virtue of Theorem 6.26, we have

$$\delta^o = \max_{||g|| \leq 1} \{g'c - L||g'S||\} = g_o'c - L||g_o'S||$$

$$= g_o' F(t_1) F^{-1}(t_o) x_o - L \int_{t_o}^{t_1} |g_o' F(t_1) F^{-1}(\tau) b(\tau)| d\tau \quad .$$

We assume that $\delta^o > 0$. Then for determination of the optimal control $u^o(\cdot)$ we have (cf. Theorem 6.26) the following relation

$$g_0'F(t_1)F^{-1}(t)b(t)u^o(t) = \min_{|u|\leq L} g_0'F(t_1)F^{-1}(t)b(t)u \quad .$$

Consequently,

$$u^o(t) = -L \text{ sign } g_0'F(t_1)F^{-1}(t)b(t) \quad . \tag{27}$$

Below it is assumed that system (16) is normal. If not, to compute $u^o(\cdot)$ we use the results of §7.8. Thus, for computation of $u^o(\cdot)$ it is necessary to find the maximum of the concave function

$$\Lambda(g) = g'c - L||g'S|| \quad . \tag{28}$$

We introduce the set

$$G = \{x: x = Su(\cdot) + F(t_1)F^{-1}(t_0)x_0, u(\cdot)\varepsilon U_\infty^L(\cdot)\} \quad .$$

Let g_1 be a vector such that

$$g_1'c > 0 \ , \ \Lambda(g_1) > 0 \ , \ ||g_1|| = 1 \quad . \tag{29}$$

The quantity $\Lambda(g_1)$ equals the minimal projection of the set G on g_1. Therefore, the set of points

$$\Pi_1 = \{x: x = \Lambda(g)g, ||g|| = 1, \Lambda(g) > 0\}$$

is a part of the pedal [116] of the boundary of G. Let $\tilde{x} = \Lambda(g_1)$, $g_1 \varepsilon \Pi_1$. We find $u_1(\cdot)$ from the condition

$$g_1'Su_1(\cdot) = \min_{u(\cdot)\varepsilon U_\infty^L(\cdot)} g_1'Su(\cdot) \tag{30}$$

and determine

$$x_1 = Su_1(\cdot) + c \quad . \tag{31}$$

It is clear that $(\tilde{x} - x_1)'g_1 = 0$. Therefore, $\Lambda(g_1) \leq ||x_1||$ (the equality $\Lambda(g_1) = ||x_1||$ is possible only under the condition $\Lambda(g_1)g_1 = x_1$). We consider the points

$$x(\alpha) = (1 - \alpha)\Lambda(g_1)g_1 + \alpha x_1 \ , \quad 0 \leq \alpha \leq 1 \quad .$$

There exists a unique α^* for which $x(\alpha^*)\varepsilon\Pi_1$. By definition $x(\alpha^*) = \Lambda(g_2)g_2$ and $g_1'(\Lambda(g_1)g_1 - \Lambda(g_2)g_2) = 0$. This means that $\Lambda(g_1) = \Lambda(g_2)g_1'g_2$. If $g_2 \neq g_1$, then

$$\Lambda(g_2) > \Lambda(g_1) \quad . \tag{32}$$

We have the inequality

$$\Lambda\left(\frac{\Lambda(g_1)g_1 + \Lambda(g_2)g_2}{||\Lambda(g_1)g_1 + \Lambda(g_2)g_2||}\right) > \min[\Lambda(g_1),\Lambda(g_2)] \ . \tag{33}$$

Inequalities (32), (33) lie at the foundations of the algorithms presented below.

1°. Let g_1 be a vector satisfying condition (29). We find a function $u_1(\cdot)$ from condition (30) and determine x_1 according to (31). We solve the equation

$$g_1'(\Lambda(g_1)g_1 - \Lambda(g(\alpha))g(\alpha)) = 0 \quad ,$$

$$g(\alpha) = (1 - \alpha)\Lambda(g_1)g_1 + \alpha x_1 \quad . \tag{34}$$

Let α^* be the solution of (34). For the second approximation, we choose the vector $g_2 = g(\alpha^*)/||g(\alpha^*)||$.

The process is then continued. As a result, we obtain sequences g_k, $\Lambda(g_k)$, $u_k(\cdot)$, x_k. The sequence $\Lambda(g_k)$ is monotonically increasing (see (32)), and is bounded from above. We assume that

$$\lim_{m\to\infty} \Lambda(g_{km}) = \Lambda^* \neq \max_{||g||=1} \Lambda(g) = \Lambda(g_0) \quad .$$

Then there exists $g \neq g_0$ for which $\Lambda(g) = \Lambda^*$. We take g as the first approximation. From (30), (31), (34) we find $\tilde{g} \neq g$ for which $\Lambda(\tilde{g}) > \Lambda(g)$, which contradicts the definition of Λ^*. Hence, $\Lambda^* = \Lambda(g_0)$. Since g_0 is the unique extremal element of the problem, $g_k \to g_0$ as $k \to \infty$, $u_k(\cdot) \to u^0(\cdot)$ in measure, and

$$x_k \to x(x_0, u^0(\cdot), t_1) \quad .$$

2^0. We construct the first and second approximations g_1, g_2 as described above. We obtain

$$g_3 = [\Lambda(g_1)g_1 + \Lambda(g_2)g_2]/||\Lambda(g_1)g_1 + \Lambda(g_2)g_2|| \quad .$$

It is now possible to follow one of two paths:

α) to find g_4 from Eq.(32),

β) to set

$$g_4 = [\Lambda(g_2)g_2 + \Lambda(g_3)g_3]/||\Lambda(g_2)g_2 + \Lambda(g_3)g_3|| \quad .$$

On succeeding steps, it is possible to average several of the previous g_i. We carry out the averaging until the inequality

$$\Lambda\left(\frac{\Lambda(g_i)g_i + \Lambda(g_{i+1})g_{i+1}}{||\Lambda(g_i)g_i + \Lambda(g_{i+1})g_{i+1}||}\right) > \max\{\Lambda(g_i), \Lambda(g_{i+1})\} \quad .$$

is satisfied.

It is not difficult to see that this process will converge for a second order system. For terminal control systems of higher order, we obtain the successive g_i from Eq.(34). Thus, Eq.(34) serves here as a test. The above method is particularly convenient for terminal control processes.

Remark. If two terms $\Lambda(g_k) = \Lambda(g_{k+1})$ are equal in the sequence $\Lambda(g_k)$, then $g_k = g_0$ and the process is completed.

3. The Method of Steepest Ascent

We again study the problem of minimizing the norm of the terminal state. Let

$$\Lambda_1(g) = [g'c - L||g'S||]/||g|| \quad .$$

It is not difficult to see that

$$\text{grad } \Lambda_1(g)\Big|_{||g_1||=1} = x_1 - \Lambda(g_1)g_1 \quad ,$$

where the point x_1 satisfies the relations (31), (30). We construct the method of steepest ascent according to the known procedure. Let $||g_1|| = 1$. We find $u_1(\cdot)$, x_1 by the rules of paragraph 2. We construct the element $g(\alpha) = g_1 + \alpha(x_1 - \Lambda(g_1)g_1)$. We determine α_1 from the condition $\Lambda_1(g(\alpha_1)) = \max\limits_{0\leq\alpha\leq 1} \Lambda_1(g(\alpha))$. We obtain $g_2 = g(\alpha_1)$ and so on.

Other variants of the method of steepest ascent are possible. From (28), we have

$$\text{grad } \Lambda(g)\Big|_{||g_1||=1} = c + Su_1(\cdot) = x_1 \quad ,$$

where the function $u_1(\cdot)$ satisfies condition (30). We set

$$g(\alpha) = [\alpha g_1 + (1 - \alpha)x_1]/||\alpha g_1 + (1 - \alpha)x_1||$$

and compute α_1 from the condition $\Lambda(g(\alpha_1)) = \max_{0 \le \alpha \le 1} \Lambda(g(\alpha))$. We obtain $g_2 = g(\alpha_1)$ and so on.

4. The Problem of Maximizing the Norm of the Terminal State

Let the initial state $x(t_o) = x_o$ of system (16) be given and fix a moment of time t_1. It is required to find a function $u^o(\cdot)$ under which

$$||x(x_o, u^o(\cdot), t_1)|| = \max_{u(\cdot) \varepsilon U_\infty^L(\cdot)} ||x(x_o, u(\cdot), t_1)|| = \Delta^o .$$

By virtue of Theorem 6.39, we have

$$\Delta^o = \max_{||g||=1} [g'c + L||g'S||] = \tilde{g}_o'c + L||\tilde{g}_o'S|| \quad . \qquad (35)$$

The optimal control is given by the relation (27) with $g_o = -\tilde{g}_o$. Clearly, $\Delta^o > 0$. The function

$$\Lambda_2(g) = g'c + L||g'S||$$

is convex in g, so that it is possible to carry out the calculation of $\tilde{g}_o$ from (35). In this case, it is impossible to guarantee that the local maximum coincides with the global. Let $g = g_1$, $||g_1|| = 1$. The quantity $\Lambda_2(g_1)$ equals the maximal projection of the points of the admissibility set G on g_1. Therefore, the set

$$\Pi_2 = \{x\colon x = \Lambda_2(g)g\ ,\quad ||g|| = 1\}$$

coincides with part of the pedal of the boundary of G. The ray directed into the subspace $\Gamma = \{g\colon g'c > 0\}$, always intersects the set Π_2. This means that for one of the vectors g_1, $-g_1$ we always have $\Lambda_2(g) > 0$, i.e. problems in the choice of the first approximation do not arise here.

We construct the succeeding approximations by the following scheme. The first approximation is the vector g_1, $||g_1|| = 1$, belonging to Γ. We find $u^1(\cdot)$ from the condition

$$g_1'Su_1(\cdot) = \max_{u(\cdot)\varepsilon U_\infty^L(\cdot)} g_1'Su(\cdot)\quad .$$

We compute $x_1 = c + Su^1(\cdot)$ and obtain

$$g(\alpha) = [(1-\alpha)\Lambda_2(g_1)g_1 + \alpha x_1]/||(1-\alpha)\Lambda_2(g_1)g_1 + \alpha x_1||\quad .$$

There exists a unique α^* for which

$$g_1'[\Lambda_2(g_1)g_1 - \Lambda_2(g(\alpha))g(\alpha)] = 0\quad . \tag{36}$$

For the second approximation we choose $g_2 = g(\alpha^*)$. If $g_2 \neq g_1$, then from (36) follows the inequality $\Lambda_2(g_1) < \Lambda_2(g_2)$. If $g_2 = g_1$, then g_1 provides the maximum for $\Lambda_2(g)$. In the event $g_2 \neq g_1$, the process continues. It is easy to see that $\Lambda_2(g_k) \to \Lambda_2(g^*)$ as $g_k \to g^*$. We also have $\Lambda_2(g^*) \geq \Lambda_2(g_k)$ for $k > k_1$.

5. Successive Approximations in the Minimal Time Problem

For Eq.(16), where the control $u(t)$ is constrained by the condition $|u(t)| \leq L$, we consider the minimal time problem (6.171). By virtue of Theorem 6.32, the optimal control is given by the switching function

$$u^o(t) = L \operatorname{sign} g_o' F^{-1}(t) b(t) \quad ,$$

where g_o is the solution of problem (6.175). Since condition (6.175) is equivalent to the equation

$$\max_{||g||=1} \left[g' x_o - L \int_{t_o}^{t_1^o} |g' F^{-1}(t) b(t)| dt \right] =$$

$$= \max_{||g||=1} [g' x_o - L||g'S||] = \tilde{g}_o' x_o - L||\tilde{g}_o' S|| \quad ,$$

the vector $\tilde{g}_o = -g_o$ may be sought by the following scheme (see also paragraph 1). We give a small number $\varepsilon > 0$, which is determined by the given accuracy of the calculation. Let t_1^1 be a quantity close to t_o. We set $g_1 = x_o/||x_o||$. Then

$$\Lambda(g_1, t_1') = g_1' x_o - L||g_1' S|| \geq \varepsilon \quad .$$

We find $u_1(\cdot)$. We have

$$g_1' S u_1(\cdot) = \max_{u(\cdot) \varepsilon U_\infty^L(\cdot)} g_1' S u(\cdot) \quad ,$$

and set $x_1 = x_o + S u_1(\cdot)$. We construct the element $g(\alpha)$:

$$g(\alpha) = [(1-\alpha)\Lambda(g_1, t_1^1)g_1 + \alpha x_1]/||(1-\alpha)\Lambda(g_1, t_1^1)g_1 + \alpha x_1||$$

and find α^* from the equation

$$g_1'[\Lambda(g_1, t_1^1)g_1 - \Lambda(g(\alpha), t_1^1)g(\alpha)] = 0 \quad .$$

The vector $g(\alpha^*)$ is chosen to be the second approximation g_2. For t_1^2, we take the largest t_1 for which $\Lambda(g_2, t_1) \geq \varepsilon$.

Such sequences g_k, t_1^k satisfy the conditions

$$g_k \to g_o \ , \quad t_1^k \to t_1^o \quad .$$

6. Obtaining the First Approximation

In several cases, obtaining a vector g_1 satisfying condition (29) turns into an independent problem of its own. In the problem of minimizing the norm $||x(t_1)||$, we recommend the following approaches in choosing the vector g_1.

1) Set $g_1 = c/||c||$ (for such a choice of g_1, the first term in (28) takes on the maximum) and solve problem (26) for t_1^1 sufficiently near to t_o. We find $\min\limits_{u(\cdot)} ||x(t_1^1)||$ with some accuracy. We set $g_1 =$ $x(t_1^1)/||x(t_1^1)||$ and solve problem (26) with $t_1^2 > t_1^1$ and so on. After a finite number of steps, we obtain a g_1 satisfying condition (29).

2) We assume $g_1 = c/||c||$ and solve problem (26) with sufficiently small L. We find an approximation to $\min\limits_{u(\cdot)} ||x(t_1)||$. We obtain $g_1 = x(t_1)/||x(t_1)||$ and

increase L. After a finite number of steps, we find a first approximation g_1 satisfying condition (29).

7. Several Recommendations for Choosing the Step Size in the Method of Steepest Ascent

We assume that the process of constructing g_i is at the kth step. We obtain $g_k = g_{k-1}(\alpha_{k-1})$ and we find $u_k(\cdot)$, x_k according to (30), (31). We introduce the element

$$g_k(\alpha) = (1 - \alpha)g_k + \alpha x_k$$

and construct the function $\mu_k(\alpha) = \Lambda(g_k(\alpha))/||g_k(\alpha)||$. We denote $x_k/||x_k||$ by $\bar{g}_k$ and find $\bar{u}_k(\cdot)$, x_k from (30), (31) with $g_1 = \bar{g}_k$. It is easy to calculate

$$\mu_k(0) = g_k' x_k \ , \ \mu_k(1) = \bar{x}_k' x_k/||x_k|| \ ,$$

$$\left.\frac{d\mu_k}{d\alpha}\right|_{\alpha=0} = x_k' x_k - [g_k' x_k]^2 \ , \ \left.\frac{d\mu_k}{d\alpha}\right|_{\alpha=1} = \bar{x}_k' \frac{g_k' x_k x_k - x_k' x_k g_k}{[x_k' x_k]^{3/2}} \ .$$

We approximate the function $\mu_k(\alpha)$ by the values $\mu_k(0)$, $\mu_k(1)$, $d\mu_k/d\alpha|_{\alpha=0,1}$, obtaining a function $\tilde{\mu}_k(\alpha)$. We compute α_k: $\tilde{\mu}_k(\alpha_k) = \max_{0\le\alpha\le1} \tilde{\mu}_k(\alpha)$. We obtain $g_{k+1} = g_k(\alpha_k)$. Hence, the size of the step is determined by the value of α_k.

Remarks. 1) The numbers x_k, $\bar{x}_k$, g_k, $\bar{g}_k$ suffice for determination of the second derivatives $d^2\mu_k/d\alpha^2|_{\alpha=0,1}$.

2) It is not difficult to see that the optimal value δ^o satisfies the inequalities

$$\Lambda(g_k) \leq \delta^o \leq ||x_k|| \;,\; \Lambda(\bar{g}_k) \leq \delta^o \leq ||\bar{x}_k|| \quad ,$$

which allow us to establish a "stopping" criterion in relation to the quantities

$$\delta = ||x_k|| - \Lambda(g_k) \qquad (\text{for } \delta^1 = ||\bar{x}_k|| - \Lambda(\bar{g}_k)) \quad .$$

Let ε be a given accuracy for the calculation. The "stopping" criterion is checked at the beginning of the following step: we find $u_{k+1}(\cdot)$, x_{k+1} and $\varepsilon_{k+1} = ||x_{k+1}|| - g'_{k+1}x_{k+1}$.

If $\varepsilon_{k+1} \leq \varepsilon$, the process is stopped. In the method based upon the solution of Eq.(34), it is possible to approximate the function

$$\phi(\alpha) = g_1'[\Lambda(g_1)g_1 - \Lambda(g(\alpha))g(\alpha)] \quad .$$

For the following iteration, we take the vector (34) with $\alpha = \alpha^*$: $\tilde{\phi}(\alpha^*) = 0$, when $\tilde{\phi}(\alpha)$ is the approximate function.

For the solution of concrete examples, we have used quadratic, cubic, and piecewise-linear approximations.

§5. Generalizations to Systems with Delays and Nonlinear Inputs

We consider the problem of minimizing the functional

$$J(u) = \phi(x(t_1))$$

($\phi(x)$ is a continuous, quasiconvex function) along

trajectories of the linear system

$$\frac{dx(t)}{dt} = A(t)x(t) + A_1(t)x(t-h(t)) + b(u(t),t) ,$$

$$x(t) = \Phi(t) , \quad t\varepsilon S_0 = \{t: t - h(t) \leq t_0\} .$$

Here u(t) is an r-dimensional piecewise-continuous control with values in a compact set U. Following §6.9, 6.10, it is possible to prove that the optimal control $u^o(t)$ satisfies the condition

$$\psi'(t)b(u^o(t),t) = \max_{u\varepsilon U} \psi'(t)b(u,t) ,$$

where $\psi(t)$ is the solution of the equation

$$\frac{d\psi(t)}{dt} = -A'(t)\psi(t) - A_1'(r(t))\psi(r(t))r'(t) , \quad t\varepsilon[t_0,t'],$$

$$\frac{d\psi(t)}{dt} = -A'(t)\psi(t) , \quad t\varepsilon[t',t_1] , \quad \psi(t_1) = -g_o ,$$

$$t' = t_1 - h(t_1) .$$

For the vector g_o, we have the relation

$$g_o's(t_1,x_o) + \int_{t_o}^{t_1} \min_{u\varepsilon U} g_o'S(t,u)dt - \max_{\phi(x)\leq\delta^o} g_o'x$$

$$= \max_{||g||=1}\left[g's(t_1,x_o) + \int_{t_o}^{t_1} \min_{u\varepsilon U} g'S(t,u)dt - \max_{\phi(x)\leq\delta^o} g'x\right],$$

where

$$s(t_1,x_o) = F(t_1,t_o)\Phi(t_o) + \int_{t_o-h(t_o)}^{t_o} F(t_1,r(\tau))A_1(r(\tau))\Phi(\tau)d\tau ,$$

$$S(t,u) = F(t_1,t)b(u,t) \quad ,$$

$$\frac{dF(t_1,\tau)}{d\tau} = -F(t_1,\tau)A(\tau) - F(t_1,r(\tau))A_1(r(\tau))r'(\tau) \quad ,$$

$F(t_1,t_1) = E, \quad F(t_1,\tau) = 0 \quad \text{for } \tau > t_1, \delta^o$ (is the root of the equation)

$$\lambda(t_1,\delta) = \max_{||g||=1} \lambda(g,t_1,\delta)$$

$$= \max_{||g||=1} [g's(t_1,x_o) + \int_{t_o}^{t_1} \min_{u \varepsilon U} g'S(t,u)dt - \max_{\phi(x) \leq \delta} g'x] = 0. \tag{37}$$

Thus, determination of the optimal control is reduced to finding the vector g_o from (37). A successive approximation scheme for solution of an analogous problem was presented in [34i] and developed in [27,40a].

From the results of §6.10, it follows that the minimal value of the functional J(u) may be found from the condition

$$\delta^o = \max_{||g||=1} \{\delta(g): \lambda(g,t_1,\delta(g)) = 0\} \quad .$$

Since the function $\lambda(g,\delta)$ is monotonically decreasing in δ, the following scheme is suggested for computation of δ^o and g_o. Let the vector g_1, $||g_1|| = 1$, satisfy the condition

$$\lambda(g_1,0) > 0 \quad .$$

We find δ_1: $\lambda(g_1,\delta_1) = 0$. Let g_2, $||g_2|| = 1$, be such that $\lambda(g_2,\delta_1) > 0$. We find $\delta_2 = \lambda(g_2,\delta_2) = 0$, and so forth. Thus, we obtain a non-decreasing sequence $\delta_1 < \delta_2 < \ldots$, which converges to the optimal δ^o. On each step, determination of δ_p in many cases is a simple operation. For example, if $\phi(x) = ||x||$, then

$$\delta_p = g_p' s(t_1, x_o) + \int_{t_o}^{t_1} \min_{u \in U} g_p' S(t,u)\,dt \quad .$$

If we assume that the operation of computing $\min_{u \in U} g'S(t,u)$ is known for each g, then the basic difficulty presented by the scheme consists of determination of the elements of the sequence g_p, $p = 1,2,\ldots$. Determination of the first element g_1 may be carried out with the help of a partitioning of the set U or a successive expansion of the interval of regulation from zero to the given length. A concrete scheme depends upon the form of the set U and, in particular cases, is described in [34i].

Let $g_1,\ldots,g_{p-1}$ be given. For determination of g_p, we use the gradient method. We introduce additional assumptions. Assume that for each g, the problems

$$\min_{u \in U} g'S(t,u) \quad , \qquad \max_{\phi(x) \leq \delta} g'x$$

have unique solutions $u(t,g)$ and $x(g)$. We set $g(\alpha) = g_p + \alpha(\bar{g} - g_p)$, where $\bar{g}$ is an arbitrary vector. We denote the value of $\lambda(g,\delta)$ on $g(\alpha)$ by $\nu(\alpha)$. We have

$$\nu(\alpha) = g'(\alpha)s(t_1,x_o) + \int_{t_o}^{t_1} \min_{u \in U} g'(\alpha)S(t,u)\,dt - \max_{\phi(x) \leq \delta} g'(\alpha)x$$

$$\leq g'(\alpha)s(t_1,x_o) + \int_{t_o}^{t_1} g'(\alpha)S(t,u(t,g))\,dt - g'(\alpha)x(g)$$

$$\leq \nu(0) + \alpha(\bar{g} - g_p)'\left[s(t_1,x_o) + \int_{t_o}^{t_1} S(t,u(t,g_p))dt - x(g_p)\right] .$$

On the other hand,

$$\nu(\alpha) \geq g_p's(t_1,x_o) + \alpha(\bar{g} - g_p)'s(t_1,x_o)$$

$$+ \int_{t_o}^{t_1} g_p'S(t,u(t,g_p))dt$$

$$+ \alpha \int_{t_o}^{t_1} (\bar{g} - g_p)'S(t,u(t,g(\alpha)))dt - g_p'x(g_p)$$

$$- \alpha(\bar{g} - g_p)'x(g(\alpha)) = \nu(0) + \alpha(\bar{g} - g_p)'$$

$$\cdot \left[s(t_1,x_o) + \int_{t_o}^{t_1} S(t,u(t,g(\alpha)))dt - x(g(\alpha))\right] .$$

By virtue of the above assumptions,

$$\int_{t_o}^{t_1} S(t,u(t,g(\alpha)))dt \to \int_{t_o}^{t_1} S(t,u(t,g_p))dt ,$$

$$x(g(\alpha)) \to x(g_p) \quad \text{for } \alpha \to 0 ,$$

$$\left.\frac{d\nu}{d\alpha}\right|_{\alpha=0} = (\bar{g} - g_p)'\left[s(t_1,x_o) + \int_{t_o}^{t_1} S(t,u(t,g_p))dt - x(g_p)\right] .$$

This means

$$\frac{\partial \lambda(g_p,\delta)}{\partial g} = s(t_1,x_o) + \int_{t_o}^{t_1} S(t,u(t,g_p))dt - x(g_p) \quad .$$

Consequently, for sufficiently small α, on the element

$$g_{p-1} = g(\alpha) = g_p + \alpha(s(t_1,x_o) + \int_{t_o}^{t_1} S(t,u(t,g_p))dt - x(g_p))$$

the condition

$$\lambda(g_{p+1},\delta) > \lambda(g_p,\delta)$$

is satisfied.

More precise estimates for the choice of α are presented in [27,40a,64i], where "stopping" criteria are also given. In conclusion, we note that the function $\lambda(g,\delta)$ is concave in g which gives the danger of a local maxima in the computation of g_o.

§6. Algorithms Based on the Theory of Games for the Computation of Optimal Controls

In application to the theory of optimal processes, the minimax theorem regards each optimal control problem as some game. Thus, we may turn to the numerical algorithms from game theory. Below, we describe only one typical algorithm.

Let it be required to minimize the norm of the terminal state $||x(t_1)||$ along trajectories of the system

$$\frac{dx}{dt} = Ax + bu \quad ,$$

generated by measurable functions $u(t)$, $t \varepsilon T = [0,t_1]$, from a convex set $U(\cdot)$ of admissible controls. From the results of §6.9, it follows that

$$\delta^o = \min_{u(\cdot)\varepsilon U(\cdot)} ||x(t_1)|| = \min_{x \varepsilon R} \max_{||g|| \leq 1} g'x = \max_{||g|| \leq 1} \min_{x \varepsilon R} g'x. \tag{38}$$

Here $R = \{x: x = x(t_1) = s(t_1,x_o) + Su(\cdot),\ u(\cdot)\varepsilon U(\cdot)\}$. Thus, the expression $\delta(u,g) = g'x$ may be considered as the payoff function of a game, the players of which have strategies g and $x(t_1) = x(u(\cdot))$. In this "game", one of the sides chooses the function, $u(\cdot)\varepsilon U(\cdot)$, having the goal to minimize the payoff, the other chooses the vector g, $||g|| \leq 1$, in an attempt to maximize the payoff. The optimal strategies $u^o(\cdot)$, g^o for each side must satisfy the inequalities $\delta(u^o,g) \leq \delta(u^o,g^o) \leq \delta(u,g^o)$ for any $u(\cdot)\varepsilon U(\cdot)$, $||g|| \leq 1$. The quantity $\delta^o = \delta(u^o,g^o)$ is the value of the game.

For the solution of problem (38), we apply the Brown-Robinson iteration method, which we consider on a continuous game. We write this method as it applies to problem (37). We present the step-by-step solution to the problem.

A. We choose an arbitrary admissible function $u_1(t)$, $t\varepsilon T$, $u_1(\cdot)\varepsilon U(\cdot)$ (if we have any a priori information about the character of the optimal controls, we must use it for the choice of the initial approximation since a successful choice may, to a great degree, influence the speed of convergence of the process. In the absence of such information, we take $u_1(\cdot) = 0$).

B. We find g_1 from the relation

$$g_1'[Su_1(\cdot) + s(t_1,x_0)] = \max_{||g||\leq 1} g'[Su_1(\cdot) + s(t_1,x_0)] .$$

C. We determine an intermediate value $\bar{u}_2(\cdot)$ of the control function giving the minimum of the expression $g_1'Su(\cdot)$:

$$g_1'S\bar{u}_2(\cdot) = \min_{u(\cdot)\varepsilon U(\cdot)} g_1'Su(\cdot)$$

D. We find the second approximation to the value $u_2(\cdot)$ by the formula

$$u_2(\cdot) = \frac{u_1(\cdot) + \bar{u}_2(\cdot)}{2} .$$

We continue this process finding the sequences $\{u_m(\cdot)\}$ and $\{g_m\}$, using the recurrence formulas

$$u_{m+1}(\cdot) = \frac{1}{m + 1}\left(\sum_{k=1}^{m} u_k(\cdot) + \bar{u}_{m+1}(\cdot)\right) , \tag{39}$$

where $\bar{u}_{m+1}(\cdot)$ is such that

$$g_m'S\bar{u}_{m+1}(\cdot) = \min_{u(\cdot)\varepsilon U(\cdot)} g_mSu(\cdot) , \tag{40}$$

$$g_{m+1} = \frac{1}{m + 1}\left(\sum_{k=1}^{m} g_k + \bar{g}_{m+1}\right) , \tag{41}$$

where

$$\bar{g}'_{m+1}[Su_{m+1}(\cdot) + s(t_1,x_o)] = \max_{||g|| \leq 1} g'[Su_{m+1}(\cdot) + s(t_1,x_o)]. \tag{42}$$

It is possible to prove that any convergent subsequences constructed from the convergent sequences above converge to the optimal strategies $u^o(\cdot)$ and g^o. For this, we have

$$\lim \frac{1}{m} \sum_{k=1}^{m} g_k'[Su_k(\cdot) + s(t_1,x_o)] = \delta^o .$$

The described scheme may be programmed for solution on a computer. At each step of the iteration process, problems (39)-(42) are solved. Problem (41) consists of the determination of the extremal element with unit norm in the dual space E_n'. This is a well known operation from functional analysis. In the case when $||x||$ is the euclidean norm, the vector g_{m+1} is obtained by norming the vector $Su_{m+1}(\cdot) + s(t_1,x_o)$, i.e.

$$\bar{g}_{m+1} = \frac{Su_{m+1}(\cdot) + s(t_1,x_o)}{||Su_{m+1}(\cdot) + s(t_1,x_o)||} .$$

Commentary on Chapter VIII

1. The problem of constructing algorithms for the numerical determination of optimal controls is one of the most pressing in the theory of optimal processes. In view of the extraordinary complexity of the problem in the general case, different authors have presented different methods for refinement of initial approximations [20,48a,49,63b,132,133]. The solution of this problem has great value for the optimization of concrete processes because the initial control is usually taken to be one found from experiments with many operators and people employed by the controlled object. Improvement of such "practical" controls by 10-15% may yield a great effect.

2. On each step, the value of the coefficient x in §1 may be given in its dependence on the value $g(x(t_1))$. The choice of this dependency is determined by the actual problem under study.

3. The value δ_1, under which the algorithm presented in §1 automatically improves the control, depends on the degree of nonlinearity of the problem in x and y and upon the structure of the functions $\phi(x)$, $g(x)$. Estimates for δ_1 may be obtained starting from the results of §6.1. In general, these estimates are unwieldy and rough. Therefore, they have not been given here.

4. For the solution of concrete problems, the functions F_1 and F_2 from §1 may be computed for different values of δ_1. This can only improve the results.

5. The quantity ε in §1 is estimated by the constant α which depends on the degree of nonlinearity (on the magnitudes of the second derivatives) of the problem in the variables x, y, u. A priori estimates of the second derivatives may be improved during the course of the computation. To estimate ε, it is useful to take into account the structure of the functional.

6. In §2, for $c_1 > 0$ it is possible to take $\varepsilon < 0$, i.e. the control (14) for sufficiently small $\varepsilon_1\varepsilon_3$ and ε_2 simultaneously reduces $\phi(x(t_1))$ and $g(x(t_1))$. Generally speaking, choice of ε depends on the concrete problem and upon the preceding steps of the refinement process.

7. The principal difference between the scheme described in §§3-5 and the scheme of §1,2 consists in the fact that in §§3-5 the operation of improving the control was carried out in the space of impulses (the variables ψ and g), while the analogous operations were carried out earlier in the space of controls. Successive approximation schemes in the space of impulses may also be found in the works [67,79].

CHAPTER IX

Toward a Theory of Optimal Processes for Discrete Systems

This chapter represents an independent part with its own definitions and notation.

§1. Necessary Conditions for Optimality in Discrete Systems

1. Introduction. Difficulties in the determination of optimal controls in serious optimization problems lead to the study of computational stability under discrete perturbations. On the other hand, there exist optimization problems for which the systems are described by relations of the type

$$x(t + 1) = f(x(t),u(t),t) \quad .$$

Investigations of optimal processes for discrete systems have been pursued in a series of works. The most general approach is contained in [11b] where the optimization problem is solved using dynamic programming. Although we have a preference for this technique compared to other methods, it does not, in general, allow us to formulate explicit necessary conditions which optimal controls must satisfy. Therefore, beginning with the work [74b], a search for necessary conditions of optimality in a form analogous to the maximum principle [109] has been instituted.

For problems with a free endpoint, in [114b] it was shown that the optimal control satisfies a maximum condition if the discrete system has the form

$$x(t + 1) = A(t)x(t) + b(u,t) \quad .$$

Doubt about the possibility of extending the maximum principle to discrete systems of general form, which was expressed in [114b], was corroborated in [216] by an example.

Results of numerours works in the theory of optimal processes for discrete systems may be separated into two groups. Results of the first group are related to attempts to obtain necessary conditions for optimality in the general case. A series of necessary conditions have been given, the basic features of which consist of the replacement of global characteristics given by the maximum principle by local characteristics of the optimal controls. Literature for works of this group may be found in [122,124c]. Results of the second group are related to the identification of systems for which the maximum principle is valid. The basic results in this direction [110,160b] consist of proofs of the maximum principle for discrete systems when the set $\{f(x,u,t), u \varepsilon U\}$ is convex. Below we study both questions of obtaining necessary conditions for discrete systems, and questions of identifying discrete systems whose optimal control satisfies a maximum condition.

2. Notations, Definitions. Let the variables of the vector $x(t) \varepsilon E_n$, describe a discrete process, having the dynamics

$$x(t+1) = f(x(t),u(t),t), \quad x(t_o) = x_o \quad ,$$
$$t \varepsilon T = [t_o, t_o + 1, \ldots, t_1 - 1] \quad , \tag{1}$$

where $u(t) \varepsilon E_r$ is the control vector, t is the number of the current step (in discrete time), and $f(x,u,t)$ is a function which is continuous in x,u,t and differentiable in x. We let $U(t)$, $t \varepsilon T$, denote a given bounded, closed set from E_r.

Let u_j be elements from $U(t)$, α_j numbers such that $\alpha_j \geq 0$, $\sum_{j=1}^{n+1} \alpha_j = 1$, $j = 1,\ldots,n+1$, $V(t)$ a set of $(r\times(n+1) + (n+1))$-dimensional vectors v with components u_j, α_j:

$$V(t) = \{(u_j,\alpha_j) : u_j \varepsilon U(t),\ \alpha_j \geq 0,\ \sum_{j=1}^{n+1} \alpha_j = 1, \quad j = 1,\ldots,n+1\}$$

We let $W(t)$ denote the convex hull of the set $U(t)$:

$$W(t)$$
$$= \{\sum_{j=1}^{r+1} \alpha_j u_j : u_j \varepsilon U(t),\ \beta_j \geq 0,\ \sum_{j=1}^{r+1} \beta_j = 1,\ j = 1,\ldots,r+1\}$$

We introduce the systems

$$y(t+1) = g(y(t),v(t),t),\ y(t_o) = y_o = x_o,\ y(t) \varepsilon E_n,\ t \varepsilon T; \tag{2}$$

$$\left.\begin{aligned} \xi(t-1) &= \frac{\partial g'(y(t),v(t),t)}{\partial y}\,\xi(t),\ \xi(t_1 - 1) = -c,\ t \varepsilon T\ , \\ g(y,v,t) &= \sum_{j=1}^{n+1} \alpha_j f(y,u_j,t), \quad v(t) \varepsilon V(t); \end{aligned}\right\} \tag{3}$$

$$z(t+1) = f(z(t),w(t),t), \quad z(t_0) = x_0, \quad z(t)\varepsilon E_n, \quad t\varepsilon T; \tag{4}$$

$$\zeta(t-1) = \frac{\partial f'(z(t),w(t),t)}{\partial z}\zeta(t), \quad \zeta(t_1 - 1) = -c, \quad w(t)\varepsilon W(t). \tag{5}$$

We let $\Omega(t)$ and $\omega(t)$ denote sets of 2n-vectors generating Eqs.(2),(3) and (4),(5), respectively, under the perturbations $v(s)\varepsilon V(s)$, $w(s)\varepsilon W(s)$, $s\varepsilon T$:

$$\Omega(t) = \{y(t),\xi(t): v(s)\varepsilon V(s), \ s\varepsilon T\} \quad ,$$

$$\omega(t) = \{z(t),\zeta(t): w(s)\varepsilon W(s), \ s\varepsilon T\} \quad .$$

Let $x\varepsilon E_n$, $\psi\varepsilon E_n$. We set

$$U_1(x,\psi,t) = \{u: \psi' f(x,u,t) = \max_{\tilde{u}\varepsilon U(t)} \psi' f(x,\tilde{u},t); \ u\varepsilon U(t)\} \quad ,$$

$$R(x,\psi,t) = \{f(x,u,t): u\varepsilon U_1(x,\psi,t)\} \quad ,$$

$$Q(x,\psi,w,t) = \{u: \psi' \frac{\partial f(x,w,t)}{\partial u} u = \max_{w\varepsilon W(t)} \psi' \frac{\partial f(x,w,t)}{\partial u} u\}.$$

We call a set $\sigma(x,X)$ a <u>star-shaped</u> neighborhood of the point x in the set X, associated with all points $y\varepsilon X$ possessing the property: for each y there exists a sequence of numbers ε_i, $\varepsilon_i > 0$, $\varepsilon_i \to 0$, such that for sufficiently large m, all $y(\varepsilon_i)\varepsilon X$, $i \geq m$, where $y(\varepsilon_i) = (1-\varepsilon_i)x + \varepsilon_i y$. If X is a convex set, then $\sigma(x,X) = X$ for any point $x\varepsilon X$. If the set X is e-convex* [167b], then all the points $(1-\mu)x + \mu y + \beta e\varepsilon X$ for which

---

* A set X is called <u>e-convex</u> if for any $x,y\varepsilon X$ and for each $\mu\varepsilon[0,1]$, there exists a number $\beta \geq 0$ such that $(1-\mu)x + \mu y + \beta e\varepsilon X$.

$\beta(x,y) = \max$, have a star-neighborhood coinciding with X. It is not difficult to construct examples of sets the points of which have $\sigma(x,X)$ consisting of only the single element x. For example, on the line $X = \{x\colon x = 1, x = -1\}$. and on the plane $X = \{x,y\colon x^2 + y^2 = 1\}$.

Let $u(t)$, $u_1(t) \varepsilon U(t)$, $t \varepsilon T$, be two controls; the vector $\Delta u(t) = u_1(t) - u(t)$ is called the increment of the control $u(t)$. We call the increment $\Delta u(t)$ special and denote it by the symbol $\Delta_{\theta,u*}u(t)$, if

$$\Delta u(t) = \begin{cases} 0 \quad , & t \neq \theta \quad , \\ u* - u(\theta) \quad , & u* \varepsilon U(\theta) \quad . \end{cases}$$

By virtue of (1), to the controls $u(t)$, $u_1(t)$ correspond trajectories $x(t)$, $x_1(t)$. The vector $x(t) = x_1(t) - x(t)$ is called the increment of the trajectory corresponding to the increment $\Delta u(t)$. To the special increment $\Delta_{\theta,u*}u(t)$ corresponds the special increment $\Delta_{\theta,u*}x(t)$ which, obviously, has the form

$$\Delta_{\theta,u*}x(t) = 0 \quad , \quad t_0 \leq t \leq \theta \quad ,$$

$$\Delta_{\theta,u*}x(\theta + 1) = f(x(\theta),u*,\theta) - f(x(\theta),u(\theta),\theta) \quad ,$$

$$\Delta_{\theta,u*}x(t + 1) = f(x(t) + \Delta_{\theta,u*}x(t),u(t),t)$$

$$- f(x(t),u(t),t), \quad \theta + 1 \leq t \leq t_1 - 1.$$

For ease of writing, we introduce the notation

$$\Delta_{u*}f(x,u,t) = f(x,u*,t) - f(x,u,t) \quad .$$

We define Δv, Δy, Δw, Δz, etc. analogously.

3. Basic Lemmas.

Lemma 1. Let the control $v(t)\varepsilon V(t)$ *and the corresponding trajectory* $y(t+1)$, $t\varepsilon T$, *be such that there exists a nonzero vector* $c\varepsilon E_n$ *under which, for all* $v^*\varepsilon V(\theta)$, $\theta\varepsilon T$, *we have the inequality*

$$c'\Delta_{\theta,v^*}y(t_1) \geq o(||\Delta_{\theta,v^*}y(t_1)||),\quad o(\alpha) \leq k\alpha^2,\ k \geq 0. \tag{6}$$

If the set $R(x,\psi,t)$ *is convex on* $\Omega(t)$, $t\varepsilon T$, *then there exists a control* $u(t)\varepsilon U(t)$, $t\varepsilon T$, *and a corresponding trajectory* $x(t)$ *of Eq.(1), for which, together with the vector* $\psi(t)\varepsilon E_n$ *from the relation*

$$\psi(t-1) = \frac{\partial f'(x(t),u(t),t)}{\partial x}\psi(t),\quad \psi(t_1-1) = -c, \tag{7}$$

for all $u^*\varepsilon U(t)$, $t\varepsilon T$, *we have the inequality*

$$\psi'(t)f(x(t),u(t),t) \geq \psi'(t)f(x(t),u^*,t)\ . \tag{8}$$

Lemma 2. Let the control $u(t)\varepsilon U(t)$, $t\varepsilon T$, *and the trajectory* $x(t)$ *of Eq.(1) be such that there exists a nonzero vector* $c\varepsilon E_n$, *under which, for all* $u^*\varepsilon U(\theta)$, $\theta\varepsilon T$, *we have the inequality*

$$c'\Delta_{\theta,u^*}x(t_1) \geq o(||\Delta_{\theta,u^*}x(t_1)||)\quad . \tag{9}$$

Then for $u(t)$, $x(t)$, *and* $\psi(t)$, *the following assertions follow from (7)*:

1) *inequality (8) is satisfied for all* $t\varepsilon T$ *and those* u^* *such that*

$$f(x(t),u^*,t)\varepsilon\sigma(f(x(t),u(t),t),f(x(t),u(t),t));$$

2) *if*

$$f(x,u,t) = A(u,t)x + b(u,t)$$

and the quantity $o(\alpha)$ in (9) is nonnegative, then for all $u^*\varepsilon U(t)$, $t\varepsilon T$, inequality (8) is satisfied.

3) *inequality (8) is satisfied for all $u^*\varepsilon U(t)$, if $t = t_1 - 1$ and the quantity $o(\alpha)$ in (9) is nonnegative.*

Lemma 3. Let $f(x,u,t)$ be differentiable in u and let the control $w(t)\varepsilon W(t)$, $t\varepsilon T$, and the corresponding trajectory $z(t)$ of Eq.(4) be such that there exists a nonzero vector $c\varepsilon E_n$, under which, for all $w^*\varepsilon W(\theta)$, $\theta\varepsilon T$, we have the inequality

$$c\Delta_{\theta,w^*}z(t_1) \geq o_1(||w^* - w(\theta)||), \quad o_1(\alpha) \leq k_1\alpha^2, \; k_1 \geq 0. \tag{10}$$

If

1) $Q(z(t),\zeta(t),w(t),t) \subset U(t)$ for $\dfrac{\partial f'(z(t),w(t),t)}{\partial u}\zeta(t) \neq 0,$

2) $w(t)\varepsilon U(t)$ for $\dfrac{\partial g'(z(t),w(t),t)}{\partial u}\zeta(t) = 0$,

then there exists a control $u(t)\varepsilon U(t)$, $t\varepsilon T$, for which $x(t)$ from (1) and $\psi(t)$ from (7) satisfy the inequality

$$\psi'(t)\frac{\partial f(x(t),u(t),t)}{\partial u}u(t) \geq \psi'(t)\frac{\partial f(x(t),u(t),t)}{\partial u}u^*, \tag{11}$$

for all $u^*\varepsilon U(t)$, $t\varepsilon T$.

Lemma 4. Let $f(x,u,t)$ be differentiable in u and let the control $u(t)$ and the trajectory $x(t)$ be such that there exists a nonzero vector $c\varepsilon E_n$ under which, for all $u^* \varepsilon U(\theta)$, $\theta \varepsilon T$, we have the inequality

$$c' \Delta_{\theta, u^*} x(t_1) \geq o_1(||u^* - u(\theta)||)$$

Then for $u(t)$, $x(t)$, and $\psi(t)$ from (7), we have

1) inequality (11) is satisfied for $u^* \varepsilon \sigma(u(t), U(t))$, $t \varepsilon T$,

2) the relation

$$\psi'(t) \frac{\partial f(x(t), u(t), t)}{\partial u} = 0 \quad ,$$

is valid if $u(t)$ is an interior point of the set $U(t)$.

Proof of Lemma 1. From the obvious identity

$$\sum_{t=t_1}^{t_1-1} \xi'(t) \Delta y(t+1)$$
$$= \sum_{t=t_o}^{t_1-1} \xi'(t-1) \Delta y(t) + \xi'(t_1 - 1) \Delta y(t_1) - \xi'(t_o - 1) \Delta y(t_o),$$

by virtue of $\Delta y(t_o) = 0$, $\xi(t_1 - 1) = -c$, we have

$$c' \Delta y(t_1) = - \sum_{t=t_1}^{t_1-1} \xi'(t) \Delta y(t+1) + \sum_{t=t_o}^{t_1-1} \xi'(t-1) \Delta y(t) \ .$$

Here we substitute the values $\Delta y(t+1)$, $\xi(t-1)$ from their defining equations

$$c' \Delta y(t_1) = - \sum_{t=t_o}^{t_1-1} \xi'(t) g(y_1(t), v_1(t), t)$$
$$+ \sum_{t=t_o}^{t_1-1} \xi'(t) g(y(t), v(t), t) + \sum_{t=t_o}^{t_1-1} \xi'(t) \frac{\partial g(y(t), v(t), t)}{\partial y} \Delta u(t) .$$

On the right side, we add the expression

$\xi'(t)g(y(t),v_1(t),t) - \xi'(t)g(y(t),v_1(t),t)$

and use the expression

$$\xi'(t)[g(y_1(t),v_1(t),t) - g(y(t),v_1(t),t)]$$

$$= \xi'(t)\frac{\partial g(y(t),v_1(t),t)}{\partial y}\Delta y(t) + o_2(||\Delta y(t)||),$$

$$o_2(\alpha) \le k_2\alpha^2 , \quad k_2 \ge 0 .$$

As a result, we obtain

$$c'\Delta y(t_1) = - \sum_{t=t_o}^{t_1-1} \xi'(t)\Delta_{v_1(t)}g(y(t),v(t),t)$$
$$- \sum_{t=t_o}^{t_1-1} \xi'(t)\frac{\partial\Delta_{v_1(t)}g(y(t),v(t),t)}{\partial y}\Delta y(t)$$
$$- \sum_{t=t_o}^{t_1-1} o_2(||\Delta y(t)||) . \qquad (12)$$

Let $\Delta v(t)$ be a special increment. Then with regard for (6), we obtain for all $v^*\varepsilon V(\theta)$, $\theta\varepsilon T$:

$$-\xi'(\theta)\Delta_{v^*}g(y(\theta),v(\theta),\theta)$$
$$- \sum_{t=\theta+1}^{t_1-1} o_2(||\Delta_{\theta,v^*}y(t)||) \ge o(||\Delta_{\theta,v^*}u(t_1)||) . \qquad (13)$$

Using the estimates for $o_2(\alpha)$, $o(\alpha)$, we have

$$\xi'(\theta)\Delta_{v^*}g(y(\theta),v(\theta),\theta)$$
$$\le k||\Delta_{\theta,v^*}y(t_1)||^2 + k_2 \sum_{t=\theta+1}^{t_1-1} ||\Delta_{\theta,v^*}u(t)||^2 .$$

From the equation for $\Delta_{\theta,v*}y(t)$, it is not difficult to obtain the estimate

$$||\Delta_{\theta,v*}y(t)|| \leq k_3||\Delta_{v*}g(y(\theta),v(\theta),\theta)||, \quad t \geq \theta + 1 .$$

Therefore, for all $v*\varepsilon V(\theta)$, $\theta\varepsilon T$, we have

$$\xi'(\theta)\Delta_{v*}g(y(\theta),v(\theta),\theta) \leq k_4||\Delta_{v*}g(y(\theta),v(\theta),\theta)||^2 . \quad (14)$$

The set $\{\Delta_{v*}g(y(\theta),v(\theta),\theta)\colon v*\varepsilon V(\theta)\}$ is convex, by definition, and contains the origin. Hence, it follows that

$$\xi'(\theta)\Delta_{v*}g(y(\theta),v(\theta),\theta) \leq 0 \quad , \quad (15)$$

for all $v*\varepsilon V(\theta)$, $\theta\varepsilon T$. To see this, assume the contrary and let θ be a moment from T, v* a control from $V(\theta)$, for which $\xi'(\theta)\Delta_{v*}g(y(\theta),v(\theta),\theta) = \eta > 0$. We construct the control v_ε:

$$\Delta_{v_\varepsilon}g(y(\theta),v(\theta),\theta) = \varepsilon\Delta_{v*}g(y(\theta),v(\theta),\theta), \quad \varepsilon > 0 .$$

Clearly, $v_\varepsilon \varepsilon V(\theta)$ and $\xi'(\theta)\Delta_{v_\varepsilon}g(y(\theta),v(\theta),\theta) = \varepsilon\eta > 0$. We substitute v_ε into (14), obtaining

$$\varepsilon\eta \leq k_4\varepsilon^2||\Delta_{v*}g(y(\theta),v(\theta),\theta)||^2 .$$

The last inequality must be satisfied for all ε, $0 \leq \varepsilon \leq 1$, which is impossible. From (15), we have

$$\xi'(\theta) \sum_{i=1}^{n+1} \alpha_i(\theta)f(y(\theta),u_i(\theta),\theta)$$

$$= \sum_{i=1}^{n+1} \alpha_i(\theta) \max_{u_i\varepsilon U(\theta)} \xi'(\theta)f(y(\theta),u_i,\theta) .$$

Thus, each $u_i(\theta)$, $i = 1,\ldots,n+1$, satisfies the condition

$$\xi'(\theta)f(y(\theta),u_i(\theta),\theta) = \max_{u\varepsilon U(\theta)} \xi'(\theta)f(y(\theta),u,\theta)$$

and, consequently, lies in the set $U_1(y(\theta),\xi(\theta),\theta)$. By assumption, the set $R(y(\theta),\xi(\theta),\theta)$ is convex. Therefore, $\sum_{i=1}^{n+1} \alpha_i(\theta)f(y(\theta),u_i(\theta),\theta) \varepsilon R(y(\theta),\xi(\theta),\theta)$, i.e., there exists a vector $u(\theta)\varepsilon U(\theta)$ such that

$$f(y(\theta),u(\theta),\theta) = \sum_{i=1}^{n+1} \alpha_i(\theta)f(y(\theta),u_i(\theta),\theta)$$

The control $u(\theta)$, $\theta\varepsilon T$, generates the trajectory $x(\theta) = y(\theta)$ and satisfies the condition

$$\xi'(\theta)f(x(\theta),u(\theta),\theta) = \max_{u\varepsilon U(\theta)} \xi'(\theta)f(x(\theta),u,\theta), \quad \theta\varepsilon T,$$

by virtue of (15). This proves Lemma 1.

Proof of Lemma 2. Inequality (14) has been proved for (2),(3); thus, it is valid for the particular case (1),(7):

$$\psi'(\theta)\Delta_{u*}f(x(\theta),u(\theta),\theta) \leq k_4||\Delta_{u*}f(x(\theta),u(\theta),\theta)||^2 ,$$

$$u^*\varepsilon U(\theta), \quad \theta\varepsilon T \quad . \tag{16}$$

Let u^* be such that $f(x(\theta),u^*,\theta)\varepsilon\sigma(f(x(\theta),u(\theta),\theta), f(x(\theta),U(\theta),\theta))$. We show that

$$\psi'(\theta)\Delta_{u*}f(x(\theta),u(\theta),\theta) \leq 0 \quad .$$

In the opposite case, there is a sequence $\varepsilon_i \to 0$, $\varepsilon_i > 0$, and a number m such that $f(x(\theta),u(\varepsilon_i),\theta) \varepsilon f(x(\theta),U(\theta),\theta)$ for $i \geq m$, $\psi'(\theta)\Delta_{u(\varepsilon_i)} f(x(\theta),u(\theta),\theta) = \varepsilon_i \eta$. Here $u(\varepsilon_i)$ is such that

$$f(x(\theta),u(\varepsilon_i),\theta) = (1-\varepsilon_i)f(x(\theta),u(\theta),\theta) + \varepsilon_i f(x(\theta),u^*,\theta),$$

$\eta = \psi'(\theta)\Delta_{u^*} f(x(\theta),u(\theta),\theta) > 0$. Clearly, $u(\varepsilon_i) \varepsilon U(\theta)$ for $i \geq m$. Substituting $u(\varepsilon_i)$ into (16) instead of u^*, we obtain the inequality $\varepsilon_i \eta \leq k_4 \varepsilon_i^2 ||\Delta_{u^*} f(\theta)||^2$, which must be satisfied for all ε_i, $i \geq m$. This is impossible since $\varepsilon_i \to 0$. This proves assertion 1).

Assertion 2) follows from (13) if the latter is written for system (1),(7) and it is considered that $o_2(||\Delta_{\theta,u^*} x(t)||) \equiv 0$ by virtue of the linearity in x of the function f(x,u,t); $o(||\Delta_{\theta,u^*} x(t_1)||) \geq 0$ by assumption.

Assertion 3) also follows from (13) if we set $\theta = t_1 - 1$ and consider that $o(||\Delta_{\theta,u^*} x(t_1)||) \geq 0$. This completely proves Lemma 2.

Proof of Lemma 3. It is not difficult to establish that identity (12) remains unchanged if we substitute y, ξ, v, by z, ζ, w, respectively. We introduce a special control increment for w(t). Then with regard for (10), we have

$$-\zeta'(\theta)\Delta_{w^*} f(z(\theta),w(\theta),\theta)$$

$$- \sum_{t=\theta+1}^{t_1-1} o_2(||\Delta_{\theta,w^*} z(t)||) \geq o_1(||w^* - w||)$$

for all $w^* \varepsilon W(\theta)$, $\theta \varepsilon T$. By virtue of the differentiability of the function f(z,u,t) in u, we obtain

$$\zeta'(\theta)\Delta_{w*}f(z(\theta),w(\theta),\theta)$$

$$= \zeta'(\theta)\ \frac{\partial f(z(\theta),w(\theta),\theta)}{\partial w}(w* - w(\theta)) + o_3(||w* - w(\theta)||)\ ,$$

$$||\Delta_{w*}f(z(\theta),w(\theta),\theta)|| \le k_5||w* - w(\theta)||,\ o_3(\theta) \le k_6\alpha^2.$$

Therefore,

$$\zeta'(\theta)\ \frac{\partial f(z(\theta),w(\theta),\theta)}{\partial w}(w* - w(\theta)) \le k_4||w* - w(\theta)||^2, \qquad (17)$$

where $k_7 = k_1 + k_6 + k_2k_3^2k_5^2(t_1 - t_0)$, and k_3 is such that

$$||\Delta_{\theta,w*}z(t)|| \le k_3||\Delta_{w*}f(z(\theta),w(\theta),\theta)|| \quad .$$

Hence, as for the proof of (15) in Lemma 1, we obtain

$$\zeta'(\theta)\ \frac{\partial f(z(\theta),w(\theta),\theta)}{\partial w}[w* - w(\theta)] \le 0 \quad . \qquad (18)$$

By the conditions of Lemma 3, $w(\theta) < U(\theta)$ if

$$\zeta'(\theta)\ \frac{\partial f(z(\theta),w(\theta),\theta)}{\partial w} = 0 \text{ and } Q(z(\theta),\zeta(\theta),w(\theta),\theta) \subset U(\theta),$$

if $\zeta'(\theta)\ \frac{\partial f(z(\theta),w(\theta),\theta)}{\partial w} \neq 0$. Therefore, there exists a function $u(\theta)\varepsilon U(\theta)$ to which corresponds the trajectory $x(t)$, $\psi(t)$ which, according to (18), satisfy the required inequality (11). Lemma 3 is proved.

Proof of Lemma 4. Assertion 1) is proved by the same scheme as assertion 1) of Lemma 2. Assertion 2) is a natural corollary of assertion 1).

4. Necessary Conditions for Optimality in Free-Endpoint Problems (problem A). We introduce the scalar function

$$H(x,\psi,u,t) = \psi' f(x,u,t) \ , \quad H_1(y,\xi,v,t) = \xi' g(y,v,t) \quad .$$

Then Eqs.(1)-(5),(7) may be written in the following form:

$$x(t+1) = \frac{\partial H(x(t),\psi(t),u(t),t)}{\partial \psi} \ , \quad x(t_0) = x_0 \ , \quad (19)$$

$$\psi(t-1) = \frac{\partial H(x,\psi,u,t)}{\partial x} \ , \quad \psi(t_1) = -c \ , \quad (20)$$

$$y(t+1) = \frac{\partial H_1(y,\xi,v,t)}{\partial \xi} \ , \quad y(t_0) = x_0 \ , \quad (21)$$

$$\xi(t-1) = \frac{\partial H_1(y,\xi,v,t)}{\partial y} \ , \quad \xi(t_1 - 1) = -c \ , \quad (22)$$

$$z(t+1) = \frac{\partial H(z,\zeta,w,t)}{\partial \zeta} \quad z(t_0) = x_0 \ , \quad (23)$$

$$\zeta(t-1) = \frac{\partial H(z,\zeta,w,t)}{\partial z} \ , \quad \zeta(t_1 - 1) = -c \ . \quad (24)$$

Let $\phi(x)$ be a scalar function defined and continuous on E_n, together with grad $\phi(x)$. We will call the control $u(t)$ $(v(t),w(t))$ admissible if $u(t)\varepsilon U(t)$ $(v(t)\varepsilon V(t)$, $w(t)\varepsilon W(t)$, respectively).

By problem A for the optimization of system (19), we understand the problem of determing an admissible (optimal) control $u(t)$, $t\varepsilon T$, under which the criterion

$$J(u) = \phi(x(t_1))$$

of the system (19) is minimized, i.e.

$$J(u) \leq J(\tilde{u}) \quad , \quad \tilde{u}(t) \varepsilon U(t) \quad , \quad t \varepsilon T \quad .$$

The first generalization of problem A (problem 1OA) is the problem of minimizing the function $\phi(y(t_1))$ along trajectories of Eq. (21) generated by the controls $v(t)\varepsilon V(t)$, $t\varepsilon T$.

Minimization of the function $\phi(z(t_1))$ along trajectories of Eq. (23), generated by the controls $w(t)\varepsilon W(t)$, $t\varepsilon T$, is called the <u>second generalization of problem A</u> (problem 2OA).

We call problem A <u>nonsingular</u> if the condition $\text{grad}\ \phi(x(t_1)) = 0$ holds along the optimal trajectory $x(t)$. Nonsingularity of problems 1OA, 2OA is defined analogously.

<u>Definitions</u>

1. The control $u(t)$, $t\varepsilon T$, satisfies the first maximum condition if, jointly with $x(t)$, $\psi(t)$ from (19), (20), for all $u^*\varepsilon U(t)$, $t\varepsilon T$, we have the inequality

$$H(x(t),\psi(t),u(t),t) \geq H(x(t),\psi(t),u^*,t) \quad . \tag{25}$$

2. The control $u(t)$, $t\varepsilon T$, satisfies the second maximum condition if for the $x(t)$, $\psi(t)$ from (19), (20) that correspond to it, for all $u^*\varepsilon U(t)$, $t\varepsilon T$, we have

$$u'(t)\ \frac{\partial H(x(t),\psi(t),u(t),t)}{\partial u} \geq (u^*)'\ \frac{\partial H(x(t),\psi(t),u(t),t)}{\partial u} . \tag{26}$$

3. The control $u(t)$, $t\varepsilon T$, satisfies the first local maximum condition if for the $x(t)$, $\psi(t)$ from (19), (20), that correspond to it, inequality (25) is satisfied for $t\varepsilon T$ and u^* such that $f(x(t),u^*,t)\varepsilon\sigma(f(x(t),u(t),t)$,

$f(x(t),U(t),t))$.

4. The control $u(t)$, $t\varepsilon T$, satisfies the second local maximum condition if for the $x(t)$, $\psi(t)$ from (19),(20) that correspond to it, inequality (26) is satisfied for $t\varepsilon T$, $u^*\varepsilon\sigma(u(t), (t))$.

Now let $u(t)$, $t\varepsilon T$, be an optimal control. Then for any $u(t)$, $t\varepsilon T$, we have

$$\phi(\tilde{x}(t_1)) - \phi(x(t_1)) \geq 0$$

and from the expansion

$$\phi(x(t_1)) = \phi(x(t_1))$$
$$+ \Delta x'(t_1) \text{ grad } \phi(x(t_1)) - o(||\Delta x(t_1)||)$$

follows

$$\Delta x'(t_1) \text{ grad } \phi(x(t_1)) \geq o(||\Delta x(t_1)||) \quad .$$

If problem A is nonsingular, for a special increment of the control $u(t)$ we obtain inequality (9). Analogously, from the optimality of the control $v(t)$, $t\varepsilon T$, follows inequality (6) for $c = \text{grad } \phi(y(t_1))$, where c is a nonzero vector if problem 10A is nonsingular. Finally, assuming $w(t)$, $t\varepsilon T$, is optimal, it is not difficult to obtain (10), where $c = \text{grad } \phi(z(t_1))$, and $c \neq 0$ in the nonsingular problem 20A.

Beginning with these remarks and Lemmas 1-4, we formulate the following theorems which contain necessary conditions for the optimality of controls in problem A for the discrete system (1).

Theorem 1. *Let problem 10A be nonsingular and let the set $R(x,\psi,t)$ be convex on $\Omega(t)$, $t\varepsilon T$. Then there exists an optimal control $u(t)$, $t\varepsilon T$, in problem A which satisfies the first maximum condition; $c = \text{grad } \phi(x(t_1))$.*

Theorem 2. *If problem A is nonsingular, then the optimal control $u(t)$, $t\varepsilon T$, satisfies: 1) the first local maximum condition, 2) the first maximum condition if*

$$f(x,u,t) = A(u,t)x + b(u,t)$$

and the function $\phi(x)$ is concave, 3) the condition

$$H(x(t),\psi(t),u(t),t) \geq H(x(t),\psi(t),u^*,t) \quad ,$$

$$t = t_1 - 1 \quad , \quad u^* \varepsilon U(t_1 - 1) \quad ,$$

if the function $\phi(x)$ is concave.

Theorem 3. *Let $f(x,u,t)$ be differentiable in u, let problem 20A be nonsingular, and assume conditions 1),2) of Lemma 3 are satisfied. Then there exists an optimal control $u(t)$, $t\varepsilon T$, for problem A which satisfies the second maximum condition.*

Theorem 4. *If the function $f(x,u,t)$ is differentiable in u and problem A is nonsingular, then the optimal control satisfies: 1) the second local maximum condition, $c = \text{grad } \phi(x(t_1))$; 2) the equality*

$$\frac{\partial H(x(t),\psi(t),u(t),t)}{\partial u} = 0 \quad ,$$

if $u(t)$ is an interior point of the set $U(t)$, $c = \text{grad } \phi(x(t_1))$.

5. Discussion of Theorems 1-4. Examples.

a) Necessary conditions for optimality in the form of the first maximum condition in problem A for the equation

$$x(t + 1) = A(t)x(t) + b(u,t)$$

and the criterion $\phi(x) = c'x$ were first obtained in [114b]. In [110], it was proved that the first maximum condition occurs for the discrete system (1) if the set $f(x,U,t)$ is convex, bounded, and closed. These results follow from Theorem 2.

The first maximum condition is the strongest of those presented in Theorems 1-4. In a given system, the presence of this property allows us to transform the original minimization problem of dimension $[t_1 - t_0]\times r$ to $t_1 - t_0$ maximization problems of dimension r. In general, the first maximum condition is not satisfied for discrete systems. This was proven in [21b] by construction of a rather complex example. The following example achieves the same goal more simply.

Example 1.

$$x(t+1) = \frac{1}{2}\, u(t), \quad y(t+1) = y(t) + x^2(t) - u^2(t), \quad t = 0,1 \quad ,$$

$$x(0) = y(0) = 0, \quad |u| \leq 1, \quad \phi(x,y) = y, \quad t_0 = 0, \quad t_1 = 2 \quad .$$

We have

$$J(u) = -\frac{3}{4}\, u^2(0) - u^2(1) \quad .$$

The control $\{u(0) = 1,\ u(1) = 1\}$ is optimal. For it, we have $\psi_2(1) = -1$, $\psi_2(0) = -1$, $\psi_1(1) = 0$, $\psi_1(0) = -1$, $H(x(0),y(0),\psi(0),u,0) = -\frac{1}{2}u + u^2$. The point $u = 1$ is not a maximum point of the latter function.

b) The first maximum condition is, in general, not satisfied for those discrete systems which arise as difference approximations to continuous systems.

Example 2.

$$x(t+1) = x(t) + \frac{1}{2}u(t) \ , \quad y(t+1) + y(t) + x^2(t) - u^2(t) \quad .$$

Under the conditions of Example 1, we obtain a map which is identical to the preceding one.

c) In [21b], the local maximum principle was presented as the basic necessary condition for optimality for discrete systems. In this case, the function $H(x,\psi,u,t)$ assumes a local maximum along the optimal control. In examples 1,2 this condition is satisfied. However, in general, this principle is not valid.

Example 3.

$$x(t+1) = 2u(t), \quad y(t+1) = y(t) + x^2(t) - u^2(t) \quad ,$$

$$x(0) = y(0) = 0, \quad |u| \leq 1, \quad \phi(x,y) = y, \quad t_0 = 0, \quad t_1 = 2.$$

We have

$$J(u) = y(2) = 3u^2(0) - u^2(1), \quad |u| \leq 1 \quad .$$

For the optimal control $\{u(0) = 0,\ u(1) = 1\}$, we obtain

$$H(x(0),\psi(0),u,0) = u^2 \quad .$$

At the point $u = 0$, corresponding to the optimal control, this function assumes its absolute minimum.

Example 4.

$$x(t+1) = u(t) \sin \frac{\pi}{2} u(t), y(t+1) = u(t) \cos \frac{\pi}{2} u(t),$$

$$z(t+1) = z(t) + x^2(t) + y^2(t) - u^2(t) \quad ,$$

$$x(0) = y(0) = z(0) = 0, \ |u| \leq 1, \ \phi(x,y,z) = z, \ t_o = 0, \ t_1 = 2.$$

It isn't difficult to compute that the control $\{u(0) = 0, u(1) - 1\}$ is optimal and the function $H(x(0), y(0), z(0), \psi_1(0), \psi_2(0), \psi_3(0), u, 0) = u^2$ assumes its absolute minimum at the point $u = 0$.

Example 5.

$$x(t+1) = v(t) \sin \frac{\pi}{2} u(t), \ y(t+1) = v(t) \cos \frac{\pi}{2} u(t),$$

$$z(t+1) = z(t) + x^2(t) + y^2(t) + u^2(t) - v^2(t) \quad ,$$

$$x(0) = y(0) = z(0) = 0, \ |u| \leq 1, \ |v| \leq 1, \ \phi(x,y,z) = z,$$

$$t_o = 0, \ t_1 = 2 \quad .$$

We calculate $J(u)$: $J(u) = z(2) = u^2(0) + u^2(1) - v^2(1)$. For the optimal control $\{u(0) = 0, v(0) = 0, u(1) = 0, v(1) = 1\}$, we have $H(x(0), \psi(0), u, 0) = u^2 - v^2$. The latter function has a saddle-point at $u = 0$, $v = 0$.

d) Sometimes, the local maximum principle is replaced by other basic necessary conditions of optimality from which, in particular, we derive the conclusion that among the points where the function $H(x,\phi,u,t)$ is not differentiable in u, are found those at which the function H assumes a local minimum. It may be thought that these are automatically non-optimal but this is not always true.

Example 6.

$$x(t+1) = 2\sqrt{|u(t)|}\ , \quad y(t+1) = y(t) + x^2(t) - |u(t)|\ ,$$

$$x(0) = y(0) = 0\ , \quad |u| \le 1\ , \quad \phi(x,y) = y\ ,$$

$$t_0 = 0\ , \quad t_1 = 2\ .$$

We obtain

$$J(u) = y(z) = 3|u(0)| - |u(1)|\ .$$

For the optimal control $\{u(0) = 0,\ u(1) = 1\}$ we have

$$H(x,\psi,u,0) = |u|\ .$$

Example 7.

$$x(t+1) = \sqrt{|u(t)|}\ \sin\frac{\pi}{2}u(t), \quad y(t+1) = \sqrt{|u(t)|}\ \cos\frac{\pi}{2}u(t),$$

$$z(t+1) = z(t) + x^2(t) + y^2(t) - |u(t)|\ ,$$

$$x(0) = y(0) = z(0) = 0\ , \quad |u| \le 1\ ,$$

$$\Phi(x,y,z) = z\ , \quad t_0 = 0\ , \quad t_1 = 2\ .$$

e. In several works, the basic necessary condtion for the optimality of discrete systems is formulated in the following form: the function H is either stationary or assumes a local maximum on optimal u(t). The preceding examples do not contradict this conclusion. However, in general it is not valid.

Example 8.

$$x(t+1) = u(t) \quad , \quad y(t+1) = y(t) + x^2(t) \quad ,$$

$$x(0) = y(0) = 0 \quad , \quad u = \pm 1 \quad , \quad \phi(x,y) = y \quad ,$$

$$t_0 = 0 \quad , \quad t_1 = 2 \quad .$$

We obtain

$$J(u) = y(2) = u^2(0) \quad .$$

For the optimal control $u(0) = 1$, we have

$$x(1) = 1 \quad , \quad \psi_2(1) = -1 \quad , \quad \psi_2(0) = -1 \quad ,$$

$$\psi_1(1) = 0 \quad , \quad \psi_1(0) = -2 \quad ,$$

$$H(x(0), y(0), \psi_1(0), \psi_2(0), u, 0) = -2u \quad .$$

The point $u = 1$, corresponding to the optimal control, is not a stationary point for the last function, nor does the function assume a local maximum at $u = 1$.

We obtain such a conclusion if we consider

Example 9.

$$x(t+1) = u(t) \quad , \quad y(t+1) = v(t) \quad ,$$

$$z(t+1) = z(t) + x^2(t) + y^2(t) \quad ,$$

$$x(0) = y(0) = z(0) = 0 \quad , \quad (u-1)^2 + (v+1)^2 \geq 8 \quad ,$$

$$(u-2)^2 + (v+2)^2 \leq 18 \quad , \quad \phi(x,y,z) = z \quad ,$$

$$t_0 = 0 \quad , \quad t_1 = 2 \quad .$$

In example 8, it is possible to see that the necessary conditions for optimality in discrete systems with quantized levels of control have the same singularities.

f) The second maximum condition, as proved in Examples 8,9, is not satisfied for all discrete systems.

g) The first and second conditions for a local maximum lose their content if the star neighborhood consists only of the optimal control (Examples 3-9).

h) Assertion 2) of Theorem 4 may be satisfied for u belonging to the boundary of the set U(t).

6. On a Particular Problem (Problem B).

By problem B, we mean the particular case of problem A when: 1) the right sides of the first n-1 equations of the system do not contain the coordinate x_n, 2) the function $f_n(x,u,t)$ has the form

$$f_n(x,u,t) = x_n + f_{1n}(x_1,\dots,x_{n-1},u,t)$$

3) the function $\phi(x) = x_n + \phi_1(x_1,\dots,x_{n-1})$.

When considering problem B, we agree to let $\bar{x}$ denote an n-vector while the vector composed of the first n-1 components of $\bar{x}$ is denoted by x. Under these conditions, it is not difficult to see that $\psi_n(t) \equiv -1$, $t \in T$. Therefore, the equations for $\psi(t)$ have the form

$$\psi(t - 1) = \frac{\partial \mathcal{E}(x(t),\psi(t),u(t),t)}{\partial x} ,$$

$$\psi(t_1 - 1) = -\text{grad } \phi_1(x(t_1)) , \tag{27}$$

$$\mathcal{E}(x,\psi,u,t) = \psi' f(x,u,t) - f_{1n}(x,u,t) . \tag{28}$$

In problem B, the sets $\Omega(t)$, $\omega(t)$ consist of those $\overline{\psi}$ for which $\psi_n = -1$. Therefore, checking of the conditions of Theorems 1,3 may be done for $\overline{\psi}$ with $\psi_n = -1$. Hence, at once it follows that the first maximum condition is satisfied for systems in which: 1) the function $f(x,u,t)$ is linear in u, 2) the function $f_{1n}(x,u,t)$ is convex in u, 3) the set U is convex.

We call the function $\alpha(x)$ strongly convex with coefficient ℓ if it is differentiable and the quantity $o_3(||\Delta x||)$ in the expansion

$$\alpha(x + \Delta x) - \alpha(x) = \Delta x' \text{ grad } \alpha(x) + o_3(||\Delta x||)$$

satisfies the condition

$$o_3(||\Delta x||) \geq \ell||\Delta x||^2 .$$

Theorem 5. Let: 1) the function $f(x,u,t)$ be differentiable in u and the quantity $o_4(||\Delta u||)$ in the expansion

$$\psi' f(x,u+\Delta u,t) - \psi' f(x,u,t) = \psi' \frac{\partial f(x,u,t)}{\partial u} \Delta u + o_4(||\Delta u||)$$

satisfy the inequality

$$o_4(||\Delta u||) \leq \ell_1||\Delta u||^2 ;$$

2) the function $f_{1n}(x,u,t)$ be strongly convex in u with coefficient ℓ; 3) the set U be convex. Then in problem B with $\ell \geq \ell_1$, the optimal control satisfies the first maximum condition.

Proof. Under the conditions of the theorm, the function $H(x,\psi,u,t)$ is concave in u. Therefore, the second maximum condtion which, by virtue of Theorem 4 must be satisfied by the optimal control, coincides with the first maximum condition.

Remarks. 1. An estimate of the quantity ℓ_1 is not difficult to obtain from estimates for ψ and the second derivative $\partial^2 f/\partial u^2$, if the latter exists.

2. The assumption on the differentiability of the functions $f(x,u,t)$, $f_{1n}(x,u,t)$ in u may be weakened (cf. below) if the idea of strong convexity and nonlinearity of the functions is introduced as in [107].

Other general schemes for isolating discrete systems in which the first maximum condition is satisfied are obtained from consideration of problem 10A. In the proof of Lemma 1, it was shown that each optimal control $u_i(t)$, $i = 1,\ldots,n+1$; $t\varepsilon T$, of this problem satisfies the maximum condition

$$\xi'(t)f(x(t),u_i(t),t) = \max_{u\varepsilon U(t)} \xi'(t)f(x,u,t) \quad .$$

Therefore, in discrete systems for which the function $\xi' f(x,u,t)$, $\{x,\xi\}\varepsilon\Omega(t)$, assumes its maximum in u at a unique point, the optimal control satisfies the first maximum condition. Hence, Theorem 5 immediately follows for $\ell > \ell_1$ since the considered function $\mathcal{E}$ is strongly concave under these conditions and assumes its maximum at a unique point. Incidentally, under these conditions the assumption about the boundedness of the set $U(t)$ is unnecessary.

In the study of problem B, it is not difficult to isolate discrete systems for which the second maximum condition is satisfied and for which $U(t)$ is not convex.

We introduce the sets

$$N_t = \left\{ -\frac{\partial f_{1n}(x,w,t)}{\partial w}, \quad z\varepsilon\Omega(t), \ w\varepsilon W(t) \right\},$$

$$Q_1(\ell_2,t) = \{\bar{w}: (\nu+\mu)'\bar{w} = \max_{w^*\varepsilon W(t)} (\nu+\mu)'w^*, \ \nu\varepsilon N_t, ||\mu|| \le \ell_2\}.$$

Theorem 6. Let: 1) problem 20B be nonsingular; 2) the functions $f(x,u,t)$, $f_{1n}(x,u,t)$ be differentiable in u; 3) $\left|\left|\zeta'\frac{\partial f(x,w,t)}{\partial w}\right|\right| \le \ell_3$: 4) $Q_1(\ell_2,t) \subset U(t)$. Then for $\ell_2 \ge \ell_3$, the optimal control in problem B satisfies the second maximum condition.

Proof. The assertion follows naturally from Theorem 3 since, under the given assumptions, the conditions of that theorem are satisfied.

Remark. The estimates in Theorem 6 are significantly simplified if

$$f_{1n}(x,u,t) = f_2(x,t) + bu, \quad f(x,u,t) = f_3(x,t) + Bu \quad .$$

In this case, the set N_t consists of the single element b, and condition 3) has the form $||\zeta'B|| \le \ell_3$. Roughly speaking, the conditions of Theorem 6 will be satisfied if U(t) is e-convex, where e is any vector in a neighborhood of b.

§2. Classes of Discrete Systems for Which the Maximum Principle is Satisfied.

From formula (12) in problem 10A, we obtain

$$\Delta J(v) = -\sum_{t=t_o}^{t_1-1} [H(y,\xi,v_1,t) - H(y,\xi,v,t)]$$

$$-\sum_{t=t_o}^{t_1-1} \Delta y'(t)\left[\frac{\partial H(y,\xi,v_1,t)}{\partial y} - \frac{\partial H(y,\xi,v,t)}{\partial y}\right]$$

$$-\sum_{t=t_o}^{t_1-1} o_2(||\Delta y(t)||) - o(||\Delta y(t_1)||) \quad .$$

For problem 10B, the formula for the increment of the function J(v) assumes the form

$$\Delta J(v) = -\sum_{t=t_o}^{t_1-1} [\mathscr{E}(y,\xi,v_1,t) - (y,\xi,v,t)]$$

$$-\sum_{t=t_o}^{t_1-1} \Delta y'(t)\left[\frac{\partial \mathscr{E}(y,\xi,v_1 t)}{\partial y} - \frac{\partial \mathscr{E}(y,\xi,v,t)}{\partial y}\right]$$

$$-\sum_{t=t_o}^{t_1-1} o_2'(||\Delta y(t)||) + \sum_{t=t_o}^{t_1-1} o''_2(||\Delta y(t)||) - o(||\Delta y(t_1)||),$$

where the quantities o_2' and o_2'' are found from the expressions

$$\xi'[g(y+\Delta y,v_1,t) - g(y,v_1,t)] = \xi'\frac{\partial g(y,v_1,t)}{\partial y}\Delta y + o_2' \quad ,$$

$$\sum_{i=1}^{n+1} \alpha_i[f_{1n}(y + \Delta y,u_{i1},t) - f_{1n}(y,u_{i1},t)]$$

$$= \sum_{i=1}^{n=1} \alpha_i\frac{\partial f_{1n}(y,u_{i1},t)}{\partial y}\Delta y + o_2'' \quad .$$

Let the function $f_{1n}(x,u,t)$ be strongly concave in x with coefficient ℓ_4, i.e. $o_2''(||\Delta y||) \leq -\ell_4||\Delta y||^2$, ℓ_4 = constant, $\ell_4 \geq 0$, and let the function $f(x,u,t)$ be weakly nonlinear in x, i.e. $o_2'(||\Delta y||) \leq \ell_5||\Delta y||^2$, ℓ_5 small. Then for $\ell_4 \geq \ell_5$, the expression $\sum_{t=t_0}^{t_1-1} o_2' + o_2''$ remains negative. Assuming, in addition, that the function $\phi(x)$ is concave $[o(||\Delta y||) \geq 0]$, we obtain

$$\Delta J(v) \leq - \sum_{t=t_0}^{t_1-1} [\mathcal{E}(y,\xi,v_1,t) - (y,\xi,v,t)]$$

$$- \sum_{t=t_0}^{t_1-1} \Delta y'(t)\left[\frac{\partial\mathcal{E}(y,\xi,v_1,t)}{\partial y} - \frac{\partial\mathcal{E}(y,\xi,v,t)}{\partial y}\right] . \qquad (29)$$

As the control $v_1(t)$, we take one of the controls $u_i(t),\alpha_i(t)$. More precisely, we assume

$$v_1(t) = \begin{cases} u_j(t),\alpha_j(t) = 1 , \quad \text{if} \\ \quad \Delta y(t)\left[\dfrac{\partial\mathcal{E}(y,\xi,u_j,\alpha_j,t)}{\partial y} - \dfrac{\partial\mathcal{E}(y,\xi,v,t)}{\partial y}\right] > 0 , \\ u_\ell(t),\alpha_\ell(t) = 1, \ \ell \text{ is one of } 1,\ldots,n+1, \text{ if} \\ \Delta y'(t)\left[\dfrac{\partial\mathcal{E}(y,\xi,u_\ell,\alpha_\ell = 1,t)}{\partial y} = 0, \ u_\ell(t)\varepsilon U(t)\right]. \end{cases}$$

For each $t\varepsilon T$, we find an index $j = j(t)$ for which the given choice of $v_1(t)$ is possible. Then

$$\Delta y'(t)\left[\frac{\partial\mathcal{E}(y,\xi,u_k,\alpha_k = 1,t)}{\partial y} - \frac{\partial\mathcal{E}(y,\xi,v,t)}{\partial y}\right]$$

$$= \Delta y''(t) \left[\frac{\partial f'(y,u_k,t)}{\partial y} \xi' - \frac{\partial f_{ln}(y,u_k,t)}{\partial y} \right.$$

$$\left. - \sum_{i=1}^{n+1} \alpha_i \frac{\partial f'(y,u_i,t)}{\partial y} \xi + \sum_{i=1}^{n+1} \alpha_i \frac{\partial f_{ln}(y,u_i,t)}{\partial y} \right] ,$$

$$k = 1,\dots,n+1 .$$

We multiply these equalities by $\alpha_k(t)$ and sum over k from k = 1 to k = n+1. On the right we obviously obtain zero. Therefore, for

$$\Delta y'(t) \frac{\partial \mathcal{E}(y,\xi,u_\ell,\alpha_\ell = 1,t)}{\partial y} \not\equiv 0$$

among

$$\Delta y'(t) \left[\frac{\partial \mathcal{E}(y,\xi,u_k,\alpha_k = 1,t)}{\partial y} - \frac{\partial \mathcal{E}(y,\xi,v,t)}{\partial y} \right] ,$$

$$k = 1,\dots,n + 1 .$$

there are found both positive and negative quantities.

We assume that v(t), $t\varepsilon T$, is the optimal control. Then (see the proof of Lemma 1) each $u_k(t)$, $t\varepsilon T$, $k = 1,\dots,n + 1$, satisfies the maximum condition and, therefore,

$$\mathcal{E}(y,\xi,u_k,\alpha_k, = 1,t) \not\equiv \mathcal{E}(y,\xi,v,t) \quad , \quad t\varepsilon T .$$

Substituting the controls $u_j(t)$, $\alpha_j(t) = 1$ into the inequality (29), we obtain $J(v) \leq 0$, which shows the optimality of the constructed controls $u_j(t)$, $\alpha_j(t) = 1$, in problem B. We formulate this result in the form of a theorem.

Theorem 7. Let 1) the problem 10B be nonsingular; 2) the function $f_{1n}(x,u,t)$ be strongly concave in x with coefficient ℓ_4; 3) the function $\phi(x)$ be concave; 4) the function f(x,u,t) be weakly nonlinear ($\ell_5 \leq \ell_4$). Then there exists an optimal control in problem B satisfying the first maximum condition.

Theorem 7 strengthens the result of assertion 2) of Theorem 2. By an analogous argument, it is not difficult to prove that in problem B, for the system

$$\bar{x}(t+1) = [A(t) + \sum_{i=1}^{r} u_i B^i(t)]\bar{x}(t) + \bar{\phi}(u,t)$$

there exists an optimal control satisfying the second maximum condition if $\phi_n(u,t)$ is strongly concave in u, and if the nonlinearity in u of the function $\phi(u,t)$ is sufficiently small ($\bar{\phi}(u,t)$ is differentiable in u).

§3. Miscellaneous Problems

1. Problems with Moving End Conditions (Problem C). In what follows, by problem AC we mean problem A with an additional condition containing a constraint on the right endpoint of the trajectory of Eq.(1):

$$p(x(t_1)) \leq 0 \quad , \tag{30}$$

where p(x) is a differentiable function. We introduce the vector $\delta_{\theta,u*}x(t)$, computed according to the equations

$$\delta_{\theta,u*}x(t) = 0, \quad t = t_0, \ldots, ; \quad \delta_{\theta,u*}x(\theta+1) = \Delta_{u*}f(x(\theta),u(\theta),\theta) ,$$

$$\delta_{\theta,u*}x(t+1) = \frac{\partial f(x(t),u(t),t)}{\partial x}\delta_{\theta,u*}x(t), \quad \theta + 1 \leq t \leq t_1 - 1 .$$

Comparing this vector with $\Delta_{\theta,u^*}x(t)$, we see that

$$||\Delta_{\theta,u^*}x(t) - \delta_{\theta,u}x(t)|| \leq \int_{t_0}^{t_1} o_5(||\Delta_{\theta,u^*}x(t)||)dt \ , \ t\varepsilon T. \tag{30}$$

Let the set $\{f(x,U(t),t)\}$ be convex, let τ_i, $i = 1,\ldots,\ell$ be any (not necessarily distinct) points of T, and let u_i^*, $i = 1,\ldots,\ell$, be any (not necessarily distincet) elements of $U(\tau_i)$. To each pair τ_i,u_i^*, there corresponds a vector $\delta_{\tau_i,u_i^*}x(t_1)$:

$$\delta_{\tau_i,u_i^*}x(t_1) = F(t_1,\tau_i)\Delta_{u_i^*}f(x(\tau_i),u(\tau_i),\tau_i) \quad ,$$

where

$$F(t_1,\tau_i) = \prod_{j=i}^{t_1-2} \frac{\partial f(x(\tau_j + 1),u(\tau_j + 1),\tau_j + 1)}{\partial x} \quad .$$

We construct a vector $u_i^*(\varepsilon_i)$, $0 \leq \varepsilon_i \leq 1$, depending on the parameter ε_i:

$$\Delta_{u_i^*(\varepsilon_i)}f(x(\tau_i),u(\tau_i),\tau_i) = \varepsilon_i\Delta_{u_i^*}f(x(\tau_i),u(\tau_i),\tau_i) \ .$$

By virtue of the convexity of $\{f(x,U(t),t)\}$, $t\varepsilon T$, each vector $u_i^*(\varepsilon_i)$, $0 \leq \varepsilon_i \leq 1$, lies in the set $U(\tau_i)$. Clearly,

$$\delta_{\tau_i,u_i^*(\varepsilon_i)}x(t_1) = \varepsilon_i F(t_1,\tau_i)\Delta_{u_i^*}f(x(\tau_i),u(\tau_i),\tau_i) \quad .$$

We introduce the control

$$u_\varepsilon^*(t) = \begin{cases} u(t) \ , & t \neq \tau_i \ , \\ u_i^*(\varepsilon_i), & t = \tau_i \ . \end{cases}$$

To this control corresponds the vector

$$\delta_\varepsilon x(t_1) = \sum_{i=1}^{\ell} \delta_{\tau_i, u_i^*(\varepsilon_i)} x(t_1)$$

$$= \sum_{i=1}^{\ell} \delta_i F(t_1, \tau_i) \Delta_{u_i^*} f(x(\tau_i), u(\tau_i), \tau_i) \quad .$$

The set of vectors $K = \{\delta_\varepsilon x(t_1): \ 0 \leq \varepsilon_i \leq 1, \ i = 1, \ldots, \ell\}$, is clearly convex, bounded, and closed. If $u(t)$, $t \varepsilon T$, is the optimal control and $p(x(t_1)) < 0$, then the set K has no interior points in common with the set $K_1 = \{\delta x: \ \delta x' \text{ grad } \phi(x(t_1)) \leq 0\}$. Therefore,

$$\delta_\varepsilon x'(t_1) \text{ grad } \phi(x(t_1)) \geq 0 \ , \ \delta_\varepsilon x(t_1) \varepsilon K \quad .$$

If under the optimal control $u(t)$, $t \varepsilon T$, we have the equality $p(x(t_1)) = 0$, then the set K has no interior point in common with the intersection of the sets K_1 and $K_2 = \{\delta x: \ \delta x' \text{ grad } p(x(t_1)) \leq 0\}$. This means that there exist constants λ, μ, $\lambda \geq 0$, $\mu \geq 0$, $\lambda + \mu = 1$, such that

$$\delta_\varepsilon x'(t_1) [\lambda \text{ grad } \phi(x(t_1)) + \mu \text{ grad } p(x(t_1))] \geq 0$$

when $\text{grad } \phi(x(t_1)) \nparallel \chi \text{ grad } p(x(t_1))$. Taking into account (31), it is possible to assert that if in problem AC the control $u(t)$ is optimal, then there exist constants λ, μ, $\lambda \geq 0$, $\mu \geq 0$, $\lambda + \mu = 1$, such that

$$\Delta_{\theta, u*} \ x'(t_1) [\lambda \text{ grad } \phi(x(t_1)) + \mu \text{ grad } p(x(t_1))]$$

$$\geq o(||\delta_{o, u*} x(t_1)||) \quad ,$$

$$(\text{grad } \phi(x(t_1)) \nparallel \chi \text{ grad } p(x(t_1))) \quad ,$$

where $\mu = 0$ if $p(x(t_1)) < 0$.

Returning to Lemma 2 (Assertion 1), we have

Theorem 8. Let the set $f(x,U(t),t)$, $t \in T$, be convex. Then for the optimal control $u(t)$ and optimal trajectory $x(t)$ of problem AC with grad $\phi(x(t_1)) = \chi$ grad $p(x(t_1))$, $-\infty < \chi < \infty$, there exist nonnegative numbers λ, μ, $\lambda + \mu = 1$, such that the control $u(t)$, $t \in T$, satisfies the first maximum condition with $\psi(t)$ from (7), where

$$c = \lambda \,\text{grad}\, \phi(x(t_1)) + \mu \,\text{grad}\, p(x(t_1)) \quad .$$

If $p(x(t_1)) < 0$, we may set $\mu = 0$.

Remark. Theorem 8 is presented only to illustrate the method. More precise results are obtained from Lemma 1,2 by the same scheme as in Theorems 1,2. The details are omitted.

We call problem B, with the additional condition (3), problem BC and retain the earlier notations for $x,\bar{x}$. In addition, we will assume that $p(\bar{x})$ does not depend upon x_n. It is not difficult to see that $\psi_n(t) \equiv -\lambda$, $t \in T$, and the equation for $\psi(t)$ has the form

$$\psi(t - 1) = \frac{\partial \mathscr{E}_{\lambda_1}(x(t),\psi(t),u(t),t)}{\partial x} ,$$

$$\psi(t_1 - 1) = \lambda \,\text{grad}\, \phi(x(t_1)) - \mu \,\text{grad}\, p(x(t_1)),$$

$$\mathscr{E}\lambda_1(x,\psi,u,t) = \psi' f(x,u,t) - \lambda_1 f_{1n}(x,u,t),$$

$$\mu p(x(t_1)) = 0 \quad , \quad \lambda,\mu \geq 0 \quad , \quad \lambda + \mu = 1 \quad .$$

In problem BC, it is already impossible to use the strong convexity of $f_{1n}(x,u,t)$ in u, since the coefficient λ_1 is not known in advance. However, the results on the first maximum condition for problem B(§2) can be transferred to problem BC if $f_{1n}(x,u,t)$ is convex in u and the function f(x,u,t) is linear in u.

Now we describe a scheme for obtaining the maximum condition in problem AC. Let the function f(x,u,t) be differentiable in u. We introduce the vector $\delta_{\theta,u*}x(t)$ by the formula

$$\delta_{\theta,u*}x(t) = 0 \quad , \quad t = t_o,\ldots,\theta \quad ,$$

$$\delta_{\theta,u*}x(\theta = 1) = \frac{\partial f(x(\theta),u(\theta),\theta)}{\partial u}(u^* - u(\theta)) \quad ,$$

$$\delta_{\theta,u*}x(t + 1) = \frac{\partial f(x(t),u(t),t)}{\partial x}\,\delta_{\theta,u*}x(t), \quad \theta + 1 \leq t \leq t_1 - 1.$$

Comparing with $\Delta_{\theta,u*}x(t)$, we obtain

$$||\Delta_{\theta,u*}x(t) - \delta_{\theta,u*}x(t)|| \leq o_6||u^* - u(\theta)|| \quad .$$

Let U(t), $t \varepsilon T$, be convex, We introduce τ_i, u_i^* as in the proof of Theorem 8. The vector $u_i^*(\varepsilon_i)$, $0 \leq \varepsilon_i \leq 1$, is constructed by the rule

$$u_i^*(\varepsilon_i) = (1 - \varepsilon_i)u(\theta) + \varepsilon_i u^*_i \quad .$$

The subsequent constructions are obvious. As a result, we obtain an assertion formulated analogously to Theorem 8 with two differences: 1) instead of the convexity of f(x(t),U(t),t), $t \varepsilon T$, we require the convexity of U(t) and the differentiability of the function f(x,u,t) in u; 2) the first maximum condition is replaced by the second maximum condition.

2. Minimax Problems. Optimal Processes for Discrete Systems with Constraints on the States. We introduce problem D, which is the problem to minimize

$$J(u) = \max_{t \varepsilon T} q(x(t+1), t+1) \quad , \tag{32}$$

along trajectories of the system (1), where $q(x,t)$ is a differentiable function. Expression (32) may be represented in the form

$$J(u) = \max_{\sum \omega(t) = 1} \sum_{t=t_o+1}^{t_1} \omega(t) q(x(t), t) \quad .$$

Therefore, if $u(t)$, $t \varepsilon T$, is the optimal control, $x(t)$ the optimal trajectory, then there corresponds a function $\omega(t)$, concentrated on the set $\sigma = \{t: \; q(x(t+1), t+1) = \max_{s \varepsilon T} q(x(s+1), s+1)\}$, $\sum \omega(t) = 1$, $\omega(t) q(x(t+1), t+1) \geq 0$. From the optimality of the control $u(t)$, it follows that

$$\sum_{t=t_o+1}^{t_1} \omega(t) q(x(t), t) \leq \sum_{t=t_o+1}^{t_1} \omega(t) q(\tilde{x}(t), t) \quad , \; \tilde{u}(t) \varepsilon U(t), \; t \varepsilon T \; ,$$

where $\tilde{x}(t)$ is the trajectory generated by the admissible control $\tilde{u}(t)$. Problem D is reduced to problem B if the coordinate x_n is determined by the rule

$$x_n(t) = \sum_{s=t_o}^{t-1} \omega(s) q(x(s), s), x_n(t_o) = 0 \quad .$$

The last equation of system (1) now assumes the form

$$x_n(t+1) = x_n(t) + \omega(t) q(x(t), t) \quad , \; t \varepsilon T \; ,$$

$$\omega(t_o) = 0 \; , \qquad x_n(t_o) = 0 \; ,$$

with the function $\phi_1(x) = q(x(t_1),t_1)\omega(t_1)$. We write the equations for $\psi(t)$, $H_\omega(x,\psi\ u,t)$ for the corresponding problem:

$$H_\omega(x,\psi,u,t) = \psi' f(x,u,t) - \omega q(x,t) ,$$
$$\psi(t - 1) = \frac{\partial H_\omega(x,\psi,u,t)}{\partial x} , \tag{33}$$

$$\psi(t_1 - 1) = -\omega(t_1) \frac{\partial q(x(t_1),t_1)}{\partial x} . \tag{34}$$

The results obtained above for problem B transfer to problem D except those results using the concept of strong concavity in x of the function $f_{1n}(x,u,t)$. However, in this case Theorem 7 is transferred to problem D if $f(x,u,t)$ is linear in x, $q(x) \geq 0$, $q(x)$ a concave function. We formulate one result.

Theorem 9. Let $f(x,U(t),t)$, $t\varepsilon T$, be a convex set. Then for the optimal control $u(t)$, $t\varepsilon T$, and the optimal trajectory $x(t)$ (grad $q(x(t),t) \not\equiv 0$), there is a function $\omega(t)$, concentrated on σ, such that the control $u(t)$, $t\varepsilon T$, satisfies the maximum condition (25) with the function $\omega(t)$ from (34).

Remark. For determination of the function $\omega(t)$, we use the inequality

$$\sum_{t=t_0+1}^{t_1} \omega(t)q(x(t),t) \geq \sum_{t=t_0+1}^{t_1} \tilde{\omega}(t)q(x(t),t) ,$$

which is valid for all $\tilde{\omega}(t)$, $\sum_{t=t_0+1}^{t_1} \tilde{\omega}(t) = 1$.

From problems of minimax we pass to the problem of optimization of discrete systems with state variable constraints. We use the nomenclature "problem AD" to mean problem A with the additional condition

$$q_1(x(t),t) \leq 0 \quad , \quad t \in T \quad .$$

We consider the minimax problem with the function q(x,t) having the form

$$q(x,t) = \begin{cases} q_1(x,t), & t \in T \quad , \\ \psi(x) - \nu, & t = t_1, \quad -\infty \leq \nu \leq \infty \quad . \end{cases}$$

The smallest ν for which the solution of the minimax problem satisfies $J(u) \leq 0$ is the minimal value of the function $\phi(x)$ in problem AD. The optimal control in problem AD coincides with its analogue in problem D. By such a scheme, we obtain necessary conditions for optimality for problem AD. Exact formulations of the theorems are easily obtained and are therefore omitted.

Remark. If it is required to satisfy the additional conditions $q(x(t_1),t_1) \leq 0$, then it is necessary to know the solution of the minimax problem with a moving right endpoint.

3. One Form of the Necessary Conditions of Optimality. Let us consider problem A with the function f(x,u,t) being differentiable in u and the set U(t) being given in the form

$$U(t) = \{u: \ q(u,t) \leq 0\}, \quad t \in T \ ,$$

where q(u,t) is differentiable in u(problem A_1). We introduce the system

$$\left.\begin{aligned} x_1(t+1) &= f_1(x_1,x_2,t) \ , \\ x_2(t+1) &= f_2(x_2,\nu,t) \ , \quad x_1(t_o) = x_o \ , \\ x_2(t_o) &= u(t_o) \ , \\ t\varepsilon T \ , \ x_1 &\varepsilon E_n \ , \ x_2 \varepsilon E_r \ , \end{aligned}\right\} \tag{35}$$

where x_1 is an n-vector, f_2, x_2 are r-vectors, $f_1(x_1,x_2,t) = f(x_1,x_2,t)$, $f_2(x_2,\nu,t)$ is continuous in all arguments together with $\partial f_2(x_2,\nu,t)/\partial x_2$, and ν is a control with $\nu \varepsilon E_m$. The region $\Xi(t)$ of admissible values of the vector ν and the function $f_2(x_2,\nu,t)$ is chosen so that the set $f_2(x_2,\Xi(t),t)$ is bounded, convex, and closed. Along trajectories of the system (45), we impose the constraint $q(x_2,t) \leq 0$, $t\varepsilon T$, and we consider the problem of minimizing

$$J(\nu) = \phi(x_1(t_1)) \quad .$$

We call this problem the preliminary generalization of problem A (problem ZOA). By virtue of the assumption on the optimal control $\nu(t)$, $t\varepsilon T$, problem ZOA satisfies the first maximum condition. Let the control $u(t_o)$ in problem A be held fixed. Then from the bounded set $U(t)$, it is possible to select the function $f_2(x_2,\nu,t)$ and the set $\Xi(t)$ such that for each $u(t)\varepsilon U(t)$, $t\varepsilon T$, there exists a $\nu(t)\varepsilon\Xi(t)$ for which $x_2(t) = u(t)$, $t\varepsilon T$. Under these conditions, we may take $x_2(t)$ as the optimal control in problem A and it in turn generates the optimal control $\nu(t)$ in problem ZOA.

We pause to detail two simple properties of f_2 and

$\nu(t)$. Let:

1) $$x_2(t+1) = \nu(t) \quad , \tag{36}$$

$$\Xi(t) = \{\nu: \sum_{i=1}^{r} (\nu^i)^2 \leq L^2\};$$

2) $$x_2(t+1) = x_2(t) + \nu(t) \quad . \tag{37}$$

We let

$$d = \max_t \; (\bar{u} - u)'(\bar{u} - u)^{1/2}$$
$$\bar{u}, u\varepsilon U(t)$$

denote the largest diameter of the set $U(t)$ and set $L \geq d$. In this case, for each $u(t)$, $t\varepsilon T$, there exists $\nu(t)$, $t\varepsilon T$, in (36),(37), such that $x_2(t) \equiv u(t)$. Now we formulate the necessary condition for optimality in problem A_1.

Theorem 10. Let $f(x,u,t)$ be differentiable in u. For the optimal control $u(t)$, $t\varepsilon T$, there is a function $\omega(t)$ concentrated on the set $\sigma = \{t: \; q(u(t),t) = 0\}$, such that $\sum_\sigma \omega(t) = 1$,

$$\sum_\sigma \omega(t)q(u(t),t) \geq \sum_\sigma \omega(t)q(u(t),t) \quad , \quad \sum_\sigma \omega(t) = 1 \quad , \tag{39}$$

and such that the control $u(t)$, $t\varepsilon T$, satisfies the maximum condition

$$\psi'(t) \frac{\partial f(x(t),u(t),t)}{\partial u} u(t) - 2\omega(t)u'(t)u(t)$$

$$\geq \psi'(t) \frac{\partial f(x(t),u(t),t)}{\partial x} u^* - 2\omega(t)u'(t)u^* \quad , \tag{40}$$

$u^*\varepsilon U(t)$, $t\varepsilon T$.

Here $\psi(t)$ satisfies Eq. (7) with

$$\psi(t_1 - 1) = -\omega(t_1) \text{ grad } \phi(x(t_1)) \quad . \tag{41}$$

Remark. The assumption that $u(t_0)$ be held fixed is inessential since system (35) may be defined at $t = t_0 - 1$.

Use of Eq. (37) to obtain necessary conditions for optimality in problem A is omitted. The maximum condition obtained in Theorem 10 is analogous in form to the second maximum condition. This is explained by the concrete representation of Eq. (36) and the set $\Xi(t)$.

Sometimes it turns out to be helpful to obtain necessary conditions for optimality by means of the fourth generalization of problem A (problem 40A), which is the problem of minimizing the function $\phi(x_1(t_1))$ along trajectories of the system (35), under the condition that the set $\Xi(t)$ be convex. The scheme of passing from problem 40A to problem A is fairly obvious in light of earlier remarks.

§4. The Quasimaximum Principle

In operations with the continuous system

$$\frac{dy}{ds} = g(y,v,s) \quad , \quad s_0 \leq s \leq s_1 \quad , \tag{42}$$

on digital computers, the operator dy/ds is, in the simplest cases, replaced by the difference $(y(s+h) - y(s))/h$ where h is a small quantized time step. Naturally, the smaller h is, the more accurate is the approximation. Setting $t = s/h$, $x(t) = y(th)$, $u(t) = v(th)$, $f(x,u,t) = g(y,v,th)$, we are led to the discrete system

$$x(t + 1) = x(t) + hf(x,u,t) \quad , \tag{43}$$

depending upon the parameter h. In general, the discrete system with parameters has the form

$$x(t + 1) = f(x,u,t,h) \quad , \tag{44}$$

where h may be a vector. In what follows, quantities and concepts connected with discrete systems with a parameter will be written with the index h. We describe relations connected with obtaining necessary conditions for optimality in problem A_h for Eq.(44). The optimal control is $u(t,h)$, $t\varepsilon T_h$. The trajectory of Eq.(44) is $x(t,h)$, $t - 1\varepsilon T_h$:

$$\psi(t - 1,h) = \frac{\partial f'(x,u,t,h)}{\partial x}\psi(t,h), \quad \psi(t_1 - 1,h) = -\text{grad}\ \phi(x(t_1,h)) \quad , \tag{45}$$

$$H(x,\psi,u,t,h) = \psi' f(x,u,t,h) \quad ,$$

$$\Delta_{\theta,u*}x(t,h) = 0 \quad , \quad t = t_0,\ldots,\theta \quad ,$$

$$\Delta_{\theta,u*}x(\theta + 1,h) = \Delta_{u*}f(x,u,\theta,y) \quad ,$$

$$\Delta_{\theta,u*}x(t + 1,h) = f(x + \Delta_{\theta,u*}x,u,t,h) - f(x,u,t,h) \quad ,$$

$$\theta + 1 \le t \le t_1 - 1 \quad ,$$

$$\Delta_{\theta,u*}J_h(u) = -\psi'(\theta,h)\Delta_{u*}f(x,u,\theta,h) - \sum_{t=\theta+1}^{t_1-1} o(||\Delta_{\theta,u*}x(t,h)||) + o_1(||\Delta_{\theta,u*}x(t_1,h)||), \tag{46}$$

$$o(||\Delta x||) \le k||\Delta x||^2, \quad o_1(||\Delta x||) \le k_1||\Delta x||^2.$$

Let $\varepsilon > 0$ be some number. We introduce the set

$$U_\varepsilon(\theta,h) = \{u^*: \ o_1(||\Delta_{\theta,u^*}x(t_1)||) - \sum_{t=\theta+1}^{t_1-1} o(||\Delta_{\theta,u^*}x(t)||) \le \varepsilon, \ y^* \varepsilon U(\theta)\} .$$

<u>Theorem 11. The optimal control u(t,h) in problem A_h for (44) satisfies the quasimaximum condition</u>

$$H(x(t,h),\psi(t,h),u(t,h),t,h) \geq H(x(t,h),\psi(t,h),u^*,t,h) - \varepsilon$$

<u>for all $y^* \varepsilon U_\varepsilon(t,h)$, $\varepsilon > 0$, $t \varepsilon T_h$.</u>

In fact, let $u(t,h) \varepsilon U(t)$, $t \varepsilon T_h$, be the optimal control but assume there exist $\theta \varepsilon T_h$, $\varepsilon > 0$, $u^* \varepsilon U_\varepsilon(\theta,h)$ for which

$$H(x(\theta,h),\psi(\theta,h),u(\theta,h),\theta,h) < H(x(\theta,h),\psi(\theta,h),y^*,\theta,h) - \varepsilon.$$

Then, from (46) we have $\Delta_{\theta,u^*}J(u) < -\varepsilon + \varepsilon = 0$, which contradicts the optimality of u(t,h). If ε,h are such that for all $\theta \varepsilon T_h$

$$U_\varepsilon(\theta,h) \supset U(\theta) \quad ,$$

then along the optimal control we necessarily have satisfied the condition (ε-maximum condition)

$$H(x(t,h),\psi(t,h),u(t,h),t,h) \geq H(x(t,h),\psi(t,h),u^*,t,h) - \varepsilon \tag{47}$$

for all $u^* \varepsilon U(t)$, $t \varepsilon T_h$. The condition reduces to the first maximum condition if h is such that

$U_\varepsilon(\theta,h) \supset U(\theta)$ as $\varepsilon \to 0$ and all $\theta \varepsilon T_h$.

Application of the quasimaximum condition for concrete determination of the optimal control is not always simple because of ineffective procedures for determing the set $U_\varepsilon(\theta,h)$. Therefore, we introduce the set

$$U_\varepsilon^1(\theta,h) = \{u^*: \quad B(\theta,u^*,h) \leq \varepsilon, u^* \varepsilon U(\theta)\} \quad ,$$

where the function $B(\theta,u^*,h)$ is chosen from the condition

$$o_1(||\Delta_{\theta,u^*} x(t_1,h)||) - \sum_{t=\theta+1}^{t_1-1} o(||\Delta_{\theta,u^*} x(t,h)||) \leq B(\theta,u^*,h).$$

The form of the function $B(\theta,u^*,h)$ depends upon the concrete conditions. For example, we consider the case of the discrete system (43). Let the solution of Eq. (42) be bounded for all measurable $v(s) \varepsilon U(s)$. Then, for sufficiently small h_o, it is possible to find a number M such that for all h, $0 \leq h \leq h_o$, the solutions of Eq. (43) satisfy the inequality

$$||x(t,h)|| \leq M \quad , \quad t \varepsilon T_h \quad .$$

For simplification of the estimate, we assume that the functions $f(x,u,t)$, $\phi(x)$ are twice-differentiable in x and

$$\left|\left|\frac{\partial f(x,u,t)}{\partial x}\right|\right| \leq M_1 \quad , \quad \left|\left|\frac{\partial^2 f}{\partial x^2}\right|\right| \leq M_2 \quad ,$$

$$\left|\left|\frac{\partial \phi(x)}{\partial x}\right|\right| \leq M_3 \quad , \quad \left|\left|\frac{\partial^2 \phi}{\partial x^2}\right|\right| \leq M_4$$

in the region $||x|| \leq M$, $u \varepsilon \bigcup_{t \varepsilon T_h} U(t)$, $t \varepsilon T_h$, $h \leq h_o$.

Then from the definition of the quanitites $o_1(||\Delta x||)$, $o(||\Delta x||)$, we have

$$|o(||\Delta x(t,y)||)| \leq \frac{h}{2} ||\psi(t,h)||M_2||\Delta x(t,h)||^2 ,$$

$$|o_1(||\Delta x(t_1,h)||)| \leq \frac{1}{2} M_4||\Delta x(t_1,h)||^2 .$$

From (45), we obtain the estimate

$$||\psi(t,h)|| \leq M_5 , \quad t \varepsilon T_h ,$$

where M_5 does not depend on h, $h \leq h_o$, and is determined by the quantities M_1, s_o, s_1, M_3. The estimate for $||\Delta_{\theta,u*}x(t,h)||$ has the form

$$||\Delta_{\theta,u*}x(t,h)|| \leq hM_6||\Delta_{u*}f(x(\theta,h),u(\theta,h),\theta)||, \quad t \varepsilon T_h ,$$

where M_6 is independent of h. As a result, for $B(\theta,u*,h)$, we may take the function

$$B(\theta,u*,h)$$

$$= \left[\frac{M_4}{2} + (s_1 - s_o)\frac{M_2M_5}{2}\right] M_6^2h^2||\Delta u*f(x(\theta,h),u(\theta,h),\theta)||^2 .$$

Hence, we see that if the set $U(\theta)$, $\theta \varepsilon T$, is bounded for any ε it is possible to find a number $h(\varepsilon)$ such that

$$U_\varepsilon^1(\theta,h) = U(\theta) \text{ for } \theta \varepsilon T_h, \ h \leq h(\varepsilon).$$

Conversly, for bounded $U(\theta)$, for each h there exists an $\varepsilon(h)$ such that the last equality is satisfied. From the definition of $B(\theta,u*,h)$, it follows that $\varepsilon(h) \sim h^2$.

The effective application of the ε-maximum condition depends upon the relationship between H and the quantity ε. If $\varepsilon \geq 2|H|$, then the ε-maximum condition will not isolate from the set $U(\theta)$, the characteristic subset of candidates for the optimal control. In general, it is difficult to estimate the order in h of the function H from below. However, for problem B_h the function $\mathcal{E}$ for small h has order not greater than one in h if $|f_{1n}(x(t,h),u(t,h),t)| \geq M_7$, for $t\varepsilon T_h$, $h \leq h_0$. It is possible to find other criteria (for example, the relation between $|f_{1n}|$ and $||\Delta_{u*}f||$), under which $\mathcal{E} \sim h$. If the function $\mathcal{E}$ possesses the indicated properties, then the effectiveness of the ε-maximum condition increases with decreasing h in system (43). We formulate these results in the form of a theorem.

Theorem 12. If $u(t,h)$, $t\varepsilon T_h$, is the optimal control for problem A_h in system (43), then for any $\varepsilon > 0$, there exists a number $h(\varepsilon)$ such that the control $u(t,h)$ for $h \leq h(\varepsilon)$, satisfies the ε-maximum condition.

For problem B_h in system (43), the effectiveness of the ε-maximum condition rises with decreasing h. In general, it is impossible to improve upon Theorem 12. In other words, there exist discrete systems of the form (43), the optimal controls for which do not satisfy the first maximum condition for $h > 0$.

Example 10.

$$x(t + 1) = x(t) + hu(t),$$

$$y(t + 1) = y(t) + h[x^2(t) - u^2(t)],$$

$$x(0) = y(0) = 0, \ |u| \leq 1, \ \phi(x,y) = y, \ t_0 = 0, \ t_1 = 2 .$$

We have $J_h(u) = -(h - h^3)u^2(0) - hu^2(1)$. For any h, $0 < h \leq 1$, the control $\{u(0,h) = 1, u(1,h) = 1\}$ is optimal. For it,

$$H_o = -2h^3u - hu^2 \quad , \quad H_o = H(x(0,h),\psi(0,h),u,0,h) \quad .$$

At the point $u = 1$, the latter function corresponding to the optimal $u(0,h) = 1$, does not assume a maximum for any h, $0 < h \leq 1$.

Let $h > 1$. Then the control $\{u(0,h) = 0, u(1,h) = 1\}$ is optimal and for it $H_o = hu^2$. For $h > 1$, this function does not assume its maximum at the point $u = 0$.

§5. Two-Parameter Nonlinear Systems. The Quasimaximum Principle.

1. Statement of the Problem. At the node $\{mh,n\tau\}$ of the rectangle $D = \{mh,n : m = 0,\ldots,M;\ n = 0,1,\ldots,N\}$, let the connection between the q-vector $x = x(m,n)$ and the r-vector $u = u(m,n)$ be given by the equation

$$R_{h\tau}(x(m,n)) = f(x,R_h,R_\tau,u,m,n) \tag{48}$$

and the initial conditions

$$x(0,n) = \gamma^n,\ n = 0,\ldots,N;$$

$$x(m,0) = \beta^m,\ m = 0,1,\ldots,M \ , \tag{49}$$

where

$$R_{h\tau}(x(m,n)) = \frac{x(m+1,n+1) - x(m+1,n) - x(m,n+1) + x(m,n)}{h\tau},$$

$$R_h = R_h(x(m,n)) = \frac{x(m+1,n) - x(m,n)}{h\tau},$$

$$R_\tau = R_\tau(x(m,n)) = \frac{x(m,n+1) - x(m,n)}{\tau}$$

while the function $f(x,T_h,R_\tau,u,m,n)$ is uniquely defined and continuous on $\Omega_1, = E_q \times E_q \times E_q \times E_r \times E_1 \times E_1$ and has continuous second derivatives in x, R_h, and R_τ on Ω_1. We assume that the admissible controls

$$u(m,n) \varepsilon U \subset E_r \quad .$$

Along trajectories (48),(49), we define the functional

$$J(u) = c'x(M,N) \quad .$$

Here c is a given vector.

<u>Problem</u>. It is required to determine $u(m,n)\varepsilon U$, $0 \leq m \leq M-1$, $0 \leq n \leq N-1$, such that

$$J(u) = \min_{\tilde{u}(m,n)\varepsilon U} J(\tilde{u}) \quad .$$

2. <u>The Formula for the Increment of the Functional</u>. Let the solution x(m,n) correspond to the control u(m,n) and let $x(m,n) + \Delta x(m,n)$ correspond to $u(m,n) + \Delta u(m,n)$.

$$\left.\begin{aligned} & R_{h\tau}(\Delta x(m,n)) \\ & \quad = f(x + \Delta x, R_h + \Delta R_h, R_\tau + \Delta R_\tau, u + \Delta u, m, n) \\ & \qquad - f(x, R_h, R_\tau, u, m, n) \quad , \\ & \Delta x(m,0) = \Delta x(0,n) = 0 \quad , \\ & \Delta R = R(x(m,n) + \Delta x(m,n)) - R(x(m,n)) = R(\Delta x(m,n)) \quad . \end{aligned}\right\} \quad (51)$$

For an arbitrary sequence of q-vectors p(m,n), we have

$$\sum_{m=0}^{M-1}\sum_{n=0}^{N-1} p'(m,n)R_{h\tau}(\Delta x(m,n))h\tau$$

$$= \sum_{m=0}^{M-1}\sum_{n=0}^{N-1} \Delta x'(m,n)\bar{R}_{h\tau}(p(m,n))h\tau$$

$$- \sum_{m=0}^{M-1} \Delta x'(m,N)\bar{R}_{h}(p(m,N-1))h$$

$$- \sum_{n=0}^{N-1} \Delta x'(M,n)\bar{R}_{\tau}(p(M-1,n))\tau$$

$$+ p'(M-1,N-1)\Delta x(M,N) \quad , \qquad (52)$$

where m = -1,0,...,M-1; n = -1,0,...,N-1,

$$\bar{R}_{h\tau}(p(m,n)) = \frac{p(m,n) - p(m-1,n) - p(m,n-1) + p(m-1,n-1)}{h\tau} ,$$

$$\bar{R}_{h} = (p(m,n)) = \frac{p(m,n) - p(m-1,n)}{h\tau} ,$$

$$\bar{R}_{\tau} = (p(m,n)) = \frac{p(m,n) - p(m,n-1)}{\tau} .$$

We set

$$H(x,R_h,R_\tau,p,u,m,n) = p'(m,n)F(x,R_h,R_\tau,u,m,n)$$

and choose p(m,n) from the equations

$$\bar{R}_{h\tau}(p(m,n)) = \frac{\partial H(x,R_h,R_\tau,p,u,m,n)}{\partial x} - \bar{R}_h \frac{\partial H(x,R_h,R_\tau,p,u,m,n)}{\partial R_h} - \bar{R}_\tau \frac{\partial H(x,R_h,R_\tau,p,u,m,n)}{\partial R_\tau} , \tag{53}$$

with boundary conditions

$$p(M-1,N-1) = -c ,$$

$$\bar{R}_h(p(m,N-1)) = -\frac{H(x,R_h,R_\tau,p,u,m,N-1)}{\partial R_\tau} , \tag{54}$$

$$\bar{R}_\tau(p(M-1,n)) = -\frac{H(x,R_h,R_\tau,p,u,M-1,n)}{\partial R_h} .$$

From (51)-(54), we obtain

$$c'\Delta x(M,N)$$

$$= \sum_{m=0}^{M-1} \sum_{n=0}^{N-1} h\tau[H(x,R_h,R_\tau,p,u+\Delta u,m,n) - H(x,R_h,R_\tau,p,u,m,n)]$$

$$- \sum_{m=0}^{M-1} \sum_{n=0}^{N-1} h\tau[H(x+\Delta x,R_h+\Delta R_h,R_\tau+\Delta R_\tau,$$

$$p,u,+\Delta u,m,n) - H(x,R_h,R_\tau,p,u+\Delta u,m,n)]$$

$$= \sum_{m=0}^{M-1} \sum_{n=0}^{N-1} h\tau\left[\Delta x'(m,n) \frac{\partial H(x,R_h,R_\tau,p,u,m,n)}{\partial x}\right.$$

$$+ \Delta R_h'(m,n) \frac{H(x,R_h,R_\tau,p,u,m,n)}{\partial R_h}$$

$$+ \Delta R'_\tau(m,n) \frac{\partial H(x,R_h,R_\tau,p,u,m,n)}{\partial R_\tau}\Bigg] .$$

Hence

$$\Delta J \equiv c'\Delta x(M,N) = -\sum_{m=0}^{M-1}\sum_{n=0}^{N-1} h\tau[H(z,p,u+\Delta u,m,n)$$

$$- H(z,p,u,m,n)] - \eta, \quad \eta = \eta_1 + \eta_2 .$$

$$\eta_1 = \sum_{m=0}^{M-1}\sum_{n=0}^{N-1} h\tau\Delta z'(m,n)$$

$$\cdot \frac{\partial H(z,p,u+\Delta u,m,n)}{\partial z} - \frac{\partial H(z,p,u,m,n)}{\partial z} ,$$

$$\eta_2 = \frac{h\tau}{2}\sum_{m=0}^{M-1}\sum_{n=0}^{N-1} \Delta z'(m,n) \frac{\partial^2 H(z+\theta\Delta z,p,u+\Delta u,m,n)}{\partial z^2} \Delta z(m,n),$$

$$0 \le \theta(m,n) \le 1 .$$

Here $z = \{x,R_h,R_\tau\}$ is a 3-dimensional vector. For the increment

$$\Delta^* u(m,n) = \begin{cases} u^* - u(k,\ell) , & m = k,\ n = \ell, \\ 0 , & m \neq k,\ n \neq \ell \end{cases}$$

we have

$$\Delta^* J = -h\tau[H(z,p,u^*,k,\ell) - H(z,p,u,k,\ell) - \eta^*] ,$$

$$\eta^* = \frac{1}{2}\sum_{m=k+1}^{M-1}\sum_{n=\ell+1}^{N-1} \Delta z'(m,n)\frac{\partial^2 H(z+\theta\Delta^* z,p,u^*,m,n)}{\partial z^2} \Delta z(m,n).$$

It is not difficult to show that

$$|\eta^*| \leq \frac{1}{2}(h\tau)^2 ||f(z,u^*,k,\ell) - f(z,u,k,\ell)||^2$$

$$\cdot \sum_{m=k+1}^{M-1} \sum_{n=\ell+1}^{N-1} L(m,n) \left|\left|\frac{\partial^2 H}{\partial z^2}\right|\right| , \quad L(m,n) = \text{const.} \qquad (55)$$

3. <u>The Quasimaximum Principle.</u> Let $f(z,u,m,n)$ be differentiable in u. We introduce the set $a(k,\ell,\alpha)$:

$$a(z,p,u(m,n),\ldots,u(M-1,N-1),k,\ell,\alpha) = \{u^*: \ |\eta^*| \leq \alpha\}$$

and the quantity $\delta_u H$:

$$H(z,p,u+\Delta u,m,n) - H(z,p,u,m,n) = \delta_u H(z,p,u,m,n) + o(||\Delta u||).$$

We say that the control u at the node $\{k,\ell\}$ satisfies the quasimaximum condition with number $\mu(k,\ell)$ and set $\omega(k,\ell)$ if: 1) $H(z,p,u,k,\ell) \geq H(z,p,u^*,k,\ell) - \mu(k,\ell)$ for all $u^* \varepsilon \omega(k,\ell)$; 2) $\delta_u H \leq 0$ if U is convex; 3) $\delta_u H = 0$ at all interior points of U.

<u>Theorem 13. The optimal control $u = \{u(m,n)\}$ in problem (50) satisfies the conditions:</u>

1) <u>the quasimaximum principle with $\mu(k,\ell) = \alpha$, $\omega(k,\ell) = a(k,\ell,\alpha) \cap U$, $k = 0,1,\ldots,M-2$; $\ell = 0,1,\ldots,N-2$;</u>

2) $H(z,p,u,k,\ell) \geq H(z,p,u^*,k,\ell)$, (56)

$u^* \varepsilon U$, $\{k = M-1; \ell = 0,\ldots,N-1\}$,

$\{k = 0,\ldots,M-1; \ell = N-1\}$.

The proof is analogous to Theorem 11.

Remark. The set $a(k,\ell,\alpha)$ may be replaced by the set $a'(k,\ell,\alpha) = \{u^*: |\bar{\eta}| \leq \alpha\}$, where $\bar{\eta}$ is some estimate from above for η^*, for example (55). From the definition of the set $a(k,\ell,\alpha)$, it follows that the local maximum principle is valid at the node $\{k,\ell\}$ if the set $a(k,\ell,0)$ contains a point different from $u(k,\ell)$.

Thus, the quasimaximum condition isolates some set U^* of points $u(k,\ell)$ which are candidates for optimality. Further narrowing of the set U^* is carried out by varying α and by the local conditions $\delta_u H \leq 0$, $\delta_u H = 0$.

4. Systems of Particular Form. Conditions for Optimality.

a) For the linear variant

$$R_{h\tau}(x(m,n)) = A(m,n)x(m,n) + B(m,n)R_h(x(m,n))$$

$$+ C(m,n)R_\tau(x(m,n)) + b(u(m,n),m,n) \tag{57}$$

of Eq.(48), we have $\eta = 0$. Therefore, the following assertion is valid.

Theorem 14. In order that a control $u \in U$ be optimal for (57), it is necessary and sufficient that the condition (56) be satisfied at the nodes $\{m,n\}$, $m = 0,\ldots,M-1$, $n = 0,\ldots,N-1$.

b) For the case

$$R_{h\tau}(x(m,n)) = A(u(m,n),m,n)x(m,n) + B(u(m,n),m,n)R_h(x(m,n))$$

$$+ C(u(m,n),m,n)R_\tau(x(m,n)) + b(u(m,n),m,n)$$

it is not difficult to see that $\eta^* = 0$.

Theorem 15. *If $u = \{u(m,n)\}$ is an optimal control, then at each node $\{k,\ell\}$, $k = 0,\ldots,M-1$; $\ell = 0,\ldots,N-1$, the relations (56) are satisfied.*

c) Let U be convex and let Eq. (48) have the form

$$R_{h\tau}(x(m,n))$$

$$= \sum_{i=1}^{r} f_i(x,R_h,R_\tau,m,n)U_i(m,n) + g(x,R_h,R_\tau,m,n) \quad .$$

Theorem 16. *Along the optimal control, conditions (56) are satisfied at all nodes of the rectangle D.*

Theorems 15 and 16 follow from the formulas for ΔJ.

§6. Application of the Methods of Functional Analysis to the Optimization of Discrete Systems

1. *Reduction of the Minimal-Time Problem to an L-Problem.* Let the motion of the controlled system be given by the difference equation [99,128,149,223]

$$x(n+1) = Ax(n) + bu(n), \quad x = \{x_1,\ldots,x_l\}, \quad b = \{b_1,\ldots,b_l\}, \quad (58)$$

where $x(n)$ is a vector in state space, A is a constant nonsingular matrix, b is some given vector, and $u(n)$ is a scalar function of the discrete argument n. The minimal-time control problem is the following: given an initial point $x(0) = \{x_1(0),\ldots,x_\ell(0)\}$, it is required to choose a control $u^o(n)$, $n = 0,1,\ldots,K^o-1$ such that the trajectory of the system (58) "reaches" the origin in the smallest possible number of steps K^o. It is assumed that the values $u(n)$ are restricted by the condition

$$|u(n)| \leq 1 \;, \; n = 0,\ldots,K^o - 1 \quad . \quad (59)$$

By the scheme of §6.11, we reduce this problem to an L-problem in a finite-dimensional space. The solution of (58) may be written as

$$x(n) = F(n)x(0) + \sum_{i=0}^{n-1} F(n - i - 1)bu(i) \quad , \tag{60}$$

(an analogue to Cauchy's formula), where $F(n)$ is the fundamental solution matrix for the homogeneous part of (58). At some moment of time K, let the trajectory of system (58) reach the origin: $x(K) = 0$. Then from (60) it follows that

$$-x(0) = \sum_{j=1}^{K} F(-j)bu(j - 1) \quad . \tag{61}$$

Introducing the notation

$$h(n) = F(-n)b \quad , \quad \eta(n) = u(n - 1), \tag{62}$$

and taking into account (61),(62), the original problem may be formulated as: find the smallest number K^o and a linear function $f(\cdot)$ (in this case, it is a linear form) such that

$$f(h^{\alpha}) = \sum_{n=1}^{K^o} h_{\alpha}(n)\eta(n) = -x_{\alpha}(0) \quad , \tag{63}$$

$$\alpha = 1,\dots,\ell \quad , \quad h^{\alpha} = \{h_{\alpha}(1),\dots,h_{\alpha}(K^o)\} \; ,$$

$$||f(\cdot)|| = \max_{n} |\eta(n)| \leq 1 \quad . \tag{64}$$

We will assume that the vectors h^{α} are linearly independent, where $K^o > \ell$. According to the basic assumptions of §6.11, a solution of (63)-(64) exists only in the event

$$\lambda(K) \geq 1 \quad , \tag{65}$$

where $\lambda(K)$ is determined from the condition

$$\lambda(K) = \min \sum_{n=1}^{K} |\alpha_1 h_1(n) + \cdots + \alpha_\ell h_\ell(n)| \qquad (66)$$

where $\sum_{i=1}^{\ell} \alpha_i x_i(0) = -1$. The optimal time K^o is determined as the smallest number K satisfying (65). The optimal control is sought in the following form. Let $\alpha^o = \{\alpha_1^o, \ldots, \alpha_\ell^o\}$ be the solution of problem (66); we call the element $\{h^o(n)\} = \{\sum_{i=1}^{\ell} \alpha_j^o h_j(n)\}$ minimizing. If for a given functional $f(\cdot)$, there exists an element $h(\cdot)$

$$|f(h)| = ||f(\cdot)|| \; ||h(\cdot)|| \quad , \qquad (67)$$

then we call this element extremal. The element $h(\cdot)$ is called normal if the functional $f(\cdot)$ is determined by (67) up to a scalar multiplier. The optimal control is determined from the condition that the minimizing element is extremal for an arbitrary choice of a minimal norm solution to (63)-(64). If the minimizing element $\{h^o(n)\}$ of problem (66) is normal, then the optimal control is determined uniquely. The vector $\{h^o(n)\}$ is normal only when

$$h^o(n) \neq 0 \quad , \quad n = 1, \ldots, K^o \quad . \qquad (68)$$

At those points where condition (68) is satisfied, the optimal control (of minimal modulus) is computed by the formula

$$u^o(n-1) = \frac{1}{\lambda(K^o)} \operatorname{sign} h^o(n) \quad . \qquad (69)$$

Below we consider cases when condition (68) is not satisfied everywhere.

2. <u>A Theorem on the Uniqueness of the Optimal Control.</u> We introduce several criteria for estimating the number of zero coordinates in the vector $\{h^o(n)\}$, $n = 1,\dots,K^o$. We consider the set of vectors

$$b, Ab, \dots, A^{\ell-1}b \ , \quad A^{\ell}b, \dots, A^{K^o-1}b \ . \tag{70}$$

We will call the optimal problem non-dengenerate if the first ℓ vectors are linearly independent. In a non-degenerate optimal problem, the vector $\{h^o(n)\}$ may not have more than $\ell-1$ successive zero coordinates. Assume the contrary. Let there exist a number k for which

$$h^o(k) = h^o(k+1) = \cdots = h^o(k+\ell-1) = 0 \ .$$

Then, by the definition of $\{h^o(n)\}$, we have

$$(\alpha^o)'h(i) = 0 \ , \quad i = k,\dots,k+\ell-1 \ ,$$

or, taking into account Eq.(62), we obtain

$$b'F'(-i)\alpha^o = 0 \ , \quad i = k,\dots,k+\ell-1 \ .$$

It is not difficult to show that $F(-i) = F(-i-1)A$. Therefore,

$$b'(A')^j F'(-k-\ell+1)\,\alpha^o = 0 \ , \ j = 0,\dots,\ell-1 \ .$$

But, by the condition that the first ℓ vectors from (70) are linearly independent, it follows that the system of homogeneous equations

$$b'(A')^j F'(-k-\ell+1)\alpha^o = 0 \ , \quad j = 0,\dots,\ell-1 \ ,$$

(where F' is the transpose of F) may have only the zero

solution $\alpha_1^o = \cdots = \alpha_\ell^o = 0$, which contradicts problem (66). The assertion is proved.

We will call the optimal problem <u>strongly non-degenerate</u> if any ℓ vectors from (7)) are linearly independent. We prove that in a strongly non-degenerate problem the vector $\{h^o(n)\}$ may not have more than $\ell-1$ zero coordinates.

We assume the contrary. If

$$h^o(i_1) = \cdots = h^o(i_\ell) = 0 \ , \ i_1 < i_2 < \cdots < i_\ell \ ,$$

then

$$b'F'(-i_k)\alpha^o = 0 \quad , \quad k = 1,\ldots,\ell \ .$$

Hence, it follows that

$$b'(A^{i_\ell - i_j})'F'(-i\ )\alpha^o = 0 \quad , \quad j = 1,\ldots,\ell \ .$$

But, by assumption any ℓ vectors from (70) are linearly independent. Thus, the last system of homogenous equations has only the zero solution which again contradicts problem (66).

We return to the problem of determining optimal controls in the case when the minimizing element of problem (66) is not normal. Let the vector $\{h^o(n)\}$ have q zero coordinates

$$h^o(i_1 - 1),\ldots,h^o(i_q - 1) \ .$$

Then the values

$$u^o(i_1 - 1),\ldots,u^o(i_q - 1) \ ,$$

cannot be determined from formula (69). We substitute the known values u(i) into (63) and isolate the terms with unknown control coordinates

$$\sum_{s=1}^{q} h_j(i_s)\eta(i_s) = -x_j(0) - \sum_{t} h_j(t)\eta(t), \; j=1,\ldots,\ell, \tag{71}$$

where $\sum_t$ is taken over the known control values. The system of equations (71) is compatible for any relation between q and ℓ. This follows from the fact that there exists a solution to the optimization problem. Moreover, from general results relating to the L-problem, it follows that all solutions of Eq.(71) satisfy inequality (64). As follows from the results of §1, the non-degeneracy of the optimal problem implies the existence and uniqueness of the optimal control for any initial state in a neighborhood of the origin for a continuous, linear system. For a discrete system, this assertion ceases to be true (cf. example 13). But the following theorem is true.

Theorem 17. In a strongly non-degenerate problem the optimal control is unique.

Actually, in this case we have no more than ℓ-1 moments of time at which the optimal control is not determined from formula (69). For the sake of definiteness, let the number of such moments be q ($q < \ell$). In the matrix $\{h_j(i_s)\}$, $s = 1,\ldots,q$; $j = 1,\ldots,\ell$, at least one minor of order q is different from zero. This follows from the fact that the vectors $\{h_j(i_s)\}$ are determined by relation (62) and our problem is strongly non-degenerate: if all minors of order q equalled zero, then among the vectors of (70), there would be one which is linearly dependent upon ℓ-1 others which is impossible. This means that relation (71) determines a unique control at the remaining q moments of time. The assertion is proved.

3. Examples

Example 11. We consider the optimal problem for the system of two equations

$$x(n + 1) = x(n) + y(n) \quad , \quad y(n + 1) = y(n) + u(n) \ .$$

We have

$$A = \begin{bmatrix} 1 & 1 \\ 0 & 0 \end{bmatrix} \quad , \qquad b = \begin{bmatrix} 0 \\ 1 \end{bmatrix} \ .$$

The problem is strongly non-degenerate since any two vectors of the sequence

$$b = \begin{bmatrix} 0 \\ 1 \end{bmatrix} \quad , \qquad Ab = \begin{bmatrix} 1 \\ 1 \end{bmatrix} , \ldots , A^{K^o - 1} b = \begin{bmatrix} K^o - 1 \\ 1 \end{bmatrix}$$

are linearly independent. From the discussion in section 2, it follows that all the numbers $\{u^o(n)\}$, except the first, are determined from formula (69). We construct the optimal control for the following points: $\{-3,0\}$, $\{-4,1/2\}$. The fundamental solution matrix has the form

$$F(n) = \begin{bmatrix} 1 & n \\ 0 & 1 \end{bmatrix} \ ;$$

hence

$$h_1(k) = -k \quad , \quad h_2(k) = 1 \ .$$

Problem (66) assumes the form: find min $\sum_{n=1}^{K} |-\alpha_1 n + \alpha_2|$ $= \lambda(k)$, where $\alpha_1 x(0) + \alpha_2 y(0) = -1$.

For the point $\{-3,0\}$, the minimal number K^o for which $\lambda(K^o) \geq 1$ is $K^o = 4$, $\lambda(K^o) = 4/3$. The minimum is assumed on two elements: $\{\alpha_1^o = 1/3,\ \alpha_2^o = 2/3\}$, $\{\alpha_1^o = 1/3,\ \alpha_2^o = 1\}$. From (69), we determine the optimal control

$$u^o(n-1) = \frac{3}{4}\ \mathrm{sign}\left[-\frac{1}{3}n + \frac{2}{3}\right], \tag{72}$$

$$u^o(n-1) = \frac{3}{4}\ \mathrm{sign}\left[-\frac{1}{3}n + 1\right]. \tag{73}$$

For the case (72), the value $u^o(1)$ is undetermined, while in case (73) the value $u^o(2)$ is indefinite. These values are found from system (71) which, in this example, has the form $\sum(-n\eta(n)) = 3$, $\sum\eta(n) = 0$. Hence, $u^o(1) = 3/4$, $u^o(2) = -3/4$. We note that the moduli of all control numbers are equal.

For the point $\{-4,1/2\}$, the solution is the same as for the point $\{-3,0\}$. These values of the optimal control are determined by the formula

$$u^o(n-1) = \frac{3}{4}\ \mathrm{sign}\left[-\frac{1}{3}n + \frac{2}{3}\right].$$

The value $u^o(1)$ is determined from the system $\sum(-n\eta(n)) = 4$, $\sum\eta(n) = 1/2$. Since the last, as shown above, is always consistent, from the second equation we find that $\eta(2) = u^o(1) = 1/4$. In this case, the quantity $u^o(1)$, determined from system (71), has modulus less than the other values of the control function.

Example 12. We solve the optimal control problem for the point $\{0,-1/c_o\}$. We find again that

$$\min \sum_{n=1}^{K^o} |-\alpha_1 n + \alpha_2| = \lambda(K^o) \text{ for } \alpha_2\left(-\frac{1}{c_o}\right) = -1 .$$

Substituting α_2 from the second equation into the first, we obtain

$$\min \sum_{n=1}^{K^o} |-\alpha_1 n + c_o| = \lambda(K^o) \quad .$$

We find the roots of the left side. They are

$$(\alpha_1)_n = \frac{c_o}{n} \quad , \; n = 1,\ldots,K^o \quad .$$

It is not difficult to see that the minimum is assumed if and only if one term reduces to zero. We calculate the value $\sum_{n=1}^{K} |-\alpha_1 n + c_o|$ at the point $(\alpha_1)_K$:

$$\sum_{n=1}^{K} |-(\alpha_1)_K n + c_o$$

$$= \left(\frac{c_o}{1} - \frac{c_o}{K}\right) 1 + \cdots + \left(\frac{c_o}{K-1} - \frac{c_o}{K}\right) (K - 1)$$

$$= (K - 1)\, c_o - \frac{c_o}{K} (1 + \cdots + (K - 1)) = \frac{K-1}{2} c_o.$$

The value $\sum_{n=1}^{K} |(-\alpha_1)_i n + c_o|$ may be computed as the sum of two quantities: $c_o(i - 1)/2$ and

$$\left(\frac{c_o}{1} - \frac{c_o}{i+1}\right)(i + 1) + \left(\frac{c_o}{i} - \frac{c_o}{i+2}\right)(i + 2)$$

$$+ \cdots + \left(\frac{c_o}{i} - \frac{c_o}{K}\right) K = \frac{(K - i)(K - i + 1)}{2i} c_o \; ,$$

that is

$$\sum_{n=1}^{K} |-(\alpha_1)_i n + c_o| = \frac{2i^2 - 2(K + 1)i + K + K^2}{2i} c_o \quad .$$

$$F(n) = \begin{Vmatrix} \left(-\frac{1}{2}\right)^n & \frac{1}{4}\left(\frac{1}{2}\right)^n - \frac{1}{4}\left(-\frac{1}{2}\right)^n \\ 0 & \left(\frac{1}{2}\right)^n \end{Vmatrix} .$$

Hence

$$\begin{Vmatrix} h_1(1) \\ h_2(1) \end{Vmatrix} = \begin{Vmatrix} 0 \\ 2 \end{Vmatrix} ,\ldots, \begin{Vmatrix} h_1(2i) \\ h_2(2i) \end{Vmatrix} = \begin{Vmatrix} 2 \\ 4 \end{Vmatrix} 4^{i-1}$$

$$\begin{Vmatrix} h_1(2i+1) \\ h_2(2i+1) \end{Vmatrix} = \begin{Vmatrix} 0 \\ 2 \end{Vmatrix} 4^i, \ldots .$$

We construct the optimal control for the point $\{-2,0\}$. From general results, it follows that it is determined by formula (69) at least for one step. The properties of (70) are such that the values of the controlling function are computed successively from formulas (69),(71).

If we solve problem (66) for this example, we obtain $K^o = 6$, $\lambda(K^o) = 21/12$, where the minimum is assumed on the element $\alpha^o = \{1/12,-1/24\}$. The optimal control has the form

$$u^o(n-1) = \frac{12}{21} \operatorname{sign}\left[\frac{1}{12} h_1(n) - \frac{1}{24} h_2(n)\right] , \quad n = 1,3,5.$$

We find the values $u^o(1)$, $u^o(3)$, $u^o(5)$, solving the system (71) which, in this case, assumes the form

$$\left.\begin{aligned} 2u^o(1) + 8u^o(3) + 32u^o(5) &= 12 , \\ 4u^o(1) + 16u^o(3) + 64u^o(5) &= 24 . \end{aligned}\right\}$$

Hence

$$u^o(5) = \frac{3}{8} - \frac{1}{16} u^o(1) - \frac{1}{4} u^o(3) \quad .$$

Thus, the optimal control depends upon two arbitrary constants $u^o(1)$ and $u^o(3)$, which corresponds to the result of section 2.

4. <u>The L-Problem of Moments on Linearly Dependent Elements</u>. In the event $K^o < \ell$ (above we assumed that $K^o \geq \ell$), direct use of the results of [4] is not possible. For optimal regulation problems, the case $K^o < \ell$ is not uncommon. This corresponds to the case of those initial states for which optimal control is possible in a number of steps less than the dimension of the state space. In this case, we reduce the problem to an ordinary L-problem preliminary to solving it in general form.

<u>Problem</u>. Given n elements $x^1,\ldots,x^n$ in the space E_m (linearly independent vectors are not required), find necessary and sufficient conditions for the numbers $c_1,\ldots,c_n$, $L(\sum c_i^2 > 0,\ L > 0)$ in order that there exist a linear functional f(x) satisfying the relations

$$f(x^i) = c_i \ , \quad ||f(\cdot)|| \leq L \quad , \quad i = 1,\ldots,n \quad . \tag{76}$$

From the elements $x^1,\ldots,x^n$ we choose a basis. Let it be formed from the elements $x^1,\ldots,x^\ell$. Then the remaining elements $x^{\ell+i}$ are expressed through the vectors $x^1,\ldots,x^\ell$:

$$x^{\ell+i} = \sum_{j=1}^{\ell} a_{ij} x^j \quad , \quad i = 1,\ldots,n-\ell \quad .$$

Substituting these values into (76), we obtain

$$f(x^i) = c_i \;, \quad f(x^{\ell+j}) = f(\sum_{k=1}^{\ell} a_{jk}x^k) = \sum_{k=1}^{\ell} a_{jk}c_k = c_{\ell+j},$$

$$i = 1,\dots,\ell; \quad j = 1,\dots,n-\ell.$$

Hence, it follows that the constants $c_{\ell+j}$, $j = 1,\dots,n-\ell$, are uniquely determined by the elements $x^1,\dots,x^n$, and the constants c_i, $i = 1,\dots,\ell$. Thus, the problem has been reduced to an ordinary L-problem for ℓ linearly independent elements $x^1,\dots,x^\ell$. If we consider the method of passage from an optimal control problem to an L-problem, then the obtained results may be expressed as: the optimal regulation time and the optimal control for a transition process, the duration of which does not exceed the dimension of the state space, correspond to points of some subspace.

5. Application of the Minimax Theorem, the Separation of Convex Sets, and the Existence of a Separating Hyperplane. It is natural to apply the L-problem to linear systems of the form

$$x(n + 1) = A(n)x(n) + b(n)u(n) \quad ,$$

in which the class of admissible controls may be given in the form of an inequality with a norm

$$U(\cdot) = \{u(\cdot): ||u(\cdot)|| \leq L\} \quad .$$

Below (§7,8), we solve a series of complex problems by such a path. Studies of the system with nonlinear inputs

$$x(n + 1) = A(n)x(n) + b(u(n),n) \qquad (77)$$

by a similar path encounter serious difficulties. For such systems, the methods of §6.9, 6.10 are effective. We recommend as an exercise to investigate a typical problem from §6.9, 6.10 by the indicated methods for the discrete system (77) in the case when $u(n) \in U$, U a bounded set such that $b(U,n)$ is convex.

Remark. In section 1, we described a procedure for finding the optimal minimal norm control. Finally, the methods of sections 2 and 3 are suitable for determing the switching controls with given levels. For this it is possible to draw on the ideas of hierarchical system criteria (§7.9).

§7. Optimal Processes in Connection with Discrete-Time Systems

1. Problem Statement. We consider systems consisting of the controlled object, the controller and, connected with them, a second object. It is assumed that for regulating the first object, we use signals from the second [82],[124b]. In such situations, the usual problems of optimal control are more complex since for their solution it is necessary to impose constraints on the quantities describing the state of the second object. Below we investigate one optimal regulation problem for such systems.

Hereafter, we let $x(n)$ and $y(n)$ denote vectors describing the states of the first and second objects, respectively, and we let $u(n)$ be the control action. Let the considered regulator system be described by the difference equations

$$x(n + 1) = Ax(n) + Bu(n) , \tag{78}$$

$$y(n + 1) = Cy(n) + Du(n) , \tag{79}$$

where $x(n)$ is a p-vector describing the first object, $y(n)$ is a q-vector describing the second object, $u(n)$ is an r-vector of controls, the matrices C,A are nonsingular, while the matrix D has at least one minor of order r different from zero.

For Eqs. (78),(79) we pose the problem. Given initial values $n = 0$, $x(0)$, $y(0) = 0$, it is required to choose the controls $u(n)$ such that the point $x(n)$ of the trajectory of system (78) with initial value $x(0)$, is transferred to the origin in minimal time. It is assumed that we allow only control actions for which the vector $y(n)$ of (79) satisfies

$$|y_i(n)| \leq 1 \ , \quad i = 1,\dots,q; \quad n \geq 0 \ . \tag{80}$$

For solution of the above problem, there are three cases: a) $r = q$, b) $r < q$, c) $r > q$. Cases a) and b) are rather easy to study. We solve the problem for the case when $q = mr$ (m in a positive integer). It will not be difficult to extend the obtained solution to case a).

The solution $u^o(n)$, $x^o(n)$, $y^o(n)$ of the posed problem will be called the optimal control and optimal trajectory.

2. <u>The Solution Method</u>. We introduce the matrix $F(n)$, defined for $n \geq 0$, being the solution of the equation

$$F(n + 1) = CF(n) \ ,$$

with the initial value $F(0) = E$. We consider the solution $y(n)$ of Eq.(79) under the condition that $y_i(0) = 0$, $i = 1,\dots,q$. It is easily verified that

$$y(n) = \sum_{s=0}^{n-1} F(n - s - 1)Du(s) \ . \tag{82}$$

We let $||u||$ denote the expression

$$||u|| = \max_{1 \le n \le M} \max_{1 \le i \le q} \left| \sum_{s=0}^{n-1} \sum_{\ell=1}^{q} \sum_{g=1}^{r} f_{i\ell}(n - s - 1) d_{\ell g} u_g(s) \right| ,$$

$$F = \{f_{i\ell}\} , \quad D = \{d_{\ell g}\} , \tag{83}$$

where M is some number. Then, condition (80) is equivalent to the inequality

$$||u|| \le 1 . \tag{84}$$

Let the matrix $\Phi(n)$ be a solution of the equation

$$\Phi(n + 1) = A\Phi(n) \tag{85}$$

with the initial condition $\Phi(0) = E$. We express the solution of Eq. (78) with the help of $\Phi(n)$. We have

$$x(n) = \Phi(n)x(0) + \sum_{s=0}^{n-1} \Phi(n - s - 1)Bu(s) . \tag{86}$$

We assume that some control u(n) transfers the trajectory x(n) to the origin at time n = M, i.e. x(M) = 0. In (86), let n = M. We multiply both sides of (86) by $\Phi(-M)$ and transfer x(0) to the left. As a result, we obtain

$$-x(0) = \sum_{s=0}^{M-1} \Phi(-s - 1)Bu(s) .$$

We set

$$\Gamma(s) = \Phi(-s)B , \quad \eta(s) = u(s - 1) . \tag{87}$$

Now we may write

$$-x_\nu(0) = \sum_{s=1}^{M} \sum_{g=1}^{r} \gamma_{\nu g}(s)\eta_g(s) \quad , \ \nu = 1,\ldots,p \ , \ \Gamma = \{\gamma_{\nu g}\}. \tag{88}$$

By virtue of (83) and (87), we may represent the constraint (84) in the form

$$||\eta|| \leq 1 \quad . \tag{89}$$

The expression $\eta(\cdot) = \sum_{s=1}^{M} \sum_{g=1}^{r} \beta_g \eta_g(s)$ is some linear functional defined on the system of functions $\beta_g(s)$, $g = 1,\ldots,r$; $s = 1,\ldots,M$; the functional $\eta(\cdot)$ is assumed to be known if the system of functions $\eta_g(s)$, $g = 1,\ldots,r$; $s = 1,\ldots,M$, is known. Now, beginning with expressions (88),(89), we give problem (78)-(80) the following formulation. Find the smallest number M and a linear functional $\eta(\cdot)$ such that the functional η on p given systems of functions $\{\gamma_{\ell_g}(s),\ g = 1,\ldots,r;\ s = 1,\ldots,M\}$, $\ell = 1,\ldots,p$, assumes the given values $-x_\ell(0)$, $\ell = 1,\ldots,p$. For this, we assume only that the functionals satisfy (89). Thus, problem (78)-(80) is reduced to an L-problem of moments.

We formulate the basic result of this section.

Theorem 18. *For problem (78)-(79), the optimal control has the form*

$$u_i^o(n) = \{[W^{\alpha_o}]^{-1}\eta^{\alpha_o}\}_i(n+1) \quad ,$$

$$n = 0,\ldots,K^o-1; \qquad i = 1,\ldots,r \quad .$$

Here the symbol $\{a\}_i(n)$ denotes the ith coordinate of the vector a at the moment n. The expressions N^{α_o}, η^{α_o}, K^o are determined by formulas (94),(98),(114),(115).

Proof. First, we show that the expression

$$||\eta|| = \max_{1\leq n\leq M} \max_{1\leq j\leq g} \left| \sum_{s=1}^{n} \sum_{\ell=1}^{q} \sum_{g=1}^{r} f_{j\ell}(n - s)d_{\ell g}\eta_g(s) \right| \qquad (90)$$

determines a norm on the system of functions $\eta_g(s)$, $g = 1,\ldots,r$; $s = 1,\ldots,M$. For this it suffices to show that three conditions are satisfied: a) $||\eta|| = 0$ if and only if $\eta_g(s) = 0$, $g = 1,\ldots,r$; $s = 1,\ldots,M$; b) $||\lambda\eta|| = |\lambda|\ ||\eta||$, if λ is a real number; c) c) $||\eta_1 + \eta_2|| \leq ||\eta_1|| + ||\eta_2||$. We demonstrate the validity of these conditions for expression (90). If $\eta_g(s) = 0$, $g = 1,\ldots,r$; $s = 1,\ldots,M$, then from (90) it follows that $||\eta|| = 0$. Let $||\eta|| = 0$. Then by virtue of (90) and(82), $y(n) = 0$, $n = 0,1,\ldots,M$. Since $y(n)$ is the solution of Eq.(79), from (87) we obtain

$$Du(n) = 0 \quad , \quad n = 0,1,\ldots,M-1 \quad . \qquad (91)$$

By assumption, at least one of the rth order minors of the matrix D is nonzero. Hence, it follows that $u_g(n) = 0$, $g = 1,\ldots,r$; $n = 0,\ldots,M-1$, is the unique solution of the obtained system of equations.

We will not establish the validity of b) and c). We pass on to obtaining the form of the controlling function $u^o(n)$ in the optimal process. We introduce the nonsingular linear transformation* S, defined on the system of functions $\eta_g(s)$, $g = 1,\ldots,r$; $s = 1,\ldots,M$. We let $[S\eta]_i(\rho)$ denote the ith coordinate of the transform $S\eta$

---

* We assume that the transformations S,W, and the transformation of (101) are nonsingular. By a minor variation in the arguments, this constraint may be dropped.

of the function η at $s = \rho$. We compute that

$$[S\eta]_i(\ ) = \sum_{s=1}^{\rho} \sum_{=1}^{q} \sum_{g=1}^{r} f_i\ (\rho - s)\ d_{\ell g}\eta_g(s)$$

$$+ \sum_{s=1}^{\rho} \sum_{\ell=1}^{q} \sum_{g=1}^{r} f_{i+r,\ell}(\rho - s) d_{\ell g}\eta_g(s)$$

$$+ \cdots + \sum_{s=1}^{\rho} \sum_{\ell=1}^{q} \sum_{g=1}^{r} f_{(m-1)r+i,\ell}(\rho - s) d_{\ell g}\eta_g(s)$$

$$= \sum_{\alpha=1}^{m} \sum_{s=1}^{\rho} \sum_{\ell=1}^{q} \sum_{g=1}^{r} f_{(\alpha-1)r+i,\ell}(\rho - s) d_{\ell g}\eta_g(s) \quad , \qquad (92)$$

$$i = 1,\ldots,r; \quad \rho = 1,\ldots,M \ .$$

Beginning with formula (92), it is possible to compute the inverse transformation S^{-1}. Let $(S^{-1})'$ denote the transpose of S^{-1}. From the theory of linear transformations, we know that

$$\sum_{s=1}^{M} \sum_{g=1}^{r} \beta_g(s)\eta_g(s) = \sum_{s=1}^{M} \sum_{g=1}^{r} [(S^{-1})'\beta]_g(s)[S\eta]_g(s) \ . \qquad (93)$$

We introduce the following notation

$$\eta_i^{\alpha}(\rho) = \sum_{s=1}^{\rho} \sum_{\ell=1}^{q} \sum_{g=1}^{r} f_{(\alpha-1)r+i,\ell}(\rho - s) d_{\ell g}\eta_g(s), \quad (94)$$

$$i = 1,\ldots,r; \ \alpha = 1,\ldots,m; \ \rho = 1,\ldots,M.$$

We represent the function $[(S^{-1})'\beta]_g(s)$ in the form

$$[(S^{-1})'\beta]_g(s) = h_g^1(s) + \cdots + h_g^m(s) \quad , \qquad (95)$$

where the function $h_g^{\alpha}(s)$ is computed from the conditions

$$\sum_{s=1}^{M} \sum_{g=1}^{r} h_g^{\alpha}(s)\eta_g^{\delta}(s) = - \sum_{s=1}^{M} \sum_{g=1}^{r} h_g^{\delta}(s)\eta_g^{\delta}(s), \quad \delta \neq \alpha. \tag{96}$$

Using expressions (95),(96), identity (93) is written in the following form:

$$\sum_{s=1}^{M} \sum_{g=1}^{r} \beta_g(s)\eta_g(s) = \sum_{\alpha=1}^{m} \sum_{s=1}^{M} \sum_{g=1}^{r} h_g^{\alpha}(s)\eta_g^{\alpha}(s) \ . \tag{97}$$

The functions $h_g^{\alpha}(s)$ may be expressed by the function $\beta_g(s)$. In matrix form, Eq.(94) has the form

$$\eta^{\alpha} = W^{\alpha}\eta \ . \tag{98}$$

Let the transformation W be nonsingular. Then

$$\eta^{\alpha} = W^{\delta}(W^{\alpha})^{-1}\eta^{\alpha} = W^{\delta\alpha}\eta^{\alpha} \ . \tag{99}$$

We introduce notation setting

$$(h^{\alpha})'\eta^{\delta} = \sum_{s=1}^{M} \sum_{g=1}^{r} h_g^{\alpha}(s)\eta_g^{\delta}(s) \ .$$

From relations (96) and (98), we have

$$(h^{\alpha})'\eta^{\delta} = (h^{\alpha})'W^{\delta\alpha}\eta^{\alpha} = -(h^{\delta})'\eta^{\alpha} \ , \quad \alpha \neq \delta \ ,$$

or

$$h^{\delta} = -(W^{\delta\alpha})'h^{\alpha} \ . \tag{100}$$

Therefore, from equality (95) we obtain

$$h^{1} = [E - \sum_{i=2}^{m} (W^{i1})']^{-1}[(S^{-1})'\beta] \ . \tag{101}$$

We compute the remaining functions h^{α}, $\alpha = 2,\ldots,m$, from formula (100). We consider the space of the system of functions $h_g^{\alpha}(s)$, $\alpha = 1,\ldots,m$; $g = 1,\ldots,r$; $s = 1,\ldots,M$, where the norm of an element h is given by the expression

$$||h|| = \sum_{\alpha=1}^{m} \sum_{g=1}^{r} \sum_{s=1}^{M} |h_g^{\alpha}(s)| \quad . \tag{102}$$

On this space we define a functional $\eta(h)$, the general form of which is

$$\eta(h) = \sum_{\alpha=1}^{m} \sum_{g=1}^{r} \sum_{s=1}^{M} h_g^{\alpha}(s)\eta_g^{\alpha}(s) \quad . \tag{103}$$

The norm of the functional η is uniquely defined from (101) and (102):

$$||\eta|| = \max_{1\leq\alpha\leq m} \max_{1\leq g\leq r} \max_{1\leq s\leq M} |\eta_g^{\alpha}(s)| \quad . \tag{104}$$

We apply the transformations S and $(S^{-1})'$ to expression (88) and, after computations of the type (93)-(101), we obtain

$$-x_{\nu}(0) = \sum_{\alpha=1}^{m} \sum_{g=1}^{r} \sum_{s=1}^{M} \gamma_{\nu g}^{\alpha}(s)\eta_g^{\alpha}(s), \quad \nu = 1,\ldots,p, \tag{105}$$

where

$$\gamma_{\nu g}^{\alpha}(s) = \{-(\overset{\alpha}{W}{}^{1})'[E - \sum_{i=2}^{M} (\overset{i}{W}{}^{1})']^{-1}[(S^{-1})'\gamma^{\nu}]\}_g(s).$$

We define the functions $\eta_g^{\alpha}(s)$ by formula (94). Relation (99) may be written in a simpler form when $\alpha = 1$:

$$\eta_i^{\delta}(s) = [\overset{\delta}{W}{}^{1}\eta^{1}]_i(s) \quad , \; i = 1,\ldots,r; \; \delta = 2,\ldots,m;$$
$$s = 1,\ldots,M \; . \tag{106}$$

The general form of linear operations by matrices on $\{\eta_i^1(s)\}$, $i = 1,\ldots,r$; $s = 1,\ldots,M$, is described by

$$\sum_{t=1}^{M} \sum_{j=1}^{r} T_{ij}(t,s)\eta_j(t) \quad .$$

Thus, relation (106) may be written as

$$\eta_i^\delta(s) - \sum_{t=1}^{M} \sum_{j=1}^{r} \tilde{T}_{ij}^\delta(t,s)\eta_j^1(t) = 0 \quad , \qquad (107)$$

$$i = 1,\ldots,r; \quad \delta = 2,\ldots,m; \quad s = 1,\ldots,M \quad .$$

We set

$$\overset{\delta}{T}{}_{ij}^{1}(t,s) = \tilde{T}_{ij}^\delta(t,s), \quad \overset{\delta}{T}{}_{ij}^{\varepsilon}(t,s) = 0 \ , \quad \text{if } \varepsilon \neq 1, \ , \qquad (108)$$

$$T_{ii}^{\delta\delta}(s,s) = -1 \quad , \quad T_{ij}^{\delta\delta}(t,s) = 0 \quad , \quad \text{if } i \neq j \ .$$

Then (107) will have the form

$$\sum_{\varepsilon=1}^{m} \sum_{t=1}^{M} \sum_{j=1}^{r} \overset{\delta}{T}{}_{ij}^{\varepsilon}(t,s)\eta_j^\varepsilon(t) = 0 \quad , \qquad (109)$$

$$i = 1,\ldots,r; \quad \delta = 2,\ldots,m; \quad s = 1,\ldots,M \ .$$

Consequently, the optimal problem (78)-(80) may be given a functional formulation. In the space of the system of functions $h_g^\alpha(s)$, $\alpha = 1,\ldots,m$; $g = 1,\ldots,r$; $s = 1,\ldots,M$, find the smallest number $M = K^0$ and a linear functional $\eta(h)$ such that the linear functional η on the p elements γ^ν, $\nu = 1,\ldots,p$, and on the $r(m-1)K^0$ elements $T_i^\delta(s)$, $i = 1,\ldots,r$; $\delta = 2,\ldots,m$; $s = 1,\ldots,K^0$, assumes the values $-x_\nu(0)$, $\nu = 1,\ldots,p$, and $0,\ldots,0$, respectively (c.f. (105),(109)). We also demand that the norm (104) of the functional satisfy the inequality

$$||\eta|| \leq 1 \quad . \tag{110}$$

Let the elements $\{\gamma^{\nu}\}$, $\{T_i^{\delta}(s)\}$ be linearly independent. Then we may apply results related to the L-problem. According to the basic assumptions, problem (105),(109), (110) has a solution if and only if

$$\lambda(M) \geq 1 \quad , \tag{111}$$

when the number $\lambda(M)$ is determined by

$$\lambda(M) = \min \sum_{\alpha=1}^{m} \sum_{g=1}^{r} \sum_{s=1}^{M} |\xi_1\gamma_{1g}^{\alpha}(s) + \cdots + \xi_p\gamma_{pg}^{\alpha}(s)$$

$$+ \sum_{i=1}^{r} \sum_{=2}^{m} \sum_{t=1}^{M} \xi_i^{\delta}(t)T_{ig}^{\delta\alpha}(s,t)| \tag{112}$$

where $\sum_{i=1}^{p} \xi_i x_i(0) = -1$. The optimal time equals the smallest number satisfying inequality (111). In order to determine the optimal control, we use the following scheme. Let $\xi_1^{\text{o}},\ldots,\xi_p^{\text{o}}$, $[\xi_i^{\delta}(t)]^{\text{o}}$, $\delta = 2,\ldots,m$; $t = 1,\ldots,M$; $i = 1,\ldots,r$, be the solution of problem (112). We consider the element

$$[\gamma_g^{\alpha}(s)]^{\text{o}} = \sum_{i=1}^{p} \xi_i^{\text{o}}\gamma_{ig}^{\alpha}(s) + \sum_{i=1}^{r} \sum_{\delta=2}^{m} \sum_{t=1}^{M} [\xi_i^{\delta}(t)]^{\text{o}}T_{ig}^{\delta\alpha}(x,t). \tag{113}$$

By virtue of one of our assumptions, the minimal norm solution $[\eta_g^{\alpha}(s)]^{\text{o}}$ of problem (105),(109),(110) has an element $[\gamma_g^{\alpha}(s)]^{\text{o}}$ as extremal, i.e.

$$|\eta\{[\gamma_g^{\alpha}(s)]^{\text{o}}\}| = ||[\eta_g^{\alpha}(s)]^{\text{o}}|| \; ||[\gamma_g^{\alpha}(s)]^{\text{o}}|| \; .$$

From this condition, we find that

$$[\eta_g^\alpha(s)]^o = \frac{1}{\lambda(K^o)} \text{ sign } [\gamma_g^\alpha(s)]^o \text{ if } [\gamma_g^\alpha(s)]^o \neq 0. \quad (114)$$

If $[\gamma_g^\alpha(s)]^o = 0$, then we obtain

$$[\eta_g^\alpha(s)]^o = \omega_g^\alpha(s) \quad , \quad (115)$$

where $\omega_g^\alpha(s)$ must satisfy the condition $|\omega_g^\alpha(s)| \leq 1$. In order to find these numbers, we substitute the values $[\eta_g^\alpha(s)]^o$, defined by relation (114), into (105),(109). From the latter equations, we will determine $\omega_g^\alpha(s)$. Here two cases are possible: the number $\omega_g^\alpha(s)$ is determined uniquely or it will depend upon several numerical parameters.

For our problem, it is not necessary to seek the values $[\eta_g^\alpha(s)]^o$ at all points. We assume that W^{α^o} is a nonsingular transformation. We find all values $[\omega_g^{\alpha^o}(s)]^o$, $g = 1,\ldots,r$; $s = 1,\ldots,M$. Now, from (98) we find the optimal control

$$\eta_g^o(s) = \{[W^{\alpha^o}]^{-1}\eta^{\alpha^o}\}_g(s) \quad .$$

3. Optimal Control of a System with Inertial Regulation. Below we consider the following problem. Given the initial states $x(0)$ and $y(0) = 0$, of the object and the regulator, determine a rule of action $u(n)$ which annihilates the initial state $x(0)$ in minimal time while obeying the constraints on the state space of the regulator.

We give a mathematical formulation of this problem. Let the state of the object be described by the equation

$$\left.\begin{aligned} x(n+1) &= Ax(n) + By(n) \quad , \\ x &= x_1,\ldots,x_p \quad , \quad y = y_1,\ldots,y_q \quad , \end{aligned}\right\} \qquad (116)$$

with the regulator given by the equation

$$y(n+1) = Cy(n) + Du(n), \quad u = \{u,\ldots,u_r\}, \quad (117)$$

where A,B,C,D are constant matrices. We assume that the given initial states are $x(0)$ and $y(0) = 0$. It is necessary to find a control $u(n)$ which transfers the point $x(0)$ along the trajectory of (116) to the origin in minimal time K^o; the coordinates of the vector $y(n)$ must satisfy the inequality

$$|y_k(n)| \leq 1, \quad k = 1,\ldots,q; \quad n \geq 0 \quad .$$

We show that the optimization problem for (116)-(117) reduces to a problem considered in sections 1,2. We introduce the functions $h_{ij}(n)$, $i = 1,\ldots,p+q$; $j = 1,\ldots,p+q$, $n \geq 0$, being the solution of the equations.

$$h_{ij}(n+1) = \sum_{k=1}^{p+q} \tilde{a}_{ik} h_{kj}(n), \; h_{ii}(0) = 1, \; h_{ij}(0) = 0, \quad i \neq j ,$$

where the number $\tilde{a}_{ik}$ is determined by the coefficients of equations (116),(117) in the form

$$\tilde{a}_{ik} = a_{ik}, \quad i = 1,\ldots,p; \quad k = 1,\ldots,p,$$

$$\tilde{a}_{i,k+p} = b_{ik}, \quad i = 1,\ldots,p; \quad k = 1,\ldots,q,$$

$$\tilde{a}_{ik} = 0, \quad i = p+1,\ldots,p+q; \quad k = 1,\ldots,p,$$

$$\tilde{a}_{i+p,k+p} = c_{ik}, \quad i = 1,\ldots,q; \quad k = 1,\ldots,q.$$

With the help of the functions $\{h_{ij}(n)\}$, the solution $x(n)$ of (116) is written in the form

$$x_i(n) = \sum_{j=1}^{p} h_{ij}(n) x_j(0) + \sum_{s=1}^{n} \sum_{j=1}^{q} \sum_{k=1}^{r} h_{ij+p}(n - s)$$

$$\cdot\, d_{jk} u_k(s - 1), \quad i = 1, \ldots, p \,. \tag{118}$$

Except for a change of notation, Eq.(118) is identical to Eq.(86). Therefore, following the arguments in sections 1,2, we obtain the solution of the problem for the system (116)-(117).

4. Example

Example 14. For illustration of the results of sections 1,2, we consider an example. Let the first object be described by the equation

$$x_1(n + 1) = \frac{1}{2} x_1(n) + u(n) \quad ,$$

while the state of the second object is determined by the equations

$$y_1(n + 1) = y_1(n) + y_2(n), \quad y_2(n + 1) = y_2(n) + u(n).$$

We assume that the initial states are such that: $x_1(0)$, $y_1(0) = 0$, $y_2(0) = 0$. It is required to reduce $x_1(0)$ to the origin in minimal time, controlling such that the coordinates $y_1(n)$ and $y_2(n)$ satisfy the constraints

$$|y_1(n)| \le 1 \quad , \quad |y_2(n)| \le 1 \quad , \quad n \ge 0 \quad .$$

This a particular case of the optimal problem considered in sections 1,2, when p = 1, q = 2, r = 1. For the given problem, the functions which are needed for determination of the optimal control, have the form

$$F(n) = \begin{bmatrix} 1 & n \\ 0 & 1 \end{bmatrix}, \quad D = \begin{bmatrix} 0 \\ 1 \end{bmatrix}, \quad \Phi(n) = \left\{ \left(\frac{1}{2}\right)^n \right\}, \quad B = \{1\} \quad ,$$

$$[S\eta](\rho) = \sum_{s=1}^{\rho} (\rho - s)\eta(s) + \sum_{s=1}^{\rho} \eta(s) \quad ,$$

$$[S^{-1}\nu](1) = \nu(1) \quad , \quad [S^{-1}(\nu)](2) = \nu(2) - 2\nu(1) \quad ,$$

$$[S^{-1}\nu](\rho) = \nu(\rho) - 2\nu(\rho - 1) - \nu(\rho - 2), \; \rho = 3,\ldots,M.$$

$$[(S^{-1})'\nu](\rho) = \nu(\rho) - 2\nu(\rho + 1) + \nu(\rho + 2), \; \rho = 1,\ldots,M-2,$$

$$[(S^{-1})'\nu](M-1) = \nu(M-1) - 2\nu(M), \; [(S^{-1})'\nu](M) = \nu(M) \; ,$$

$$\eta^1(n) = \sum_{s=1}^{n} \eta(s) \equiv [W^1\eta](n);$$

$$\eta^2(n) = \sum_{s=1}^{n} (n - s)\eta(s) \equiv [W^2\eta](n) \quad .$$

Hence

$$\eta(n) = [W^1)^{-1}\, {}^1](n) \equiv \begin{cases} \eta^1(1) \quad , \quad n = 1 \quad , \\ \eta^1(n) - \eta^1(n-1) \quad , \; n = 2,\ldots,M \end{cases}$$

$$\eta^2(n) = [W^{21}\eta^1](n) \equiv \begin{cases} 0 \quad , \qquad n = 1 \quad , \\ \sum_{s=1}^{n-1} \eta^1(s), \quad n = 2,\ldots,M. \end{cases} \qquad (119)$$

The operator $(W^{21})'$ has the form

$$h^2(n) = [(W^{21})'h^1](n) = \begin{cases} \sum_{i=n+1}^{M} h^1(i), & n = 1,\ldots,M-1. \\ 0 & n = M. \end{cases}$$

We compute the functions $\gamma_{11}^{\alpha}(s)$:

$$\gamma_{11}^{1}(M) = 2^M, \quad \gamma_{11}^{1}(M-1) = -2^{M-1}, \quad \gamma_{11}^{1}(M-2) = 3 \cdot 2^{M-2}, \ldots$$

$$\gamma_{11}^{2}(M) = 0, \quad \gamma_{11}^{2}(M-1) = -2^{M}, \quad \gamma_{11}^{2}(M-2) = -2^{M-1}, \ldots$$

From formula (119), we obtain the elements $T_{11}^{2}(t,s)$ entering into expression (109): $T_{11}^{21}(t,s) = 1$ for $t = 1,\ldots,s-1$; $T_{11}^{21}(t,s) = 0$ for $t = s,\ldots,M$; $T_{11}^{22} = -1$ for $t \neq s$; $T_{11}^{22}(t,s) = 0$ if $t = s$. For our example, expression (112) has the form

$$\lambda(M) = \min \sum_{\alpha=1}^{2} \sum_{s=1}^{M} \left| \xi\gamma_{11}^{\alpha}(s) + \sum_{i=1}^{M} \xi_i T_{11}^{2\alpha}(s,i) \right|$$

for $\xi x(0) = -1$. Let $x(0) = 12$. We calculate $\lambda(M)$ for this value. It turns out that the smallest number M satisfying the inequality $\lambda(M) \geq 1$, equals 3. Thus, $\lambda(K^{o} = e) = 1$,

$$\xi^{o} = -1/12, \quad \xi_1^{o} = 1/3, \quad \xi_2^{o} = 3/4, \quad \xi_3^{o} = -1/6 \quad .$$

Hence, we obtain the form of the extremal element (113)

$$[\gamma^{\alpha}(s)]^{o} = -\frac{\gamma_{11}^{\alpha}(s)}{12} + \frac{1}{3} T_{11}^{2\alpha}(s,1) + \frac{3}{4} T_{11}^{2\alpha}(s,2) - \frac{1}{6} T_{11}^{2\alpha}(s,3) \quad .$$

By virtue of formula (114), we have

$$[\eta^{\alpha}(s)]^{o} = \text{sign } [\gamma^{\alpha}(s)]^{o} \quad .$$

We set $\alpha=1$. Then $[\gamma^{1}(1)]^{o} = 0$, $[\gamma^{1}(2)]^{o} = 2/12$, $[\gamma^{1}(3)]^{o} = -8/12$. Therefore,

$$[\eta^{1}(1)]^{o} = c, \quad [\eta^{1}(2)]^{o} = 1, \quad [\eta^{1}(3)]^{1} = -1 \ , \tag{120}$$

where c is a number which must be determined from the condition that the trajectory of $x_1(n)$ hit the origin. From relations (119) and (120), we obtain

$$\eta^{o}(1) = c, \quad \eta^{o}(2) = 1 - c, \quad \eta^{o}(3) = -2 \quad . \tag{121}$$

In our case, formula (88) has the form: $-12 = 2c + 4(1 - c) - 16$. Hence, we find that $c = 0$, while, since $u(i) = \eta(i+1)$, from (121) we obtain the optimal control $u^{o}(0) = 0$, $u^{o}(1) = 1$, $u^{o}(2) = -2$.

§8. A General Optimal Control Problem Connected with Discrete-Time Systems

1. Problem Statement.* We consider the regulator system described by the equations

$$x(n + 1) = Ax(n) + Bu(n) \quad , \tag{122}$$

$$y(n + 1) = Cy(n) + Du(n) \quad , \tag{123}$$

---

* This problem was posed by A.A. Fel'dbaum [124b, p.620].

where** $x = \{x_1, \ldots, x_p\}$; $y = \{y_1, \ldots, y_q\}$, $u = \{u_1, \ldots, u_q\}$, and A,B,C,D are constant matrices, of which A,D,C are nonsingular.*** The sizes of the matrices A,B,C,D are $p \times p$, $p \times q$, $q \times q$, $q \times q$, respectively. We consider the following problem: let the initial values $n = 0$, $x(0)$, $y(0) = 0$ be given. It is required to choose a control $u(n)$ which transfers the state $x(n)$ of system (122) in minimal time $n = K^o$ from $x(0)$ to the origin. The control $u(n)$ must lie in some class of functions. We will assume that the admissible controls are those which, together with the vector function $y(n)$, are bounded. Namely,

$$|u_i(n)| \leq 1, \quad |y_i(n)| \leq 1, \quad i = 1, \ldots, q; \; n \geq 0. \tag{124}$$

We will call the solution $u^o(n)$, $x^o(n)$, $y^o(n)$ of the formulated problem the optimal control and optimal trajectories.

2. The Solution Method. We introduce the matrix functions $\Phi(n) = \{\phi_{ij}(n)\}$ and $F(n) = \{f_{ij}(n)\}$ as the solutions of the equations

$$\left.\begin{aligned} \Phi(n+1) &= A\Phi(n), \quad F(n+1) = CF(n), \\ \Phi(0) &= E, \qquad F(0) = E. \end{aligned}\right\} \tag{125}$$

Taking into account the initial condition $y(0) = 0$, the solution $y(n)$ of Eq.(123) may be represented in the

---

* The problem considered below may be solved by the same methods when rank D < 1.

following form

$$y(n) = \sum_{s=0}^{n-1} F(n - s - 1)Du(s) \quad . \tag{126}$$

In §7, it was shown that the expression

$$\max_{1\le n\le M} \max_{1\le j\le q} \sum_{s=0}^{n-1} \sum_{\ell=1}^{q} \sum_{g=1}^{q} f_{j\ell}(n - s - 1)d_{\ell g}u_g(s), \ \{d_{\ell g}\} = D,$$

determines the norm of the vector function u(n), $n = 0,\ldots,M-1$. We denote this norm by the symbol $||u||_1$; we let $||u||_2$ denote the norm defined by the expression $\max_{1\le n\le M} \max_{1\le g\le q} |u_g(n-1)|$. Let

$$||u|| = \max \{||u||_1, ||u||_2\} \ . \tag{127}$$

Clearly, condition (124) is equivalent to the inequality

$$||u|| \le 1 \quad . \tag{128}$$

We write the solution of Eq.(122) with the help of the function $\Phi(n)$. We have

$$x(n) = \Phi(n)x(0) + \sum_{s=0}^{n-1} \Phi(n - s - 1)Bu(s) \quad .$$

Let u(n) be a control which, at time n=M, transfers the trajectory x(n) to the origin for the first time. Then from the last formula we obtain

$$-x_i(0) = \sum_{s=0}^{M-1} \sum_{j=1}^{q} [\Phi(-s - 1)B]_{ij}u_j(s), \quad i = 1,\ldots,p,$$

where the symbol $[Q]_{ij}$ denotes the elements of the matrix

Q. We set

$$\eta(s) = u(s-1), \quad \{\gamma^1(s),\ldots,\gamma^p(s)\} = \phi(-s)B ,$$

$$\eta'\gamma^i = \sum_{s=1}^{M} \sum_{j=1}^{q} \gamma_j^i(s)\eta_j(s) \quad .$$

Then, for determination of the function $\eta(s)$ we obtain the system

$$-x_i(0) = \eta'\gamma^i \quad , \quad i = 1,\ldots,p \quad . \tag{129}$$

We let R' denote the space of the system $\{\eta_j(s)\}$ with norm (127). Let R'' = R be the dual space of R'. It is the space of the system $\{\gamma_j(s)\}$. The general form of a linear functional η over R is given by the expression $\eta(\gamma) = \eta'\gamma$. Therefore, the original problem may be formulated as: find the smallest number $M = K^o$ and a linear functional η, $||\eta|| \leq 1$, which assumes the values $-x_i(0)$ on the given elements $\gamma^i \varepsilon R$.

Thus, the problem has been reduced to an L-problem. For practical application of the results, the solution of the L-problem must be known by an explicit expression of the norm in R. Below, we describe a method which allows us to by-pass calculation of the norm in R.

We consider a transformation S, defined in the space R' by the expression

$$S\eta = \eta + Q\eta = (E+Q)\eta \quad ,$$

where the transformation Q is given by

$$[Q\eta]_i(\rho) = \sum_{s=0}^{\rho=1} \sum_{j=1}^{q} [F(\rho-s-1)D]_{ij}\eta_j(s+1) \quad , \tag{130}$$

$$i = 1,\ldots,q; \quad \rho = 1,\ldots,M \; .$$

We assume that the operator S has an inverse S^{-1}. We let $(S^{-1})'$ denote the operator satisfying the identity

$$\gamma'\eta = [(S^{-1})'\gamma]S\eta \quad .$$

In the space R', we introduce two elements η^1 and η^2, setting

$$\eta^1 = \eta \quad , \quad \eta^2 = Q\eta^1 \quad . \tag{131}$$

We decompose the element $(S^{-1})'$ of the space R into h^1 and h^2 in the following way:

$$(S^{-1})'\gamma = h^1 + h^2 \quad , \quad (h^1)'\eta^2 = -(h^2)'\eta^1 \quad . \tag{132}$$

From the identity

$$(h^1)'\eta^2 = (h^1)'Q\eta^1 = [Q'h^1]'\eta^1 = -(h^2)'\eta^1$$

we obtain $h^2 = -Q'h^1$. We assume that the operator $(E - Q')^{-1}$ exists. From the relations (131),(132), it is possible to obtain a formula for determination of the elements h^1 and h^2. We have

$$(S^{-1})'\gamma = h^1 - Q'h^1 = (E - Q')h^1 \quad ,$$

$$h^1 = (E - Q')(S^{-1})'\gamma \quad .$$

Consequently,

$$h^1 = (E - Q')^{-1}(E + Q')^{-1}\gamma = (E - (Q')^2)^{-1}\gamma \quad ,$$

$$h^2 = -Q'(E - (Q')^2)^{-1}\gamma \quad .$$

If we decompose the given elements γ^i in such a way (cf.(129)) into elements h_i^1, h_i^2, $i = 1,\ldots,p$, then instead of the relation (129), we obtain

$$-x_i(0) = (h_i^1)'\eta^1 + (h_i^2)'\eta^2 \quad . \tag{133}$$

Now, instead of the space R, we consider the space R_1 consisting of the functions $h = \{h^1,h^2\}$. The norm $||h||$ of elements of this space is defined as:

$$||h|| = ||h^1|| + ||h^2|| \quad .$$

The general form of a linear functional over R_1 is described in the form $\beta(h) = (h^1)'\eta^1 + (h^2)'\eta^2$, where $\beta = \{\eta^1,\eta^2\}$ is an element of the space R_1'. Hence we obtain a norm $||\beta||$ in the dual space R_1':

$$||\beta|| = \max \; \{||\eta^1||, \; ||\eta^2||\} \quad ,$$

i.e. $||\beta||$ coincides with expression (127). By virtue of the transformations made above, we keep in mind that among the functionals from R_1', it is possible to use only those functionals $\beta = \{\eta^1,\eta^2\}$ consisting of components η^1,η^2 which satisfy the second equality from (131). We transform this equality to a form more suitable for what follows. Using the explicit form (13) of the operator Q, instead of the second relation from (131) we obtain the expression

$$\sum_{s=1}^{\rho} \sum_{j=1}^{q} [F(\rho - s)D]_{ij}\eta_j^1(s) - \eta_i^2(\rho) = 0 \; , \tag{134}$$

$$i = 1,\ldots,q; \quad \rho = 1,\ldots,M \; .$$

In the space R_1, we introduce into consideration the qM elements $\{K_\rho^i, N_\rho^i\}$, $i = 1,\ldots,q$; $\rho \doteq 1,\ldots,M$, and we set

$$\left.\begin{aligned} &[K_\rho^i]_j(s) = [F(\rho - s)D]_{ij}, \; s = 1,\ldots,\rho; \; j = 1,\ldots,q; \\ &[K_\rho^i]_j(s) = 0, \; s = \rho + 1,\ldots,M; \\ &[N_\rho^i]_j(s) = -1, \quad s = \rho; \quad j = i; \\ &[N_\rho^i]_j(s) = 0, \quad s = \rho; \quad i = j. \end{aligned}\right\} \quad (135)$$

Expression (134) may be written in the form

$$[K_\rho^i]'\eta^1 + [N_\rho^i]'\eta^2 = 0, \; i = 0,\ldots,q; \; \rho = 1,\ldots,M. \quad (136)$$

Let the elements $\{h_i^1, h_i^2\}$, $\{K_\rho^j, N_\rho^j\}$, $i = 1,\ldots,p$; $j = 1,\ldots,q$; $\rho = 1,\ldots,M$, be linearly independent. Beginning with relations (133) and (136), the problem posed in section 1 may be formulated in the form of the following functional problem. Find the smallest number $M = K^0$ and a linear functional $\beta = \{\eta^1, \eta^2\}$, $||\beta|| \leq 1$, which assumes the values $\{-x_i(0)\}$ on the elements $\{h_i^1, h_i^2\}$, $i = 1,\ldots,p$, while on the elements $\{K_\rho^j, N_\rho^j\}$ (cf.(135)), $j = 1,\ldots,q$; $\rho = 1,\ldots,M$, its value is zero.

The solution of the last problem has already been cited repeatedly. We compute the number

$$\lambda(M) = \min_{\xi'x(0)=-1} \left\{ \sum_{s=1}^{M} \sum_{j=1}^{q} \left[\left| \sum_{i=1}^{p} \xi_i [h_i^1]_j(s) + \sum_{\rho=1}^{M} \sum_{i=1}^{q} \sigma_\rho^i [K_\rho^i]_j(s) \right| + \left| \sum_{i=1}^{p} \xi_i [h_i^2]_j(s) + \sum_{\rho=1}^{M} \sum_{i=1}^{q} \sigma_\rho^i [N_\rho^i]_j(s) \right| \right] \right\} . \quad (137)$$

We find the smallest number $M = K^o$ satisfying the inequality $\lambda(M) \geq 1$. The number K^o is the optimal time. We let $\xi_i^o, \sigma_\rho^{j\,o}$, $i = 1,\ldots,p$; $j = 1,\ldots,q$; $\rho = 1,\ldots,M$, denote the solution of problem (137). We form the minimizing element $\psi = \{\psi^1, \psi^2\}$ of problem (137), the coordinates ψ^1, ψ^2 of which we compute as:

$$\psi_j^1(s) = \sum_{i=1}^{p} \xi_i^o [h_i^1]_j(s) + \sum_{\rho=1}^{M} \sum_{i=1}^{q} \sigma_\rho^{i\,o} [K_\rho^i]_j(s) \quad ,$$

$$\psi_j^2(s) = \sum_{i=1}^{p} \xi_i^o [h_i^2]_j(s) + \sum_{\rho=1}^{M} \sum_{i=1}^{q} \sigma_\rho^{i\,o} [N_\rho^i]_j(s) \quad .$$

The optimal control $u^o(n)$ and the optimal trajectory (minimal in norm) are found from the formulas

$$\left.\begin{aligned} u_j^o &= \frac{1}{\lambda(K^o)} \operatorname{sign} \psi_j^1(s+1) \text{ for } \psi_j^1(s+1) \neq 0, \\ &\qquad s = 0,\ldots,K^o-1 \,, \end{aligned}\right\} \tag{138}$$

$$\begin{aligned} y_j^o(s) &= \frac{1}{\lambda(K^o)} \operatorname{sign} \psi_j^2(s) \text{ for } \psi_j^2(s) \neq 0, \\ &\qquad s = 1,\ldots,K^o \quad . \end{aligned}$$

Formulas (138), together with the relations (133), give the solution of the problem.

3. An Example.

Example 15. We assume that the state of the first object may be described by the equations

$$x_1(n+1) = x_1(n) + x_2(n), \quad x_2(n+1) = \frac{1}{2} x_2(n) + \frac{1}{4} u(n). \tag{139}$$

For the second object, let the equation have the form

$$y(n + 1) = \frac{1}{2} y(n) + \frac{1}{3} u(n), \; y(0) = 0 \quad . \tag{140}$$

For the minimal time, it is necessary to drive the states $x_1(0)$, $x_2(0)$ to the origin, under the conditions that during the process we do not violate the inequalities

$$|u(n)| \leq 2 , \quad |y(n)| \leq \frac{1}{3} , \quad n \geq 0 \quad . \tag{141}$$

The substitution $v = u/2$, $y_1 = 3y$, allows us to write expression (139)-(141) so that

$$x_1(n + 1) = x_1(n) + x_2(n), \; x_2(n + 1) = \frac{1}{2} x_2(n) + \frac{1}{2} v(n) \; ,$$

$$y_1(n + 1) = \frac{1}{2} y_1(n) + 2v(n) \quad ,$$

$$|v(n)| \leq 1 \quad , \quad |y_1(n + 1)| \leq 1 \quad , \quad n \geq 0 \quad .$$

We compute $\Phi(-n)$, $F(n)$ (cf.(125)):

$$\Phi(-n) = \begin{bmatrix} 1 & 2 - 2^{n+1} \\ 0 & 2^n \end{bmatrix} \quad , \qquad F(n) = \{2^{-n}\} \quad .$$

The elements $\gamma^1(s)$, $\gamma^2(s)$ equal $\gamma^1(s) = 1-2^s$, $\gamma^2(s) = 2^{s-1}$. The operator Q is determined by the expression

$$[Q\eta](\rho) = \sum_{s=1}^{\rho} 2^{s-\rho+1} \eta(s) \quad .$$

Therefore,

$$[Q'\gamma](\rho) = \sum_{s=\rho}^{M} 2^{\rho-s+1} \gamma(s) \quad .$$

Further, we have

$$[S\eta](\rho) = [\eta + Q\eta](\rho) = \eta(\rho) + \sum_{s=1}^{\rho} 2^{s-\rho+1}\eta(s) \quad ,$$

$$[S^{-1}\eta](\rho) = \eta(\rho) - \frac{2}{3}\sum_{s=1}^{\rho} \sigma^{s-\rho}\eta(s) \quad ,$$

$$[(S^{-1})'\gamma](\rho) = \gamma(\rho) - \frac{2}{3}\sum_{s=\rho}^{M} \sigma^{\rho-s}\gamma(s) \quad .$$

Let $\eta^1 = \eta$, $\eta^2(\rho) = \sum_{s=1}^{\rho} 2^{s-\rho+1}\eta^1(s)$. We compute the operator

$$[(E - Q')^{-1}](\rho) = \gamma(\rho) + \sum_{s=\rho}^{M} (-2)^{\rho-s+1}\gamma(s) \quad ,$$

and we find the operator

$$(E - Q')^{-1}(E + Q')^{-1} \quad , \quad Q'(E - Q')^{-1}(E + Q')^{-1} \quad .$$

For determination of the elements $h_i^1(s)$, $h_i^2(s)$, we have the formulas

$$h_i^1(\rho) = \gamma^i(\rho) - \frac{1}{3}\sum_{s=\rho}^{M} [\sigma^{\rho-s} + 3(-2)^{\rho-s}]\gamma^i(s) \quad ,$$

$$h_i^2(\rho) = \sum_{t=\rho}^{M} [(-2)^{\rho-t} - \frac{1}{3}\sigma^{\rho-t}]\gamma^i(t) \quad .$$

We write several values ot the quantities $K_\rho(s)$, $N_\rho(s)$:

$K_1(1) = 2$, $K_1(2) = 0$, $K_1(3) = 0$, $K_2(1) = 1$, $K_2(2) = 2$, $K_2(3) = 0$, $K_3(1) = \frac{1}{2}$, $K_3(2) = 1$, $K_3(3) = 2$;

$N_1(1) = -1, \quad N_1(2) = 0, \quad N_1(3) = 0, \quad N_2(1) = 0,$

$N_2(2) = -1,$

$N_2(3) = 0, \quad N_3(1) = 0, \quad N_3(2) = 0, \quad N_3(3) = -1.$

Using these values, we find $\lambda(M)$ for the initial conditions $x_1(0) = 1/4$, $x_2(0) = 0$. The smallest number $M = K^o$ satisfying the inequality $\lambda(M) \geq 1$, equals 2; $\lambda(2) = 1$. The solution of problem (137) for this case is the numbers $\xi_1^o = -4$, $\xi_2^o = -6$, $\sigma_1^o = -1/3$, $\sigma_2^o = 0$. Therefore, from formulas (138) we obtain

$$v^o(0) = -\frac{1}{2}, \quad v^o(1) = \frac{1}{4}, \quad y_1^o(1) = -1, \quad y_1^o(2) = 0 ,$$

$$x_1^o(1) = \frac{1}{4} , \quad x_1^o(2) = 0 , \quad x_2^o(1) = -\frac{1}{4} , \quad x_2^o(2) = 0 .$$

Commentary on Chapter IX

1. The proof used in §1 is cumbersome in several of the details of the construction. These results may be simplified if we transfer the methods of Chapter VI directly to discrete systems.

2. The quasimaximum principle, studied in [34j], arises in connection with unusual situations which are complicated in the theory of optimal processes for discrete systems. The importance of discrete systems for the optimization of any control system is incontestable. The preparation of an optimization problem for solution on a digital computer requires passing to recurrance (or difference) relations. This transference generates many analytic questions associated with the continuous model of the process. It brings in the powerful methods of dynamic programming which, in the discrete case, require no analytic assumptions. It appears that this latter scheme for the solution of optimization problems makes all the studies centered around the maximum principle unnecessary. However, its realization encounters serious difficulties. It is known that the computational requirements of dynamic programming are very great [13,122]

and do not allow us to routinely solve systems of order greater than three. Finally, in particular cases the order of the considered system may be increased by several times but the principal difficulties remain. The attractive feature of dynamic programming is the possibility to compute controls insuring an absolute minimum for the criterion function where, with an increasing number of constraints on the state and control, the computing problem becomes simpler. This has generated different variants of dynamic programming with artificially introduced constraints. A limiting simplification of the problem is in the method of local variations [130] which, in essence, is already not the method of dynamic programming. An obstacle is the "curse of dimensionality" [13] which comes about in the general case where it is necessarily encountered if we want to insure finding the optimal control. Thus, the optimization problem turns out to be posed just as if it is to be solved using the necessary conditions of optimality. The effectiveness of these approaches is studied in [122]. In a series of cases (particularly for systems of high order), the use of the necessary conditions of optimality becomes preferable to dynamic programming.

In §1 we gave a brief history of questions about the necessary conditions for optimality in discrete systems. The number of works completed in this area is immense. A majority of them characteristically converge to a treatment of the effectiveness of the Pontryagin maximum principle for discrete systems, which was first proved in [19] for continuous systems. After it was discovered [21b] that this goal was, in general, unattainable, it was undertaken to strengthen the proof of the maximum principle in a weakened form (local maxima, stationarity). Generally, the new conditions did not converge to the maximum principle if the discrete system converged (with a smaller step (period) of discretization) to the continuous. In this situation, the quasimaximum principle was presented as a necessary condition for optimality in discrete systems.

3. The problem of moments for the optimization of discrete systems was first applied in [74b], where the solution of a linear two-point boundary-value minimal time problem was carried out for the minimal control norm.

REFERENCES

The following abbreviations are employed throughout the bibliography:

ARC	=	Automation and Remote Control
DAN USSR	=	Proceedings of the Soviet Academy of Sciences
Iz. AN USSR	=	Izvestia Academy of Sciences of the USSR
BMMP	=	Journal of Computational Mathematics and Mathematical Physics

1. Isaacs, R. Differential Games, Mir Co., Moscow, 1967.
2. Aleksandrov, P. S. Combinational Topology, Gostekhizdat Moscow, 1947.
3. Al'brecht, E. G., and Krasovskii, N. N. "On the Observability of Nonlinear Control Systems in the Neighborhood of a Given Motion," ARC, vol. 25, No. 7, 1964.
4. Achiezer, N., and Krein, M. "On Some Questions in the Theory of Moments," Kharkov, 1930.
5. Babunashvili, T. G. "The Synthesis of Optimal Linear Systems," DAN USSR, vol. 155, No. 2, 1964.
6. Bagaeva, N. Ya., and Moiseev, N. N. "On One Approach to the Numerical Solution of Optimal Control Problems," DAN USSR, vol. 153, No. 4, 1963.
7. Baranov, A. Yu., Trychaev, R. I., and Chomeniok, V. V. "Foundations of the Imbedding Method in Variational Problems," ARC, vol. 20, No. 7, 1967.
8. Barbashin, E. A.
   a) "Toward a Theory of Generalized Dynamical Systems," Moscow U., vol. 2, No. 135, 1949.
   b) "On Estimating the Mean-Square Deviation From a Given Trajectory," ARC, vol. 21, No. 7, 1960.
   c) "On Estimating the Maximum Deviation From a Given Trajectory," ARC, vol. 21, No. 10, 1960.
9. Bedrov, Ya. A., and Kanarev, L. E. "The Method of Successive Synthesis of Minimal Time Control," Iz. AN USSR, Tech-Cybernetics, No. 4, 1965.

10. Beiko, I. V. and Karpenko, M. R. "The Solution of Nonlinear Optimal Problems by the Method of Successive Approximations," DAN USSR, No. 12, 1964.

11. Bellman, R.
   a) Stability Theory of Differential Equations, IL. 1954.
   b) Dynamic Programming, IL, 1960.
   c) Adaptive Control Processes, Nauka, 1964.

12. Bellman, R., I. Glicksberg, and O. Gross. "Some Questions in the Mathematical Theory of Control Processes," IL, 1962.

13. Bellman, R., and Dreyfus, S. Applied Dynamic Programming, Nauka, 1965.

14. Bellman, R., and Kalaba, R. "Reduction of Dimensionality for Dynamic Programming and Control Processes," Tech. Mechanics, ser. D., vol. 93, No. 1, 1961.

15. Bellman, R., and Cooke K. L. Differential-Difference Equations, Mir, 1967.

16. Bliss A. Lectures on the Calculus of Variations, IL, 1950

17. Bolonkin, O. O. "Impulsive, Singular, and Chattering Regimes in Problems of Flight Dynamics," in Complex Control Systems, Kiev, Nauka, 1965.

18. Boltyanskii, V. G.
   a) "The Maximum Principle in the Theory of Optimal Processes," DAN USSR, vol. 119, No. 6, 1959.
   b) "Optimal Processes with Parameters," DAN USSR, No. 10, 1959.
   c) "Sufficient Conditions for Optimality and the Foundations of Dynamic Programming," Iz. AN USSR, ser. Mathematics, vol. 20, No. 3, 1964.

d) Mathematical Methods of Optimal Control, Nauka, 1966.

19. Boltyanskii, V. G., Gamkrelidze, R. V., and Pontryagin, L. S. "Toward a Theory of Optimal Processes," DAN USSR, vol. 110, No. 1, 1956.

20. Bryson, A. "The Solution of Optimal Programming Problems by the Method of Steepest Ascent," Appl. Mechanics, No. 2, 1962.

21. Butkovskii, A. G.

a) "Optimal Processes in Distributed-Parameter Systems," ARC, vol. 22, No. 1, 1961.

b) "On Necessary and Sufficient Conditions for Optimality for Impulse Control Systems," ARC, vol. 24, No. 8, 1963.

c) "The Method of Moments in the Theory of Optimal Control of Distributed-Parameter Systems," ARC, vol. 24, No. 9, 1963.

d) Optimal Control of Distributed-Parameter Systems Nauka, 1965.

22. Butkovskii, A. G. and Lerner, A. Ya. "On the Optimal Control of Distributed-Parameter Systems," DAN USSR, vol. 134, No. 4, 1960.

23. Butkovskii, A. G., and Poltavskii, L. N. "Finite Control of Linear Systems with Lumped Parameters," ARC, vol. 20, No. 9, 1967.

24. Buzakas, V. I. "Singular Solutions of the Maximum Principle in the Optimal Control of Systems with Variable Structure," in Optimal Control Systems, Nauka, Moscow, 1967.

25. Bykov, Ya. V. "On Several Problems in the Theory of Integro-Differential Equations," Frunze, Kirghiz, 1957.

26. Vapnyarshkii, I. B. "Existence Theorems For the Optimal Control Problems of Bolza, Some of their Applications, and Necessary Conditions for the Optimality of Chattering and Singular Regimes," BMMP, vol. 7, No. 2, 1967.
27. Vasil'ev, O. V. "The Gradient Method for the Solution of a Class of Optimal Control Problems," BMMP, vol. 7, No. 1. 1967.
28. Vasil'ev, O. V., and Kirillova, F. M. "On Optimal Processes in Two-Parameter Discrete Systems," DAN USSR, vol. 175, No. 1, 1967.
29. Velichenko, V. V. "On Optimal Control Problems for Equations with Different Right Sides," ARC, vol. 27, No. 7, 1966.
30. Vinokurov, V. R. "Optimal Control of Processes Described by Integral Equations: I-III, Iz. Buzov Mathematika, Nos. 7-9, 1967.
31. Volin, Yu. M., and Ostrovskii, G. M. "On One Optimal Problem," ARC, vol. 25, No. 10, 1964.
32. Gabasov, R.
    a) "Questions on the Uniqueness of Optimal Control in Discrete Systems," DAN USSR, Energy and Automation, No. 5, 1962.
    b) "Optimal Processes with Constraints on the Period," DAN USSR, vol. 144, No. 4, 1962.
    c) "Toward Optimal Processes in Connection with Systems of Discrete Type," ARC, vol. 23, No. 7, 1962.
    d) "On One Problem in the Theory of Optimal Processes," ARC, vol. 28, No. 8, 1967.
    e) "Toward the Optimization of a Class of Dynamic Systems," DAN Byel. SSR, vol. 12, No. 3, 1968.

f) "On Necessary Conditions for Optimality for Systems Described by Partial Differential Equations," DAN Byel. SSR, vol. 12, No. 7, 1968.

g) "On the Optimality of Singular Controls," Diff. Eqs., vol. 4, No. 6, 1968.

h) "Toward a Theory of Optimal Processes in Discrete Systems," BMMP, vol. 8, No. 4, 1968.

i) "Toward a Theory of Controllability of Dynamical Systems," Diff. Eqs., vol. 4, No. 9, 1968.

j) "On Necessary Conditions of Optimality for Singular Controls," Iz. AN USSR, Tech. Cyber., No. 5, 1968.

33. Gabasov, R., and Hindis, V. B. "Toward Optimal Processes in Linear Systems with Two Control Constraints," ARC, vol. 26, No. 6, 1965.

34. Gabasov, R., and Kirillova, F. M.

a) "Toward Optimal Processes in Connected Systems," ARC, vol. 24, No. 6, 1963.

b) "On the Optimal Control of Connected Systems of Discrete Type," ARC, vol. 24, No. 7, 1963.

c) "Toward Optimal Control Problems," Iz. AN USSR, Tech. Cyber., No. 1, 1964.

d) "Application of the Theory of Linear Inequalities to Optimal Control Problems," Annot. Dok.II, All-Union Conf. on Theoret. and Applied Mechanics, Nauka, 1964.

e) "On One Approach to the Solution of Several Optimal Control Problems," ARC, vol. 25, No. 3, 1964.

f) "On the Solution of Several Problems in the Theory of Optimal Processes," ARC, vol. 25, No. 7, 1964.

g) "Optimization of Convex Functionals Along Trajectories of Linear Systems," DAN USSR, vol. 156, No. 5, 1964.

h) "Statistical Problems in the Optimal Control of Linear Systems," DAN USSR, vol. 164, No. 1, 1965.

i) "Construction of Successive Approximations for Several Optimal Control Problems," ARC, vol. 27, No. 2, 1966.

j) "On the Question of Extending the Pontryagin Maximum Principle to Discrete Systems," ARC, vol. 27, No. 11, 1966.

k) "Statistical Problems in the Optimal Control of the Terminal State of a Linear System," in Optimal Control Systems, Nauka, 1967.

35. Gabasov, R., and Naumov, S. T. "On Necessary Conditions for Optimality in Dynamical Systems," DAN Byel. SSR, vol. 13, No. 5, 1968.

36. Gabasov, R., and Churakov, S. V.

a) "An Optimal Control Problem in Systems with an Aftereffect," Diff. Eqs., vol. 2, No. 10, 1966.

b) "On the Existence of Optimal Controls in Systems with a Delay," Diff. Eqs., vol. 3, No. 12, 1967.

c) "Necessary Conditions for Optimality in Systems with a Delay," ARC, vol. 29, No. 1, 1968.

d) "Toward Necessary Conditions for Optimality in Systems with a Delay," DAN Byel. SSR, vol. 12, No. 1, 1968.

e) "Sufficient Conditions for Optimality in Systems with a Delay," ARC, vol. 29, No. 2, 1968.

37. Galiullin, A. S. "On Dynamic Programming Problems," Trud. Lumumba U., vol. 5, 1964.

38. Gamkrelidze, R. V.

a) "Toward a Theory of Optimal Processes in Linear Systems," DAN USSR, vol. 116, No. 1, 1957.

b) "The Theory of Minimal-Time Processes in Linear Systems," Iz. AN USSR, ser. Math., vol. 22, No. 4, 1958.

c) "Toward a General Theory of Optimal Processes," DAN USSR, vol. 123, No. 2, 1958.

d) "Optimal Control Processes Under State Constraints," Iz. AN USSR, ser. Math., vol. 24, No. 3, 1960.

e) "On Optimal Chattering Regimes," DAN USSR, vol. 143, No. 6, 1962.

f) "Toward a Theory of the First Variation," DAN USSR, vol. 161, No. 1, 1965.

39. Gantmacher, F. The Theory of Matrices, Gostekhizdat, 1953.

40. Hindes, V. B.

a) "On the Problem of Minimizing a Convex Functional of the Terminal State of a Linear Control System," BMMP, vol. 6, No. 6, 1960

b) "On Singular Controls in Optimal Systems," Iz. Vuzov Mathematika, No. 7, 1967.

41. Girsanov, I. V. "Minimax Problems in the Theory of Diffusion Processes," DAN USSR, vol. 136, No. 4, 1961.

42. Gnoenskii, L. S. "On a Pursuit Problem," Appl. Math. and Mech., vol. 26, No. 5, 1962.

43. Gul'ko, F. B., and Kogan, B. Ya. "Methods of Optimal Control with Prediction," Proc. 2nd IFAC Congress, vol. 2, Nauka, 1965.

44. Gurin, L. S. "Optimization in Stochastic Models," BMMP, vol. 4, No. 2, 1964.

45. Gurman, V. I.

a) "On Optimal Singular Control Processes," ARC, vol. 26, No. 5, 1965.

b) "Methods for Studying One Class of Optimal Chattering Regimes," ARC, vo. 26, No. 7, 1965.

46. Dunford, N., and Schwartz, J. Linear Operators, IL, 1962.

47. Degtyarev, G. L., and Sirazetdinov, T. G. "On the Optimal Control of One-Dimensional Distributed -Parameter Processes," ARC, vol. 28, No. 11, 1967.

48. Dem'yanov, V. F.

a) "The Minimization of Functions on Bounded Sets," Kybernetika, No. 6, 1965.

b) "On the Solution of some Minimax Problems," Kybernetika, No. 6, 1966.

49. Dem'yanov, V. F., and Rubinov, A. M. "The Minimization of Smooth, Convex Functionals on Convex sets," Vestnik Leningrad U., No. 19, No. 4, 1964.

50. Dubovitskii, A. Ya., and Miliutin, A. A.
a) "Some Optimal Problems for Linear Systems," ARC, vol. 24, No. 12, 1963.
b) "Extremum Problems in the Presence of Constraints," BMMP, vol. 5, No. 3, 1965.
c) "The Second Variation in Extremal Problems with Constraints," DAN USSR, vol. 160, No. 1, 1965.

51. Dynkin, E. B. "Controllable Stochastic Sequences," Th. Prob. and Applic., vol. 10, No. 1, 1965.

52. Egorov, A. I.
a) "Optimal Processes in Distributed-Parameter Systems and Several Problems in the Theory of Invariance," Iz. AN USSR, ser. Math., vol. 29, No. 6, 1965.
b) "Necessary Conditions of Optimality for Distributed-Parameter Systems," Math. Sbornik, vol. 69, No. 3, 1966.

53. Egorov, Yu. V.
a) "Several Problems in the Theory of Optimal Control," BMMP, Vol. 3, No. 5, 1963.
b) "Necessary Conditions for Optimal Control in a Banach Space," Math. Sbornik, vol. 64, No. 1, 1964.

54. Egorov Yu. V., and Milliutin, A. A. "On Sufficient Conditions for a Strong Extremum in the Class of Curves with a Bounded Derivative," DAN USSR, vol. 195, No. 5, 1964.

55. Zelikin, M. I., and Tynyanskii, N. T. "Deterministic Differential Games," Uk. Math.J., vol. 20, No. 4, 1965.

56. Ivanova, G. P. "On Existence Theorems in the Calculus of Variations," DAN USSR, vol. 170, No. 2, 1966.

57. Isaev, V. K., and Sonin, V. V. "On One Nonlinear Optimal Control Problem," ARC, vol. 23, No. 9, 1962.

58. Kalman, R. E.
   a) "On the General Theory of Control Systems," Proc. 1st IFAC congress, Iz. AN USSR 1961.
   b) "When is a Linear Control System Optimal?," Proc. Amer. Institute Mech. Eng., ser. D, vol. 86, No. 1, 1964.

59. Karasev, I. P. "On the Existence of Domains of Attainability," Diff. Eqs., vol. 3, No. 12, 1967.

60. Karlin, S. Mathematical Methods is Game Theory, Programming, and Economics, Mir, Moscow, 1964.

61. Katkovnik, V. Ya., and Poluektov, R. A. Multidimensional Discrete Control Systems, Nauka, Moscow, 1966.

62. Kelendzheridze, D. L. "Toward a Theory of Optimal Pursuit," DAN USSR, vol. 138, No. 3, 1961.

63. Kelly, H.
   a) "Necessary Conditions for Singualr Extremals Based Upon the Second Variation," Rocket Tech. and Astronautics, vol. 2, No. 8, 1964.
   b) "The Gradient Method," in Optimization Methods with Application to Spaceflight, Nauka, Moscow, 1965.

64. Kirillova, F. M.
   a) "On the Well-Based Statement of One Optimal Regulation Problem," Izv. Vuzov, Mathematica, No. 4, 1959.
   b) "On Passage to the Limit in the Solution of an Optimal Control Problem," Appl. Math. Mech., vol. 24, No. 2, 1960.

c) "Toward Optimal Control Processes," Proc. Ural Polytech. Institute, Sverdlovsk, Sbornik No. 133, 1961.

d) "On the Problem of the Existence of Optimal Trajectories of Nonlinear Systems," Iz. Vuzov, Mathematica, No. 2, 1961.

e) "On the Continuous Dependence of the Solution of an Optimal Control Problem on the Initial Data and Parameters," Uspekhi Math. Nauk., vol. 17, No. 4, 1962.

f) "Several Questions in the Theory of Optimal Control," Iz. Vuzov, Mathematica, No. 3, 1962.

g) "On the Problem of the Existence of Optimal Controls for Linear Systems with Stochastic Pertubations," Siberian Math. J., No. 1, 1964.

h) "Optimal Control is a Statistical Problem," Diff. Eqs., vol. 2, No. 11, 1966.

i) "On One Direction in the Theory of Optimal Processes," ARC, vol. 28, No. 11, 1967.

65. Kirillova, F. M., and Poletaeva, I. A. "On Several Pursuit Problems," Proc. Int'l. Cong. Math. Nauka, Moscow, 1966.

66. Kirillova, F. M., Churakova, S. V.

a) "On the Problem of the Controllability of Linear Systems with Aftereffects," Diff. Eqs., vol. 3, No. 3, 1967.

b) "Relative Controllability of Linear Dynamical Systems with a Delay," DAN USSR, vol. 174, No. 6, 1967.

67. Kirin, N. E.

a) "On One Numerical Method in Linear Minimal-Time Problems," Comp. Methods, No. 2, Leningrad U., 1963.

b) "Toward the Solution of General Linear Minimal-Time Problems," ARC, vol. 25, No. 1, 1964.

68. Kizhevnikov, Yu. V. "Toward Optimization in the Mean of Linear Systems with Stochastic Parameters," in Optimal Automatic Control Systems, Nauka, Moscow, 1967.

69. Kolmogorov, A. N., Mishenko, E. F., and Pontryagin, L. S. "On One Probabilistic Problem in Optimal Control," DAN USSR, vol. 145, No. 5, 1962.

70. Kolosov, G. E., and Stratonovich, R. L. "Asymptotic Methods for the Solution of Statistical Problems of Optimal Control of Quasiharmonic Systems," ARC, vol. 29, No. 2, 1967.

71. Kopp. R., and Moyer, G. "Necessary Conditions for the Optimality of Singular Extremals," Rocket Tech. and Astronautics, vol. 3, No. 9, 1965.

72. Kramer D. "On the Control of Linear Systems with a Delay," in Sb. Mechanika, No. 4, 1963.

73. Krasovskii, A. A. "The Statistical Theory of Transfer Processes in Control Systems," Nauka, Moscow, 1968.

74. Krasovskii, N. N.

a) "Toward a Theory of Optimal Control," ARC, vol. 18, No. 11, 1957.

b) "On One Optimal Control Problem," Appl. Mth. Mech., vol. 21, No. 5, 1957.

c) "On an Optimal Control Problem for Nonlinear Systems," Appl. Mth. Mech., vol. 23, No. 2, 1959.

d) "Toward Sufficient Conditions for Optimality," Appl. Mth. Mech., vol. 23, No. 3, 1959.

e) "On Problems of the Existence of Optimal Trajectories," Izv. Buzov, Mathematika, No. 6, 1959.

f) "Toward a Theory of Optimal Control," Appl. Mth. Mech., vol. 23, No. 4, 1959.

g) "On Optimal Control Under Stochastic Perturbations," Appl. Mth. Mech., vol. 27, No. 4, 1963.

h) "On Mean-Square Optimal Stabilization under Stochastic Damping Perturbations," Appl. Mth. Mech., vol. 25, No. 5, 1961.

i) "On a Pursuit Problem," Appl. Mth. Mech., vol. 27, No. 2, 1963.

j) "On the Stabilization of Unstable Motions by Additional Forces with Incomplete Feedback," Appl. Mth. Mech., vol. 27, No. 4, 1963.

k) "On the Problem of Damping Linear Systems with Minimal Inertial Controls," Appl. Mth. Mech., vol. 29, No. 2, 1965.

l) "Toward a Theory of Controllability and Observability of Linear Dynamical Systems," Appl. Mth. Mech., vol. 28, No. 1, 1964.

m) "Optimal Processes in Systems with a Delay," in Optimal Systems, Statistical Methods, Proc. 2nd IFAC Congress, vol. 2, Nauka, 1965.

n) The Theroy of Controlled Motion, Nauka, Moscow, 1968.

75. Krasovskii, N. N., and Kurzhanskii, A. B. "On the Question of the Observability of Systems with a Delay," Diff. Eqs., vol. 2, No. 3, 1966.

76. Krasovskii,, N. N., Repin, Yu. M., and Tret'yakov, V. E. "On Several Gaming Situations in the Theory of Control Systems," Iz. AN USSR, Tech. Cybernetics, No. 4, 1965.

77. Krotov, V. F.

a) "Difference Solutions of Variational Problems," Izv. Vuzov, Mathematika, No. 5, 1960.

b) "On Difference Solutions in Variational Problems," Izv. Vuzov, Mathematika, No. 2, 1961.

c) "Methods for the Solution of Variational Problems Based Upon Sufficient Conditions for an Absolute Minimum I-III," ARC, vol. 23, No. 12, 1962, vol. 24, No. 5, 1963, vol. 25, No. 7, 1964.

78. Krotov, V. F., and Gurman, V. I. "On Optimal Chattering Regimes in Variational Problems of Flight Dynamics," in Studies in the Dynamics of Flight, No. 1, Izd. Machine Construction, 1965.

79. Krylov, I. A., and Chernous'ko, F. L. "On the Method of Successive Approximations for the Solution of Optimal Control Problems," BMMP, vol. 2, No. 6, 1962.

80. Kulikowski, R. Optimal and Adaptive Processes in Automatic Control Systems, Nauka, Moscow, 1967.

81. Kurzhanskii, A. B.

a) "On the Construction of an Optimal Control by the Method of Moments to Minimize Mean-Square Errors," ARC, vol. 25, No. 5, 1964.

b) "On the Computation of Optimal Controls in Systems with Incomplete Information," Diff. Eqs., vol. 1, No. 3, 1965.

82. Larichev, O. I. "The Optimal Control of One Class of Multiconnected Systems," Iz. AN USSR, Tech. Cyber., No. 5, 1964.

83. La Salle, J. P. "The Principle of Optimal Relay Controls," Proc. Ist. IFAC Congress, vol. 2, USSR Acad. Sci., 1961.

84. Levitin, E. S., and Polyak, B. T. "Minimization Methods in the Presence of Constraints," BMMP, vol. 6, No. 5, 1966.

85. Leitmann, G. "On Optimal Rocket Trajectories," Appl. Mth. Mech., vol. 25, No. 6, 1961.

86. Lerner, A. Ya. "On the Limit of Minimal-Time Control Systems," ARC, vol. 15, No. 6, 1954.

87. Letov, A. M.
    a) "The Analytic Construction of Regulators I-V," ARC, vol. 21, No. 4-6, vol. 22, No. 4, 1961, vol. 23, No. 11, 1962.
    b) "Problems for Scientific Investigation in the Domain of Automatic Control," ARC, vol. 27, No. 8, 1966.

88. Litovchenko, I. A.
    a) The Theory of Optimal Systems. A Result of Science. Series "Mathematical Analysis, Probability Theory, Control," 1962.
    b) "Optimization of Systems with Step Constraints on the Controls," ARC, vol. 26, No. 8, 1965.

89. Lawden, D. F. "Optimal Trajectories for Spaceflight, Mir, Moscow, 1966.

90. Luzin, N. N. "On the Study of Matrix Theory in Differential Equations," ARC, No. 5, 1940.

91. Luzin, N. N., and Kuznetsov, P. I. "Toward Absolute Invariance and Invariance Up to ε in the Theory of Differential Equations," DAN USSR, vol. 51, Nos. 4-5, 1946, vol. 80, No. 3, 1951.

92. Lur'e, A. I., Tvoitskii, V. A. "The Mayer-Bolza Problems and Optimal Control Processes," Proc. IV All-Union Math. Conf., vol. II, Nauka, 1964.

93. "The Mayer-Bolza Problem for Multiple Integrals in the Optimal Behavior of Distributed -Parameter Systems," Appl. Mth. Mech., vol. 27, No. 5, 1963.

94. Miele, A. "An Extension of the Theory of Optimal Programmed Motion to the Expenditure of Fuel for Horizontal Flight," in Studies in the Optimization of Rocket Flight, Oboroniz, 1959.

95. Mihaelevich, V. S. and Shov, N. Z. "On Numerical Methods for the Solution of Multistage Planning and Economic Problems," Sci.-Math. Material of the Econ. Math. Seminar, VTs. AN USSR, No. 1, Kiev, 1962.

96. Miklin, S. G. Integral Equations and their Application, Gostekhizdat, 1949.

97. Mischenko, E. F., and Pontryagin, L. S.

a) "On a Statistical Problem in Optimal Control," Izv. AN USSR, sev. Math., vol. 25, No. 4, 1961.

b) "Linear Differential Games," DAN USSR, vol. 174, No. 1, 1967.

98. Moiseev, N. N.

a) "Dynamic Programming Methods in the Theory of Optimal Control, I-II," BMMP, vol. 4, No. 3, 1964, vol. 5, No. 1, 1965.

b) "Numerical Methods Using Variation in the State Space," Proc. Int'l. Math. Congress, Mir. 1968.

99. Moroz, A. I. "The Synthesis of Minimal-Time Controls for Linear, Discrete, Third-Order Objects, I-II," ARC, vol. 26, Nos. 2,3,8, 1965.

100. Nguen, Tchan' Bang. "On the Controllability of Quasilinear Systems," Appl. Math. Mech., vol. 31, No. 1, 1967.

101. Neustadt, L. V. "The Synthesis of Minimal-Time Systems."

102. Nemyetskii, V. V., Stepanov, V. V. The Qualitative Theory of Differential Equations, Gostekhizdat, 1950.

103. Nikol'skii, S. M. "Linear Equations in Normed Linear Spaces," Izv. AN USSR, Ser. Math., No. 7, 1943.

104. Ozhiganova, I. A.

a) "Toward a Theory of Optimal Control of Systems with a Delay," Proc. Seminar on Diff. Eqs. with a Deviating Argument, Lumumba U., vol. 2, 1963.

b) "On Conditions of Invariance for One Linear Problem with a Delay," Ibid, vol. 3, 1965.

105. Pittel', B. G. "On Several Optimal Control Problems, I-II," ARC, vol. 24, Nos. 9,11, 1963.

106. Poletaeva, I. A. "Optimal Systems with Constraints on the Mean-Square Error," ARC, vol. 27, No. 6, 1966.

107. Polyak, B. T. "Existence Theorems and the Convergence of Minimizing Sequences for Extremal Problems in the Presence of Contraints," DAN USSR, vol. 166, No. 2, 1966.

108. Pontryagin, L. S.
   a) "Optimal Control Processes," Uspekhi Math. Nauka., vol. 14, No. 1, 1959.
   b) "On Some Differential Games," DAN USSR, vol. 156, No. 4, 1964.
   c) "Toward a Theory of Differential Games," Uspekhi Math. Nauka., vol. 21, No. 4, 1966.

109. Pontryagin, L. S., Boltyanskii, V. G., and Gamkrelidze, R. V., and Mischenko, E. F. The Mathematical Theory of Optimal Processes, Fizmatgiz, Moscow, 1961.

110. Propoi, A. I. On the Maximum Principle for Discrete Control Systems," ARC, vol. 26, No. 7, 1965.

111. Pshenichnyi, B. N. "Convex Programming in Normed Spaces," Kibernetika, No. 5, 1965.

112. Riesz, F., and Sz. Nagy, B. Lectures on Functional Analysis, IL, Moscow, 1954.

113. Robbins, H. "The Optimality of Active Thrust Regions of Rocket Trajectories," Rocket Tech. and Astronautics, vol. 3, No. 6, 1965.

114. Rozonoer, L. I.
   a) "On Sufficient Conditions for Optimality," DAN USSR, vol. 127, No. 3, 1959.
   b) "The Pontryagin Maximum Principle in the Theory of Optimal Systems, I-III," ARC, vol. 20, Nos. 10-12, 1959.
   c) "A Variational Approach to the Problem of Invariance, I-II," ARC, vol. 24, Nos. 6-7, 1963.

115. Roitenberg, Ya. N. Some Problems of Controlled Motion, Fizmatgiz, Moscow, 1963.
116. Savelov, A. A. Plane Curves, Fizmatgiz, Moscow, 1960.
117. Simakova, E. N. "Differential Games," ARC, vol. 27, No. 11, 1966.
118. Sirazetdinov, T. G. "Toward a Theory of Optimal Processes with Distributed Parameters," ARC, vol. 25, No. 4, 1964.
119. Sobolev, S. L. Equations of Mathematical Physics, Nauka, Moscow, 1966.
120. Stratonovich, R. L. "The Latest Developments of Dynamic Programming Methods and their Application for the Synthesis of Optimal Systems. Proc. 2nd IFAC Congress, vol. 2, Nauka, Moscow, 1965.
121. Troitskii, V. A.
  a) "The Mayer-Bolza Problem in the Calculus of Variations and the Theory of Optimal Systems," Appl. Mth. and Mech., vol. 25, No. 4, 1961.
  b) "On Variational Problems for the Optimization of Control Processes," Appl. Mth. and Mech., vol. 26, No. 1, 1962.
122. Fan, Lyan'- Tsen', and van', Chu-Sen'. The Discrete Maximum Principle, Mir, Moscow, 1967.
123. Fan', Tsi. "Minimax Theorems," in Infinite Games of Conflict, Fizmatgiz, Moscow, 1963.
124. Fel'dbaum, A. A.
  a) "Optimal Processes in Automatic Control Systems," ARC, vol. 14, No. 5, 1953.
  b) Computational Stability in Automatic Systems, Fizmatgiz, Moscow, 1959.

c) Foundations of the Theory of Optimal Control Systems, Nauka, Moscow, 1966.

125. Fillipov, A. F. "On Several Questions in the Theory of Optimal Control," Vest. MGU, sev. Math., Mech. Astron., Phy., Chem., No. 2, 1959.

126. Fraser, R., Duncan, V., Collar, A. Matrix Theory and its Applications, IL, Moscow, 1950.

127. Kharatishvili, G. L. Optimal Processes with a Delay, Metsniereba, Tbilisi, 1966.

128. Tsypkin, Ya. Z. "Optimal Processes in Impulse Control Systems," Izv. AN USSR, Ener. and Auto., No. 4, 1960.

129. Chang, S. S. L. The Synthesis of Optimal Control Systems, Mashinastroenic Moscow, 1964.

130. Chernous'ko, F. L. "The Method of Local Variations for the Numerical Solution of Variational Problems," BMMP, vol. 5, No. 4, 1965.

131, Sharl', Zh. "The Maximum Principle for Discrete Systems," in the translation of Mechanica, vol. 105, No. 4, 1965.

131. Shatrovskii, L. I. "On a Numerical Method for the Solution of Optimal Control Problems," BMMP, vol. 2, No. 3, 1962.

133. Eneev, T. M. "Application of the Gradient Method in Problems of Optimal Control Theory," Cosmic Studies, vol. 4, No. 5, 1966.

134. Arrow, K. J., Hurwicz, L., and Uzawa, H. Studies in Linear and Nonlinear Programming, IL, Moscow, 1962.

135. Antosiewicz, H. A. "Linear Control Systems," Arch. Rational Mech. and Anal., vol. No. 4, 1963.

136. Astrom, K. A. "Optimal Control of Markov Processes with Incomplete State Information," J. Math. Anal. and Appl., vol. 10, No. 1, 1965.

137. Balakrishnan, A. V.
   a) "Optimal Control Problems in Banach Spaces," J. Soc. Industr. and Appl. Math. Control, vol. 3, No. 1, 1965.
   b) "On the Controllability of Nonlinear Systems," Proc. Nat. Acad. Sci. USA, vol. 55, No. 3, 1966.

138. Bellman, R., Glicksberg, I. and Gross, O. "On the 'Bang-Bang' Control Problem," Quart. Appl. Math., Vol. 14, No. 1, 1956.

139. Berkovitz, L. D.
   a) "Variational Methods in Problems of Control and Programming," J. Math. Anal. and Appl., vol. 3, No. 1, 1961.
   b) "A Survey of Differential Games," Math. Theory Control, New York - London, Acad. Press, 1967.

140. Bittner, L. "Lineare, Zeitoptimale Prozesse bei konvexem Steuer- und Phasengebiet," Abhandl. Dtsch. Acad. Wiss. Berlin, Kl. Math., Phys. und Techn., No. 2, 1966.

141. Bryson, A. E., Denham, W. F., Carroll, F. J., and Mikami, K. "Determination of Lift or Drag Programs to Minimize Re-entry Heating," J. Aerospase Sci. vol. 29, No. 4, 1962.

142. Bushaw, D. W. Experimantal Towing Tank. Stevens Institute of Technology, Report 469, Hoboken, N. J., 1953.

143. Chyung, D. H., and Lee, E. B. Optimization Theory Optimal Systems with Time Delays. Third Congress IFAC, Abstracts, London, 1966.

144. Canon, M., Cullum, J., and Polak, E. "Constrained Minimization Problems in Finite-Dimensional Spaces," J. Soc. Industr. and Appl. Math. Control, Vol. 4, No. 3, 1966.

145. Cesari, L. "An Existence Theorem in Problems of Optimal Control," J. Soc. Industr. and Appl. Math., Ser. A3, No. 1, 1965.

146. Chen, Chi-Tsong. "Output Controllability of Composite Systems," IEEE Trans. Automat. Control, Vol. 12, No. 2, 1967.

147. Conti, R. Sui Problema Della Controllabilita di un Sistema Lineare. Atti. Accad. Naz. Lincei, Rend. Cl. sci. fis. mat. e natur., t. 37, No. 3,4, 1964.

148. Datko, R. "An Implicit Function Theorem with an Application to Control Theory," Michigan Math. J., Vol. 11, No. 4, 1964.

149. Desoer, C. A., and Wing, J. "An Optimal Strategy Saturating Sampled-Data System," IRE Trans. Automat. Control, Vol. 6, No. 1, 1961.

150. Eaton, J. H.

a) "An Iterative Solution to Time-Optimal Control," J. Math. Anal. and Appl., Vol. 5, No. 2, 1962.

b) "Identification for Control Purposes," IEEE Internat. Convent. vol. 15, No. 3, 1967.

151. Falb, P. L. "Infinite Dimensional Control Problems," J. Math. Anal. and Appl., Vol. 9, No. 1, 1964.

152. Fancher, P. "Iterative Computational Procedures for an Optimum Control Problem," IEEE Trans. Automat. Control, Vol. 10, No. 3, 1965.

153. Farison, J. B. "Parameter Identification for a Class of Linear Discrete Systems," IEEE Trans. Automat. Control, Vol. 12, No. 1, 1967.

154. Fattorini, H. O.
   a) "On Complete Controllability of Linear Systems," J. Diff. Equat., Vol. 3, No. 3, 1967.
   b) "Controllability of Higher Order Linear Systems," Math. Theory Control, New York - London, Acad. Press, 1967.

155. Fleming, W. H. "Stochastic Lagrange Multipliers," Math. Theory Control, New York - London, Acad. Press, 1967.

156. Gamkrelidze, R. V. "On Some Extremal Problems in the Theory of Optimal Control," J. Soc. Industr. and Appl. Math. Control, Vol. 3, No. 11, 1965.

157. Gilbert, E. G. "Controllability in Multivariable Control Systems," J. Soc. Industr. and Appl. Math. Control, A1, No. 2, 1963.

158. Goldstein, A. A. "Convex Programming and Optimal Control," J. Soc. Industr. and Appl. Math., A3, No. 1, 1965.

159. Goodman, G. S. "On a Theorem of Scorza - Dragoni and its Application to Optimal Control," Math. Theory Control, New York and London, Acad. Press, 1967.

160. Halkin, H.

a) "Liapunov's Theorem on the Range of a Vector Measure and Pontryagin's Maximum Principle," Arch. Rat. Mech. and Anal., Vol. 10, No. 4, 1962.

b) "Optimal Control for Systems Described by Difference Equations," Advances Control Syst., Vol. 1, New York - London, Acad. Press, 1964.

c) "Nonlinear Nonconvex Programming in an Infinite Dimensional Space. Math. Theory Control, New York - London, Acad. Press, 1967.

161. Halkin, H., Jordan, B. W., Polak, E., and Rosen, J. B., "Theory of Optimum Discrete Time Systems." Third Congress IFAC, Abstracts, London, 1966.

162. Halkin, H., and Neustadt, L. W. "General Necessary Conditions for Optimization Problems." Proc. Acad. Sci. USA, Vol. 56, No. 4, 1966.

163. Harvey, C. A., and Lee, E. B. "On the Uniqueness of Time-Optimal Control for Linear Processes," J. Math. Anal. and Appl. Vol. 5, No. 2, 1962.

164. Hermes, H. "The Equivalence and Approximation of Optimal Control Problems," J. Different Equat. Vol. 1, No. 4, 1965.

165. Hestenes, M. R. "On Variational Theory and Optimal Control Theory," J. Soc. Industr. and Appl. Math., A3, No. 1, 1965.

166. Ho, B. L., and Kalman, R. E. "Effective Construction of Linear State-Variable Models from Input/Output Functions," Regelungstechnik, Bd. 14, No. 12, 1966.

167. Holtzman, J. M.

a) "Convexity and the Maximum Principle for Discrete Systems," IEEE Trans. Automat. Control, Vol. 11, No. 1, 1966.

b) "On the Maximum Principle for Non-linear Discrete-Time systems," IEEE Trans. Automat. Control, Vol. 11, No. 2, 1966.

168. Holtzman, J. M, and Halkin, H. "Directional Convexity and the Maximum Principle for Discrete Systems," J. Soc. Industr. and Appl. Math. Control, Vol. 4, No. 2, 1966.

169. Hopkin, A. M. "A Phase-Plane Approach to the Compensation of Saturating Servomechanisms," Trans. AIEE, part I, Vol. 70, 1951.

170. Hsin, Chu. "A Remark on Complete Controllability," J. Soc. Industr. and Appl. Math., A3, No. 3, 1965.

171. Hwang, C. L., and Fan, L. T. "A Discrete Version of Pontryagin's Maximum Principle.," Operat. Res., Vol. 15, No. 1, 1967.

172. Jackson, R., and Horn, F. "On Discrete Analogues of Pontryagin's Maximum Principle," Internat. J. Control, Vol. 1, No. 4, 1965.

173. Jacobs, M. Q. "Remarks on Some Recent Extensions of Filippov's Implicit Functions Lemma," J. Soc. Industr. and Appl. Math, Control, Vol. 5, No. 4, 1967.

174. Johnson, C. D., and Gibson, J. E. "Singular Solutions in Problems of Optimal Control," IEEE Trans. Automat. Control, Vol. AC-8, No. 1, 1963.

175. Jordan, B. W., and Polak, E. "Theory of a Class of Discrete Control Systems," J. Electron. and Control, Vol. 17, No. 6, 1964.

176. Kalman, R. E. The Theory of Optimal Control and the the Calculus of Variations. RIAS, Baltimore, Maryland, RIAS, Tech. Rept. 61-3, 1961.

177. Katz, S. "A Discrete Version of Pontryagin's Maximum Principle," J. Electron. and Control, Vol. 13, No. 2, 1962.

178. Kelley, H. J.

a) "A Transformation Approach to Singular Subarcs in Optimal Trajectory and Control Problems," J. Soc. Industr. and Appl. Math., Ser. A2, 1964.

b) "A Second Variation Test for Singular Extremals," AIAA J. Vol. 2, No. 8, 1964.

179. Kharatishvili, G. L. "A Maximum Principle in Extremal Problems with Delays," Math. Theory Control, New York - London, Acad. Press, 1967.

180. Kirillova, F. M.

a) "On the Application of Functional Analysis to the Theory of Optimal Processes," J. Soc. Industr. and Appl. Math. Control, Vol. 5, Ser. A, No. 1, 1967.

b) The Application of Functional Analysis to Problems of Pursuit. Proc. Conf. Math. Theory Control, USA Acad. Press, 1967.

181. Knudsen, H. K. "An Iterative Procedure for Computing Time-Optimal Controls," IEEE Trans. Automat. Control, AC-9, No. 1, 1964.

182. Kokotovic, P., and Heller, J. "Direct and Adjoint Sensitivity Equations for Parameter Optimization," IEEE Trans. Automat. Control, Vol. 12, No. 5, 1967.

183. Kushner, H. J. "On Stochastic Extremum Problems: Calculus," J. Math. Anal. and Appl., Vol. 10, No. 2, 1965.

184. Larson, R. E., and Peschon, J. "A Dynamic Programming Approach to Trajectory Estimation," IEEE Trans. Automat. Control, Vol. 11, No. 3, 1966.

185. La Salle, J. P. "Time Optimal Control Systems," Proc. Nat. Acad. Sci. USA, Vol. 45, No. 4, 1959.

186. Lee, E. B. "A Sufficient Condition in the Theory of Optimal Control," J. Soc. Industr. and Appl. Math. Control, A1, No. 3, 1963.

187. Lee, E. B., and Markus, L. "Optimal Control of Nonlinear Processes," Arch. Rational Mech. Anal., Vol. 8, No. 1, 1961.

188. Luenberger, D. G. "Observers for Multivariable Systems," IEEE Trans. Automat. Control, Vol. 11, No. 2, 1966.

189. Malanowski, K., and Rolewicz, S. "Zastosowania Metody Piaszczyzn Podpierajacych do Wyznaczania Sterowania Czasowooptymalnego," Arch. Automat. i telemech., Vol. 10, No. 2, 1965.

190. McShane, E. J.
   a) "On Multipliers for the Lagrange Problem," Amer. J. Math., 61, 1939.
   b) "Optimal Controls, Relaxed and Ordinary," Math. Theory Control, New York - London, Acad. Press, 1967.

191. Miele, A. "Generalized Variational Approach to the Optimum Thrust Programming for the Vertical Flight of a Rocket," Zeit. Flug-Wissenschaften, Bd. 6, No. 3, 1958.

192. Mischenko, E. G. "On a Certain Problem for Parabolic Differential Equations Connected with Optimal Pursuit," J. Soc. Industr. and Appl. Math., A3, No. 1, 1965.

193. Mitter, S. K. "Theory of Inequalities and the Controllability of Linear Systems," Math. Theory Control, New York - London, Acad. Press, 1967.

194. Neustadt L. W.

a) "Time Optimal Control Systems with Position and Integral Limits," J. Math. Anal. and Appl., Vol. 3, No. 3, 1961.

b) "The Existence of Optimal Controls in the Absence of Convexity Conditions," J. Math. Anal. and Appl., Vol. 7, No. 1, 1963.

c) "An Abstract Variational Theory with Applications to a Broad Class of Optimization Problems. I. General Theory," Soc. Industr. and Appl. Math. Control, Vol. 4, No. 3, 1966.

195. Oguztoreli, M. N. Time-Lag Control Systems, New York and London, Acad. Press, 1966.

196. Ohap, R. F., and Stubberud, A. R. "A Technique for Estimating the State of a Nonlinear System," IEEE Trans Automat. Control, Vol. 10, No. 2, 1965.

197. Okamura, K. "Some Mathematical Theories of the Penalty Method for Solving Optimum Control Problems," J. Soc. Industr. and Appl. Math. A2, No. 3, 1965.

198. Oldenburger, R. "Theorie und Anwendung Optimaler Nichtlinearer Regelungen." Regelungstechnik, Bd. 11, No. 4, 1963.

199. Paiewonsky, B. H., Woodraw, P. G., Brunner, W., and Halbert, P. "Synthesis of Optimal Controllers Using Hybrid Analog-Digital Computers," Comput. Methods in Opt. Problems, Acad. Press. New York, 1964.

200. Pearson, J. D.
a) "On the Duality Between Estimation and Control," J. Soc. Industr. and Appl. Math. Control, Vol. 4, No. 4, 1966.
b) "On Controlling a String of Moving Vehicles," IEEE Trans. Automat. Control, Vol. 12, No. 3, 1967.

201. Pearson, J. D., Jr., and Sridhar, R. "A Discrete Optimal Control Problem," IEEE Trans. Automat. Control. Vol. 11, No. 2, 1966.

202. Polak, E. "Minimal Time Control of a Discrete System with a Nonlinear Plant," IEEE Trans. Automat. Control, Vol. AC-8, No. 1, 1963.

203. Ragg, B. C. "Necessary Conditions for the Optimal Control of a System with Time-Varying Transport Lags," IEEE Trans. Automat. Control, Vol. 11, No. 4, 1966.

204. Reynolds, P. A. and Cadzow, J. A. "Solution of an Optimization Problem for Linear Discrete Systems Through Ordinary Calculus," IEEE Trans. Automat. Control, Vol. 11, No. 4, 1965.

205. Roxin, E. "The Existence of Optimal Controls," Michigan Math. J., Vol. 9, No. 2, 1962.

206. Sancho, N. G. F. "Optimization of Linear Stochastic Control Systems Operating Over a Finite Time Interval," Intern. J. Control, Vol. 3, No. 4, 1966.

207. Sarachik, P. E. "Identification of the Steady State Operator for Discrete Selfoptimizing Systems," IEEE Trans. Autom. Contr. Vol. 10, No. 1. 1965.